KORN
KORN
KORN
KORN
KORN
KORN
KORN
KORN

Life is Peachy

2.9.73

2.22.98
8.3.25

4:54
4 1 5 74
74 0
9:

148
120
28

1111

304

1111

4:02
4 1.

12 55

1111.

4 0 7
3:45
52

(44)

15
8

4:15
325
40

111

1111

1111

Mathematics
With Business Applications

Third Edition

Walter H. Lange
The University of Toledo

Temoleon G. Rousos
The University of Toledo

Robert D. Mason
Professor Emeritus/The University of Toledo

Macmillan/McGraw-Hill

New York, New York   Columbus, Ohio   Mission Hills, California   Peoria, Illinois

Imprint 1994

Formerly published under the title *Business Mathematics*.

Send all inquiries to:
GLENCOE DIVISION
Macmillan/McGraw-Hill
936 Eastwind Drive
Westerville, OH 43081

ISBN 0-02-800124-9 (Student Edition)
ISBN 0-02-800125-7 (Teacher's Wraparound Edition)

Printed in the United States of America.

4 5 6 7 8 9 10 11 12 13 14 15 AGH/LP 00 99 98 97 96 95 94

TABLE OF CONTENTS

TABLE OF CONTENTS

TABLE OF CONTENTS

TABLE OF CONTENTS

TABLE OF CONTENTS

TABLE OF CONTENTS

TABLE OF CONTENTS

TABLE OF CONTENTS

TABLE OF CONTENTS

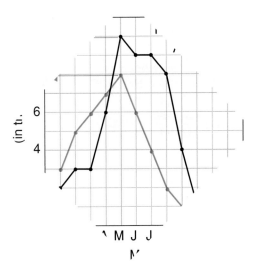

TABLE OF CONTENTS

TABLE OF CONTENTS

Spreadsheet Applications

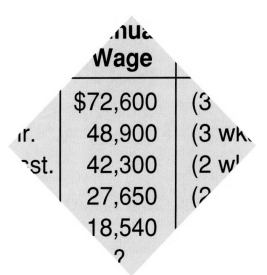

MEMO TO STUDENTS

To: The Student
From: The Authors

This is your textbook. It is specially designed to help you become a knowledgeable consumer and business person. This text contains information and resources to help you achieve these goals.

Each lesson begins with a brief description of the topic and highlights the key formula for you.

The **Example** *shows the topic in an applied situation. It provides you with a step-by-step method to solve for the answer. If you have any trouble or need to review, there is helpful information in the Reference Files at the back of the book. The references at the top of the example tell you just which items to study.*

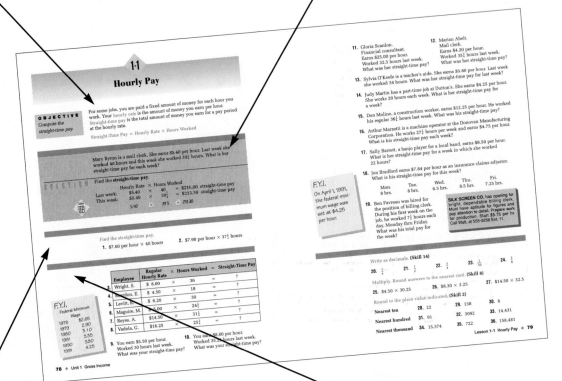

The **Self-Check** *lets you practice the concept and use the formula on your own before you begin the lesson problems.*

After you work through the Example and the Self-Check, complete the problems assigned by your instructor. These problems are arranged systematically to apply gradually what you have just studied. Answers to selected problems are found at the back of your book. Use the answers to check your work.

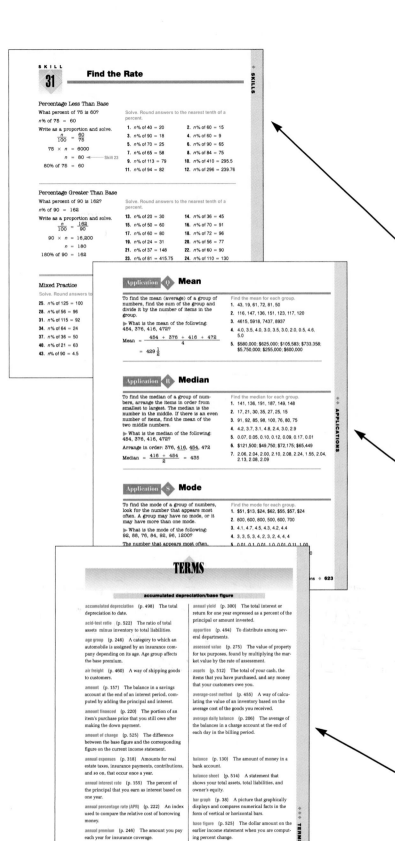

A wise consumer and business person keeps important information in orderly files for easy reference. The tabbed Reference Files at the back of your book contain all the Skills, Applications, and Terms you need for this course.

The *Skills File* contains 32 basic computational skills arranged in order of difficulty. This file helps you review and practice arithmetic, if needed. Study the model problems, then complete the exercises. For more practice, do the set of mixed problems at the bottom of the page.

The *Applications File* will help you recall other math-related topics that are applied in the consumer and business situations. The applications are coded from A to Z.

The *Terms* file contains clear, precise definitions of ideas covered in your textbook. The page references show where the topic may be found in your book.

PREFACE

A knowledge of basic mathematics is absolutely essential for survival in today's world. Mathematics as a survival tool is implemented in almost every phase of personal and business life. How effectively one is able to use that important tool may determine how well one is able to prepare for the future, manage one's personal and business resources, and benefit from one's efforts as a consumer, worker, or business person.

The main purpose of *Mathematics with Business Applications,* Third Edition, is to assist students in learning to use mathematics effectively as a tool in their personal and business lives. After students have completed their study of the textbook and related ancillaries, they will be able to

1. Understand terminology relating to personal and business mathematics applications.
2. Apply basic math skills to the solution of both personal and business applications.
3. Use common mathematics formulas to solve a variety of personal and business mathematics problems.

Organization of the Textbook

While retaining the sound organization and approach of the first and second editions *Mathematics with Business Applications,* Third Edition focuses even more strongly on the review, remediation, and reinforcement of basic math skills. An expanded workshop section offers methods students can use to approach the solving of problems in a more logical manner. Computer activities, accompanied by an optional template diskette, provide the opportunity for students to use the computer in solving selected problems. Additional mini-simulations in the workbook add a touch of realism and heighten student interest.

The textbook is organized into four sections: Part One, Basic Math Skills Workshop: A Review of Fundamentals; Part Two, Personal Application Mathematics; Part Three, Business Applications; and Reference Files.

Basic Math Skills Workshop: A Review of Fundamentals

Basic math skills are an essential element of functional literacy. As an aid to students who need to review basic math skills or who need reinforcement of those skills, we have expanded the basic skills workshops.

Sixteen new workshops have been added to the third edition for a new total of thirty workshops. Each workshop includes a description of the covered skill, a variety of examples, and numerous applications and problems that allow students to apply each specific skill. These workshops provide students with a foundation of basic skills they will need during their study of Parts Two and Three of the textbook.

PREFACE

Personal Application Mathematics

In Part Two of the textbook, Personal Application Mathematics, students examine the personal applications of basic math skills used by workers and consumers. These applications include the calculation of gross income and net income, the use of both checking and savings accounts, cash purchases, charge purchases, loans, the cost of owning and operating an automobile, housing costs, insurance and investments, and budgets.

Business Applications

In Part Three, Business Applications, students apply mathematics fundamentals to realistic business situations as they pertain to eleven different departments of a large business. Representative business departments covered include personnel, production, purchasing, sales, marketing, warehousing and distribution, services, accounting, accounting records, financial, and corporate planning.

Parts Two and Three of the textbook are organized into units covering a specific type of application or a business department. Each unit is composed of numbered lessons. Each lesson begins with a clear, concise statement of the objective to be met followed by a brief explanation of the personal or business application to be examined in the lesson. The basic mathematical formula to be used in the lesson is followed by a

worked-out, sample problem that illustrates in detail how to solve the application. Table, phrase, and word problems—all graded in order of difficulty—give the students a variety of opportunities to apply the lesson formula. Many lessons contain "A Brief Case" problem, which challenges students to use critical thinking skills as they apply the lesson formula to a real world situation. Each lesson concludes with a set of basic math review problems that help students maintain or improve their basic math skills.

Each unit is followed by review activities; a feature on spreadsheet applications; and a one-page unit test. A career feature is included at the end of each odd-number unit. Also, five mini-simulations are spaced throughout the textbook to provide interesting and realistic opportunities for students to apply their math skills.

Reference Files

Extensive reference files are included at the end of the student textbook to help students who may need to review basic skills or may want to use the files as study aids in completing individual lessons in the textbook. The reference files contain a Skills File of thirty-two numbered computational math skills with practice problems; an Applications File of twenty-six common practice applications; and a Terms File of key terms used in business and consumer situations. These skills, applications, and terms

PREFACE

are keyed to the examples within the textbook lessons. Students who need further explanation or practice can turn to the reference files and complete the problems found there to reinforce lesson content or basic math skills.

In addition to the preceding files, the textbook also contains several tables that can be used in solving related problems, as well as the answers to the Self-Check and odd-number problems in the text lessons.

Supporting Materials

In addition to the student textbook, the program includes the following: a new teacher's wraparound edition of the textbook; a problems and simulations workbook with additional practice, review, and reinforcement; a teacher's edition of the workbook; a teacher's resource book that contains additional test material, a solution key to all the problems in the textbook, and answers to the Spreadsheet Application problems as well as instructions for using the template diskette to solve the spreadsheet activities in the textbook.

Acknowledgments

We wish to acknowledge the valuable contributions made to this textbook by teachers, students, colleagues, friends, and collaborators throughout its various stages of revision. In particular, we would like to express our appreciation to Professors James Arbaugh, Clayton L. Ziegler, and Robert A. Siddens of the University of Toledo for their expert advice in accounting, engineering, and transportation. We would also like to thank Carla Christy and Danuta T. Lange, who checked solutions and helped prepare the manuscript.

We would like to acknowledge the significant contributions of Robert D. Mason, Professor Emeritus, The University of Toledo—College of Business, with whom we co-authored previous editions of this textbook.

Walter H. Lange

Temoleon G. Rousos

P A R T

Basic Math Skills Workshop:
A Review of Fundamentals

As you use this book, you will need to know certain mathematical fundamentals. To be sure you are prepared, here are lessons for review and practice. Each workshop uses realistic business problems to show how math is used in different job settings.

Writing Numbers as Words and Rounding Numbers

Look up Skills 1 and 2 on pages 582–584 for more practice.

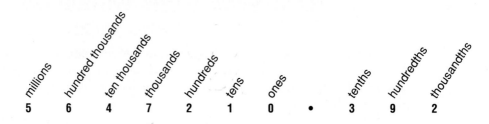

5   6   4   7   2   1   0   •   3   9   2

The place-value chart shows the value of each digit in the number 5,647,210.392. For example, the digit 7 is in the thousands place and has a value of 7000. The digit 9 is in the hundredths place and has a value of $\frac{9}{100}$, or 0.09. The place-value chart can help you write numbers.

EXAMPLE	SOLUTION (read as or written as)
536	five hundred thirty-six
72,051	seventy-two thousand fifty-one
9.227	nine and two hundred twenty-seven thousandths
$45.63	forty-five and sixty-three one hundredths dollars
	or forty-five and $\frac{63}{100}$ dollars
	or forty-five dollars and sixty-three cents

✓ SELF-CHECK Complete the problems, then check your answers in the back of the book.

Write in words.

1. 241   *two hundred + fourty-one*
2. 7.317   *seven + three hundred seventeen*
3. $761.13   *seven hundred sixty one thousand + thirteen hundreaths*

Place value is also used in rounding numbers. If the digit to the right of the place value you want to round is 5 or more, round up by adding 1 to the number in the place value and then change all the digits to the right of the place value to zeros. If the number is 4 or less, round down by changing all the numbers to the right of the place value to zeros.

EXAMPLE	Round 7862 to the nearest hundred.

SOLUTION	
7862	**A.** Find the digit in the hundreds place. It is 8.
7862	**B.** Is the digit to the right 5 or more? Yes.
7900	**C.** Add 1 to the hundreds place. Change the digits to the right to zeros.

EXAMPLE	Round 0.637 to the nearest tenth.

SOLUTION	
0.637	**A.** Find the digit in the tenths place. It is 6.
0.637	**B.** Is the digit to the right 5 or more? No.
0.6	**C.** Do not change the tenths digit. Drop the digits to the right.

Dollar and cents amounts are often rounded to the nearest cent, or the hundredths place. Begin with the digit to the right of the hundredths place, or the thousandths place.

EXAMPLE	**SOLUTION**
$13.6341	$13.63
$112.4171	$112.42
$7918.325	$7918.33

✓ SELF-CHECK Complete the problems, then check your answers in the back of the book.

Round to the nearest cent.

4. $21.277 *$21.30*   5. $967.461

6. $138.7836 *138.80*   7. $647.555

PROBLEMS

Write as numbers.

8. five thousand six hundred seventy-two *$5672,*

9. sixty-five and $\frac{75}{100}$ dollars

10. one hundred four and forty-nine thousandths *104.049*

11. ninety-five and $\frac{7}{100}$ dollars

Write in word form.

12. 216,798   13. $15.43   14. 214.093   15. 300.5   16. $9.81

12) two hundred sixteen thousand, seven hundred + ninety eight

14) two hundred fourteen + ninety three thousandths

16) Nine dollars + eighty one cents.

Round to the nearest place value shown.

Ten thousand	**17.** 416,719	**18.** 514,917	**19.** 960,817	**20.** 5,149,601
Thousand	**21.** 312,410	**22.** 42,614	**23.** 867,501	**24.** 7,523,614
Hundred	**25.** 41,515	**26.** 3156	**27.** 17,771	**28.** 21,910
Ten	**29.** 7291	**30.** 694	**31.** 21,945	**32.** 30,740
One	**33.** 3.341	**34.** 516.17	**35.** 42.716	**36.** 147.88
Tenth	**37.** 21.217	**38.** 86.554	**39.** 4718.24	**40.** 717.97178
Hundredth	**41.** 5.2167	**42.** 15.518	**43.** 31.79861	**44.** 14.52499

Round 15,748,516 to the place value given.

45. millions **46.** ten millions **47.** thousands
48. hundreds **49.** ten thousands **50.** hundred thousands

Round to the nearest place value shown.

Cent	**51.** $14.7151	**52.** $49.592	**53.** $72.195	**54.** $94.1249
Ten cents	**55.** $119.864	**56.** $8.7199	**57.** $7.0451	**58.** $89.975
One dollar	**59.** $151.912	**60.** $510.47	**61.** $623.97	**62.** $53.099
Ten dollars	**63.** $3145	**64.** $1789.75	**65.** $515.11	**66.** $710.00
Hundred dollars	**67.** $19,876	**68.** $21,050	**69.** $4132	**70.** $74,656

71. In his job as an accountant for the Tarwood Manufacturing Co., Steve Alexander needs to write numbers as words. Write each check amount in words.

a. $7.17 **b.** $9.75 **c.** $42.91 **d.** $51.80
e. $217.23 **f.** $531.49 **g.** $9751.63 **h.** $1270.82

Tarwood Manufacturing Company   1403
2168 Main Street
Baltimore, Maryland 21233

July 14 19 ——   54-168 / 114

PAY TO THE ORDER OF *Giant Supply Co., Inc.* $ 917.82
Nine hundred seventeen and 82/100 ——————— DOLLARS

Derry Bank & Trust Company

FOR _____   *Steve Alexander*

⑈001403⑈ ⑆011401685⑆ ⑈0802 0450⑈

72. Amy Cole records the attendance data for the sporting events at the Glass Bowl. Round each number to the amount indicated.

	Football Game	Number	Nearest Thousand	Nearest Hundred	Nearest Ten
a.	Austin Peay	18,971	?	?	?
b.	Bowling Green	25,687	?	?	?
c.	Ohio University	20,119	?	?	?
d.	Western Michigan	24,567	?	?	?
e.	Central Michigan	19,424	?	?	?
f.	Ball State	21,972	?	?	?
g.	Kent State	19,876	?	?	?
h.	Miami	23,947	?	?	?

73. To get an overall sales picture of first-quarter sales, the marketing executive wants the sales figures in round numbers. Round the sales of each company to the nearest thousand dollars.

FIRST-QUARTER SALES, 19--

	Company	Sales
a.	Able Computers	$617,571
b.	Crawford Electronics	719,417
c.	Diamond Research, Inc.	271,710
d.	Laser Systems, Inc.	914,620
e.	Meltex Graphics	535,498
f.	Perfect Software	419,501
g.	Quality Communications	178,071
h.	Superior Limited	571,719
i.	Total Surveys, Inc.	423,412

74. John Brown, an inventory clerk, often rounds inventory figures for easier handling. Round the numbers from the inventory list to the nearest ten.

CORN INVENTORY

	Description	Quantity	Rounded
a.	Golden Cross	119 lb	?
b.	Golden Beauty	221 lb	?
c.	Early Glo	476 lb	?
d.	X-tra Sweet	219 lb	?
e.	White Sunglo	431 lb	?
f.	Silver Queen	578 lb	?
g.	Honey-Cream	364 lb	?
h.	Country Gtlm	597 lb	?

WORKSHOP 2

Comparing Numbers

Look up Skill 1 on pages 582–583 for more practice.

To compare numbers to see which one is greater, first find out if one of the numbers has more digits to the left of the decimal point. For example, 100 is greater than 99.62 because 100 has three digits to the left of the decimal point and 99.62 has only two. If both numbers have the same number of digits to the left of the decimal point, compare each digit, beginning at the left. One way is to write the value of each digit.

EXAMPLE   Compare 5972 and 5983. Which number is greater?

SOLUTION
$5972 = 5000 + 900 + 70 + 2$
$5983 = 5000 + 900 + 80 + 3$
            ▲   ▲   ▲
         same   same   80 is greater than 70, so 5983 is
                              greater than 5972.

You may not have to write the value of each digit. Just compare each digit, starting at the left.

EXAMPLE   Compare 237.51 and 237.29. Which number is greater?

SOLUTION
2   3   7 • 5   1
2   3   7 • 2   9
▲   ▲   ▲   ▲
same same same   5 is greater than 2, so 237.51 is
                      greater than 237.29.

✔ SELF-CHECK   Complete the problems, then check your answers in the back of the book.

Which number is greater?

1. 8891 and 8889   **2.** 759.39 and 759.80

3. 3064 and 3079   **4.** 56.633 and 56.629

When you are comparing decimals, you may have to write zeros to the right of the decimal point so that the numbers being compared have the same number of decimal places.

EXAMPLE   Compare 5.17 and 5.149 (without writing the value of each digit). Which number is greater?

SOLUTION   5.170 ◄—Write in the zero.
                    5.149
                       ▲
                    7 is greater than 4, so 5.17 is greater than 5.149.

✔ SELF-CHECK   Complete the problems, then check your answers in the back of the book.

Which number is greater?

5. 284.4 and 284.396   **6.** 0.06 and 0.006   **7.** 62.3 and 62.313

Which number is greater?

8. 26 or 29　　　　　**9.** 545 or 554　　　　**10.** 4201 or 4210

11. 3943 or 3934　　　**12.** 2.65 or 2.56　　　**13.** 81.2 or 80.9

14. 0.696 or 0.695　　　**15.** 0.4 or 0.04　　　**16.** 3.54 or 4

17. 0.1 or 0.9　　　　　**18.** 0.03 or 0.003　　**19.** 2.234 or 2.244

20. $8.79 or $8.97　　　**21.** $329.02 or $329.20　**22.** $13.11 or $11.13

23. "29 miles to the gallon in highway driving. 24.9 miles to the gallon in city driving." Which mileage is greater?

24. "Still available for the play-offs: 3428 box seats, 3128 grandstand seats." Which number is greater?

Write the numbers in order from lowest to highest.

25. 1.37, 1.36, 1.39　　　　　　　**26.** 5.11, 5.09, 5.10

27. 7.18, 7.38, 7.58　　　　　　　**28.** 5.86, 5.95, 5.81

29. 15.321, 15.231, 15.270　　　　**30.** 40.004, 40.04, 40.4

31. 121.012, 121.021, 121.210　　**32.** 365.15, 365.51, 365.490

33. 0.1234, 0.1342, 0.1423　　　　**34.** 0.2539, 0.4139, 0.3239

35. 10.100, 11.11, 11.01　　　　　**36.** 57.57, 57.571, 57.75

37. 3.1, 3.062, 3.09　　　　　　　**38.** 8.001, 8.02, 8.0135

39. Janet Swick works part-time shelving books in the school library. Arrange the following library call numbers from lowest to highest.

　a. Science: 513.12, 519.03, 532.626, 571.113, 587.41

　b. Literature: 94.79, 32.615, 11.7, 67.192, 34.9

　c. Religion: 46.94, 18.7, 15.04, 71.21, 26.311

　d. Language: 22.5, 67.21, 48.275, 38.9, 93.047

40. Fourteen books were returned in the last hour. To help Janet shelve them faster, put them in numerical order from lowest to highest.

824.62, 392.4, 329.213, 419.52, 723.6, 122.51, 918.7, 749.02, 728.31, 927.263, 918.712, 826.401, 615.02, 131.06

41. Janet also files the cards from checked-out books, according to the card numbers. Between which two cards would she file the checked-out books?

　a. Card number of the checked-out book is 874.192.
　　Filed cards are: 872.41, 873.15, 877.142, 879.190

　b. Card number of the checked-out book is 332.75.
　　Filed cards are: 309.8, 311.75, 332.075, 332.749, 333.54

42. Use the tax table at the right to complete the following problem.

Amount	At least	But less than
$197.50	$195	$200
a. $193.40	?	?
b. $178.30	?	?
c. $180.00	?	?
d. $181.50	?	?
e. $195.01	?	?

FEDERAL WITHHOLDING TAX
WEEKLY Payroll Period

And the wages are:

At least	But less than
$170	$175
175	180
180	185
185	190
190	195
195	200

43. Shipping charges often depend on the dollar amount of the order, as shown in the table. Find the cost of shipping each order.

a. $9.90 **b.** $14.35
c. $8.54 **d.** $15.00
e. $12.44 **f.** $21.22
g. $45.00 **h.** $6.54
i. $11.70 **j.** $16.15

SHIPPING AND HANDLING

Up to $8.00 . . . . Add $1.95
$8.01 to $11.00 . . . . Add $2.45
$11.01 to $15.00 . . . . . . Add $2.95
$15.01 to $20.00 . . . . . . Add $3.45
$20.01 to $30.00 . . . . . . Add $3.95
Over $30.00 . . . . Add $4.45

44. The minimum payment required on a charge account can be determined from a minimum payment schedule. Find the minimum payment for each balance.

a. $75.90 **b.** $322.82
c. $520.00 **d.** $782.29
e. $9.30 **f.** $427.20
g. $800.01 **h.** $26.13
i. $330.85 **j.** $700.00

MINIMUM PAYMENT SCHEDULE

New Balance	Minimum Monthly Payment
$.01 to $ 300.00	$25.00
$300.01 to $ 400.00	$30.00
$400.01 to $ 500.00	$35.00
$500.01 to $ 600.00	$40.00
$600.01 to $ 700.00	$45.00
$700.01 to $ 800.00	$50.00
$800.01 to $ 900.00	$55.00
$900.01 to $1000.00	$60.00

45. Olympic scores are often given in decimals. In racing events, the lowest time wins. List the order of finish for each event.

a. Men's 1500 m

Baumann	3:15.52
Ngugi	3:11.70
Kunze	3:15.73

b. Women's 1500 m

Samoienko	4:00.30
Ivan	3:53.96
Baikauskaite	4:00.24

c. Men's 400 m relay

Spain	38.19 s
France	38.40 s
Britain	38.28 s

In some events, the highest number wins. Determine the winner for each of following events.

d. Men's discus

Oubartas	221 ft $4\frac{1}{2}$ in
Schult	225 ft $9\frac{1}{4}$ in
Danneberg	221 ft $0\frac{1}{2}$ in

e. Women's shot put

Meisu	69 ft 1 in
Lisovskaya	72 ft $11\frac{1}{2}$ in
Neimke	69 ft $1\frac{1}{2}$ in

f. Men's archery

S. Korea	986 pts.
Britain	968 pts.
USA	972 pts.

Adding Decimals

Look up Skill 5 on page 587 for more practice.

When adding decimals, write the addition problem in vertical form. Be sure to line up the decimal points. Write a decimal point in the answer directly below the decimal points in the problem. Then add as you would whole numbers.

EXAMPLE   26.94 + 31.71 + 19.47

SOLUTION   Line up decimals.

```
            26.94      Add.    26.94
            31.71              31.71
            19.47            + 19.47
                               78.12
```

✔ SELF-CHECK   Complete the problems, then check your answers in the back of the book.

1. 17.56 + 11.73 + 12.47 = ?   **2.** 31.71 + 22.46 + 41.97 = ?
3. 53.15 + 41.71 + 79.84 = ?   **4.** 41.98 + 74.71 + 81.96 = ?

When adding amounts with different numbers of decimal places, you may want to write zeros in the empty decimal places. Put a decimal point in any whole number included in the problem.

EXAMPLE   41.51 + 117.483 + 7 + 19.6

SOLUTION   Line up decimals.   Write zeros.   Add.

```
            41.51               41.510           41.510
           117.483             117.483          117.483
             7.                   7.000            7.000
            19.6                 19.600         + 19.600
                                                185.593
```

Adding amounts of money is just like adding decimals. The decimal point separates the dollars and cents. Remember to put a dollar sign in the total.

EXAMPLE $43.78 + $6.93

SOLUTION   Line up decimals. $43.78   Add. $43.78
                                      6.93   + 6.93
                                                     $50.71

✔ SELF-CHECK   Complete the problems, then check your answers in the back of the book.

5. 114.7 + 31.83 + 51.678 = ?   **6.** 13.7 + 6.012 + 17.412 + 6 = ?
7. $14.71 + $21.84 + $7.51 = ?   **8.** $75 + $14.40 + $21.72 = ?

PROBLEMS

```
 9.    36.74      10.    78.91      11.    715.17      12.    513.88
     + 51.27            +  4.17           + 612.94           + 721.93
```

13.	71.172 + 14.835	**14.**	64.574 + 76.915	**15.**	3.171 + 7.847	**16.**	7.974 + 4.157

17.	71.91 84.76 + 48.53	**18.**	44.76 53.12 + 87.94	**19.**	7.4189 4.5567 + 3.7124	**20.**	17.471 35.683 + 71.457

21.	417.9 2.171 + 43.42	**22.**	119.71 14.861 + 9.5	**23.**	2.178 47.5 + 917.	**24.**	61.178 191.41 + 7.786

25.	7.456 517.1 + 120.8	**26.**	7180.9 15.754 + 617.9	**27.**	14.79 3.1451 + 976.8	**28.**	0.42 817. + 41.7181

29.	$17.98 14.31 + 19.47	**30.**	$161.91 72.87 + 54.36	**31.**	$17.56 71. + 41.84	**32.**	$517. 14.97 + 416.34

33. 17.47 + 34.71 + 56.78 + 15.07   **34.** 0.17 + 17.94 + 13 + 147.7

35. 14.6 + 19.314 + 4.71 + 264.5176   **36.** 4.917 + 6 + 4.37 + 15.971

37. $1.98 + $71.49 + $0.49 + $50   **38.** $7.79 + $0.89 + $412.37 + $7

39. $71.84 + $2.79 + $143.54 + $71   **40.** $7.98 + $4.14 + $71.84 + $0.47

APPLICATIONS   Find the net deposit for each bank deposit slip by adding the deposits.

41.

		DOLLARS	CENTS
CASH	CURRENCY	17	00
	COINS		
CHECKS	LIST SEPARATELY 89-1	114	79
	89-2	17	08
	SUBTOTAL		
	⬦ LESS CASH RECEIVED		
	TOTAL DEPOSIT	?	

42.

		DOLLARS	CENTS
CASH	CURRENCY		
	COINS	4	75
CHECKS	LIST SEPARATELY 85-12	17	85
	85-10	71	94
	85-11	114	47
	SUBTOTAL		
	⬦ LESS CASH RECEIVED		
	TOTAL DEPOSIT	?	

43.

		DOLLARS	CENTS
CASH	CURRENCY	7	00
	COINS		
CHECKS	LIST SEPARATELY 95-76	314	78
	98-11	37	14
	95-13	212	71
	SUBTOTAL	?	
	⬦ LESS CASH RECEIVED		
	TOTAL DEPOSIT	?	

Complete the sales receipts by finding the subtotal and the total.

44.

DATE 3/14/-	AUTH. NO 42	IDENTIFACTION	CLERK JR	REG/DEPT.	☑TAKE ☐SEND
QTY	CLASS	DESCRIPTION	PRICE	AMOUNT	
1		dress shirt		24	95
1		sweater		49	98

a. The issuer of the card identified on this item is authorized to pay the amount shown as TOTAL upon proper presentation. I promise to pay such TOTAL together with any other charges due thereon subject to and in accordance with the agreement governing the use of such card.

CUSTOMER SIGNATURE X *Betty Clark*

	SUBTOTAL	?
	TAX	4 50
b. | SALES SLIP | TOTAL | ? |

45.

DATE 6/1/-	AUTH. NO 86430	IDENTIFACTION	CLERK DL	REG/DEPT.	☑TAKE ☐SEND
QTY	CLASS	DESCRIPTION	PRICE	AMOUNT	
1		Knit shirt		32	98
2		Pairs Socks	3.19 ea.	6	38
1		Can Tennis Balls		2	19

a. The issuer of the card identified on this item is authorized to pay the amount shown as TOTAL upon proper presentation. I promise to pay such TOTAL together with any other charges due thereon subject to and in accordance with the agreement governing the use of such card.

CUSTOMER SIGNATURE X *Ed Mack*

	SUBTOTAL	?
	TAX	3 32
b. | SALES SLIP | TOTAL | ? |

46. The personnel office of Color Sales, Inc., received this partially completed travel expense report for Beatrice Casey. Find the missing totals. First add the total expenses for each category horizontally. Then add all expenses for each day vertically. Add the Daily Totals horizontally. The sum of the vertical totals should equal the sum of the horizontal totals.

Color Sales, Inc.
419 N. Second St.
Allen, NY 01212

Travel Expense Report

Name _Beatrice Grey_   Position _Sales Representative_

Destination _Philadelphia, PA_

Purpose of Trip _Regional Sales Meeting_

Date(s) of Trip _12/5-9_

	Expense Item	Sun.	Mon. 12-5	Tue. 12-6	Wed. 12-7	Thur. 12-8	Fri. 12-9	Sat.	TOTALS
a.	Hotel		104.74	104.74	104.74	104.74			?
b.	Breakfast			8.14	9.16	8.71	11.14		?
c.	Lunch			9.17	8.24	6.54	7.19		?
d.	Dinner		21.74	19.91	22.41	25.15			?
e.	Tips		5.00	6.00	6.00	6.00	2.00		?
	Entertainment								
f.	Taxi		10.00				12.00		?
	Public trans.								
	Parking								
	Tolls								
	Car rental								
g.	Airfare		240.00						?
h.	Telephone		6.50	7.15		9.50	4.70		?
	DAILY TOTALS		?	?	?	?	?	?	Total Before Mileage

    i.   j.   k.   l.   m.   n.

47. These 5 contestants at the gymnastic tryouts received the scores listed below. Find the total for each contestant.

	Judge:	A	B	C	D	E	F	G	H	Total
	Contestant									
a.	Mark	4.7	4.9	5.1	5.0	4.9	5.3	4.7	4.8	?
b.	Steve	6.1	6.3	6.0	5.9	6.4	6.3	6.0	5.8	?
c.	Tom	9.8	9.6	9.5	10.0	9.7	9.6	10.0	9.4	?
d.	John	9.5	9.4	9.6	9.3	9.2	9.6	9.5	9.4	?
e.	Brad	8.7	8.8	9.1	8.5	8.7	8.8	8.9	9.0	?

Subtracting Decimals

Look up Skill 6 on page 588 for more practice.

When subtracting decimals, write the subtraction problem in vertical form. Be sure to line up the decimal points. Write a decimal point in the answer directly below the decimal points in the problem. Then subtract as you would with whole numbers.

EXAMPLE

85.29 − 34.72 ➡

SOLUTION

Line up decimals.
85.29
34.72
➡

Subtract.
85.29
− 34.72
50.57

When subtracting amounts with different numbers of decimal places, you may want to write zeros in the empty decimal places. Put a decimal point in any whole number included in the problem.

EXAMPLE

838.5
− 39.248
➡

SOLUTION

Put in zeros and subtract.
838.500
− 39.248
799.252

EXAMPLE

47.49
− 9
➡

SOLUTION

Put in zeros and subtract.
47.49
− 9.00
38.49

✔ SELF-CHECK Complete the problems, then check your answers in the back of the book.

1. 87.86 − 34.25 = ?
3. 675.4 − 65.32 = ?

2. 125.9 − 87.6 = ?
4. 76.76 − 8 = ?

Subtracting amounts of money is just like subtracting decimals. The decimal point separates the dollars and cents. Remember to put a dollar sign in the answer.

EXAMPLE

$89.59
− 56.37
➡

SOLUTION

$89.59
− 56.37
$33.22

EXAMPLE

$572.
− 89.64
➡

SOLUTION

$572.00
− 89.64
➡

$572.00
− 89.64
$482.36

✔ SELF-CHECK Complete the problems, then check your answers in the back of the book.

5. $47.85
− 31.07

6. $998.41
− 86.35

7. $679.13
− 346.98

8. $114.
− 74.95

PROBLEMS

9. 94.7
− 31.4

10. 98.6
− 88.5

11. 19.87
− 8.54

12. 7.93
− 2.03

13. 49.64
− 10.34

14. 96.13
− 12.37

15. 38.065
− 33.426

16. 68.111
− 9.648

17.	18.	19.	20.
28.37 − 3.067	28.36 − 4.239	6.1536 − .1985	6.9909 − 3.9999

21.	22.	23.	24.
87.492 − 31.875	43.686 − 20.320	754.8 − 34.632	895.1 − 33.47

25.	26.	27.	28.
420. − 12.78	201. − 9.764	3.2 − .459	36. − .357

29.	30.	31.	32.
$99.85 − 32.16	$75.67 − 28.30	$953.22 − 287.32	$506.37 − 243.70

33.	34.	35.	36.
$347. − 82.97	$275. − 85.12	$55,553.65 − 38,872.58	$47,005.45 − 3257.64

37. 335.4 − 217.9 **38.** 148.1 − 132.5 **39.** 5.21 − 0.71

40. 348.1 − 25.38 **41.** 23.57 − 0.6421 **42.** 537 − 46.11

43. $61.92 − $33.01 **44.** $75.99 − $31.49 **45.** $8 − $2.63

46. $434.66 − $51.43 **47.** $3479.31 − $2616.16 **48.** $6000 − $4333.83

APPLICATIONS

49. You work as a cashier in a restaurant. Compute the correct change for each of the following orders.

	Customer's Order	Customer Gives You	Change
a.	$ 6.94	$ 7.00	?
b.	$ 9.12	$10.00	?
c.	$16.97	$17.00	?
d.	$ 3.42	$ 5.42	?
e.	$ 5.01	$ 5.01	?
f.	$23.11	$25.00	?
g.	$41.97	$42.07	?
h.	$27.42	$30.00	?
i.	$20.13	$21.13	?
j.	$150.84	$151.00	?

Complete the computational part of the following bank deposit tickets.

50.

		DOLLARS	CENTS
CASH	CURRENCY	2	00
	COINS		
CHECKS	LIST SEPARATELY 1-43	22	76
	1-44	27	80
a.	SUBTOTAL	?	
	LESS CASH RECEIVED	10	00
b.	TOTAL DEPOSIT	?	

51.

		DOLLARS	CENTS
CASH	CURRENCY		
	COINS	3	50
CHECKS	LIST SEPARATELY 57-12	42	50
	57-10	86	55
a.	SUBTOTAL	?	
	LESS CASH RECEIVED	20	00
b.	TOTAL DEPOSIT	?	

52.

		DOLLARS	CENTS
CASH	CURRENCY	13	00
	COINS	6	41
CHECKS	LIST SEPARATELY 85-76	155	40
	88-11	25	35
		15	75
a.	SUBTOTAL	?	
	LESS CASH RECEIVED	45	00
b.	TOTAL DEPOSIT	?	

53.

	DOLLARS	CENTS
CURRENCY		
COINS	17	25
LIST SEPARATELY 50-50	16	40
50-75	35	83
45-30	27	11
a. SUBTOTAL	?	
LESS CASH RECEIVED	40	00
b. TOTAL DEPOSIT	?	

54.

	DOLLARS	CENTS
CURRENCY	137	00
COINS	12	35
LIST SEPARATELY 66-40	15	27
60-50	19	38
65-70	214	51
a. SUBTOTAL	?	
LESS CASH RECEIVED	75	00
b. TOTAL DEPOSIT	?	

55.

	DOLLARS	CENTS
CURRENCY	21	00
COINS	37	75
LIST SEPARATELY 40-10	43	33
45-50	37	40
42-13	437	65
a. SUBTOTAL	?	
LESS CASH RECEIVED	75	00
b. TOTAL DEPOSIT	?	

56. A partially completed budget comparison statement for the Roadway Bicycle Club is shown below.

ROADWAY BICYCLE CLUB   BUDGET COMPARISON STATEMENT
For the year 19--

ITEM	ESTIMATED	ACTUAL	DIFFERENCE	OVER/UNDER?
Receipts				
Dues	3 6 0 0 00	5 2 0 0 00	?	?
Bike-a-thon	4 0 0 0 00	3 2 4 00	?	?
Raffle	1 0 0 0 0 00	1 1 4 2 00	?	?
T-shirt sales	2 7 5 00	2 3 1 60	?	?
Total Receipts	?	?		

a. Find the club's total estimated and total actual receipts.

b. Complete the difference column by subtracting the Actual column from the Estimated column. Label the difference as being "over" or "under" the estimate.

57. The marketing department for Jackson Sporting Goods Company prepared a comparison sheet that shows sales projections and actual sales for ten products. Find the difference between the projected sales and the actual sales by subtracting actual sales from projected sales.

	Jackson Sporting Goods Company		
	Product	Sales Projections	Actual Sales	Difference
a.	Electric fishing motors	$ 3000.00	$ 2749.67	?
b.	Fishing tackle	$ 32,000.00	$ 31,897.40	?
c.	Wilderness boots	$ 6500.00	$ 5607.15	?
d.	Boat moccasin	$ 950.00	$ 571.28	?
e.	Enamel campware sets	$ 400.00	$ 339.67	?
f.	Geodesic dome tent	$ 6400.00	$ 5379.58	?
g.	Backpacking tent	$ 2500.00	$ 1999.40	?
h.	Sleeping bags	$ 6300.00	$ 5919.75	?
i.	Sleeping bag pads	$ 800.00	$ 589.20	?
j.	Folding cot	$ 740.00	$ 731.52	?

Multiplying Decimals

Look up Skill 8 on page 590 for more practice.

When multiplying decimals, multiply as if the decimal numbers were whole numbers. Then count the total number of decimal places to the right of the decimals in the factors. This number will be the number of decimal places in the product.

EXAMPLE	SOLUTION

```
    18.1   ← factor          18.1   ←      1 decimal place
 ×  0.35   ← factor       ×  0.35   ←    + 2 decimal places
     905                      905
     543                      543
    6335   ← product       6.335   ←      3 decimal places
```

If the product does not have enough digits to place the decimal in the correct position, you will need to write zeros. Start at the right of the product in counting the decimal places and write zeros at the left.

EXAMPLE	SOLUTION

```
     0.72                    0.72   ←      2 decimal places
  ×  0.03                 ×  0.03   ←    + 2 decimal places
     216                   0.0216   ←      4 decimal places
```

✔ SELF-CHECK Complete the problems, then check your answers in the back of the book.

1.	**2.**	**3.**	**4.**
24.7	41.8	0.78	0.74
× 0.33	× 2.14	× 0.11	× 0.08

When multiplying amounts of money, round the answer off to the nearest cent. Remember to put a dollar sign in the answer.

EXAMPLE	SOLUTION	

```
   $3.35                  $3.35   ←    2 places      $3.35 × 4.5 = $15.075
 ×   4.5               ×    4.5   ←  + 1 place                  = $15.08
   15075              $15.075   ←    3 places     rounded to nearest cent
```

When multiplying by 10, 100, or 1000, count the number of zeros and then move the decimal point to the right the same number of spaces.

EXAMPLE 1	SOLUTION	

6.7 × 10   6.7 × 10 = 6.7 = 67.   10 has 1 zero; move decimal 1 place.

EXAMPLE 2	SOLUTION	

5.24 × 100   5.24 × 100 = 5.24 = 524.   100 has 2 zeros; move decimal 2 places.

EXAMPLE 3	SOLUTION	

9.4718 × 1000   9.4718 × 1000 = 9.4718   1000 has 3 zeros; move decimal 3 places.

= 9471.8

Complete the problems, then check your answers in the back of the book.

5. $4.15 × 8.5   **6.** 71.4 × 10   **7.** 41.861 × 100   **8.** 3.1794 × 1000

PROBLEMS

9. 41.3
  × 0.2

10. 78.4
  × 0.3

11. 84.8
  × 0.25

12. 74.5
  × 0.15

13. 4.31
  × 0.71

14. 51.7
  × 0.72

15. 97.8
  × 0.31

16. 51.7
  × 0.67

17. 0.62
  × 0.06

18. 0.86
  × 0.08

19. 0.41
  × 0.02

20. 0.74
  × 0.08

21. 0.51
  × 0.06

22. 0.93
  × 0.07

23. 0.635
  × 0.005

24. 0.051
  × 0.003

25. 4.171
  × 0.04

26. 7.823
  × 0.09

27. 3.471
  × 0.63

28. 9.875
  × 0.42

29. 0.071
  × 0.002

30. 0.207
  × 0.034

31. 0.973
  × 1.918

32. 0.573
  × 6.423

33. $2.25
  × 20

34. $5.15
  × 85

35. $5.85
  × 3.5

36. $3.35
  × 7.6

37. $18.74
  × 2.3

38. $87.98
  × 8.7

Multiply by 10   **39.** 31.7   **40.** 5.71   **41.** 6.517   **42.** 0.9186

Multiply by 100   **43.** 32.85   **44.** 41.786   **45.** 3.987   **46.** 412.42

Multiply by 1000   **47.** 72.716   **48.** 7.1956   **49.** 0.415   **50.** 916.518

Multiply by 10,000   **51.** 6.9178   **52.** 3.42876   **53.** 94.7189   **54.** 723.471

APPLICATIONS

55. Find the total price for each item and the total order amount.

	Quantity	Description	Price Each	Total Price
a.	3	Notebooks	$1.29	?
b.	2	Automatic pencils	3.98	?
c.	2	Pads of 3 × 5 cards	1.49	?
d.	1	Pen	9.98	?
e.	4	Packs of paper	2.19	?
f.	3	Boxes of paper clips	3.09	?
g.	1	Stapler with staples	7.49	?
h.		Total		?

56. Find the total price of each item, the total list price, the 10% trade discount, the net price, the 2% cash discount, the cash price, the 6% sales tax, and the final price.

	Quantity	Item Number	Unit Price	Total Price
a.	32	2Z-143	$47.85	?
b.	15	5R-719	$14.73	?
c.	75	6S-427	$ 3.87	?
d.		Total List Price		?
e.		− Trade Discount (× 0.1)		?
f.		Net Price		?
g.		− Cash Discount (× 0.02)		?
h.		Cash Price		?
i.		+ Sales Tax (× 0.06)		?
j.		Final Price		?

57. Find the gross pay for each employee by multiplying the hourly rate times the number of hours. Find social security by multiplying the gross pay by 0.0765, FIT by multiplying by 0.25, SIT by multiplying by 0.045. Round each deduction to the nearest cent. Find the total deductions and subtract from the gross pay to find the net pay.

	Employee	Hourly Rate	Number of Hours	Gross Pay	Deductions			Total Deductions	Net Pay
					Social Security	FIT	SIT		
a.	Faye White	$7.85	40	?	?	?	?	?	?
b.	John Zielinski	$8.50	36	?	?	?	?	?	?
c.	Bill Adams	$8.75	40	?	?	?	?	?	?
d.	Jane Boyce	$9.70	35	?	?	?	?	?	?
e.	Al Tone	$9.70	40	?	?	?	?	?	?
f.			Total	?	?	?	?	?	?

Dividing Decimals

Look up Skills 9, 10, and 11 on pages 591, 592, and 593 for more practice.

When dividing decimals, first check if there is a decimal point in the divisor. If there is, move the decimal point to the right to make the divisor a whole number. Then move the decimal point in the dividend to the right the same number of places you moved the decimal point in the divisor. Write the decimal point in the quotient directly above the decimal point in the dividend. Then divide as with whole numbers.

$$\text{divisor} \rightarrow 3 \overline{)693} \quad \begin{matrix} \leftarrow \text{quotient} \\ \leftarrow \text{dividend} \end{matrix}$$

with 231 above the bar as quotient

EXAMPLE

$23.78 \div 5.8$

or

$5.8 \overline{)23.78}$

SOLUTION

$5.8 \overline{)23.78}$ ➡

$$\begin{array}{r} 4.1 \\ 58 \overline{)237.8} \\ -232 \\ \hline 5\ 8 \\ -5\ 8 \end{array}$$

Add zeros to the right of the decimal point in the dividend if needed.

EXAMPLE

$0.147 \div 0.42$

or

$0.42 \overline{)0.147}$

SOLUTION

$0.42 \overline{)0.147}$ ➡

$$\begin{array}{r} 0.35 \\ 42 \overline{)14.70} \\ -12\ 6 \\ \hline 2\ 10 \\ -2\ 10 \end{array}$$ ← zero added

✓ SELF-CHECK Complete the problems, then check your answers in the back of the book.

1. $35.96 \div 5.8$

2. $12.9 \overline{)55.341}$

3. $1.47 \div 0.05$

4. $0.052 \overline{)1.872}$

When the dividend is an amount of money, remember to place the dollar sign in the quotient and round the answer to the nearest cent.

EXAMPLE

$24 \overline{)\$47.56}$

SOLUTION

$$\begin{array}{r} \$1.981 \\ 24 \overline{)\$47.560} \end{array}$$

$\$47.56 \div 24 = \$1.98$ ← rounded to nearest cent

When dividing by 10, 100, or 1000, count the number of zeros in 10, 100, or 1000 and move the decimal point to the left the same number of places.

EXAMPLE 1	SOLUTION	
9.3 ÷ 10	9.3 ÷ 10 = 09.3 = 0.93	10 has 1 zero; move decimal 1 place.

EXAMPLE 2	SOLUTION	
742.64 ÷ 100	742.64 ÷ 100 = 742.64 = 7.4264	100 has 2 zeros; move decimal 2 places.

EXAMPLE 3	SOLUTION	
13,436.1 ÷ 1000	13,436.1 ÷ 1000 = 13436.1 = 13.4361	1000 has 3 zeros; move decimal 3 places.

✔ SELF-CHECK Complete the problems, then check your answers in the back of the book.

5. $16.32 ÷ 12   **6.** 7.9 ÷ 10

7. 138.9 ÷ 100   **8.** 9862.8 ÷ 1000

PROBLEMS

9. $5.3\overline{)9.54}$ **10.** $3.2\overline{)11.2}$ **11.** $2.6\overline{)24.18}$ **12.** $3.2\overline{)40.32}$

13. $4.8\overline{)35.52}$ **14.** $2.2\overline{)70.84}$ **15.** $46\overline{)354.2}$ **16.** $9.9\overline{)564.3}$

17. $6.02\overline{)51.17}$ **18.** $3.12\overline{)39.936}$ **19.** $3.87\overline{)10.836}$ **20.** $3.04\overline{)29.336}$

21. $0.004\overline{)0.2928}$ **22.** $0.007\overline{)0.2415}$ **23.** $0.108\overline{)0.5616}$ **24.** $0.101\overline{)0.70498}$

Divide. Round answers to the nearest hundredth or to the nearest cent.

25. $4.3\overline{)7.871}$ **26.** $5.9\overline{)5.343}$ **27.** $7.36\overline{)88.34}$ **28.** $2.95\overline{)177.9}$

29. $0.88\overline{)46.98}$ **30.** $0.97\overline{)602.7}$ **31.** $14.32\overline{)8.027}$ **32.** $31.44\overline{)9.109}$

33. $31\overline{)\$82.42}$ **34.** $28\overline{)\$23.98}$ **35.** $78\overline{)\$678.40}$ **36.** $43\overline{)\$177.79}$

37. $62\overline{)\$267.32}$ **38.** $57\overline{)\$793.28}$ **39.** $961\overline{)\$9803.88}$ **40.** $817\overline{)\$4635.79}$

Divide by 10	**41.** 9.3	**42.** 14.42	**43.** 726.81	**44.** 0.034
Divide by 100	**45.** 429.8	**46.** 133.39	**47.** 8462.65	**48.** 0.746
Divide by 1000	**49.** 5896.9	**50.** 321.29	**51.** 22.098	**52.** 5.432
Divide by 10,000	**53.** 63,652.18	**54.** 3879.19	**55.** 415.49	**56.** 64.858

57. Ken Kowlinski drove 495 miles on 15.0 gallons of gas. How many miles per gallon did he get?

58. A 64-ounce bottle of detergent costs $3.49. What is the cost per ounce?

59. A 4.25-kilogram roast costs $17.39. What is the cost per kilogram?

60. As a clerk and stockperson at the Quick Stop Carryout, you are required to compute the cost per item (to the nearest cent) for the following items.

	Item	Cost per Case	Number per Case	Cost per Item
a.	Orange juice, 64 oz	$ 17.94	6	?
b.	Wild rice	$ 40.56	24	?
c.	Cake mix	$ 30.96	24	?
d.	Mushrooms, 4 oz	$ 6.09	12	?
e.	Cat food, 6 oz	$ 19.20	48	?
f.	Tuna fish, 6.5 oz	$ 25.36	36	?
g.	Batteries, 9 V	$108.00	60	?
h.	Cream cheese, 8 oz	$ 39.60	40	?
i.	Raisins, 15 oz	$ 31.20	24	?

61. Compute the miles per gallon (to the nearest tenth) for the types of vehicles leased by the Pilloid Plastics Company the first week in May.

	Type of Vehicle	Miles	Gallons of Fuel	Miles per Gallon
a.	Standard	665.4	40.3	?
b.	Intermediate	371.3	19.9	?
c.	Compact	407.0	16.8	?
d.	Subcompact	511.6	17.3	?
e.	Passenger van	291.5	22.1	?
f.	Motor home	423.7	64.2	?

62. The unit price of an item can be found by dividing the selling price by the weight. Determine the unit price of each item to the nearest thousandth.

	Item	Selling Price	Weight	Unit Price (3 decimal places)
a.	Vegetable oil	$1.79	32 oz	?
b.	Orange juice	$1.29	12 oz	?
c.	Olives	$1.39	7 oz	?
d.	Crackers	$1.59	16 oz	?
e.	Tea bags	$2.39	100 ctns	?
f.	Onion soup	$0.99	1.25 oz	?
g.	Dog food	$8.49	25 lb	?
h.	Pie spice	$1.69	1.125 oz	?

Fraction to Decimal, Decimal to Fraction

Look up Skill 14 on page 596 for more practice.

Any fraction can be renamed as a decimal and any decimal can be renamed as a fraction.

To rename a fraction as a decimal, use division. Think of the fraction bar in the fraction as meaning "divide by." For example, $\frac{5}{8}$ means "5 divided by 8." After the 5, write a decimal point and as many zeros as are needed. Then divide by 8.

EXAMPLE

Change $\frac{5}{8}$ to a decimal.

SOLUTION

$$\frac{5}{8} \longrightarrow \begin{array}{r} 0.625 \\ 8\overline{)5.000} \\ -48 \\ \hline 20 \\ -16 \\ \hline 40 \\ -40 \end{array}$$

EXAMPLE

Change $\frac{2}{5}$ to a decimal.

SOLUTION

$$\frac{2}{5} \longrightarrow \begin{array}{r} 0.4 \\ 5\overline{)2.0} \\ -20 \end{array}$$

If a fraction does not divide out evenly, divide to one more decimal place than you are rounding to.

EXAMPLE

Change $\frac{3}{7}$ to a decimal rounded to the nearest hundredth. (Divide to the thousandths place.)

SOLUTION

$$\frac{3}{7} \longrightarrow \begin{array}{r} 0.428 = 0.43 \\ 7\overline{)3.000} \\ -28 \\ \hline 20 \\ -14 \\ \hline 60 \\ -56 \\ \hline 4 \end{array}$$

$$\frac{3}{7} = 0.43 \text{ (rounded)}$$

EXAMPLE

Change $\frac{5}{6}$ to a decimal rounded to the nearest thousandth. (Divide to the ten thousandths place.)

SOLUTION

$$\frac{5}{6} \longrightarrow \begin{array}{r} 0.8333 = 0.833 \\ 6\overline{)5.0000} \\ -48 \\ \hline 20 \\ -18 \\ \hline 20 \\ -18 \\ \hline 20 \\ -18 \\ \hline 2 \end{array}$$

$$\frac{5}{6} = 0.833 \text{ (rounded)}$$

✔ SELF-CHECK Complete the problems, then check your answers in the back of the book.

Change the fractions to decimals. Round to the nearest thousandth. (Divide to the ten thousandths place.)

1. $\frac{7}{8}$ **2.** $\frac{3}{5}$ **3.** $\frac{7}{9}$ **4.** $\frac{2}{3}$

To rename a decimal as a fraction, name the place value of the digit at the far right. This is the denominator of the fraction.

$0.79 = \frac{79}{100}$

$0.003 = \frac{3}{1000}$

9 is in the hundredths place, so the denominator is 100.

3 is in the thousandths place, so the denominator is 1000.

Note that the number of zeros in the denominator is the same as the number of places to the right of the decimal point. The fraction should always be written in lowest terms.

$0.05 = \frac{5}{100} = \frac{1}{20}$

$4.625 = 4\frac{625}{1000} = 4\frac{5}{8}$

✔ SELF-CHECK Complete the problems, then check your answers in the back of the book.

Change the decimals to fractions reduced to lowest terms.

5. 0.4   **6.** 0.12   **7.** 1.125   **8.** 7.82

PROBLEMS Change the fractions to decimals. Round to the nearest thousandth.

9. $\frac{1}{5}$ **10.** $\frac{3}{4}$ **11.** $\frac{5}{7}$ **12.** $\frac{7}{15}$

13. $\frac{9}{20}$ **14.** $\frac{1}{6}$ **15.** $\frac{4}{9}$ **16.** $\frac{17}{120}$

17. $\frac{17}{40}$ **18.** $\frac{3}{8}$ **19.** $\frac{7}{10}$ **20.** $\frac{13}{50}$

21. $\frac{9}{25}$ **22.** $\frac{1}{2}$ **23.** $\frac{5}{12}$ **24.** $\frac{117}{200}$

Change the fractions to decimals. Round to the nearest hundredth.

25. $\frac{1}{4}$ **26.** $\frac{4}{5}$ **27.** $\frac{1}{8}$ **28.** $\frac{5}{9}$

29. $\frac{6}{7}$ **30.** $\frac{1}{3}$ **31.** $\frac{3}{10}$ **32.** $\frac{7}{16}$

33. $\frac{3}{100}$ **34.** $\frac{17}{25}$ **35.** $\frac{13}{14}$ **36.** $\frac{71}{75}$

Change the fractions to decimals. Round to the nearest thousandth.

37. $\frac{1}{7}$ **38.** $\frac{8}{9}$ **39.** $\frac{7}{12}$ **40.** $\frac{11}{13}$

41. $\frac{4}{15}$ **42.** $\frac{11}{16}$ **43.** $\frac{17}{30}$ **44.** $\frac{13}{40}$

45. $\frac{27}{50}$ **46.** $\frac{173}{200}$ **47.** $\frac{491}{500}$ **48.** $\frac{171}{1000}$

Change the decimals to fractions reduced to lowest terms.

49. 0.25   **50.** 0.6   **51.** 0.5   **52.** 0.375

53. 0.3   **54.** 0.17   **55.** 0.45   **56.** 0.75

57. 0.125   **58.** 0.48   **59.** 0.755   **60.** 0.05

61. 0.1875   **62.** 0.65   **63.** 0.325   **64.** 0.68

65. 0.34   **66.** 0.2125   **67.** 0.9375   **68.** 0.95

69. 1.1   **70.** 14.35   **71.** 7.08   **72.** 4.5

73. 5.117   **74.** 10.085   **75.** 3.4375   **76.** 2.85

77. 21.975   **78.** 57.44   **79.** 43.34375   **80.** 76.313

81. $30\frac{1}{4}$ yards equal one square rod. Write $30\frac{1}{4}$ as a decimal.

82. A basketball player made 163 free throws in 200 attempts. Write $\frac{163}{200}$ as a decimal. What was the player's free throw average?

83. Three months is $\frac{1}{4}$ of a year. Write $\frac{1}{4}$ as a decimal.

84. The price of a stock is $16\frac{1}{4}$. Write $16\frac{1}{4}$ as a decimal.

85. One cubic foot is about 0.8 bushel. Write 0.8 as a fraction in lowest terms.

APPLICATIONS

86. The Major Indoor Lacrosse League standings are given. Convert the decimals to fractions reduced to lowest terms.

MAJOR INDOOR LACROSSE LEAGUE

	National Division	W	L	Pct.	GB
a.	Detroit	3	0	1.000	—
b.	Pittsburgh	1	1	.500	$1\frac{1}{2}$
c.	New England	1	2	.333	2

	American Division	W	L	Pct.	GB
d.	New York	1	1	.500	—
e.	Philadelphia	1	2	.333	$\frac{1}{2}$
f.	Baltimore	1	2	.333	$\frac{1}{2}$

87. Stock prices are quoted as dollars and fractions of a dollar. Change the stock prices to dollars and cents. Round to the nearest cent.

	Stock	Price
a.	Alcoa	$52\frac{1}{2}$
b.	BnkAm	$22\frac{7}{8}$
c.	Chrysler	$12\frac{5}{8}$
d.	E Kodak	$45\frac{1}{4}$
e.	JP Ind	$8\frac{3}{8}$
f.	Ford	$26\frac{3}{8}$
g.	GenMtrs	$39\frac{1}{2}$
h.	Wendys	$5\frac{5}{8}$

88. The stock market summary lists certain key measures in decimal form. Change the decimals to fractions reduced to lowest terms.

STOCK MARKET SUMMARY

	Dow Jones Average	Decimal
a.	30 Industrials	2643.07
b.	20 Transportation	1036.32
c.	15 Utilities	206.49
d.	65 Stocks	954.21
e.	20 Bonds	92.18
f.	10 Pub util bonds	93.85
g.	10 Ind bonds	90.08
h.	Commodity futures	124.69

89. Individual bowling averages in The Blade Classic League are carried to the nearest hundredth. Convert the decimals to fractions reduced to lowest terms.

	Name	Average
a.	Kevin Taber	222.08
b.	Dan Koles	216.32
c.	Eric Roberts	216.01
d.	Terry Jacobs	212.29
e.	Jim Ferguson	210.05
f.	Steve Hold	208.42
g.	Paul Tyler	208.25

Percent to Decimal, Decimal to Percent

Look up Skills 26 and 28 on pages 608 and 610 for more practice.

Percent is an abbreviation of the Latin words *per centum*, meaning "by the hundred." So percent means "divide by 100." A percent can be written as a decimal. To change a percent to a decimal, first write the percent as a fraction with a denominator of 100, then divide by 100.

| EXAMPLE | Change 42% to a decimal. | | EXAMPLE | Change 19.4% to a decimal. |

SOLUTION

$42\% = \frac{42}{100} = 0.42$

SOLUTION

$19.4\% = \frac{19.4}{100} = 0.194$

When dividing by 100, you can just move the decimal point two places to the left. So when you write a percent as a decimal, you are moving the decimal point two places to the left and dropping the percent sign (%). If necessary, use zero as a placeholder.

EXAMPLE   **SOLUTION**

A. 42%   42%   = 42.   = 0.42 ◄── Drop % sign.

B. 37.2%   37.2% = 37.2 = 0.372 ── Move decimal 2 places.

C. 435%   435% = 435. = 4.35

D. 5%   5%   = 05.   = 0.05

Insert a zero as a placeholder.

✔ SELF-CHECK   Complete the problems, then check your answers in the back of the book.

Change the percents to decimals.

  1. 25%   **2.** 37.5%   **3.** 142%   **4.** 9%

To write a decimal as a percent, move the decimal point two places to the right and add a percent sign (%).

EXAMPLE   **SOLUTION**

A.   0.42   0.42   = 0.42   = 42% ◄── Add % sign.

B.   0.05   0.05   = 0.05   = 5% ── Move decimal 2 places.

C.   0.5   0.5   = 0.50   = 50%

D.   7.5   7.5   = 7.50   = 750%

E.   0.005   0.005 = 0.005 = 0.5%

Complete the problems, then check your answers in the back of the book.

Change the decimals to percents.

5. 0.85 **6.** 0.07 **7.** 0.3 **8.** 1.55 **9.** 0.004

Write as decimals.

10. 54% **11.** 34% **12.** 49% **13.** 60%

14. 23% **15.** 87% **16.** 71% **17.** 99%

18. 52.1% **19.** 97.6% **20.** 47.3% **21.** 81.8%

22. 20.6% **23.** 34.4% **24.** 60.9% **25.** 73.6%

26. 252% **27.** 817% **28.** 798% **29.** 376%

30. 485% **31.** 508% **32.** 2578% **33.** 1783%

34. 8% **35.** 3% **36.** 2% **37.** 9%

38. 1.1% **39.** 4.7% **40.** 5.5% **41.** 7.2%

42. 3.05% **43.** 8.03% **44.** 8.0765% **45.** 5.0921%

Write as percents.

46. 0.84 **47.** 0.13 **48.** 0.67 **49.** 0.35

50. 0.75 **51.** 0.24 **52.** 0.15 **53.** 0.97

54. 0.03 **55.** 0.05 **56.** 0.08 **57.** 0.01

58. 0.016 **59.** 0.042 **60.** 0.0251 **61.** 0.0965

62. 0.001 **63.** 0.005 **64.** 0.0032 **65.** 0.0061

66. 4.125 **67.** 9.371 **68.** 4.5 **69.** 7.2

70. 1.11 **71.** 2.98 **72.** 3.004 **73.** 7.000

74. 0.004 **75.** 0.009 **76.** 9 **77.** 46

78. 0.5 **79.** 0.1 **80.** 750.4 **81.** 1083

82. "Coats on sale, 33% off." Write 33% as a decimal.

83. "Cost of food up 112.3%." Write 112.3% as a decimal.

84. "42 out of 100 attempts." Write $\frac{42}{100}$ as a percent.

85. "All furniture 30%–50% off." Write 30% and 50% as decimals.

86. "Batteries on sale, 25% off." Write 25% as a decimal.

87. The percent changes in retail sales were reported as a decimal in the October issue of *Modern Merchandise* magazine. Change the decimals to percents.

RETAIL SALES		
	Month	**Change**
a.	February	0.016
b.	March	0.002
c.	April	0.015
d.	May	0.045
e.	June	0.042
f.	July	0.013
g.	August	0.019

88. Erica Stubenhofer is an investment counselor for Foxboro Investment Company. When working with clients, she uses this chart. Change the percents to decimals.

	Taxable Income	**Marginal Rate**
a.	$0–7,300	1.0%
b.	7,301–17,300	2.0%
c.	17,301–27,300	4.0%
d.	27,301–37,900	6.0%
e.	37,901–47,900	8.0%
f.	47,901 and up	9.3%

89. The commission rate schedule for a stockbroker is shown. Change the percents to decimals.

COMMISSION RATE SCHEDULE		
	Dollar Amount	**% of Dollar Amount**
	Stocks:	
a.	$0 – $2,499	1.7%, minimum $30
b.	$2,500 – $4,999	1.3%, minimum $42
c.	$5,000 – $9,999	1.1%, minimum $65
d.	$10,000 – $14,999	0.9%, minimum $110
e.	$15,000 – $24,999	0.7%, minimum $135
f.	$25,000 – $49,999	0.5%, minimum $175
	$50,000 and above	negotiated

90. During the National Basketball Association season, the teams had these won–lost records. The Pct. column shows the percent of games won, expressed as a decimal. Change the decimals to percents.

EASTERN CONFERENCE

Atlantic Division

		W	L	Pct.	GB
a.	Boston	29	9	.763	—
b.	Philadelphia	22	17	.564	$7\frac{1}{2}$
c.	New York	17	21	.447	12
d.	Washington	17	21	.447	12
e.	New Jersey	12	26	.316	18
f.	Miami	11	29	.275	19

Central Division

		W	L	Pct.	GB
g.	Chicago	28	11	.718	—
h.	Detroit	28	12	.700	
i.	Milwaukee	27	13	.675	$1\frac{1}{2}$
j.	Atlanta	24	15	.615	4
k.	Indiana	15	24	.385	13
l.	Charlotte	12	26	.316	$15\frac{1}{2}$
m.	Cleveland	12	26	.316	$15\frac{1}{2}$

WESTERN CONFERENCE

Midwest Division

		W	L	Pct.	GB	
San Antonio	27	10	.730	—	n.
Utah	26	13	.667	2	o.
Houston	20	19	.513	8	p.
Dallas	13	24	.351	14	q.
Minnesota	13	24	.351	14	r.
Orlando	10	31	.244	19	s.
Denver	9	30	.231	19	t.

Pacific Division

	W	L	Pct.	GB	
Portland	34	7	.829	—	u.
L.A. Lakers	27	11	.711	$5\frac{1}{2}$	v.
Phoenix	25	12	.676	7	w.
Golden State	21	17	.553	$11\frac{1}{2}$	x.
Seattle	17	19	.472	$14\frac{1}{2}$	y.
L.A. Clippers	14	27	.341	20	z.
Sacramento	10	26	.278	$21\frac{1}{2}$	aa.

Finding a Percentage

Look up Skill 30 on page 612 for more practice.

Finding a percentage means finding a percent of a number. To find a percent of a number, you change the percent to a decimal, then multiply it by the number.

EXAMPLE 1 20% of 95 is what number?

SOLUTION $20\% \times 95 = n$ In mathematics, *of* means "times" and *is* means "equals." Let n stand for the unknown number.

$0.20 \times 95 = n$ Change the percent to a decimal.
$19 = n$ Multiply.

20% of 95 $= 19$ Write the answer.

EXAMPLE 2 The delivery charge is 7% of the selling price of $140.00. Find the delivery charge.

EXAMPLE 3 The student had 85% correct out of 60 questions. How many answers were correct?

SOLUTION
$7\% \times \$140.00 = n$
$0.07 \times \$140.00 = n$
$\$9.80 = n$
$7\% \times \$140.00 = \$9.80$
delivery charge

SOLUTION
$85\% \times 60 = n$
$0.85 \times 60 = n$
$51 = n$
$85\% \times 60 = 51$
correct

✔ SELF-CHECK Complete the problems, then check your answers in the back of the book.

1. Find 40% of 70. **2.** Find 25% of 120.
3. Find 5% of 30. **4.** Find 145% of 200.

PROBLEMS Find the percentage.

5. 35% of 70 **6.** 60% of 95 **7.** 20% of 54
8. 40% of 216 **9.** 42% of 335 **10.** 75% of 815
11. 32% of 315 **12.** 64% of 320 **13.** 50% of 419
14. 8% of 50 **15.** 4% of 95 **16.** 6% of 48
17. 3% of 217 **18.** 2% of 371 **19.** 9% of 912
20. 5% of 41.6 **21.** 1% of 76.4 **22.** 8% of 64
23. 130% of 600 **24.** 175% of 615 **25.** 195% of 860
26. 420% of 470 **27.** 425% of 746 **28.** 650% of 416
29. 315% of 2170 **30.** 515% of 3350 **31.** 715% of 5218

32. 6.5% of 30   **33.** 8.25% of 75   **34.** 1.67% of 64

35. 5.5% of 136   **36.** 7.45% of 234   **37.** 4.5% of 584

38. 24.5% of 419   **39.** 32.5% of 591   **40.** 56.25% of 691

41. 71.2% of 875   **42.** 84.6% of 340   **43.** 99.9% of 742

Round answers to the nearest cent.

44. 5.5% of $60   **45.** 6% of $70   **46.** 7.75% of $30

47. 6.5% of $420   **48.** 7.5% of $160   **49.** 4.25% of $470

50. 8.25% of $76   **51.** 4.5% of $36   **52.** 7.455% of $246

53. 1.75% of $71.80   **54.** 1.25% of $117.45   **55.** 1.85% of $48.79

56. 0.8% of $24,000   **57.** 0.25% of $174,000   **58.** 2.0% of 96,500

APPLICATIONS

59. Allison Welles works for Klein Department Store as a salesclerk. When items are on sale, she computes the amount saved. She also subtracts the amount saved from the regular price to find the sale price. These insulated guide boots are marked down 30%. Complete the computations for her.

	Height	Color	Order No.	Was	Amount Saved	Sale Price
a.	8"	Brown	699 B 7011	$29.99	?	?
		Black	699 B 7012			
b.	10"	Brown	699 B 7013	$33.99	?	?
		Black	699 B 7014			
c.	12"	Brown	699 B 7015	$39.99	?	?
		Black	699 B 7016			

60. Use the chart at the right to find the postage and handling charges for the amounts listed below.

a. $3.00   **b.** $6.00
c. $20.00   **d.** $40.00
e. $4.60   **f.** $9.80
g. $18.50   **h.** $196.40
i. $2.89   **j.** $7.49
k. $19.94   **l.** $94.49

TO FIGURE POSTAGE AND HANDLING
. . . add to order

Sale items, total order:
  $3.00 or less, add 75¢
  $3.01–$10, add 20% of total
  $10.01–$25, add 15% of total
  Over $25, add 10% of total

61. Student Bert Trace received these test scores. How many answers were correct on each test?

	Subject	Test Score	Number of Items
a.	Math	90%	60
b.	English	80%	50
c.	Science	75%	44
d.	Spanish	85%	40
e.	Government	90%	120

62. Schedule X of the tax rate schedule is used if the filing status is single. Find the tax for each of the incomes listed.

a. $16,500

b. $24,640

c. $74,916

d. $43,150

SCHEDULE X Taxable Income			
Over	But not over	Your tax is	of amount over
$0	$17,850	15%	$0
17,850	43,150	$2677.50 + 28%	17,850
43,150	89,560	9761.50 + 33%	43,150
89,560	See further instructions.		

63. Betty Einstein is a stockbroker. She charges a percent of the dollar amount of stocks sold. Use the rate schedule and find how much she charges on these sale amounts. Round to the nearest cent.

COMMISSION RATE SCHEDULE	
Dollar Amount	% of Dollar Amount
Stocks:	
$0–$2499	2%, minimum $30
$2500–$4999	1.7%, minimum $42
$5000–$9999	1.4%, minimum $65
$10,000–$14,999	1%, minimum $110
$15,000–$24,999	0.8%, minimum $135
$25,000–$49,999	0.6%, minimum $175
$50,000 and above	negotiated

a. $1400

b. $3000

c. $8400

d. $11,700

e. $3640

f. $18,670

g. $41,948

h. $25,148

i. $12,147.75

j. $56,148.95

64. Sales taxes are found by multiplying the tax rate times the selling price of the item. The total purchase price is the selling price plus the sales tax. Find the sales tax and total purchase price for each selling price. Round to the nearest cent.

	Selling Price	Tax Rate	Sales Tax	Total Purchase Price
a.	$ 12.40	4%	?	?
b.	19.49	5%	?	?
c.	2.19	6%	?	?
d.	74.79	6.5%	?	?
e.	119.49	7.25%	?	?
f.	96.79	8.25%	?	?
g.	44.99	7.455%	?	?
h.	21.19	4.625%	?	?

Average (Mean)

Look up
Application Q
on page 625 for
more practice.

The average, or mean, is a single number used to represent a group of numbers. The average, or mean, of two or more numbers is the sum of the numbers divided by the number of items added.

EXAMPLE 1 Find the average of 7, 9, 4, 6, and 4.

Add to find the total.

SOLUTION
$$\frac{7 + 9 + 4 + 6 + 4}{5} = \frac{30}{5} = 6$$

Divide by the number of items.

EXAMPLE 2 Find the average of 693, 367, 528, and 626.

SOLUTION
$$\frac{693 + 367 + 528 + 626}{4} = \frac{2214}{4} = 553.5$$

EXAMPLE 3

SOLUTION

Find the average of 5.7, 6.3, 4.2, 5.8, and 3.4. Round to the nearest tenth.

$$\frac{5.7 + 6.3 + 4.2 + 5.8 + 3.4}{5} = \frac{25.4}{5} = 5.08 = 5.1$$

EXAMPLE 4

SOLUTION

Find the average of $17, $24, $38, $23, $19, and $26. Round to the nearest dollar.

$$\frac{\$17 + \$24 + \$38 + \$23 + \$19 + \$26}{6} = \frac{\$147}{6} = \$24.50 = \$25$$

✔ SELF-CHECK Complete the problems, then check your answers in the back of the book.

Find the average for each set of numbers. Round to the nearest cent.

1. 3, 6, 2, 5, 8, 6

2. 3.2, 1.8, 6.5, 8.1, 5.9

3. 134, 126, 130

4. $25, $37, $49, $53, $42, $42

PROBLEMS Find the average for each group.

5. 7, 8, 9, 12, 14

6. 4, 6, 10, 12, 8

7. 70, 85, 90, 75

8. 44, 86, 35, 95

9. 197, 108, 116

10. 225, 432, 321

11. 776, 709, 754, 733

12. 526, 387, 431, 388

13. 2.4, 3.5, 4.7, 2.9, 8.4

14. 6.8, 5.6, 3.4, 2.5, 4.7

15. 1.4, 2.5, 4.8, 3.7, 2.0, 3.9

16. 8.1, 5.3, 3.8, 7.9, 4.6, 3.2

17. $14, $12, $16, $14, $15

18. $64, $38, $92, $51, $65

19. $86, $75, $82, $87, $64, $70

20. $44, $44, $40, $42, $42, $40

Find the average for each group. Round to the amount indicated.

21. 7.8, 6.3, 8.3, 4.9, 7.7, 6.9, 5.1 (nearest tenth)

22. 9.2, 7.6, 8.2, 5.9, 9.5, 7.8 (nearest tenth)

23. 31.7, 33.9, 36.1, 33.8 (nearest tenth)

24. 4.37, 3.74, 4.90, 5.74 (nearest hundredth)

25. 34.87, 42.90, 46.21, 36.34, 39.89 (nearest hundredth)

26. $37.50, $44.50, $39.65, $34.25, $15.61, $11.22 (nearest cent)

27. $450,000; $386,000; $425,000; $372,000 (nearest thousand dollars)

28. Ben Agars had bowling scores of 175, 132, and 142. What was his average?

29. Marcia McComber recorded her pulse rate on 4 occasions as folows: 68, 85, 77, and 82. What was her average pulse rate?

30. Rachel Kelley's tips from being a bellhop were $4.00, $2.00, $3.50, $1.00, $4.00, $2.00, $1.00, and $3.00. What was her average tip?

31. During a 6-day period in June, you earned an average of $25 a day for mowing lawns. What were your total earnings?

32. Last year, Andre Barsotti's telephone bills averaged $35.45 a month. What was his total bill for the year?

33. Hung Lee had an average grade of 92 on his first 4 business math tests. He had a 95 and a 98 on the next 2 tests. What is his average grade for the 6 tests?

34. Christy Houck recorded her math test scores this quarter. What is her average?

Test Number	1	2	3	4	5	6	7	8
Score	87	75	98	95	82	77	78	88

35. What does she need on the next test to have an average of 86?

36. If there are a total of 10, 100-point tests for the quarter, is it possible for her to raise her average to 90?

37. Erik Pitts earns extra money each spring by rototilling gardens. His brother, John, went along to record the time spent doing each part of the job. Complete the chart in order to find the average time for each part of the job and the average time for each job. Round to the nearest minute.

	Task	Job 1	Job 2	Job 3	Job 4	Job 5	Average
a.	Unload tiller	3	2	2	3	3	?
b.	Till garden	55	40	65	45	50	?
c.	Clean up tiller	6	10	11	8	6	?
d.	Load tiller	5	4	3	5	6	?
e.	Travel time	16	20	25	12	8	?
f.	Average	?	?	?	?	?	

38. As captain of his bowling team, Burton McKaig has to complete this form after each match. Help him by computing the total and the average for each bowler. He also computes the total and the team average for each game. Round to the nearest whole number.

	Bowler	Game 1	Game 2	Game 3	Total	Average
a.	McKaig	145	171	163	?	?
b.	Pegorsch	220	215	203	?	?
c.	Hoskinson	175	160	172	?	?
d.	Goldberg	214	210	190	?	?
e.	Hallauer	136	181	172	?	?
f.	Total	?	?	?		
g.	Team average	?	?	?		

39. During a recent vacation trip, Denise and Dennis Hogan purchased 10 gallons of gasoline for $12.79, 11 gallons for $14.29, 8 gallons for $10.63, and 7 gallons for $9.51. What was the average cost per gallon of gasoline?

40. While on a business trip, you purchased 11.2 gallons of gasoline at $1.299 per gallon, 8.6 gallons at $1.319 per gallon, 7.3 gallons at $1.369 per gallon, and 10.3 gallons at $1.429 per gallon. What was the average cost per gallon of gasoline?

41. Jeri Keefer wanted to estimate the number of words in her 12-page term paper for her history class. She picked out 8 lines and counted the number of words per line: 12, 9, 7, 11, 13, 8, 10, and 12. Next she counted the lines on four pages: 28, 26, 27, and 27.

    a. What was the average number of words per line?

    b. What was the average number of lines per page?

    c. Use the averages to estimate the number of words in the term paper.

Elapsed Time

Look up
Application F
on page 619 for
more practice.

To find elapsed time, subtract the earlier time from the later time.

EXAMPLE Find the elapsed time for Shirley Miller who worked from:

A. 2:15 p.m. to 10:30 p.m.   **B.** 6:45 a.m. to 12:56 p.m.

SOLUTIONS

$$
\begin{array}{r}
10:30 \\
-\ 2:15 \\
\hline
8:15
\end{array}
= 8 \text{ hours 15 minutes} \\
\text{written as 8 h:15 min}
$$

$$
\begin{array}{r}
12:56 \\
-\ 6:45 \\
\hline
6:11
\end{array}
= 6 \text{ h:11 min}
$$

You cannot subtract 45 minutes from 30 minutes unless you borrow an hour and add it to the 30 minutes. Remember that 1 hour = 60 minutes.

EXAMPLE Find the elapsed time from 1:45 p.m. to 8:30 p.m.

SOLUTION

$$
\begin{array}{rcccl}
8:30 &=& 7:30 \ + &:60 = & 7:90 \text{ borrowed 1 hour} \\
-\ 1:45 &=& -\ 1:45 & = & -\ 1:45 \\
\hline
& & & & 6:45 = 6 \text{ h:45 min}
\end{array}
$$

✔ SELF-CHECK Complete the problems, then check your answers in the back of the book.

Find the elapsed time for a person who worked from:

1. 4:30 p.m. to 11:45 p.m.   **2.** 6:30 a.m. to 11:45 a.m.

3. 4:15 p.m. to 10:10 p.m.   **4.** 7:43 a.m. to 10:40 a.m.

To find elapsed time when the time period spans 1 o'clock, add 12 hours to the later time before subtracting.

EXAMPLE 1 Find the elapsed time from 8:30 a.m. to 4:40 p.m.

SOLUTION

$$
\begin{array}{rcccl}
4:40 &=& 4:40 \ + \ 12:00 &=& 16:40 \\
-\ 8:30 &=& & & -\ 8:30 \\
\hline
& & & & 8:10 = 8 \text{ h:10 min}
\end{array}
$$

EXAMPLE 2 Find the elapsed time from 7:50 p.m. to 3:34 a.m.

SOLUTION

$$
\begin{array}{rccccl}
3:34 &=& 15:34 &=& 14:34 \ + \ :60 &=& 14:94 \\
-\ 7:50 &=& -\ 7:50 &=& -\ 7:50 & = & -\ 7:50 \\
\hline
& & & & & & 7:44 = \\
& & & & & & 7 \text{ h:44 min}
\end{array}
$$

✔ SELF-CHECK Complete the problems, then check your answers in the back of the book.

Find the elapsed time for a person who worked from:

5. 8:00 a.m. to 5:10 p.m.   **6.** 11:15 p.m. to 7:30 a.m.

7. 7:37 a.m. to 3:20 p.m.   **8.** 10:07 p.m. to 6:00 a.m.

PROBLEMS

Find the elapsed time.

9. From 1:15 p.m. to 5:30 p.m.   **10.** From 2:12 p.m. to 10:25 p.m.

11. From 5:40 a.m. to 11:55 a.m.   **12.** From 4:25 a.m. to 11:50 a.m.

13. From 3:30 a.m. to 10:20 a.m.   **14.** From 1:30 p.m. to 10:15 p.m.

15. From 6:35 p.m. to 10:12 p.m.   **16.** From 5:40 a.m. to 11:14 a.m.

17. From 8:00 a.m. to 4:30 p.m.   **18.** From 9:15 a.m. to 6:25 p.m.

19. From 7:35 a.m. to 11:28 a.m.   **20.** From 2:50 a.m. to 11:05 a.m.

21. From 6:20 p.m. to 11:05 p.m.   **22.** From 1:37 a.m. to 9:28 a.m.

23. From 2:50 p.m. to 10:35 p.m.   **24.** From 1:48 p.m. to 9:33 p.m.

25. From 12:15 a.m. to 7:30 a.m.   **26.** From 12:30 p.m. to 8:45 p.m.

27. From 8:15 a.m. to 4:50 p.m.   **28.** From 8:45 a.m. to 5:10 p.m.

29. From 9:00 a.m. to 5:00 p.m.   **30.** From 9:30 a.m. to 5:00 p.m.

31. From 8:30 a.m. to 5:15 p.m.   **32.** From 8:17 a.m. to 5:13 p.m.

33. From 6:30 a.m. to 2:25 p.m.   **34.** From 6:47 a.m. to 3:28 p.m.

35. From 6:45 p.m. to 1:22 a.m.   **36.** From 10:57 a.m. to 6:12 p.m.

37. From 9:37 p.m. to 4:11 a.m.   **38.** From 6:46 p.m. to 2:14 a.m.

APPLICATIONS

39. Sam Watson worked from 9:45 a.m. to 6:12 p.m. How long did he work?

40. Helen Angell took a bus that left Detroit at 9:25 a.m. and arrived in Chicago at 3:20 p.m. How long was the trip? Disregard time zones.

41. Determine the elapsed time for each flight. Disregard time zones.

	From	To	Departure Time	Arrival Time	Elapsed Time
			AIRLINE SCHEDULE		
a.	Toledo	Detroit	7:30 a.m.	8:12 a.m.	?
b.	Chicago	Houston	11:30 a.m.	2:45 p.m.	?
c.	Los Angeles	Detroit	8:35 a.m.	1:07 p.m.	?
d.	New York	Cleveland	12:30 p.m.	2:56 p.m.	?
e.	Boston	Detroit	11:15 p.m.	1:10 a.m.	?
f.	Atlanta	Miami	12:47 p.m.	2:15 p.m.	?
g.	Pittsburgh	St. Louis	12:15 a.m.	3:10 a.m.	?

Reading Tables and Charts

Look up Application C on page 617 for more practice.

To read a table or chart, find the *column* containing one of the pieces of information you have. Look across the *row* containing the other piece of information. Read down the column and across the row. Read the information you need where the column and row cross.

ANY FRACTION OF A POUND OVER THE WEIGHT SHOWN TAKES THE NEXT HIGHER RATE							
WEIGHT NOT TO EXCEED	RATE CHART TO GROUND ZONES						
	2	3	4	5	6	7	8
1 lb	$1.25	$1.28	$1.32	$1.36	$1.42	$1.48	$1.55
2 lb	1.34	1.40	1.47	1.55	1.67	1.79	1.93
3 lb	1.43	1.52	1.63	1.75	1.92	2.11	2.32
4 lb	1.52	1.64	1.78	1.94	2.18	2.42	2.70
5 lb	1.61	1.75	1.93	2.14	2.43	2.74	3.09
6 lb	1.70	1.87	2.09	2.33	2.68	3.05	3.47
7 lb	1.79	1.99	2.24	2.53	2.94	3.37	3.86
8 lb	1.88	2.11	2.40	2.72	3.19	3.68	4.24
9 lb	1.97	2.23	2.55	2.92	3.44	4.00	4.63
10 lb	2.05	2.34	2.70	3.11	3.69	4.31	5.01
11 lb	2.14	2.46	2.86	3.31	3.95	4.63	5.40
12 lb	2.23	2.58	3.01	3.50	4.20	4.94	5.78
13 lb	2.32	2.70	3.17	3.70	4.45	5.26	6.17
14 lb	2.41	2.82	3.32	3.89	4.71	5.57	6.55
15 lb	2.50	2.93	3.47	4.09	4.96	5.89	6.94

EXAMPLE What is the cost to ship a 10-lb package to Zone 4?

SOLUTION
a. Find the Zone 4 column. **b.** Find the 10-lb row.
c. Read across the 10-lb row to the Zone 4 column. The cost is $2.70.

✓ SELF-CHECK Complete the problems, then check your answers in the back of the book.

Find the cost to ship each package to the indicated zone.

1. 1 lb, Zone 2 **2.** 1 lb, Zone 8 **3.** 10 lb, Zone 3 **4.** 13.2 lb, Zone 4

To classify an item, find the row that contains the known data. Then read the classification from the head of the column.

SIZE CHART—MEN'S SIZES									
Suits and Sportcoats	Order size	36	37	38	39	40	42	44	46
Sizes 36 to 46 Order by chest size	If chest is (inches)	35–36	36–37	37–38	38–39	39–40	41–42	43–44	45–46
Be sure waist will fit comfortably.	And waist is (inches)	28–31	29–32	30–33	31–34	32–35	34–37	36–39	38–41
Jackets	Order size	36		38		40	42	44	46
Sizes 36 to 46 Order by chest size	If chest is (inches)	34 1/2–36		36 1/2–38	38 1/2–40	40 1/2–42	42 1/2–44	44 1/2–46	

EXAMPLE What size suit should a man with a 41-inch chest order?

SOLUTION
a. Find the Suits and Sportcoats section.
b. Find the row If chest is (inches). **c.** Read across the row to 41–42.
e. Read the number at the head of the column (42).
A man with a 41-inch chest should order a size 42 suit.

Complete the problems, then check your answers in the back of the book.

Determine what size garment should be ordered.

5. Suit, chest size 38 inches.　　　　**6.** Jacket, chest size 41 inches.

Use the shipping chart on page 35 to find the cost to ship each package to the indicated zone.

7. 6 lb, Zone 3　　　　**8.** 2 lb, Zone 8　　　　**9.** 11 lb, Zone 3

10. 9 lb, Zone 5　　　　**11.** 12 lb, Zone 4　　　　**12.** 9 lb, Zone 2

13. 6.5 lb, Zone 2　　　　**14.** 9.75 lb, Zone 4　　　　**15.** 11.2 lb, Zone 3

16. 2.6 lb, Zone 5　　　　**17.** 1.35 lb, Zone 3　　　　**18.** 0.75 lb, Zone 7

19. Using the same shipping chart, determine the maximum weight a package can weigh.

	a.	b.	c.	d.	e.	f.
Shipping to Zone	2	4	5	8	7	3
Shipping Cost	$1.34	$2.86	$3.70	$5.01	$5.57	$2.11
Maximum Weight	?	?	?	?	?	?

20. Use the size chart on page 35 to determine what size garment should be ordered.

	a.	b.	c.	d.	e.	f.
Chest Size (inches)	37	42	45	36	40	38
Waist Size (inches)	30	35	39	32	35	34
Suit Order Size	?	?	?	?	?	?
Jacket Order Size	?	?	?	?	?	?

21. What size suit should a man with a 36-inch chest and a 31-inch waist order?

22. What size suit should a man with a 39-inch chest and a 33-inch waist order?

Use the New England Weather chart to answer the following.

NEW ENGLAND WEATHER					
Station	**Today**	**High**	**Low**	**Tomorrow**	**High**
Bangor	Cloudy	36	24	Snow	32
Bedford	Snow	31	27	"	31
Boston	"	33	29	"	33
Brockton	"	33	28	"	32
Framingham	"	31	26	"	29
Gloucester	"	33	30	"	33

23. What was the high temperature in Boston today?

24. What was the low temperature in Bedford today?

25. What will be the high temperature in Framingham tomorrow?

26. In which city was a low temperature of 28 degrees recorded?

27. In which city was the highest temperature recorded for today?

Use the federal income tax table to find the following.

FEDERAL WITHHOLDING TAX—WEEKLY Payroll Period, Employee MARRIED

And the wages are—		And the number of withholding allowances claimed is—							
At least	But less than	0	1	2	3	4	5	6	7
		The amount of income tax to be withheld shall be—							
$480	$490	$63	$56	$50	$44	$38	$32	$25	$19
490	500	64	58	52	45	39	33	27	21
500	510	66	59	53	47	41	35	28	22
510	520	67	61	55	48	42	36	30	24
520	530	69	62	56	50	44	38	31	25
530	540	70	64	58	51	45	39	33	27
540	550	72	65	59	53	47	41	34	28
550	560	73	67	61	54	48	42	36	30
560	570	75	68	62	56	50	44	37	31
570	580	76	70	64	57	51	45	39	33
580	590	78	71	65	59	53	47	40	34
590	600	79	73	67	60	54	48	42	36
600	610	81	74	68	62	56	50	43	37
610	620	82	76	70	63	57	51	45	39
620	630	84	77	71	65	59	53	46	40

	28.	29.	30.	31.	32.	33.
Income	$481.50	$558.75	$612.31	$485.03	$522.00	$520.00
Allowances	1	3	0	2	4	2
Amount Withheld	?	?	?	?	?	?

	Number of Allowances	Tax Withheld	Income	
			At least	But less than
34.	2	$56	?	?
35.	3	$63	?	?
36.	0	$79	?	?
37.	1	$61	?	?
38.	6	$30	?	?
39.	4	$45	?	?

Use the tax table above to find the amount of tax withheld.

40. Ralph Adams earns $483 and claims 2 allowances.

41. Barbara Knighten earns $562.29 and claims 1 allowance.

42 . Luther Whittier earns $498.60 and claims 2 allowances.

43. Stacey Hagley earns $571.30 and claims 1 allowance.

44. Verna Denbridge earns $602.19 and claims 3 allowances.

45. Shirley Sagamore earns $596.78 and claims 0 allowances.

Constructing Graphs

Look up
Applications M
and N on pages
622–623 for more
practice.

A **bar graph** is a picture that displays and compares numerical facts in the form of vertical or horizontal bars. To construct a vertical bar graph, follow these steps:

a. Draw the vertical and horizontal axes.

b. Scale the vertical axis to correspond to the given data.

c. Draw one bar to represent each quantity.

d. Label each bar and the vertical and horizontal axes.

e. Title the graph.

METROPOLITAN STATISTICAL AREAS Population (in millions)	
Chicago, IL	8.1
San Francisco, CA	5.9
Detroit, MI	4.6
Washington, DC	3.6

EXAMPLE   Construct a vertical bar graph of the given data.

SOLUTION

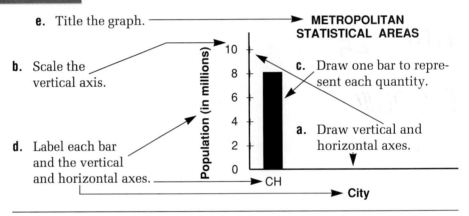

A **line graph** is a picture used to compare data over a period of time. It is an excellent way to show trends (increases or decreases). To construct a line graph, follow these steps:

a. Draw the vertical and horizontal axes.

b. Scale the vertical axis to correspond to the given data.

c. Label the axes.

d. Place a point on the graph to correspond to each item of data.

e. Connect the points from left to right.

f. Title the graph.

COMPUTER CLASSES Enrollment	
1987	15
1988	20
1989	30
1990	45
1991	50

EXAMPLE   Construct a line graph of the given data.

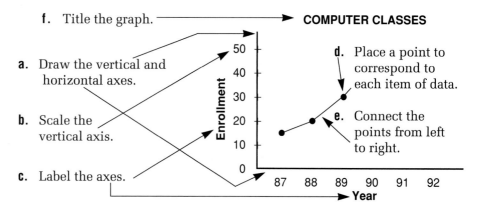

SOLUTION

f. Title the graph. ——————————→ **COMPUTER CLASSES**

a. Draw the vertical and horizontal axes.

b. Scale the vertical axis.

c. Label the axes.

d. Place a point to correspond to each item of data.

e. Connect the points from left to right.

Enrollment (vertical axis: 0, 10, 20, 30, 40, 50)

Year (horizontal axis: 87, 88, 89, 90, 91, 92)

✓ SELF-CHECK Complete the problems, then check your answers in the back of the book.

1. Complete construction of the vertical bar graph started in the Example.

2. Complete construction of the line graph started in the Example.

PROBLEMS   Construct vertical bar graphs of the given data.

3.

CITY GOVERNMENT EMPLOYMENT National Summary (in thousands)	
Schools	355
Hospitals	133
Highways	131
Police	427
Fire	236
Parks & Rec.	152

4.

HEBAN'S DEPARTMENT STORE Total Sales by Department (in thousands)	
Sports	145
Housewares	82
Men's Clothing	120
Women's Clothing	112
Appliances	75
Electronics	130

5. Read the vertical bar graph.

   a. In what year was the most money spent on leveraged buyouts? How much?

   b. In what year was the least money spent on leveraged buyouts? How much?

   c. What was the total value in 1991?

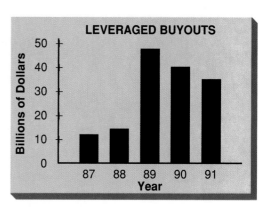

LEVERAGED BUYOUTS

Billions of Dollars (0, 10, 20, 30, 40, 50) — Year (87, 88, 89, 90, 91)

Construct line graphs for the given data.

6.

ACME INDUSTRIES STOCK	
Monthly	Average
Jan.	$2.00
Feb.	$2.50
Mar.	$2.75
Apr.	$3.25
May	$4.00
June	$4.50

7.

TAX RETURNS FILED (in millions)	
1980	90
1981	93
1982	94
1983	95
1984	96
1985	96
1986	99
1987	100
1988	101
1989	102
1990	102

8. Read the line graph.

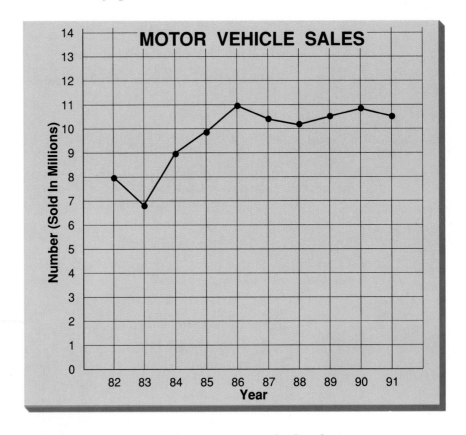

a. Which year shows the highest motor vehicle sales?

b. Which year shows the lowest motor vehicle sales?

c. Indicate the time span that showed an increase 3 years in a row. How much was the increase?

d. Compare the decrease from 1981 to 1982 with the decrease from 1985 to 1986.

e. Estimate the sales for 1992.

Units of Measure

Here are the abbreviations and conversions for units of measure in the U.S. Customary System.

Length	Volume	Weight
12 inches (in) = 1 foot (ft)	2 cups (c) = 1 pint (pt)	16 ounces (oz) =
3 ft = 1 yard (yd)	2 pt = 1 quart (qt)	1 pound (lb)
5280 ft = 1 mile (mi)	4 qt = 1 gallon (gal)	2000 lb = 1 ton (t)

Here are the symbols and conversions for units of measure in the metric system.

Length	Volume
1000 millimeters (mm) = 1 meter (m)	1000 milliliters (mL) = 1 liter (L)
100 centimeters (cm) = 1 m	**Mass**
1000 m = 1 kilometer (km)	1000 grams (g) = 1 kilogram (kg)

To convert from one unit of measure to another, use the conversion lists above.

When converting to a smaller unit, multiply.

EXAMPLE   Convert 6 ft to inches.   Convert 2 m to centimeters.

SOLUTION   Use 12 in = 1 ft   Use 100 cm = 1 m
6 ft: 6 × 12 = 72   2 m: 2 × 100 = 200
6 ft = 72 in   2 m = 200 cm

When converting to a larger unit, divide.

EXAMPLE   Convert 12 pt to quarts.   Convert 8400 g to kg.

SOLUTION   Use 2 pt = 1 qt   Use 1000 g = 1 kg
12 pt: 12 ÷ 2 = 6   8400 g: 8400 ÷ 1000 = 8.4
12 pt = 6 qt   8400 g = 8.4 kg

✔ SELF-CHECK   Complete the problems, then check your answers in the back of the book.

Convert:

1. 9 ft to inches

2. 0.15 L to milliliters

3. 24 ft to yards

4. 350 cm to meters

PROBLEMS   Do the following conversions.

5. 9 yd to feet

6. 14 gal to quarts

7. 7 lb to ounces

8. 3 ft to inches

9. 6 lb to ounces

10. 4 L to milliliters

11. 3.8 km to meters

12. 16 pt to cups

13. 3.2 kg to grams

14. 96 in to yards

15. 9 qt to gallons

16. 42 oz to pounds

17. 14 qt to gallons   **18.** 33 oz to pounds   **19.** 2000 g to kilograms

20. 90 cm to meters   **21.** 3300 mL to liters   **22.** 450 cm to meters

23. 72.1 kg to grams   **24.** 3.4 L to milliliters   **25.** 723 g to kilograms

26. 11.316 mL to liters   **27.** 18 cm to millimeters   **28.** 383.2 cm to meters

29. 1 yd 4 in to inches   **30.** 4 ft 2 in to inches   **31.** 5 qt 1 pt to pints

32. 3 lb 7 oz to ounces   **33.** 2 gal 2 qt to quarts   **34.** 2 yd 2 ft 2 in to inches

35. 3 gal 2 qt 1 pt to pints   **36.** 2 m 142 cm 45 mm to millimeters

APPLICATIONS

37. How many quarts will a 5-gallon plastic bag hold?

38. How many milliliters will a one-liter bottle hold?

39. How many cups of coffee does a 2-quart coffeepot hold?

40. How many cups of hot chocolate will a 2-gallon thermos jug hold?

41. How many inches long is a $6\frac{2}{3}$-yard roll of aluminum foil?

42. Cottage cheese is sold in 1-pint containers. How many pints must be bought to have enough for a recipe that calls for 5 cups?

43. James Hartman knows that his jogging stride is about 1 meter long. The jogging track he uses is 3.9 kilometers long. How many strides does it take him to go around the track once?

44. The cafeteria receives 32 cases of milk each day. Each case contains 24 half-pint cartons. How many gallons of milk are received each day?

45. A soft drink is sold in 355 mL cans. How many liters are in a six-pack?

46. John Patroulus, a pastry chef, baked a walnut cake weighing 2.4 kilograms. How many 75-gram servings can be cut from the cake?

47. Joan Wagoner ordered baseboard molding for the rooms of a new house. Joan needs to complete this chart to determine the total number of feet of molding needed. How much molding is needed?

	Length	Width	2 Lengths	+	2 Widths	=	Perimeter
	12 ft	9 ft	24 ft	+	18 ft	=	42 ft
a.	10 ft	8 ft	20 ft	+	16 ft	=	?
b.	18 ft	24 ft	?	+	?	=	?
c.	10 ft 6 in	9 ft 4 in	?	+	?	=	?
d.	11 ft 3 in	7 ft 8 in	?	+	?	=	?
e.	11 ft 4 in	13 ft 2 in	?	+	?	=	?
f.	17 ft 8 in	12 ft 9 in	?	+	?	=	?
g.	8 ft 5 in	7 ft 9 in	?	+	?	=	?
					Total	=	?

Estimation: Rounding

Rounding is often used to estimate an answer. First, round the numbers to the highest place value. Then perform the indicated computation. If all the numbers do not have the same number of digits, round the numbers to the highest place value of the smaller number.

ESTIMATE

$$\begin{array}{r} 76.761 \\ -\ 39.302 \end{array} \qquad \begin{array}{r} 80 \\ -\ 40 \\ \hline 40 \end{array}$$

The answer is about 40.
By computation, it is 37.459.

ESTIMATE

$$6.1\overline{)23.79} \qquad 6\overline{)24}^{\,4}$$

The answer is about 4.
By computation, it is 3.9.

ESTIMATE

$$\begin{array}{r} 5.65 \\ \times\ 4.45 \end{array} \qquad \begin{array}{r} 6 \\ \times\ 4 \\ \hline 24 \end{array}$$

The answer is about 24.
By computation, it is 25.1425.

ESTIMATE

$$\begin{array}{r} \$265.88 \\ +\ 32.47 \end{array} \qquad \begin{array}{r} \$270 \\ +\ 30 \\ \hline \$300 \end{array}$$

The answer is about $300.
By computation, it is $298.35.

ESTIMATE

$$2\tfrac{3}{4} \times 8\tfrac{1}{4} \qquad 3 \times 8 = 24$$

The answer is about 24.
By computation, it is $22\tfrac{11}{16}$.

ESTIMATE

$$33\% \text{ of } \$62 \qquad \tfrac{1}{3} \times \$60 = \$20$$

The answer is about $20.
By computation, it is $20.46.

✔ SELF-CHECK   Complete the problems, then check your answers in the back of the book.

First round the numbers to the highest place value, estimate the answer, then perform the computation.

1. $\begin{array}{r} 43.986 \\ -\ 27.491 \end{array}$ 2. $7.1\overline{)44.872}$ 3. $\begin{array}{r} 63.91 \\ \times\ 9.43 \end{array}$ 4. $5\tfrac{1}{8} \times 4\tfrac{2}{3}$

PROBLEMS   First estimate, then perform the indicated computation.

5. $\begin{array}{r} 5965 \\ +\ 1824 \end{array}$ 6. $\begin{array}{r} 7.791 \\ +\ 2.151 \end{array}$ 7. $\begin{array}{r} 15.86 \\ -\ 13.72 \end{array}$ 8. $\begin{array}{r} \$72.75 \\ -\ 6.47 \end{array}$ 9. $\begin{array}{r} 9.34 \\ \times\ 7.92 \end{array}$

10. $\begin{array}{r} \$28.40 \\ \times\ 5.20 \end{array}$ 11. $6.8\overline{)48.52}$ 12. $21.3\overline{)57.723}$ 13. $2\tfrac{3}{4} \times 1\tfrac{1}{2}$

14. $12\tfrac{1}{2} \times 2\tfrac{3}{5}$ 15. $\tfrac{1}{4} \times 12\tfrac{1}{2}$ 16. 31% of 18

17. 70% of $49.95   18. 27% of $12   19. 52% of 160

20. You purchase items for $39.45, $17.55, and $32.53. Estimate the total, then calculate.

21. Forty-two people charter a bus for $388. Estimate the amount each person pays, then calculate.

22. A living room measures $16\frac{1}{4}$ ft by $12\frac{1}{2}$ ft. Estimate the area in square feet, then calculate.

23. Blue jeans are on sale for 33% off the regular price of $26.95. Estimate the savings, then calculate.

APPLICATIONS   First estimate, then find the total for these grocery store receipts.

24.
```
SUPER VALUES MARKET
FACE TISSUE      1.15
COFFEE           3.95
FACE TISSUE      1.15
FACE TISSUE      1.15
CORN CHIPS       2.19
BACON            1.58
RYE CRACKERS     1.39
SMALL LINKS      1.98
SMALL LINKS      1.98
SMALL LINKS      1.98
CAT FOOD         3.60
DETERGENT        1.88
DETERGENT        1.08
        TOTAL      ?
```

25.
```
SUPER VALUES MARKET
CORN CHIPS       2.19
VANILLA          4.19
BNL SHK HAM     25.33
ORANGE JUICE     5.48
CASING LINKS     3.65
KIELBASA         3.18
KIELBASA         2.43
PICKLES          1.89
CEREAL           1.73
MAYO             1.69
ALUMINUM FOIL    1.09
CHEDDAR          2.99
MACARONI          .65
STEAK            6.23
RIPE OLIVES      1.29
SPAN OLIVES       .99
SPAN OLIVES       .99
SPAN OLIVES       .99
VINEGAR          1.49
        TOTAL      ?
```

26.
```
SUPER VALUES MARKET
BABY CLAMS       1.49
RICE             1.03
CHESTNUT          .97
CHEESE           3.65
SPROUTS           .95
CRANAPLE         1.23
SALAD DRSG        .89
HERB DRSG         .89
HERB DRSG         .89
SALAD DRSG        .89
HALF HALF PT      .85
WAX PAPER         .89
ITAL SAUS        2.37
ITAL SAUS        2.98
MINI CHP         1.95
SAUSAGE 2 LB     4.35
MINTS             .35
GUM               .45
CRANPLE          1.23
        TOTAL      ?
```

Use the menu from the Columbian House to estimate the bill for the following orders.

27. Two persons

1 Soup	1 Chocolate
1 Fruit Cup	Torte
1 Baked Ham	1 Apple Strudel
1 Red Snapper	1 Milk
	1 Herbal Tea

28. Party of six

4 Soup	3 Chocolate
2 Fruit Cup	Torte
1 Baked Ham	3 Walnut
3 Prime Rib	Dream Cake
1 Red Snapper	2 Milk
1 Filet Mignon	2 Soft Drinks
	2 Coffee

COLUMBIAN HOUSE

Appetizers
Soup Romaine $1.25
Fruit Cup 1.55
Shrimp Cocktail 5.95
Entrees
Vegetarian Plate 5.95
Baked Ham 6.50
Red Snapper 7.95
Roast Prime Rib 13.95
Filet Mignon 12.75
Desserts
Chocolate Torte 3.25
Apple Strudel 2.75
Walnut Dream Cake 3.50
Beverages
Coffee 95
Milk .75
Herbal Tea 75
Soft Drinks 1.00

Estimation: Front End

Estimation is a very valuable tool in business mathematics. It can be used as a quick method of checking the reasonableness of a calculation, or when an exact answer is not needed. It is important to know how to estimate. One way to estimate a sum is to add the front-end digits.

ESTIMATE			ESTIMATE	
6477	6		$8.60	9 (by rounding)
2142	2		3.19	3
+ 1321	+ 1		+ 0.65	+ 0
	9			12

The answer is about 9000.
By computation, it is 9940.

The answer is about $12.00.
By computation, it is $12.44.

Each estimate in the examples above is less than the correct sum. A closer estimate can be found by adjusting the sum of the front-end digits.

ESTIMATE			ESTIMATE	
5|67	about 100		$3.85 ← about $1.00	
3|35	←		4.44	
1|24	about 100		1.13 about $1.00	
+ |84	←		+0.65 ←	

about   900   +   200 = 1100

The answer is about 1100.
By computation, it is 1110.

about $8.00   + $2.00 = $10.00

The answer is about $10.00.
By computation, it is $10.07.

✔ SELF-CHECK Complete the problems, then check your answers in the back of the book.

Estimate by first adding the front-end digits and then by adjusting the sum of the front-end digits. Perform the computations.

1.	389	**2.**	5623	**3.**	$ 3.45
	467		221		1.49
	15		9879		9.72
	+ 240		+ 1061		+ 0.35

PROBLEMS

Estimate by adding the front-end digits. Perform the computations.

4.	4427	**5.**	7178	**6.**	112	**7.**	25.5	**8.**	$64.50
	3274		4298		448		86.6		43.60
	+ 1245		+ 5370		515		7.7		6.90
					+ 324		+ 0.6		+ 53.20

Estimate by adjusting the sum of the front-end digits. Perform the computations.

9.	335	**10.**	285	**11.**	$39.37	**12.**	$15.95	**13.**	$340.05
	660		315		7.49		54.20		21.65
	74		544		23.75		12.57		47.32
	+ 126		+ 361		+ 4.35		+ 21.54		+ 53.79

14. Estimate first, then calculate the total attendance for six home soccer games. September 12, 3187; September 19, 2234; October 5, 2108; October 17, 3421; October 24, 6790; and November 2, 3907.

15. Estimate first, then calculate the cost of these grocery items: tissues $1.29, coffee $6.99, orange juice $1.45, T-bone steak $6.79, oatmeal $1.99, cocoa mix $2.89, and a jar of salad dressing $1.99.

16. Estimate first, then calculate the cost of these options on a new car: two-tone paint $147.30, air-conditioning $578.50, oil gauge $39.50, tinted glass $75.75, automatic transmission $340.00.

APPLICATIONS   First estimate, then find the total for each receipt.

17.
```
SUPER VALUE STORES
GREEN ONIONS   .39
DAIRY         1.65
CHEESE        1.69
SPAGHETTI      .99
5# SUGAR      1.39
LETTUCE LEAF   .95
DETERGENT     2.95
FRUIT          .89
FRUIT          .89
        TOTAL    ?
```

18.
```
SUPER VALUE STORES
2% MILK        1.39
DIET COLA       .99
WALNUTS        3.78
VINEGAR        2.69
FABRIC SOFT    2.29
PEANUT BUTTER  4.79
AVOCADOS       1.59
CASHEW PARTS   2.99
COFFEE         1.99
WHITE BREAD     .49
FRUIT PUNCH    1.19
TORTILLA CHPS  2.49
CRACKERS       1.99
TOOTHPASTE     2.09
         TOTAL    ?
```

19.
```
ACME HARDWARE
GLUE         13.93
FILM          5.09
TAPE MEAS     9.97
SNOW BRUSH    2.29
EXTEN COR    11.59
AA BATT       1.25
AA BATT       1.25
AA BATT       1.25
AA BATT       1.25
LIGHT SW      3.39
POWER STR    15.95
WIRE PLI      3.99
       TOTAL    ?
```

First estimate, then find the total deposit.

20.

CASH	CURRENCY	74	00
	COINS	322	05
CHECKS		191	80
		201	20
		121	00
Total From Other Side			
SUBTOTAL			
◊ LESS CASH RECEIVED			
TOTAL DEPOSIT			?

21.

CASH	CURRENCY	40	00
	COINS	121	43
CHECKS		37	20
		73	40
Total From Other Side		158	25
SUBTOTAL			
◊ LESS CASH RECEIVED			
TOTAL DEPOSIT			?

22.

CASH	CURRENCY	2317	00
	COINS	819	95
CHECKS		501	21
		9213	76
		8324	05
Total From Other Side		913	17
SUBTOTAL			
◊ LESS CASH RECEIVED			
TOTAL DEPOSIT			?

23. Helen Hazelton used a calculator to solve the following problems. If she entered each number properly into the calculator, her answers should be correct. Estimate each answer and decide if Helen's answers are correct.

	Problem	Her Answer on Calculator	Your Estimate	Is She Right? Yes/No
a.	577 + 321 + 225 + 70	1193	?	?
b.	$8.60 + $4.90 + $0.50	$13.50	?	?
c.	12.3 + 11.2 + 9.8 + 7.6 + 8.4	42.3	?	?
d.	32 + 53 + 41 + 77 + 58 + 90	351	?	?
e.	31.26 + 44.51 + 31.07 + 28.46	189.3	?	?
f.	34.671 + 9.902 + 1.009 + 0.103	45.685	?	?
g.	1.19 + 0.9 + 1.39 + 0.08 + 1.09	12.75	?	?
h.	$345.70 + $12.35 + $75.95 + $5.45	$489.45	?	?

Estimation: Compatible Numbers

Sometimes a reasonable estimate can be arrived at by changing the numbers in the problem to numbers that can be computed easily. These are called compatible numbers.

DIVIDE	ESTIMATE
$36{,}414 \div 9$	$9\overline{)36{,}000}$ with quotient 4000

The estimate is easy since 36,000 is close to 36,414 and is compatible with 9. The answer is about 4000. By computation, it is 4046.

DIVIDE	ESTIMATE
$798 \div 42$	$40\overline{)800}$ with quotient 20

The estimate is easy since 800 is close to 798 and 40 is close to 42. 800 and 40 are compatible. The answer is about 20. By computation, it is 19.

FIND	ESTIMATE
$\frac{1}{3} \times 8\frac{3}{4}$	$\frac{1}{3} \times 9 = 3$

The estimate is easy since 9 is close to $8\frac{3}{4}$ and is compatible with $\frac{1}{3}$. The answer is about 3. By computation, it is $2\frac{11}{12}$.

FIND	ESTIMATE
25.5% of $\$420$	$\frac{1}{4} \times 400 = 100$

The estimate is easy since 25.5% is about $\frac{1}{4}$ and $420 is about 400. 400 and $\frac{1}{4}$ are compatible. The answer is about $100. By computation, it is $107.10.

✔ SELF-CHECK   Complete the problems, then check your answers in the back of the book.

Estimate using compatible numbers, then perform the computations.

1. $661.74 \div 82$ **2.** $8763.3 \div 321$ **3.** $35\% \times 926$

PROBLEMS

For problems 4–21, estimate using compatible numbers. Perform the computations.

4. $8824 \div 8$ **5.** $4879 \div 7$ **6.** $6095 \div 5$

7. $642 \div 6$ **8.** $896 \div 32$ **9.** $24{,}564 \div 575$

10. $8766.63 \div 81$ **11.** $\$6447.6 \div 7.2$ **12.** $\frac{1}{3} \times 8\frac{1}{2}$

13. $\frac{5}{7} \times 44\frac{1}{2}$ **14.** 50% of $\$430$ **15.** 65% of $\$75$

16. Delbert Rowell drove his tractor trailer rig 84,572.5 miles in 6 months. Estimate how many miles he drove each month.

17. Anna Nethery earned $47,500 this year as a stock analyst. Estimate how much she earns each month.

18. A walking path is $20\frac{1}{2}$ miles long. You walk $\frac{1}{3}$ of the path by noon. Estimate the distance walked.

19. Harriet Murdock saves 33% of her paycheck each week. Last week her check was for $247.95. Estimate the amount saved.

20. During the high school state basketball finals, 75% of the 22,000 tickets were sold. Estimate how many tickets were not sold.

21. The hotel-motel tax in Pittsburgh is 9.5% of the price of a room. If a room cost $112.50 per night, estimate the tax for 3 nights.

APPLICATIONS

22. Louis Watts was hired as the accounting clerk/bookkeeper at an annual salary of $12,500. Estimate his monthly salary.

ACCOUNTING CLERK/ BOOKKEEPER $12–16K Salary. Knowledge of accounts payable/receivable. Career Position!

ACCOUNT EXECUTIVE/ MARKETING: $23–26K Salary (Fee paid). Degreed! 2 yrs. Marketing/ promotional experience! Kathleen 555-2222.

23. Ed Cooper learned at the job interview that he would be paid an annual salary of $24,600 as account executive/marketing. Estimate his monthly salary.

BOOKKEEPER-FEE PAID. $15,000. IBM computerized bookkeeping to trial balance. Prefer complete charge small company experience. Free parking. BOARDROOM PERSONNEL, 626 Madison, 555-4271

24. Betty Borman took a job as a bookkeeper at an annual salary of $15,000. Estimate her weekly salary.

Manufacturing Outlet
No Experience
Start pay for new employees is up to
$350 PER WEEK
To apply call Mon. or Tues. 9 am to 4 pm
555-3580

25. Linda Jackson took a job at a manufacturing outlet. She earns $350 a week. Estimate her daily income if she works 5 days a week. Estimate her annual salary.

DATA TYPIST: $700–$800/ mo. Entry level! Type 45 wpm, phones. Call Polly 555-2284.

26. Jianguo Wang is a data typist earning $725 per month. Estimate his annual salary.

OFFICE CLERK: $8–9K/ year. Great benefits! Filing, light typing. Call Sue 555-7670.

27. Suppose you take a job as an office clerk earning $8,000 to $9,000 a year. Estimate your weekly salary range.

30% OFF MFG. LIST 3.30
LADIES' ANKLETS
Cotton/nylon blend. White and fashion colors.

28. Amy DiGrazia saw the sale ad for ladies' anklets. Estimate the savings.

25% OFF OUR REG. 59.99–119.99
ALL CORDLESS TELEPHONES
SALE 44.99 TO 89.99 Choose from a large selection of quality brand names.

29. Gonzalo Carlos purchased a cordless telephone that had a regular price of $71.50. If he received 25% off, estimate how much he saved.

40% OFF OUR REG. 5.99–75.99
ALL 14K GOLD FILLED JEWELRY
SALE 3.59 TO 45.59 Earrings, lockets, pendants, more. Men's and ladies' styles.

30. Lynn Hatch purchased some jewelry that had a regular price of $19.75. If she received 40% off the regular price, estimate how much she paid for the jewelry.

Estimation: Clustering

Estimation by clustering is another way of projecting what an answer will be. When the numbers to be added are close to the same quantity, the sum can be found by clustering.

ESTIMATE

$4.85	All of the
5.15	numbers cluster
4.89	around $5.00, so
+ 5.17	$5 × 4 = $20.

The answer is about $20.00.
By computation, it is $20.06.

ESTIMATE

$15.95 ⎫
16.50 ⎬ about $48.00
15.75 ⎭
+ 7.95   + 8.00
        about $56.00

The answer is about $56.00.
By computation, it is $56.15.

Clustering cannot always be used, but when it can be used, it is usually a fast method since it can be done mentally.

✔ SELF-CHECK Complete the problems, then check your answers in the back of the book.

Estimate the sums by clustering, then compute the sums.

	1.	2.	3.
	563	$27.95	$1.59
	598	31.42	1.79
	559	30.25	2.21
	+ 612	+ 29.47	+ 0.75

PROBLEMS

Estimate the sums by clustering. Compute the actual amount.

4.	5.	6.	7.	8.
763	525	$53.39	$36.20	88,026
781	496	55.24	39.67	91,521
773	512	49.26	41.78	87,842
+ 810	+ 530	+ 49.97	+ 43.59	+ 94,819

9.	10.	11.	12.	13.
7.95	16.05	$2.29	$25.95	547.89
7.87	15.95	2.39	26.30	546.90
8.13	15.50	2.25	25.70	789.32
8.25	15.50	2.40	37.95	804.90
+ 3.67	+ 5.50	+ 8.95	+ 38.50	+ 811.66

14. School supplies costing: $6.95, $7.25, $7.45, $6.65, and $6.79.

15. Groceries costing: $2.89, $3.15, $3.29, $2.67, $3.25, and $3.35.

16. Work clothes costing: $25.95, $23.95, $26.50, $39.95, and $41.25.

17. School supplies costing: $1.09, $0.99, $0.89, $2.29, $1.99, and $2.15.

18. Clothes costing: shirt $17.95, shoes $39.95, six pairs of socks $18.25, tie $16.95, and belt $16.95.

19. Sylvanus Schaibley drove the miles indicated: Mon. 367, Tues. 390, Wed. 405, Thurs. 386, Fri. 402, and Sat. 396. Estimate first, then calculate the total miles for the week.

20. Lincoln McClure had the following sales: First Qtr. $52,900, Second Qtr. $88,900, Third Qtr. $91,980, and Fourth Qtr. $89,830. Estimate first, then calculate the total sales for the year.

21. Esther Radner had the following long-distance phone charges: $6.15, $2.15, $5.98, $1.81, $5.87, $1.89, and $11.71. Estimate first, then calculate the total long-distance charges.

APPLICATIONS

22. During an extended stay at the Union Square Hotel, Mary Anderson sent the items indicated to be dry-cleaned. Estimate first, then calculate the total.

23. Mary's total dry-cleaning bill is subject to a 6% sales tax, and Mary gave the bellhop a tip equal to 10% of the dry-cleaning charges. Estimate the bill rounded to the nearest one dollar. What is the total bill?

		HOTEL VALET GUEST DRY CLEANING		
No.	Ladies'	Clean & Press	Press Only	Charges
a. 2	Suits	8.00	5.00	
b. 2	Capris & Slacks	4.00	3.00	
c. 1	Blazers	4.50	3.00	
d. 5	Blouses	4.50	3.00	
e. 1	Dresses	7.75	5.00	
f. 2	Skirts	4.50	3.00	
	Sweaters	4.50	3.00	
	Coats	8.00up	5.00	
g. 1	Coats-Car	7.50up	4.00	
h. 1	Jumpsuits	8.50up	6.50	
	Scarves	2.50up	1.50	
	Belts			
l.		**TOTAL**		**?**
Rm. No. _____				

24. Greg Leininger used a calculator to solve the following problems. If he entered each number properly into the calculator, his answers should be correct. Estimate each answer and decide if Greg's answers are correct.

	Problem	His Answer on Calculator	Your Estimate	Is He Right? Yes/No
a.	672 + 703 + 725 + 130	2230	?	?
b.	$9.80 + $9.95 + $10.35 + $10.01	$40.11	?	?
c.	1.23 + 1.95 + 7.8 + 7.6 + 8.4	24.98	?	?
d.	82 + 75 + 79 + 34 + 29+ 32	299	?	?
e.	44.20 + 44.51 + 30.07 + 29.35	148.13	?	?
f.	625.1 + 615.2 + 12.35 + 10.75	1263.4	?	?
g.	9.89 + 2.9 + 9.49 + 3.1 + 2.75	24.73	?	?
h.	12,341 + 25,452 + 13,021 + 12,981	61,231	?	?

Problem Solving:
Using the Four-Step Method

The problem-solving process consists of several interrelated actions. The solutions to some problems are obvious and require very little effort. Others require a step-by-step procedure. Using a procedure such as the four-step method should help you in solving word problems.

The Four-Step Method

Step 1:	Understand	What is the problem? What is given? What are you asked to do?
Step 2:	Plan	What do you need to do to solve the problem? Choose a problem-solving strategy.
Step 3:	Work	Carry out the plan. Do any necessary calculations.
Step 4:	Answer	Is your answer reasonable? Did you answer the question?

EXAMPLE A small office building is being remodeled. It will take 3 plumbers 7 days to install all the pipes. Each plumber works 8 hours a day at $24 per hour. How much will it cost for the plumbers?

SOLUTION

Step 1:	Given	3 plumbers, 7 days, 8 hours, $24 per hour
	Find	The cost per day for 1 plumber. The cost per day for 3 plumbers. The cost of 3 plumbers for 7 days.
Step 2:	Plan	Find the cost per day for 1 plumber, then multiply by the number of plumbers, and then multiply by the number of days.
Step 3:	Work	8 hrs per day × $24 per hr = $192 per day for 1 plumber 3 plumbers × $192 per day for 1 plumber = $576 per day for 3 plumbers 7 days × $576 per day for 3 plumbers = $4032 for 3 plumbers for 7 days
Step 4:	Answer	It will cost $4032 for 3 plumbers for 7 days.

✔ SELF-CHECK Complete the problem, then check your answer in the back of the book.

1. It takes 2 finish carpenters 8 days to do the work. Each finish carpenter earns $26.50 per hour and works $7\frac{1}{2}$ hours per day. How much will it cost for the finish carpenters?

Identify the plan, work, and answer for each problem.

2. Henry Pitulski makes a car payment of $214.50 every month. His car loan is for 5 years. How much will he pay in 5 years?

3. Barb Allen and Joe Casey spent a total of $115.75 on their prom date. Dinner cost $45.87. How much did everything else cost Barb and Joe?

4. Elaine Wong purchased 2 sweaters at $24.99 each, a belt for $14.49, slacks for $19.79, shoes for $54.49, and 5 pairs of socks at $3.99 a pair. How much did Elaine spend?

5. A builder is building 5 new homes. It will take 4 electricians 2 days to wire each home. The electricians work 8 hours per day and earn $27.45 per hour. How much will it cost for the electricians?

6. A builder is building 5 new homes. Each home has a foyer measuring 9 feet by 18 feet. Wood parquet floors for each foyer cost $46.80 per square yard. What is the cost of the wood parquet floors for the foyers in all 5 homes?

7. Nadine Huffman charges $1.75 per page for typing rough drafts and an additional 50¢ per page for changes and deletions. A manuscript had 212 pages, of which 147 pages had changes and deletions. What was the total cost of typing the manuscript?

8. Nick Elsass is paying $10.50 per week for a compact disc player. The total cost of the CD player was $504. How long will it take Nick to pay for the CD player?

9. Karen Johnson rode her 27" bicycle to the store and back. The store is 1 mile from Karen's home. Approximately how many rotations did Karen's bicycle wheels make in going to the store and back? (Hint: The circumference of a circle is approximately 3.14 times the diameter.)

10. Ben Cornell and Tom Ingulli drove to Chicago, a distance of 510 miles. Their car gets 17 miles per gallon of gasoline. Gasoline costs them 96¢ per gallon. How much did Ben and Tom spend for gasoline on their trip?

11. Nancy Greer drove due north for 3 hours at 46 miles per hour. From the same spot, Ellen Young drove due south for 2 hours at 43 miles per hour. How far apart were they after their trip?

12. Peter Norris bought 3 boxes of cereal at $1.79 each, a roll of paper towels for 78¢, and 2 pounds of margarine at 74¢ a pound. How much change would Peter get back from $10?

13. Bread is sliced from a 30 cm loaf into equal slices 2 cm thick. All the slices are then toasted for a large breakfast requiring 180 pieces of toast. However, 1 out of every 5 pieces of toast is too well done for use. How many loaves of bread are needed?

WORKSHOP 20

Problem Solving: Identifying Information

Before you begin to solve a word problem, first read the problem carefully and answer these questions:

- What are you asked to find?

- What facts are given?

- Are enough facts given? Do you need more information than the problem provides?

Some word problems provide more information than is needed to solve the problem. Others cannot be solved without additional information. Identifying what is wanted, what is given, and what is needed allows you to organize the information and plan your solution.

EXAMPLE 1 Mary Ising is a systems analyst for EDP Consultants, Inc. She earns $17.25 per hour. She is married and claims 2 withholding allowances. Last week she worked 40 hours at the regular rate and 4 hours at the weekend rate. She is 28 years old. Find her gross pay last week.

SOLUTION **A.** Wanted: Mary Ising's gross pay last week

B. Facts given: $17.25 hourly rate
                        40 hours worked at regular rate
                        4 hours worked at weekend rate

C. Facts needed: Weekend rate

This problem cannot be solved.

EXAMPLE 2 John Skaggs, age 28, runs 6 miles every day. How many miles does John run in a week?

SOLUTION **A.** Wanted: Number of miles run in 1 week

B. Facts given: Runs 6 miles every day

C. Facts needed: None

This problem can be solved. Multiply the number of miles run per day (6) by the number of days in 1 week (7). The answer is 42 miles.

✔ SELF-CHECK Complete the problem, then check your answer in the back of the book.

1. Tonia Walsh bought a new car with a $2000 down payment and monthly payments of $274.50. How much did Tonia pay, in total, for her new car?

Identify the wanted, given, and needed information. If enough information is given, solve the problem.

2. The Camp Store is having a sale on camping equipment. It has 2-person tents for $79.49, cookstoves for $27.45, and cooking sets for $24.79. How much does a lantern and a tent cost?

3. The D & J Fruit Farm pays pickers 45¢ per pound to pick blueberries. The berries are packed in pint baskets and sold to grocery stores for 95¢ per pint. How many pint baskets are needed for 300 pounds of berries?

4. Don Diamond paid $84 each way to fly round-trip from Detroit to Pittsburgh. Bob Tucker paid $139 for the round-trip fare. Who paid more? How much more?

5. Edith Fairmont paid $126 for 3 tickets to a stage play. She paid for the tickets with three $50 bills. How much change did she receive?

6. Find the cost of 3 tablecloths, each 68 inches long and 52 inches wide. Each tablecloth costs $21.95.

7. Adam Larson paid for 2 watermelons with a $10 bill. He received $2.10 in change. What did the watermelons cost per pound?

8. Food for the party cost $44.95. Party supplies cost $17.48. Tula Drake and her friends have agreed to share the total cost of food and supplies equally. How much will each pay?

9. Lou Hart is 6 feet 2 inches tall and weighs 195 pounds. He grew 3 inches in the past year. How tall was Lou 1 year ago?

10. Alice Petroll has finished 25 of the 30 mathematics problems on her test. It is now 11:50 a.m. The 1-hour test started at 11:00 a.m. What is the average number of minutes she can spend on each of the remaining problems?

11. A tennis racket and a can of balls cost a total of $59.89. What is the cost of the tennis racket?

12. A bottle of cider costs 95¢. The cider costs 65¢ more than the bottle. How much does the cider cost?

13. Richard Anderson sells magazine subscriptions and receives a weekly salary of $145. He also receives a $3 bonus for each subscription that he sells. Last week his gross pay was $202. How many subscriptions did Richard sell last week?

14. In shopping for the latest recording of her favorite artist, Cheryl Nowokoski found that the cost of the compact disc was $9.20 more than the cost of the record album, and the cost of the cassette was $1.84 less than the cost of the record album. How much more than the cost of the cassette was the compact disc?

15. Assume you are driving on a 2-mile circular racetrack. For the first half of the track, you average 30 miles per hour. What speed must you maintain for the second half of the racetrack to average 60 miles per hour?

Problem Solving:
Using More Than One Operation

Some problems require several operations to solve. After deciding which operations to use, you must decide the correct order in which to perform them.

EXAMPLE 1 The cash price of a new car is $18,750. Arthur Dennis cannot pay cash so he is making a down payment of $2750 and 60 monthly payments of $325 each. How much more does it cost to buy the car this way?

SOLUTION

A. Given:
Cash price of $18,750
$2750 down + 60 payments of $325 each

B. Multiply: To get total of payments
$60 \times \$325 = \$19,500$

Add: $2750 to total of payments
$\$2750 + \$19,500 = \$22,250$

Subtract: Total payments from cash price of car
$\$22,250 - \$18,750 = \$3500$

It costs $3500 more to buy the car this way. In this example, the order of operations is very important; that is, to first multiply, then add, then subtract.

EXAMPLE 2 Nancy Paris bought 3 notebooks costing $3.98 each. She gave the cashier a $20 bill. How much change did she receive if there was no sales tax?

SOLUTION

A. Given: Bought 3 notebooks at $3.98 each, no sales tax
Gave cashier $20.00

B. Multiply: To get total cost
$3 \times \$3.98 = \$11.94$

Subtract: To find change
$\$20.00 - \$11.94 = \$8.06$

Nancy received $8.06 in change.

✔ SELF-CHECK Complete the problem, then check your answer in the back of the book.

1. Spent $7.95, $15.20, and $12.47 on entertainment. Entertainment budget is $50. How much is left in the entertainment budget?

PROBLEMS Give the sequence of operations needed to solve the problems, then solve.

2. Donna Preski works 8 hours a day, 5 days a week. So far this year, she has worked 680 hours. How many weeks has she worked?

3. The band boosters sell cider and doughnuts at home football games. Last week they sold 318 cups of cider at 50¢ per cup and 12 dozen doughnuts at 40¢ per doughnut. What were the total sales?

4. Patsy Cole paid monthly electric bills of $51.72, $47.75, and $53.21. Her electric budget is $150 for 3 months. Is she over or under her budget? By how much?

5. Victor Haddad sold 8 pumpkins for $2.75 each, 9 for $2 each, 24 for $1.50 each, and 15 for $1 each. He receives 35¢ for each pumpkin sold plus a $10 bonus if his sales total $75 or more. How much did he receive?

6. Debra Smith worked through 174 pages of a 408-page computer training manual. It took her 2 days to work through the remaining pages. If she worked through the same number of pages each day, how many pages did she work each day?

7. Mark Quincy worked 40 hours for $3.95 per hour. He worked 5 hours for $5.93 an hour. How much money did Mark earn?

8. The Parkers spent $42.78, $45.91, and $41.15 in 3 visits to the grocery store. Their food budget is $150. How much money do they have left to spend for food?

9. The temperature in the production department is 21 degrees Celsius at 12 noon. If the temperature increases 1.5 degrees Celsius every hour, what will the temperature be at 5 p.m.?

10. Flora Sturgeon walks 4 miles round-trip 3 times a week to work. How far will she walk in one year?

11. John Piotrowski assembled a total of 642 circuit boards in 3 days of work. During the first 2 days, he assembled 211 and 208 circuit boards, respectively. How many did he assemble the last day?

12. In a one-month sales contest, Ernie Johnkovich earned 4 two-point certificates, 2 three-point certificates, and 6 one-point certificates. How many points did he earn for the month?

13. Seventeen hundred tickets costing $5 each were sold for a scholarship fund-raiser. One prize of $2000, three prizes of $1000, and five prizes of $250 were given away. How much money did the scholarship fund-raiser make?

14. Tom Kramer, sales leader for the past month, earned 47 one-point certificates. If he earned a total of 68 points, how many three-point certificates did he earn?

15. Paula Cohn bought 3 T-shirts for $7.50 each and a sweatshirt for $22.95. How much change did she receive from a $50 bill?

16. Rich Spangler saved $212. After he earned an additional $124, he spent $149 for a small color TV, $35 for a rugby shirt, and $79 for a pair of sneakers. How much money did Rich have left?

Problem Solving:
Using Estimation

An important part of problem solving is determining the reasonableness of an answer. Checking an answer doesn't mean that you must recalculate it. Quite often, it is sufficient simply to determine if your answer makes sense. Estimation can be used to check the reasonableness of an answer. Some problems may ask for just an estimate.

EXAMPLE 1 Three cans of juice cost $1.19, six cans of soda cost $1.49, and six peaches cost $0.95. About how much will it cost for one of each item?

SOLUTION

$1.19 ÷ 3 is about $0.40
$1.49 ÷ 6 is about $0.25
$0.95 ÷ 6 is about $0.15

Total is about $0.80

EXAMPLE 2 Sandra Kaselman used her calculator to find 25% of $198.50. The result is shown at the right. Is her answer reasonable? Why or why not?

$4962.50

SOLUTION Her answer is not reasonable.
25% is equal to $\frac{1}{4}$, and $\frac{1}{4}$ of $198.50 is about $50.
It looks as if she multiplied by 25, not 25%.

✓ SELF-CHECK Complete the problem, then check your answer in the back of the book.

1. Tom Lucas estimated the cost of 10 gallons of gas at 93¢ a gallon to be $930. Is his estimate reasonable? What error did he make?

PROBLEMS In problems 2–9, determine the reasonableness of the estimate. If it is not reasonable, state what error was made.

	Problem	Estimate	Reasonable	Error, If Any
2.	20% of $496.98	$100	?	?
3.	5 gallons of gas at 89¢ a gallon	$450	?	?
4.	$14,203.00 ÷ 200	$700	?	?
5.	98.7 × 516	50,000	?	?
6.	Tip of 15% on $19.47	$3	?	?
7.	Sales tax of 5% on $1014.74	$5	?	?
8.	$15.00 per sq yd carpeting for 2' × 5' area	$150	?	?
9.	$49.79 − $19.49	$70	?	?

10. A picture frame requires 36 inches of frame molding at $5.99 a foot. About how much will the molding cost?

11. Ursula VanMeer bought 100 shares of stock at 18\frac{1}{2}$ per share. One year later, she sold all her shares at 23\frac{1}{4}$ per share. About how much did she make on her stock?

12. A new Vacation Plus mini van costs $24,897. About how much would 4 of these mini vans cost?

13. If the sales tax rate is 5%, about how much sales tax will be due on a $24,897 mini van?

14. If a 10% down payment is required to finance the purchase of a $24,897 mini van, about how much money would you need for a down payment?

15. Ben Grasser is buying a new car with a total purchase price including interest of $14,987. He plans to make a $3000 down payment and to finance the rest for 5 years. About how much will his monthly payments be?

16. Molly Noonan purchased 6 dozen cupcakes for the 25 children attending a birthday party. All the cupcakes were eaten. About how many cupcakes did each child eat?

17. Peter Cummings purchased the following school supplies on sale: a $2.79 notebook for $1.99, a $1.49 ballpoint pen for 99¢, a $1.99 pack of notebook paper for 99¢, and a $3.49 automatic pencil for $2.79. Estimate the total savings.

18. It is about 1300 feet around the bicycle test track. How many times must you ride around the track to ride about 1 mile?

19. Fruit baskets containing 6 apples, 4 oranges, and 2 grapefruits are on sale for $2.98. If you have 148 apples, 121 oranges, and 64 grapefruits, about how many fruit baskets can be made?

Problem Solving: Constructing a Table

Constructing a table can be a good way of solving some problems. By organizing the data into a table, it is easier to identify the information that you need. A table is useful in classifying information.

EXAMPLE 1 Meredith McCall is a car salesperson. For each new car she sells, she earns 10 bonus points; for each used car she sells, she earns 5 bonus points. She earned 125 bonus points by selling 16 cars last week. How many of each type of car did she sell?

SOLUTION Construct a table in order to evaluate the possibilities.

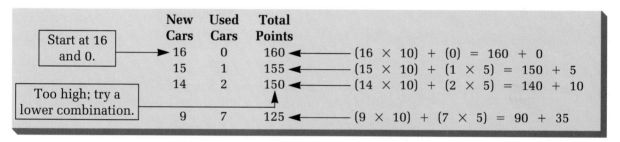

	New Cars	Used Cars	Total Points	
Start at 16 and 0.	16	0	160	$(16 \times 10) + (0) = 160 + 0$
	15	1	155	$(15 \times 10) + (1 \times 5) = 150 + 5$
	14	2	150	$(14 \times 10) + (2 \times 5) = 140 + 10$
Too high; try a lower combination.	9	7	125	$(9 \times 10) + (7 \times 5) = 90 + 35$

Meredith sold 9 new cars and 7 used cars.

EXAMPLE 2 Roger Bitter has a total of 20 coins consisting of dimes and quarters. The total value of the coins is $4.70. How many of each coin does he have?

SOLUTION Construct a table in order to evaluate the possibilities.

Number of		Value of		
Quarters	Dimes	Quarters	Dimes	Total Value
20	0	$5.00	$ 0	$5.00
19	1	4.75	0.10	4.85
18	2	4.50	0.20	4.70

Roger has 18 quarters and 2 dimes.

✔ SELF-CHECK Complete the problem, then check your answer in the back of the book.

1. A total of 11 vehicles consisting of unicycles (1 wheel) and bicycles (2 wheels) went by. Eighteen wheels were counted. How many of each were there?

2. Lieutenant Sampson is a recruitment officer for the Marines. For each high school graduate he recruits, he gets 5 points; for each college graduate, 11 points. He earned 100 points last week by recruiting 14 high school and college graduates. How many of each did he recruit?

3. Nine cycles were produced using 21 wheels. How many bicycles and how many tricycles were produced?

4. Third National Bank charges a monthly service fee of $5 plus 25¢ per check. The bank is changing its policy to a $6 service charge and 15¢ per check. A bank officer said you will save money with the new system. How many checks must you write each month in order to save money?

5. Wanda Cross has a total of 40 coins consisting of nickels and quarters with a total value of $6. How many of each coin does she have?

6. The Theatre Club sold a total of 415 tickets. The adult tickets cost $5 and the children's tickets cost $3. If $1615 was collected, how many adult tickets were sold?

7. There are 56 stools in the storeroom. Some stools have 3 legs and some have 4 legs. If there are 193 legs, how many 4-legged stools are in the storeroom?

8. Steve Swartz has 4 dimes and 3 nickels. List the amounts of all the exact-change telephone calls Steve could make using 1 or more of these coins.

9. How many different ways can you make change for a quarter?

10. Pam Young has exactly 20 dimes, 20 nickels, and 20 pennies. Find all the ways Pam can choose 22 coins whose total value is $1 if she must use at least 1 coin of each type.

11. Frank DeGeorge has 69¢ in coins. Bob White asked Frank for change for a half-dollar. Frank tried to make change but found that he didn't have the coins to do so. What coins did Frank have if each coin was less than a half-dollar?

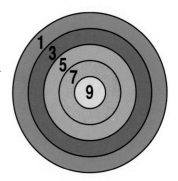

12. Carol Meeks was playing darts. She threw 6 darts, all of which hit the target shown. Which of the following scores could be hers? 2, 19, 58, 28, 33, 37

Problem Solving: Looking for a Pattern

Some problems can be solved more easily if the information is first organized into a list or table. Then the list or table can be examined to see if a pattern exists. A pattern may not "jump out" at you, but you may be able to discover a pattern after manipulating the information.

EXAMPLE   Joe Cobb has 15 coins consisting of dimes and quarters. The total value of the coins is $2.55. How many of each coin does he have?

SOLUTION   Given: 15 coins consisting of dimes and quarters

The table displays different combinations of dimes and quarters with the total number of coins equaling 15.

Number of Dimes	15	14	13	12	11
Number of Quarters	0	1	2	3	4
Total Value	$1.50	$1.65	$1.80	$1.95	$2.10

Look for a pattern.

You could continue the table, but it is easier if you see the pattern. Each time you take away a dime and add a quarter, the total value increases by $0.15. The difference between $2.55 and $1.50 is $1.05, and $1.05 ÷ $0.15 = 7. Therefore, subtract 7 from 15 and conclude that there are 8 dimes and 7 quarters.

Check: The total value of 7 quarters and 8 dimes is:

$(7 \times \$0.25) + (8 \times \$0.10) = \$1.75 + \$0.80 = \$2.55$

✔ SELF-CHECK   Complete the problems, then check your answers in the back of the book.

Write the next 3 numbers for the established pattern.

1. 1, 3, 9, 27, . . .   **2.** 2, 4, 7, 11, . . .

PROBLEMS   In problems 3–10, look for a pattern and then write the next 3 numbers.

3. 10, 16, 22, 28, . . .

4. 2, 4, 8, 16, . . .

5. 30, 27, 24, 21, . . .

6. 8, 4, 2, 1, . . .

7. 1, 4, 9, 16, . . .

8. 2, 3, 5, 9, 17, 33, . . .

9. 1, 4, 13, 40, 121, 364, . . .

10. 1, 1, 2, 3, 5, 8, 13, . . .

11. Sy Mah has 20 coins consisting of dimes and quarters. Their total value is $3.80. How many of each coin does Sy have?

12. Peggy Mays had 185 tickets to the school play. Adult tickets sold for $4 each and children's tickets sold for $2 each. The total value was $594. How many of each ticket did she sell?

13. Three-sided numbers, such as 3, are so named because dots can be used to form a triangle with an equal number of dots, such as 3, on each side. What three-sided number has 12 dots on a side?

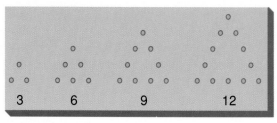

14. On the first day of school, your teacher agrees to allow 1 minute of "fun" at the end of the first day, 2 minutes on the second day, 4 minutes on the third day, 8 minutes on the fourth day, and so on. How much time will you have for "fun" at the end of 10 days?

15. At 9 a.m., there were 7 students in the computer room. At 9:30 a.m., 2 students left, and at 10 a.m., 1 student arrived. At 10:30 a.m., 2 students left, and at 11 a.m., 1 student arrived. This pattern continued with 2 students leaving at half past the hour and 1 student arriving on each hour. At what time did the computer room first become empty?

16. Bob and Ray Hunt were responsible for total lawn care of the factory grounds. The first week, Bob mowed half the lawn. The next week, he mowed two-thirds as much as he had the first week. The third week, he mowed three-fourths as much as he had the second week, and so on. The tenth week, he mowed ten-elevenths as much as he mowed the ninth week. How much of the lawn did Bob mow the tenth week?

17. There are 2 rectangular storage rooms whose sides are whole numbers and whose area and perimeter are the same number. What are their dimensions?

18. A 5-pound bag of lawn food sells for $3.25 and a 3-pound bag sells for $2.29. You need 17 pounds of lawn food. What is the least amount you can pay and buy at least 17 pounds?

19. When it is 12 o'clock, the hands on the face of the clock overlap. How many more times will the hands overlap as the clock runs until it is again 12 o'clock? (Count the second time it is 12 o'clock but not the first.)

20. Each year, before the U.S. Supreme Court opens, each of the 9 justices shakes hands with every other justice. How many handshakes are involved in this ceremony?

Problem Solving: Using Guess and Check

One way to solve a problem is by using guess and check, or trial and error. Guessing at a solution doesn't mean making a blind guess in the hopes that it is correct. It means making an informed guess and then checking it against the conditions stated in the problem to determine how to make a better guess. The process is repeated until the answer is found.

EXAMPLE 1 You need 80 sandwich buns. You can buy 8 for $1.19 or 12 for $1.69. What do you buy to obtain at least 80 sandwich buns at the lowest cost?

SOLUTION Guess and check until you find the lowest cost. Keep your information organized by using a table.

Number		Total Buns	Cost		Total Cost
12/pkg.	8/pkg.		12/pkg.	8/pkg.	
7	0	84	(7 × $1.69) +	0	$11.83
6	1	80	(6 × $1.69) +	(1 × $1.19)	$11.33
5	3	84	(5 × $1.69) +	(3 × $1.19)	$12.02
4	4	80	(4 × $1.69) +	(4 × $1.19)	$11.52
3	6	84	(3 × $1.69) +	(6 × $1.19)	$12.21
2	7	80	(2 × $1.69) +	(7 × $1.19)	$11.71
1	9	84	(1 × $1.69) +	(9 × $1.19)	$12.40
0	10	80	0	+ (10 × $1.19)	$11.90

Six packages of 12 sandwich buns and one package of 8 sandwich buns cost $11.33, the least amount, and result in 80 sandwich buns.

EXAMPLE 2 The factory building is 4 times as old as the equipment. Three years from now, the factory building will be 3 times as old. How old is the equipment?

SOLUTION

Current Age		Three Years From Now		Check
Equipment	Building	Equipment	Building	
7	28	10	31	No
8	32	11	35	No, wrong direction
6	24	9	27	Yes

The equipment is now 6 years old.

✔ SELF-CHECK Complete the problem, then check your answer in the back of the book.

1. The product of 3 consecutive whole numbers is 504. What are the numbers?

2. The planet Tetriad only has creatures with 3 legs (triads) or 4 legs (tetrads). Astronauts Peter North and Sally Clark could not bear to look at these ugly creatures, so they kept their eyes on the ground. On their first day on Tetriad, they counted 81 legs. How many Triads and Tetrads did they meet?

3. It costs 19¢ to mail a postcard and 29¢ to mail a letter. Sam Checkers wrote to 15 friends and spent $3.55 for postage. How many letters and how many postcards did he write?

4. Paula McDale earns $8 per hour Monday through Friday and $12 per hour on weekends. One week, she worked 49 hours and earned $404. How many hours did she work on the weekend?

5. Arrange the 4 dominoes below into a domino donut so that all sides equal the same sum (not necessarily the same sum as that of the example at the right).

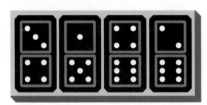

Example

6. Ti Sun wants to fence off a storage area for surplus lumber. She has 96 meters of new fencing to put along an existing fence. What are the dimensions that will give Ti the largest storage area?

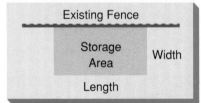

7. In the square at the right, a rule applies from top to bottom and from left to right. Find the rule and figure out the missing number.

8. Batteries come in packs of 3 or 4. If your class needs 30 batteries, how many different ways are there of buying exactly 30 batteries? Which combination do you think would be the cheapest?

9. The ages of 3 delivery vans total 22 years. The Ford is the oldest and is 10 years old. The Dodge is 6 years younger than the Ford. The third delivery van is a Chevy. What are the ages of the vans?

10. The product of 3 consecutive whole numbers is 120. What are the numbers?

11. Arrange the numbers 5 through 13 in a magic square whose rows, columns, and diagonals add up to the same number.

Problem Solving:
Working Backward

Problems that involve a sequence of events or actions can sometimes be solved by working backward. If the final result of the problem is given, start your solution with that result and work backward to arrive at the beginning conditions of the problem.

EXAMPLE 1   Each year, a delivery van is worth three fourths of its value from the previous year. A van is now worth $9000. What was its value last year?

SOLUTION   **A.**   What is the van worth now? $9000

B.   How does last year's value relate to this year's value?

$\frac{3}{4}$ of last year's value $=$ this year's value

$\frac{3}{4} x = \$9000$

C.   Solve: $x = \$9000 \div \frac{3}{4}$

$x = \$12,000$

The van was worth $12,000 last year.

EXAMPLE 2   Central Bakery baked some cookies and put one half of them away for the next day. Then Central Bakery divided the remaining cookies evenly among its 3 sales outlets so that each outlet received 40 dozen. How many cookies did Central Bakery bake?

SOLUTION   Each of the 3 sales outlets received 40 dozen cookies. Thus, Central Bakery divided a total of 120 dozen cookies. The 120 dozen cookies represent one half of what was baked; therefore, they baked 240 dozen cookies.

✔ SELF-CHECK   Complete the problem, then check your answer in the back of the book.

1.   A water lily doubles itself in size each day. It takes 30 days from the time the original plant is placed in a pond until the surface of the pond is completely covered with lilies. How long does it take for the pond to be half covered?

PROBLEMS

2.   Checker Cab Co. charges a flat fee of $3.75 plus $0.30 for every $\frac{1}{4}$ mile driven. Dick Lewis paid a driver a total of $7.95 for a trip from the airport to his office. How many miles did he travel?

3. A recipe for 24 medium-sized pancakes requires 2 eggs. Eggs are sold in cartons of 12 eggs. The band boosters plan to serve 240 people an average of 3 pancakes each. How many cartons of eggs are needed?

4. Two barrels, A and B, contain unspecified amounts of cider, with A containing more than B. From A, pour into B as much cider as B already contains. Then from B, pour into A as much cider as A now contains. Finally, pour from A into B as much cider as B presently has. Both barrels now contain 80 liters of cider. How many liters of cider were in each barrel at the start of the process?

5. Your company automobile is now worth $12,000. You read an article that indicates that each year an automobile is worth 80% of what its value was the previous year. What was the value of the company automobile last year?

6. Tiffany Cole starts at point A and enters a fun house. She pays $2 to get in and loses half of the money in her possession while she is in the fun house. She then pays $1 when she exits at B. She goes to the next entrance (C), pays $2 to get in, loses half of her money, and pays $1 to exit at D. This is repeated until she exits at H and gives her last $1 to get out. How much money did she start with?

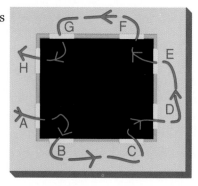

7. Merril Bakery baked some cookies in their new convection oven. Stores #1, #2, #3, and #4 sold a total of 4 dozen cookies each. The remainder were put in storage. Later, another dozen cookies were sold and one third of what was left was sent to stores #5 and #6. Those two stores sold 2 dozen cookies each and had 9 dozen left between them. How many cookies did Merril Bakery bake?

8. Sixty-four players are entered in the company single elimination tennis tournament. How many matches must be played to determine the winner?

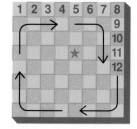

9. If you were to continue the number pattern until you got to the star, what number would you put in the star's square?

10. Three girls played a game in which two of them won and one lost on each play. The girl who lost had to double the points that each winner had at that time by subtracting from her own points. The girls played the game 3 times, each winning twice and losing once. At the end of the 3 plays, each girl had 40 points. How many points did each girl have to start?

11. There were 8 women and 16 men at the last board of directors meeting. Every few minutes, 1 man and 1 woman (a couple) left the meeting together. How many couples must leave before there are exactly 5 times as many men as women left at the meeting?

Problem Solving: Writing an Equation

A word problem can be translated into an equation that is solved by performing the same mathematical operation (adding, subtracting, multiplying, or dividing) to both sides. Solving the equation then leads to the solution of the problem.

To set up the equation, look for words in the problem that suggest which of the four mathematical operations to use.

Words	Symbol	Operation
The total, how many in all, the sum, plus	+	Addition
The difference, how much more, how much smaller, minus	−	Subtraction
The total for a number of equal items, the product	×	Multiplication
The number left over, the quotient	÷	Division

EXAMPLE 1 In 40 hours at your regular rate of pay plus 5 hours of double time (twice your regular rate of pay), you earn $425. What is your regular rate of pay?

SOLUTION Use the letter x to stand for your regular rate of pay.

$$40x + 5(2x) = \$425.00$$
$$40x + 10x = \$425.00$$
$$50x = \$425.00$$
$$x = \$8.50 \text{ (Divide each side by 50.)}$$

EXAMPLE 2 A rectangle with a perimeter of 48 mm is 20 mm long. What is the width of the rectangle?

SOLUTION Let w equal the width of the rectangle.

$$w + 20 + w + 20 = 48$$
$$2w + 40 = 48 \text{ (Subtract 40 from both sides.)}$$
$$2w = 8 \text{ (Divide both sides by 2.)}$$
$$w = 4 \text{ mm wide}$$

✔ SELF-CHECK Complete the problem, then check your answer in the back of the book.

1. The sum of 2 consecutive numbers is 23. What is the smaller number?

PROBLEMS

2. One brand of computer scanner can read 74 documents per hour while a second scanner can read 92 documents per hour. How many hours will it take to read 747 documents?

3. A bottle and a cork cost $1.10. The bottle costs $1.00 more than the cork. How much does each cost?

4. A robotic delivery unit travels 54 meters in traveling completely around the edge of a rectangular mail room. If the rectangle is twice as long as it is wide, how long is each side?

5. A football field is 100 yards long and has a distance around of 308 yards. How wide is it?

6. David and Liz Northrup make monthly payments of $727.20 on their $80,000 mortgage. They will have paid $138,160 in interest when their mortgage is paid off. For how many years is their mortgage?

7. Edith Harris had gross pay of $464.86 last week. She earns $5.65 per hour plus a 3% commission on all sales. She knows she worked 40 hours last week but can't remember her total sales. What were her total sales?

8. Steve Sutton earns $7.60 per hour plus double time for all hours over 40 per week. How much did Steve earn for working 46 hours last week?

9. Wilson Davis has 3 Guernsey cows and 2 Holstein cows that give as much milk in 4 days as 2 Guernsey and 4 Holstein cows give in 3 days. Which kind of cow is the better milk producer, the Guernsey or the Holstein?

10. The Karis Tool & Die Company building is 5 times as old as the equipment. The building was 24 years old when the equipment was purchased. How old is the equipment?

11. The sum of 3 consecutive odd numbers is 27. What are the 3 numbers?

12. Universal Inc. stock sells for 17\frac{1}{4}$ a share. Future Discount Brokers charges a flat fee of $40 for every transaction. How many shares could you buy for $730?

13. Write an equation expressing the relationship between A and B given in the table at the right. What would B equal when A is 40?

A	1	2	3	4	5	...
B	1	4	7	10	13	...

14. In problem 13, write an equation for A in terms of B. What would A equal when B is 253?

15. Harry, Jerry, and Darrel have a combined weight of 599 pounds. Jerry weighs 13 pounds more than Harry, while Harry weighs 5 pounds more than Darrel. How much does each man weigh?

Problem Solving: Making a Venn Diagram

Some problems can be solved with a diagram. Making a diagram can be an effective way of showing the information in a problem. In this workshop, we will use diagrams called Venn diagrams. They can be used to show the relationship among several groups of people, animals, or objects.

EXAMPLE 1 Of 53 employees surveyed, 43 have degrees in management, 20 have degrees in marketing, and 10 in both. How many have degrees in marketing but not management?

SOLUTION Draw two intersecting circles, one for management and one for marketing. Work out from the middle region.

A. Fill in the number in both classes (10).
B. Fill in the remaining number in management: 43 − 10 = 33
C. Fill in the remaining number in marketing: 20 − 10 = 10
D. Add the numbers:
   33 + 10 + 10 = 53
Ten employees have degrees in marketing but not management.

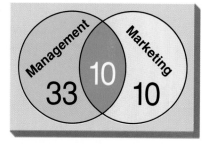

EXAMPLE 2 Two hundred people returning from a trip to Europe were asked which countries they had visited. One hundred forty-eight had been to England, 116 had been to France, and 96 had been to Spain. Eighty-two had been to England and France, 71 had been to France and Spain, 56 had been to England and Spain, and 44 had visited all 3 countries.

• How many had visited France but not England or Spain?
• How many had not visited any of these 3 countries?

SOLUTION **A.** Seven people visited France only. This was found by the following procedure:

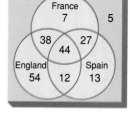

(1) Draw a Venn diagram as shown.
(2) Forty-four visited all 3 countries.
(3) Twelve visited England and Spain only. (56 − 44)
(4) Twenty-seven visited France and Spain only. (71 − 44)
(5) Thirty-eight visited England and France only. (82 − 44)
(6) Thirteen visited Spain only. [96 − (12 + 44 + 27)]
(7) Seven visited France only. [116 − (38 + 44 + 27)]
(8) Fifty-four visited England only. [148 − (38 + 44 + 12)]
B. Five people did not visit any of the 3 countries.
   [200 − (54 + 7 + 13 + 38 + 27 + 12 + 44)]

Complete the problem, then check your answer in the back of the book.

1. Of 12 classmates, 6 went to the game, 7 went to the dance, and 4 went to both. How many went to neither?

2. There are 197 delegates at the International Trade Association. Eighty-five of them speak Spanish, 74 speak French, and 15 speak both French and Spanish.
   a. How many speak Spanish but not French?
   b. How many speak French but not Spanish?
   c. How many speak neither French nor Spanish?

3. A survey of 150 people revealed that 121 people watch the early evening TV news, 64 watch the noon news, and 47 watch both. How many do not watch either one?

4. A survey of 120 college junior business students produced these results: 40 read *Business Journal*, 48 read the local paper, 70 read the campus paper, 25 read *Business Journal* and the local paper, 28 read the local paper and the campus paper, 21 read the campus paper and *Business Journal*, and 18 read all three papers.
   a. How many do not read any of the papers?
   b. How many read *Business Journal* and the local paper but not the campus paper?

5. Of 20 students eating at a restaurant, 14 ordered salad, 10 ordered soup, and 4 ordered both salad and soup. How many did not order either?

6. One hundred employees were asked which sports they play. Fifty play football, 48 play basketball, 54 play baseball; 24 play both football and basketball, 22 play both basketball and baseball, 25 play both football and baseball; and 14 play all 3 sports.
   a. How many play basketball only?
   b. How many play football and baseball but not basketball?
   c. How many play none of the 3 sports?

7. In a marketing survey, 500 people were asked their 2 favorite colors. Red was chosen by 380, blue by 292, and 10 chose neither red nor blue.
   a. How many like both colors?
   b. How many like red but not blue?

Problem Solving: Drawing a Sketch

Some word problems, particularly those that involve lengths, widths, and dimensions, can be simplified if you draw a sketch. Sketches and diagrams can also help you keep track of information in multistep problems.

EXAMPLE 1 A 30-cm piece of pipe is cut into 3 pieces. The second piece is 2 cm longer than the first piece, and the third piece is 2 cm shorter than the first piece. How long is each piece?

SOLUTION

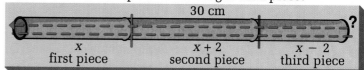

$$x + (x + 2) + (x - 2) = 30$$
$$3x = 30$$
$$x = 10 \text{ cm} \quad \text{length of first piece}$$
$$x + 2 = 12 \text{ cm} \quad \text{length of second piece}$$
$$x - 2 = 8 \text{ cm} \quad \text{length of third piece}$$

EXAMPLE 2 A 2-volume set of classics is bound in 1/4-inch covers. The text in each volume is 3 inches thick. The 2 volumes are side by side on the shelf. A bookworm travels from inside the front cover of Volume I to the inside back cover of Volume II. How far does the bookworm travel?

SOLUTION

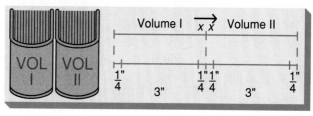

Bookworm travels: $\frac{1}{4}" + \frac{1}{4}" = \frac{1}{2}"$

The bookworm travels $\frac{1}{2}$ inch from inside the front cover of Volume I to inside the back cover of Volume II.

✔ SELF-CHECK Complete the problem, then check your answer in the back of the book.

1. Use an 18-inch, 10-inch, and 7-inch length of board to mark off a length of 15 inches.

PROBLEMS

2. Tom Brown leaves his house and jogs 8 blocks west, 5 blocks south, 3 blocks east, 9 blocks north, 6 blocks east, 12 blocks south, 3 blocks west, and then stops to rest. Where is he in relation to his house?

3. Allison Kelly bicycles to work. She travels 3 miles north of her house, turns right and bikes 2 miles, turns left and bikes 3 miles, and turns left and bikes 5 miles. At this point, where is she in relation to her home?

4. The rectangular area allotted to the shoe department of a department store has a perimeter of 96 feet. The length of the rectangle is 8 feet longer than the width. What is the width of the rectangle?

5. A barn has dimensions of 50 feet by 60 feet. A cow is tethered to 1 corner of the barn with a 60-foot rope. The cow always stays outside the barn. How many square feet of grazing land can the cow reach?

6. A 60-foot piece of fencing is cut into 3 pieces. The second piece is 3 feet longer than the first piece, and the third piece is 9 feet longer than the first piece. Find the lengths of the pieces.

7. Assuming that each corner must be tacked, what is the least number of tacks that you need to display eight 8-inch by 10-inch photographs?

8. Making identical cuts, a lumberjack can saw a log into 4 pieces in 12 minutes. How long would it take to cut a log of the same size and shape into 8 pieces?

9. Master Chemical Company has containers with capacities of 4 L, 7 L, and 10 L. How could you use these containers to measure exactly 1 L?

10. Three book volumes are arranged as shown. The thickness of each cover is 0.2 cm. The text in each volume is 3 cm thick. What is the distance from the first page of Volume I to the last page of Volume III?

11. A dog is on a 12-foot leash that is tied to the corner of a 10-foot by 15-foot shed. The dog always stays outside the shed. How many square feet of ground can the dog reach?

12. Ohio Airlines is to provide service between cities as shown on the map. The airline employed 5 new people to sit in the control tower at each of the 5 cities. The people are Carol, Connie, Clare, Charles, and Cedric. The 2 people in the cities with connecting routes will be talking to each other a great deal, so it would be helpful if these people were friends. The pairs of friends are: Charles-Connie, Carol-Cedric, Charles-Carol, Clare-Cedric, and Carol-Clare. Place the 5 people in the 5 cities so that the ones in connecting cities are friends.

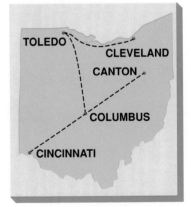

Problem Solving: Restating the Problem

To solve a problem, you may find it helpful to restate the problem in a different way. A difficult problem can be restated as a simpler problem and become easier to solve. Or, a problem may involve a series of simpler problems that will lead to the solution of the original problem.

EXAMPLE 1   A single elimination tournament is to be held among 124 contestants. Single elimination means that you are out of the tournament if you have 1 loss. How many matches will have to be played to determine a winner? (Hint: A similar problem was encountered in the "Working Backward" workshop.)

SOLUTION   Given that it is a single elimination tournament, every player but the winner will have to lose 1 match. Since 123 contestants will have to lose 1 match, 123 matches will have to be held.

EXAMPLE 2   A 27-inch bicycle makes 100 revolutions per minute. A stone stuck in the treads will travel how many feet in 1 hour?

SOLUTION   Restate the problem to follow a series of simpler problems.

A. How many inches will the stone travel in 1 revolution?
$C = \pi d$ or 3.14×27 inches $= 84.78$ inches

B. How many inches will the stone travel in 1 minute?
100 rpm $\times$ 84.78 inches per revolution $= 8478$ inches

C. How many inches will the stone travel in 1 hour?
8478 inches per minute $\times$ 60 minutes per hour $= 508,680$ inches

D. How many feet will the stone travel in 1 hour?
508,680 inches $\div$ 12 inches per foot $= 42,390$ feet

✔ SELF-CHECK   Complete the problem then check your answer in the back of the book.

1. A 27-inch bicycle wheel makes 100 revolutions per minute. A stone stuck in the treads will travel how many miles in 1 hour? (Round answer to the nearest mile.)

PROBLEMS

2. Your automobile's gas mileage is 27.5 miles per gallon. Last month your automobile was driven 1595 miles, and a gallon of gasoline cost 96¢. What did it cost to buy gasoline for your automobile last month?

3. Vanna Burzinski works in the stockroom at Keynote Auto Supply. Vanna always stacks crates using the triangular pattern shown. The top row always has 1 crate. Each row always has 2 fewer crates than the row before it.
   a. How many crates are in a stack that has 20 rows?
   b. How many crates are in a stack if the bottom row has 29 crates?

4. Rachel Murphy is having a party. The first time the doorbell rings, 3 guests arrive. Each time the doorbell rings after that, a group arrives that has 2 more guests than the preceding group.

   a. How many guests arrive on the seventh ring?
   b. If the doorbell rings 10 times, how many guests came to the party?

5. Note that $1 + 3 = 4$, $1 + 3 + 5 = 9$, and $1 + 3 + 5 + 7 = 16$. What is the sum of the first 25 odd numbers?

6. Forty-eight teams are entered in a double elimination tournament. Double elimination means that a team is out of the tournament if it loses two games.
   a. How many games will have to be played to determine a winner?
   b. Why are there 2 possible answers?

7. A train traveling at 60 miles per hour takes 4 seconds to enter a tunnel and another 50 seconds to pass completely through the tunnel.
   a. What is the length of the train?
   b. What is the length of the tunnel?

8. There are 26 teams in the National Football League. To conduct their annual draft, teams in each city must have a direct telephone line to each of the other cities.
   a. How many direct telephone lines must be installed to accomplish this?
   b. How many direct telephone lines must be installed if the league expands to 30 teams?

9. Grandpa Moyer wanted to leave his 17 horses to his 3 grandsons. Rick was to get $\frac{1}{2}$ the horses, Mike was to get $\frac{1}{3}$, and Peter was to get $\frac{1}{9}$. How could he accomplish this?

10. Two cyclists start toward each other from points 25 miles apart. One cyclist travels 15 mph, while the other cyclist travels 10 mph. A trained bird leaves the shoulder of 1 cyclist and travels to the shoulder of the other cyclist and back and forth until the cyclists meet. If the bird flies at 40 mph, how far will the bird fly?

11. Three men agree to share a hotel room and pay a total of $60 when they check in. Later, the clerk finds that he should have charged them only $55. He sends the bellhop up with a $5 refund for the room, but because the bellhop cannot divide $5 evenly by 3, he returns only $3 to the men and keeps $2. Therefore, each man paid only $19 as his share of the room for a total of $57. The bellhop kept the other $2 for a total of $59. What happened to the extra $1?

2

Personal Application Mathematics

Part 2. Personal Application Mathematics, covers the personal applications of basic math skills that you will use in your roles as worker and knowledge-able consumer. By learning how to handle your money wisely, you can use your income to meet day-to-day living expenses and still save and invest part of your income for use in the future. Eleven consumer topics are covered in Part 2.

The following definitions provide an overview of the important areas that affect how you earn, spend, and save your money.

Gross Income The total amount of money you earn for work done during a pay period, before any deductions are taken.

Net Income Your take-home pay after all tax withholdings and personal deductions have been taken from your gross income.

Checking Accounts Funds deposited in a bank from which you can write checks to pay for goods and services instead of paying by cash.

Savings Accounts Funds deposited in a bank for which you earn interest.

Cash Purchases Items or services that you buy at regular or sales prices.

Charge Accounts and Credit Cards Systems of buying in which you receive goods and services now and spread the payments, along with finance charges, if any, over a period of time.

Loans Money that is borrowed and must be repaid along with interest owed.

Automobile Transportation 1. The costs of buying, insuring, and maintaining your own automobile. 2. The costs of leasing an automobile or renting from a rental agency.

Housing Costs The costs of purchasing, owning, and maintaining your own home.

Insurance and Investments 1. Financial protection purchased for your dependents. 2. Sums of money paid out to gain future income for yourself.

Planning Activities related to planning for business decisions; includes inflation, GNP, consumer price index, and budget analysis.

Recordkeeping Maintaining accurate accounts of your expenses as tools for managing your money.

1

Gross Income

Your *income,* the money that you earn on a job, may be determined in one of several ways. Income may be computed on hours worked, or *hourly rate,* the number of items produced, or *piecework,* or *commission.* Commission may be a specified amount of money you are paid for selling a product, or it may be a percent of the selling price. You may also be paid on a *graduated commission,* which increases as your sales increase. Many workers are paid a fixed annual *salary,* which they receive *weekly, biweekly, semimonthly,* or *monthly.*

O U T L I N E

1-1 Hourly Pay

1-2 Overtime Pay

1-3 Weekly Time Card

1-4 Piecework

1-5 Salary

1-6 Commission

1-7 Graduated Commission

◆ Reviewing the Basics

◆ Unit Test

◆ A Spreadsheet Application: Gross Income

◆ Career Wise: Salesperson

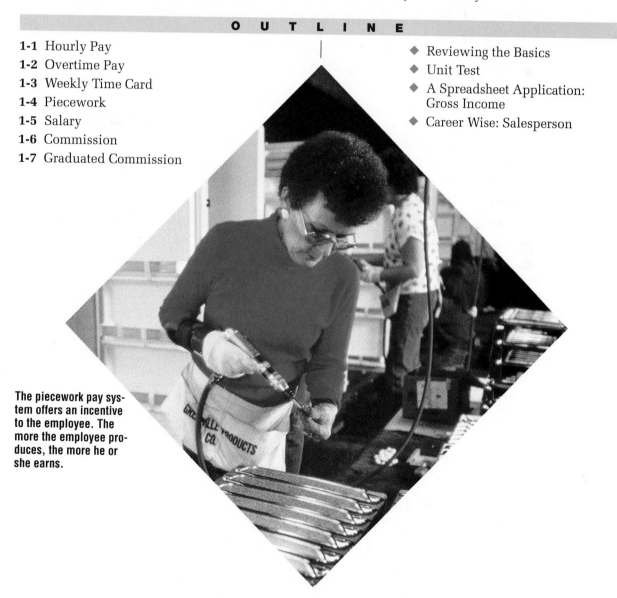

The piecework pay system offers an incentive to the employee. The more the employee produces, the more he or she earns.

Hourly Pay

OBJECTIVE
Compute the straight-time pay.

For some jobs, you are paid a fixed amount of money for each hour you work. Your **hourly rate** is the amount of money you earn per hour. **Straight-time pay** is the total amount of money you earn for a pay period at the hourly rate.

Straight-Time Pay = Hourly Rate × Hours Worked

EXAMPLE *Skills* 14, 8, 2 *Application* A *Term* Straight-time pay

Mary Byron is a mail clerk. She earns $5.40 per hour. Last week she worked 40 hours and this week she worked $39\frac{1}{2}$ hours. What is her straight-time pay for each week?

S O L U T I O N Find the **straight-time pay**.

	Hourly Rate	×	Hours Worked		
Last week:	$5.40	×	40	= $216.00	straight-time pay
This week:	$5.40	×	$39\frac{1}{2}$	= $213.30	straight-time pay

$$5.40 \quad \boxed{\times} \quad 39.5 \quad \boxed{=} \quad 213.30$$

✔ SELF-CHECK Complete the problems, then check your answers in the back of the book.

Find the straight-time pay.

1. $7.60 per hour × 40 hours

2. $7.90 per hour × $37\frac{1}{2}$ hours

PROBLEMS

Employee	Regular Hourly Rate	× Hours Worked	= Straight-Time Pay
3. Wright, S.	$ 8.00	× 36	= ?
4. Reardon, E.	$ 4.50	× 18	= ?
5. Levitt, R.	$ 6.20	× 30	= ?
6. Maguire, M.	$ 5.00	× $24\frac{1}{2}$	= ?
7. Reyes, A.	$14.50	× $31\frac{1}{4}$	= ?
8. Vadola, G.	$16.25	× $25\frac{3}{4}$	= ?

F.Y.I.
Federal Minimum
Wage
1978	$2.65
1979	2.90
1980	3.10
1981	3.35
1990	3.80
1991	4.25

9. You earn $5.50 per hour. Worked 30 hours last week. What was your straight-time pay?

10. You earn $8.00 per hour. Worked 35.25 hours last week. What was your straight-time pay?

11. Gloria Scanlon.
Financial consultant.
Earns $25.00 per hour.
Worked 32.5 hours last week.
What was her straight-time pay?

12. Marian Abelt.
Mail clerk.
Earns $4.20 per hour.
Worked $35\frac{3}{4}$ hours last week.
What was her straight-time pay?

13. Sylvia O'Keefe is a teacher's aide. She earns $5.80 per hour. Last week she worked 34 hours. What was her straight-time pay for last week?

14. Judy Martin has a part-time job at Dutton's. She earns $4.25 per hour. She works 20 hours each week. What is her straight-time pay for a week?

15. Don Moline, a construction worker, earns $12.25 per hour. He worked his regular $36\frac{1}{4}$ hours last week. What was his straight-time pay?

16. Arthur Marzetti is a machine operator at the Donovan Manufacturing Corporation. He works $27\frac{1}{4}$ hours per week and earns $4.75 per hour. What is his straight-time pay each week?

17. Sally Barrett, a banjo player for a local band, earns $8.50 per hour. What is her straight-time pay for a week in which she worked 22 hours?

18. Jon Bradford earns $7.84 per hour as an insurance claims adjustor. What is his straight-time pay for this week?

Mon.	Tue.	Wed.	Thu.	Fri.
8 hrs.	8 hrs.	6.5 hrs.	8.5 hrs.	7.25 hrs.

F.Y.I.

On April 1, 1991, the federal minimum wage was set at $4.25 per hour.

19. Ben Favreau was hired for the position of billing clerk. During his first week on the job, he worked $7\frac{1}{2}$ hours each day, Monday thru Friday. What was his total pay for the week?

SILK SCREEN CO. has opening for bright, dependable billing clerk. Must have aptitude for figures and pay attention to detail. Prepare work for production. Start $5.75 per hr. Call Walt, at 555-9256 Ext. 11.

MAINTAINING YOUR SKILLS   Look up the skills in parentheses if you need help or more practice.

Write as decimals. **(Skill 14)**

20. $\frac{1}{4}$ **21.** $\frac{1}{2}$ **22.** $\frac{3}{4}$ **23.** $\frac{1}{10}$ **24.** $\frac{3}{8}$

Multiply. Round answers to the nearest cent. **(Skill 8)**

25. 4.50×30.25 **26.** 8.30×3.25 **27.** 14.50×32.5

Round to the place value indicated. **(Skill 2)**

Nearest ten	**28.** 22	**29.** 138	**30.** 8
Nearest hundred	**31.** 91	**32.** 3092	**33.** 14,431
Nearest thousand	**34.** 15,374	**35.** 722	**36.** 158,481

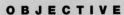

1-2

Overtime Pay

OBJECTIVE

Compute the straight-time, overtime, and total pay.

When you work more than your regular hours, you may receive overtime pay. The overtime rate may be $1\frac{1}{2}$ times your regular hourly rate. This is called time and a half. You may receive double time, 2 times your regular hourly rate, for overtime on Sundays or holidays.

Overtime Pay = Overtime Rate × Overtime Hours Worked

Total Pay = Straight-Time Pay + Overtime Pay

EXAMPLE   *Skill* 5   *Application* A   *Term* Overtime pay

Bill Learner is paid $8.20 an hour for a regular 40-hour week. His overtime rate is $1\frac{1}{2}$ times his regular hourly rate. This week Bill worked his regular 40 hours plus 10 hours of overtime. What is his total pay?

SOLUTION

A. Find the **straight-time pay.**

Hourly Rate	×	Regular Hours Worked	
$8.20	×	40	= $328.00 straight-time pay

B. Find the **overtime pay.**

Overtime Rate	×	Overtime Hours Worked	
(1.5 × $8.20)	×	10	= $123.00 overtime pay

C. Find the **total pay.**

Straight-Time Pay	+	Overtime Pay	
$328.00	+	$123.00	= $451.00 total pay

8.20 ✕ 40 = 328 M+ 1.5 ✕ 8.20 ✕ 10 = 123 M+ RM 451

✔ SELF-CHECK   Complete the problems, then check your answers in the back of the book.

Find the total pay.

1. $9.00 per hour for 40 hours.
Time and a half for 6 hours.

2. $11.50 per hour for 40 hours.
Time and a half for 7 hours.

PROBLEMS

	3.	4.	5.	6.	7.	8.
Hourly Pay (40 hrs.)	$6.00	$10.60	$4.80	$5.55	$17.00	$19.25
Straight-Time Pay	?	?	?	?	?	?
Overtime Rate	$1\frac{1}{2}$	$1\frac{1}{2}$	$1\frac{1}{2}$	2	2	$1\frac{1}{2}$
Overtime Hours	8	5	4	8	13	6
Overtime Pay	?	?	?	?	?	?
Total Pay	?	?	?	?	?	?

9. Steven Kellogg, machine operator. **10.** Dorothy Kaatz, programmer.
Regular hourly rate of $9.60.   Regular hourly rate of $9.15.
Time and a half for overtime.   Double time for overtime.
Worked 37 regular hours.   Worked 40 regular hours.
Worked 8 hours overtime.   Worked 12 hours overtime.
What is his straight-time pay?   What is her straight-time pay?
What is his overtime pay?   What is her overtime pay?
What is his total pay?   What is her total pay?

11. Cindy Haskins is paid $7.00 an hour for a regular 35-hour week. Her overtime rate is $1\frac{1}{2}$ times her regular hourly rate. This week Cindy worked her regular 35 hours plus 8 hours of overtime. What is her total pay?

12. As a maintenance engineer, Bonnie Zoltowski earns $7.85 an hour plus time and a half for weekend work. Last week she worked her regular 40 hours plus 16 hours of overtime on the weekend. What was her total pay for the week?

13. Judy Sweeney is a carryout cashier. She earns $4.50 an hour for a regular $36\frac{1}{2}$-hour week. She earns double time for work on Sundays. Last week Judy worked her regular hours plus $7\frac{1}{4}$ hours on Sunday. What was her total pay for the week?

> **CARRYOUT CASHIER**
>
> $4.50/ hr.
> Double time on weekends

14. Bill McCrate is a cable TV installer. He is paid $7.57 an hour for a 40-hour week and time and a half for overtime. What is Bill's total pay for a week in which he worked 61 hours?

15. George Keller earns a regular hourly rate of $8.64. He earns time and a half for overtime work on Saturdays and double time for overtime on Sundays. This week he worked 40 hours from Monday through Friday, 8 hours on Saturday, and 7 hours on Sunday. What is his total pay for the week? *164.25*

MAINTAINING YOUR SKILLS   Look up the skills in parentheses if you need help or more practice.

Add. (Skill 5)

16. $42.50	**17.** $160.45	**18.** $10.62	**19.** 56.7
+ 24.76	+ 86.90	9.34	0.007
		+ 0.45	121.33
			+ 1568.125

355.20

Multiply. (Skills 8, 14)

20. $(1\frac{1}{2} \times \$8.40) \times 7$ **21.** $(2 \times \$18.40) \times 12$ **22.** $(1\frac{1}{4} \times \$6.00) \times 9$

23. $(1.5 \times \$12.50) \times 8$ **24.** $(2 \times \$4.50) \times 6$ **25.** $(1.25 \times \$10.00) \times 10$

Weekly Time Card

OBJECTIVE
Compute the total hours on a weekly time card.

When you work for a business that pays you on an hourly basis, you are usually required to keep a time card. A **weekly time card** shows the time you reported for work and the time you departed each day of the week. The hours worked are computed for each day. The daily hours are added to arrive at the total hours for the week. The hours worked each day must be rounded to the nearest quarter hour.

Total Hours = Sum of Daily Hours

EXAMPLE   *Skills* 2, 16   *Applications* E, F   *Term* Weekly time card

Gail Stough is required to keep a weekly time card. What are her daily hours for 9/18? What are her total hours for the week?

EMPLOYEE TIME CARD	DATE	IN	OUT	IN	OUT	HOURS
	9/15	8:15	12:15	1:00	5:00	8
NAME: GAIL STOUGH	9/16	8:30	12:00	12:35	5:20	8 1/4
DEPT.: CREDIT	9/17	8:28	12:05	12:30	4:30	7 1/2
Gail Stough	9/18	8:15	1:45	2:15	4:59	?
	9/19	8:30	12:20	12:50	5:00	8
EMPLOYEE SIGNATURE					TOTAL HOURS	

SOLUTION

A. Find the hours worked on 9/18.
  (1) Time between 8:15 a.m. and 1:45 p.m.
  $(1:45 + 12:00) - 8:15$
  $13:45 \quad - 8:15$ 5h:30min

  (2) Time between 2:15 p.m. and 4:59 p.m.
  $4:59 - 2:15$ 2h:44min
  5h:30 min + 2 h:44min 7h:74min
  Rounded to the nearest quarter hour = $8\frac{1}{4}$ hours

B. Find the **total hours.**
Sum of daily hours

$8 + 8\frac{1}{4} + 7\frac{1}{2} + 8\frac{1}{4} + 8$

$8 + 8\frac{1}{4} + 7\frac{2}{4} + 8\frac{1}{4} + 8 = 39\frac{4}{4} = 40$ total hours

✔ SELF-CHECK   Complete the problems, then check your answers in the back of the book.

1. Karl West worked from 8:00 to 11:45 and from 12:30 to 4:15. Find the total hours.

2. Rod Abet worked from 7:30 to 11:55 and from 1:00 to 4:50. Find the total hours.

Eddie Irwin works as a computer operator in the data processing department at The Needham Medical Center. He is required to keep a weekly time card. What are his hours for each day? What are his total hours for the week?

EMPLOYEE TIME CARD NEEDHAM MEDICAL CENTER			NAME: Eddie Irwin DEPARTMENT: Data Processing		
Date	In	Out	In	Out	Total
3. 12/13	8:15	12:00	12:30	4:30	?
4. 12/14	8:30	12:30	1:10	5:10	?
5. 12/15	8:35	12:50	1:30	5:00	?
6. 12/16	8:15	12:10	12:45	4:25	?
7. 12/17	8:25	1:15	1:50	5:30	?
8.				Total Hours	?

Lucy Wallace works for Appliance Parts, Inc. Use her time card below to find her total hours worked each day and her total hours for the week.

EMPLOYEE TIME CARD APPLIANCE PARTS, INC.			NAME: Lucy Wallace DEPARTMENT: Parts		
Day	Date	In	Out	In	Out	Total
9. Mon.	6/7	7:00	11:00	12:00	4:20	?
10. Tue.	6/8	7:10	11:30	12:00	5:15	?
11. Wed.	6/9	6:50	11:15	11:55	4:20	?
12. Thu.	6/10	6:58	10:45	11:25	3:50	?
13. Fri.	6/11	7:15	11:49	12:30	5:02	?
14. Sat.	6/12	7:05	12:45	—	—	?
15.					Total Hours	?

16. Gail Stough earns $6.40 per hour. Use her time card on page 82 to find her total pay for the week.

17. Eddie Irwin is paid $7.25 for a regular 37-hour week. His overtime rate is $1\frac{1}{2}$ times his regular hourly rate. Use his time card above to find his total pay for the week.

18. Lucy Wallace is paid $7.80 for a regular 40-hour week. Her overtime rate is $1\frac{1}{2}$ her regular hourly rate. Use her time card above to find her total pay for the week.

19. You are scheduled to work from 7:00 a.m. to 11:00 a.m. and from 12:00 noon to 4:00 p.m. You are not allowed to work overtime. Your hourly rate is $7.50. Find the hours worked and the total pay for the week.

Day	In	Out	In	Out	Hrs.
Sun.	////		////		
Mon.	7:00	11:00	12:00	4:04	?
Tue.	7:05	11:10	11:55	3:59	?
Wed.	7:01	10:59	12:05	3:57	?
Thu.	6:59	10:55	12:01	4:03	?
Fri.	6:55	11:01	12:02	3:49	?
Sat.					

20. A portion of Kim Fong's time card is shown below. Find the number of hours he worked each day.

Day	In	Out	In	Out	Hrs.
Sun.	1:00	6:15	////		?
Mon.	1:00	5:00	6:00	9:10	?
Tue.	10:02	3:04	4:00	8:15	?
Wed.	////	////	////		?
Thu.	11:05	4:04	5:10	9:25	?
Fri.	1:30	5:25	6:15	10:08	?
Sat.	9:50	2:05	3:00	6:25	?

21. In problem 20, how many hours did Kim work on Saturday and Sunday? How many hours did he work Monday through Friday? If he earns $7.40 per hour plus time and a half for work on Saturday and Sunday, what is his total pay for the week?

MAINTAINING YOUR SKILLS Look up the skills in parentheses if you need help or more practice.

Add. **(Skill 16)**

22. $\frac{1}{4} + \frac{1}{4} + \frac{1}{2}$ **23.** $\frac{3}{4} + \frac{1}{8}$ **24.** $\frac{3}{8} + \frac{1}{4} + \frac{1}{4}$ **25.** $\frac{5}{8} + \frac{1}{3}$

Round to the nearest quarter hour. **(Application E)**

26. 8:09   **27.** 7:47   **28.** 11:55   **29.** 3:39   **30.** 5:23

Round to the place value indicated. **(Skill 2)**

Nearest thousand	**31.** 39,972	**32.** 21,944	**33.** 68,498
Nearest hundred	**34.** 842	**35.** 257	**36.** 3580

Piecework

OBJECTIVE
Compute the total pay on a piecework basis.

Some jobs pay on a piecework basis. You receive a specified amount of money for each item of work that you complete.

Total Pay = Rate per Item × Number Produced

EXAMPLE   *Skill* 8   *Application* A   *Term* Piecework

Ramon Hernandez works for National Cabinet Company. He is paid $8.00 for each cabinet he assembles. Last week he assembled 45 cabinets. This week he assembled 42. What is his pay for each week?

SOLUTION   Find the **total pay.**

	Rate per Item	×	Number Produced			
Last week:	$8.00	×	45	=	$360.00	total pay
This week:	$8.00	×	42	=	$336.00	total pay

✔ SELF-CHECK   Complete the problems, then check your answers in the back of the book.

Find the total pay.

48.76

1. $3.20 per item × 140 items produced
2. 15¢ per item × 1494 items produced

PROBLEMS

38.06 13.26

262.70

50

50

	Employee	Rate per Item	×	Number Produced	=	Total Pay
3.	Wo, E.	$2.05	×	180	=	?
4.	Washington, G.	$0.63	×	360	=	?
5.	Forest, J.	$1.55	×	174	=	?
6.	Gibley, T.	$0.95	×	610	=	?
7.	Jackman, L.	$9.50	×	87	=	?
8.	Mohammed, H.	$0.28	×	831	=	?
9.	Lewis, L.	$3.34	×	79	=	?
10.	Thompson, R.	$0.97	×	818	=	?

11. Hanley Fellhour.
    Beauty shop operator.
    Rate per haircut is $10.50.
    Gave 60 haircuts.
    Find his total pay.

12. Ellen Kolazinski.
    Strawberry picker.
    Rate per quart is $0.25.
    Picked 1053 quarts.
    Find her total pay.

13. Paul Aymes.
    Chrome plater.
    Rate per item is $1.25.
    Plated 321 items.
    What is his total pay?

14. Jules Gartner.
    Shirt silk screener.
    Rate per shirt is $0.45.
    Silk screened 388 shirts.
    What is his total pay?

15. Carol Ying assembles calculators at Central Electronics. She is paid $0.45 for each calculator she assembles. What is her pay for a day in which she assembles 134 calculators?

16. Leah Elliot runs a carpet cleaning service. She charges $15.95 per room. On Monday, she did 3 rooms in one house, 2 in another, and 4 in a third. Find her total cleaning charges.

17. Jeremy Sullivan delivers newspapers for the Dispatch. He receives 4.2¢ per paper, 6 days a week, for the daily paper, and 25¢ for the Sunday paper. He delivers 124 daily papers each day and 151 Sunday papers each week. What is his total pay for the week?

Use the table to find the charges for the typing services in problems 18–20.

18. A 12-page, single-spaced report plus 2 copies of each page.

19. A 35-page, double-spaced term paper plus 1 copy of each page.

20. A financial report with 14 single-spaced pages, 31 double-spaced pages, and 6 space-and-a-half pages. Plus 6 copies of each page.

Spacing	Rate per Page
Single-space	$2.25
Double-space	1.25
Space and a half	1.75
Duplicate copies: $0.15 per page	

21. Rex Moore operates a Quick Stop oil change and tune-up service. He charges $25.95 per oil change, $52.50 to tune a 4-cylinder engine, $58.50 to tune a 6-cylinder engine, and $64.50 to tune an 8-cylinder engine. What are the charges for a week in which he did 35 oil changes, tuned five 4-cylinder engines, seven 6-cylinder engines, and two 8-cylinder engines?

MAINTAINING YOUR SKILLS   Look up the skills in parentheses if you need help or more practice.

Multiply. Round answers to the nearest cent. **(Skill 8)**

22. $0.23 × 89   **23.** $1.10 × 240   **24.** $0.06 × 4192   **25.** 6¢ × 906

26. $5.20 × 23   **27.** $0.04 × 3200   **28.** $0.66 × 350   **29.** 4.7¢ × 731

Multiply. **(Skill 8)**

30. 1.17	**31.** 15.876	**32.** 242	**33.** 456	**34.** 693.25
× 100	× 100	× 0.04	× 1000	× 482

1-5

Salary

OBJECTIVE
Compute the salary per pay period.

A **salary** is a fixed amount of money that you earn on a regular basis. Your salary may be paid weekly, biweekly, semimonthly, or monthly. Your annual salary is the total salary you earn during a year. There are 52 weekly, 26 biweekly, 24 semimonthly, and 12 monthly pay periods per year.

$$\text{Salary per Pay Period} = \frac{\text{Annual Salary}}{\text{Number of Pay Periods per Year}}$$

EXAMPLE   *Skill* 11   *Application* K   *Term* Salary

Beth Huggins is a computer programmer. Her annual salary is $22,560. What is her monthly salary? What is her weekly salary?

SOLUTION   Find the **salary per pay period.**

	Annual Salary	÷	Number of Pay Periods per Year	
Monthly:	$22,560.00	÷	12	= $1880.00 monthly salary
Weekly:	$22,560.00	÷	52	= $433.85 weekly salary
	22560	÷	52	= 433.84615

✔ SELF-CHECK   Complete the problems, then check your answers in the back of the book.

1. Find the biweekly salary.
$42,900 ÷ 26 = ?

2. Find the semimonthly salary.
$18,200 ÷ 24 = ?

PROBLEMS

	Employee	Pay Period	Annual Salary	÷ Pay Periods per Year	= Salary per Pay Period
3.	K. O. Sanchez	Semimonthly	$ 15,600 ÷	24	= ?
4.	B. J. Molley	Weekly	$ 16,900 ÷	52	= ?
5.	O. L. Farrell	Monthly	$ 16,500 ÷	12	= ?
6.	T. B. Reston	Semimonthly	$ 34,650 ÷	?	= ?
7.	R. R. Halston	Weekly	$ 66,598 ÷	?	= ?
8.	K. C. Ying	Biweekly	$132,475 ÷	?	= ?

9. You are a legal clerk.
Annual salary is $12,500.
What is your weekly salary?

10. You are a store manager.
Annual salary is $28,320.
What is your semimonthly salary?

11. Geraldine Piela is a sales clerk. Her annual salary is $17,040. What is her semimonthly salary?

12. Gary Wilder is a claims adjustor. His annual salary is $21,840. What is his biweekly salary?

13. Susan Snell was just hired as a mechanical engineer for the Howwel Company. Her starting salary is $32,400 per year. What is her monthly salary? What is her weekly salary?

14. Morgan Behnke works as a court reporter. Her annual salary is $19,380. What is her weekly salary? What is her biweekly salary?

15. Ben Rodebaugh is a medical lab assistant. His annual salary is $17,400. Ben is paid on a monthly basis. What is his monthly salary?

16. Juan Rodriguez qualifies for the technical services position described in the advertisement. If he is hired, what will his weekly salary be?

F.Y.I.

Average Starting Salaries

Engineering	$32,304
Accounting	27,408
Sales/ marketing	27,828
Business administration	26,496
Mathematics	26,712
Computer science	29,100

17. Louis Rahn is currently earning an annual salary of $15,090 at Budgett Electronics. He has been offered a job at Delta Tech at an annual salary of $16,660. How much more would Louis earn per week at Delta Tech than at Budgett Electronics?

18. When Paul Sellers first started working at Custom Computers, he was paid a biweekly salary of $615. Custom Computers is now converting to a new payroll system and will be paying its employees on a semimonthly basis. What will Paul's semimonthly salary be?

19. Sune Fung earns a weekly salary of $350 at Howard's Department Store. Next month she will be promoted from assistant buyer to buyer. In her new position, she will be paid $895 semimonthly. How much more per year will Sune earn as a buyer than as assistant buyer?

20. **A Brief Case** Assume that your present job pays a monthly gross salary of $1560. You are offered a new position that pays $8.60 per hour with $1\frac{1}{2}$ for all hours over 40 per week. How many hours of overtime per week would you need to earn the same amount per week as your present job?

MAINTAINING YOUR SKILLS Look up the skills in parentheses if you need help or more practice.

Divide. Round answers to the nearest hundredth. **(Skill 11)**

21. $50)\overline{\$14,290}$ **22.** $8.6)\overline{41.62}$ **23.** $14.7)\overline{191.3}$ **24.** $12)\overline{5473.2}$

25. $0.032)\overline{17.9}$ **26.** $26)\overline{21,860}$ **27.** $52)\overline{69,146}$ **28.** $24)\overline{10,549.9}$

Find the number of pay periods. **(Application K)**

29. Weekly for 2 years

30. Semimonthly for 3 years

31. Biweekly for 4 years

1-6

Commission

OBJECTIVE
Compute the straight commission and determine the gross pay.

A **commission** is an amount of money that you are paid for selling a product or service. Your **commission rate** may be a specified amount of money for each sale or it may be a percent of the total value of your sales. If the commission is the only pay you receive, you work on **straight commission**.

Commission = Total Sales × Commission Rate

EXAMPLE *Skills* 30,2 *Application* A *Term* Commission

Milton Arps sells real estate at a $7\frac{1}{2}\%$ straight commission. Last week his sales totaled $90,000. What was his commission?

SOLUTION   Find the **commission**.

Total Sales	×	Commission Rate	
$90,000.00	×	$7\frac{1}{2}\%$	
$90,000.00	×	0.075	= $6750.00 straight commission

90000 [×] .075 [=] 6750 or 90000 [×] 7.5 [%] 6750

✔ SELF-CHECK   Complete the problems, then check your answers in the back of the book.

Find the commission.

3H7.08

1. $9400 × 8% commission rate   **2.** $143,400 × $5\frac{1}{2}$ % commission rate

314.38
3H72.08

Instead of working only on commission, you may be guaranteed a minimum weekly or monthly salary. The commission you earn during a week or month is compared with your minimum salary. Your gross pay is the greater of the two amounts.

Gross Pay = Salary or Commission

EXAMPLE *Skill* 30 *Application* A *Term* Commission

Owen Theil is guaranteed a minimum salary of $250 a week or 7% of his total sales, whichever is greater. What is his gross pay for a week in which his total sales were $3614?

SOLUTION   **A.** Find the **commission**.

Total Sales	×	Commission Rate	
$3614.00	×	7%	= $252.98 commission

B. Find the **gross pay**.
Salary or Commission
$250.00 or $252.98   Gross pay is $252.98 (greater amount).

3. Minimum salary: $160.
Commission: $5\frac{1}{2}\%$ on $2900.

4. Minimum salary: $2000.
Commission: $6\frac{1}{4}\%$ on $34,000.

PROBLEMS

	Position	Total Sales	×	Commission Rate	=	Commission
5.	Real estate sales	$98,000	×	8%	=	?
6.	Computer sales	$18,100	×	12%	=	?
7.	Major appliance sales	$ 9,598	×	16%	=	?
8.	Dress sales	$ 1,311	×	9%	=	?
9.	Computer supplies sales	$ 929	×	15%	=	?
10.	Siding contract sales	$ 754	×	$12\frac{1}{2}\%$	=	?
11.	Auto sales	$68,417	×	$3\frac{1}{2}\%$	=	?

	Salesperson	Minimum Monthly Salary	Total Monthly Sales	×	Commission Rate	=	Monthly Commission	Gross Pay
12.	Anne Moser	$2,100	$28,000	×	8.0%	=	?	?
13.	Peter Zinn	$1,600	$23,000	×	6.5%	=	?	?
14.	Vern Yoder	$3,140	$31,000	×	9.25%	=	?	?
15.	Rene Vershum	$ 850	$10,400	×	3.5%	=	?	?
16.	Virgil Ulrich	$1,200	$45,000	×	5.5%	=	?	?
17.	Robin Coy	$3,410	$29,100	×	12.1%	=	?	?

18. Roger Tussing.
3% commission on sales.
$9500 in sales for a week.
What is his commission?

19. Sam Taylor.
$6\frac{1}{2}\%$ commission on sales.
$4226 in sales for a week.
What is his commission?

20. Theresa Britsch.
Minimum weekly salary
    is $225.
Rate of commission is 6.5%.
Weekly sales are $3420.
What is the commission?
What is the gross pay?

21. Clair Ming.
Minimum weekly salary
    is $160.
Rate of commission is 7.5%.
Weekly sales are $2420.
What is the commission?
What is the gross pay?

22. Marie Busack sells used cars for Bonded Auto. She receives a straight commission of 4% of the selling price of each car. What commission will she receive for selling a $19,420 van?

23. Bruce Clay sells real estate. His commission is $5\frac{3}{4}$% of the price of every house he sells. What is his commission when he sells a $145,450 house?

24. Auto mechanics at Cline's Garage receive a straight commission of 22% of the service income they bring in each week. What is the commission for a mechanic who brings in $1465 service income in a week?

25. John Navarro is a salesperson in the appliance department at Morris Appliance, Inc. He is guaranteed a minimum salary of $185 per week or 5.5% of his total sales, whichever is greater. What is his gross pay for a week in which his total sales were $3422?

26. Maude Eggert sells cosmetics for Soft Touch, Inc. She is guaranteed a salary of $790 a month or $7\frac{1}{4}$% of her total sales, whichever is greater. What is her gross pay for a month in which her total sales were $10,984?

27. Irma DeWolfe is paid either a commission or $4.75 per hour plus time and a half overtime for all hours over 8 per day, whichever is greater. Her commission consists of $5\frac{1}{2}$% of sales. Find her gross pay for a week in which she worked 6 hours on Monday, 8 hours on Tuesday, 10 hours on Wednesday, 6 hours on Thursday, 10 hours on Friday, and 9 hours on Saturday. Her total sales for the week were $4100.

28. A Brief Case   Some jobs pay a commission plus a bonus at the end of the year. The bonus may be a percent of the salesperson's total commission for the year.

a. Madelyn Carr is a sales representative. She receives a 7% commission on all sales. At the end of the year, she receives a bonus of 5% of her commission. What is her total pay for a year in which she had sales totaling $412,454?

b. What would Madelyn's total pay be if her sales were $316,250?

MAINTAINING YOUR SKILLS   Look up the skills in parentheses if you need help or more practice.

Write as a decimal. Round answers to the nearest hundredth. (Skill 14)

29. $2\frac{1}{8}$ **30.** $5\frac{1}{2}$ **31.** $6\frac{3}{4}$ **32.** $4\frac{1}{3}$ **33.** $9\frac{1}{4}$

Find the percentage. Round to nearest cent or tenth. (Skill 30)

34. 4% of $1250   **35.** $8\frac{1}{2}$ % of $4300   **36.** $7\frac{1}{4}$ % of $8200   **37.** 9.2% of $3600

38. 7.3% of $120   **39.** 8.92% of 1380   **40.** 39% of 281   **41.** 5.7% of 9140

Round answers to the nearest tenth. (Skill 2)

42. 0.081   **43.** 0.608   **44.** 0.92   **45.** 0.821   **46.** 0.22

1·7

Graduated Commission

OBJECTIVE

Compute the total graduated commission.

Your commission rate may increase as your sales increase. A graduated commission offers a different rate of commission for each of several levels of sales. It provides an extra incentive to sell more.

Total Graduated Commission = Sum of Commissions for All Levels of Sales

---

EXAMPLE   *Skills* 5, 28, 30   *Application* A   *Term* Graduated Commission

Irene Tomas sells appliances at Twin City Sales. She receives a graduated commission of 4% on her first $1000 of sales, 6% on the next $2000, and 8% on sales over $3000. Irene's sales for the past month totaled $9840. What is her commission for the month?

SOLUTION

Find the **sum of commissions for all levels of sales.**

		Sales	× Commission Rate		
A. First $1000:		$1000	×	4%	= $ 40.00
B. Next $2000:		$2000	×	6%	= $120.00
C. Over $3000:	($9840 − $3000)	×	8%		
		$6840	×	8%	= $547.20
			Total Graduated Commission		$707.20

1000 ⊠ 4 % 40 M+ 2000 ⊠ 6 % 120 M+ 9840 − 3000 = 6840 ⊠
8 % 547.2 M+ RM 707.2

✔ SELF-CHECK   Complete the problems, then check your answers in the back of the book.

1. Commission: 10% on first $5000; 15% over $5000. Find the total graduated commission on $22,000.

2. Commission: 5% first $2000; 8% over $2000. Find the total graduated commission on $7740.

PROBLEMS

	3.	**4.**
Amount of Sales	$3,400	$2,500
Commission		
First $1000: 10%	?	?
Over $1000: 15%	?	?
Total Commission	?	?

	5.	**6.**
Amount of Sales	$3,900	$8,250
Commission		
First $1500: 6%	?	?
Next $2000: 8%	?	?
Over $3500: 10%	?	?
Total Commission	?	?

7. Bill Weston, part-time sales.
20% commission on first $500.
25% on next $1000.
30% on sales over $1500.
$1940 total sales.
What is his total commission?

8. Mary Draper, sporting goods sales.
5% commission on first $6000.
$7\frac{1}{2}$% on next $6000.
10% on amount over $12,000.
$14,640 total sales.
What is her total commission?

9. Jean Gray sells office supplies. She receives a graduated commission of 4% on her first $2000 of sales and $8\frac{1}{2}$% on all sales over $2000. Jean's sales for the past week totaled $3925. What is her commission for the week?

10. Charles Beaudry sells computer hardware for a computer firm. He is paid a 4% commission on the first $6000 of sales, 6% on the next $10,000, and 8% on sales over $16,000. What is his commission on $24,550 in sales?

11. Ralph King demonstrates cookware at the National Food Fair. He is paid $6.00 each for the first 10 demonstrations in one day and $7.50 for each demonstration over 10. What is Ralph's commission for a day in which he makes 21 demonstrations?

12. Odessa Dilulo demonstrates home fire alarm systems. She is paid $10.00 per demonstration for the first 10 demonstrations given during a week and $15.50 for each demonstration over 10. Also, for every sale, she gets a bonus of $15.00. What is her commission for a week in which she gives 18 demonstrations and makes 6 sales?

13. A Brief Case   Some sales positions pay a commission only if sales exceed a sales quota. The salesperson is rewarded only for having sales beyond an expected amount.

Joyce Doyle is a sales trainee. She is paid $550 per month plus a commission of 7.5% on all sales over a quota of $9000 per month. What is her total pay for a month in which she has sales totaling $10,650?

MAINTAINING YOUR SKILLS   Look up the skills in parentheses if you need help or more practice.

Add. **(Skill 5)**

14. $123.72	**15.** $567.89	**16.** $1350.23	**17.** 0.007
112.69	34.69	946.00	1.384
+ 45.23	+ 431.73	+ 18.11	+ 0.569

Write as a decimal. **(Skill 28)**

18. $5\frac{1}{2}$ %   **19.** $10\frac{1}{4}$ %   **20.** 15%   **21.** 97%   **22.** 342%

Find the percentage. **(Skill 30)**

23. $5\frac{1}{4}$ % of $2000   **24.** 7% of $4560   **25.** $15\frac{1}{2}$ % of $3500

Reviewing the Basics

Skills

(Skill 2)

Round to the nearest cent.

1. $45.672　　**2.** $469.3591　　**3.** $0.045　　**4.** $53.744　　**5.** $172.255

Solve.

(Skill 5)

6. $672.01 + $45.32　　**7.** 0.004 + 1.5 + 19.03　　**8.** $32.51 + $95.98

(Skill 8)

9. $45.00 × 0.055　　**10.** 11.563 × 100　　**11.** 0.007 × 0.03

(Skill 11)

12. $23,660 ÷ 26　　**13.** $55,920 ÷ 12　　**14.** 27.3 ÷ 0.05

(Skill 16)

15. $5\frac{1}{4} + 7\frac{1}{2} + 6\frac{3}{4}$　　**16.** $5\frac{1}{2} + 2\frac{2}{3} + 7\frac{5}{6}$　　**17.** $8\frac{1}{2} + 2\frac{1}{4}$

(Skill 30)

18. $15\frac{1}{2}$ % of $650　　**19.** 110% of $310　　**20.** $\frac{1}{4}$ % of $24

Write as a decimal.

(Skill 14)

21. $\frac{1}{2}$　　**22.** $\frac{1}{4}$　　**23.** $\frac{3}{4}$　　**24.** $\frac{8}{10}$　　**25.** $\frac{1}{8}$

(Skill 28)

26. 11.5%　　**27.** 1.25%　　**28.** 0.125%　　**29.** 385%　　**30.** $10\frac{1}{4}$ %

Applications

(Application E)

Round to the nearest quarter hour.

31. 8 hours 35 minutes　　**32.** 7 hours 50 minutes

Find the number of occurrences.

(Application K)

33. Weekly for 2 years　　**34.** Monthly for 3 years

35. Quarterly for 4 years　　**36.** Semimonthly for 2 years

Terms

Write a correct definition for each term. Refer to the terms file.

37. Straight-time pay　　**38.** Piecework pay

39. Salary　　**40.** Commission

41. Graduated commission　　**42.** Overtime pay

Refer to your reference files at the back of the book if you need help.

Unit Test

Lesson 1-1

1. Judy Enright, a painter, earns $8.60 an hour during a regular workweek. Last week she worked her regular 40 hours. This week she worked 32.5 hours. What is her straight-time pay for each week?

Lesson 1-2

2. Resmund Downs works as a hotel clerk. He earns $5.60 per hour for a 40-hour week. His overtime rate is $1\frac{1}{2}$ times his regular hourly rate. This week he worked his regular 40 hours plus $8\frac{3}{4}$ hours of overtime. What is his total pay?

Lesson 1-3

3. Stella Gordon fills out a weekly time card. What are her total hours for the week? (Round to the nearest quarter hour.)

Day	In	Out	In	Out	Hrs.
Sun.					
Mon.	8:30	12:00	12:35	5:35	?
Tue.	8:25	12:15	1:00	5:30	?
Wed.	8:28	12:45	1:15	4:25	?
Thu.	8:15	1:08	1:45	5:32	?
Fri.	8:50	12:30	1:00	5:08	?
Sat.					
				Total Hours	?

Lesson 1-4

4. Duane Smith is paid on a piecework basis. He is paid $14.50 for each chair he assembles. This week he assembled 32 chairs. What is his pay for the week?

Lesson 1-5

5. Frank DuByne is a production manager for National Plastics. He earns an annual salary of $32,420. What is his semimonthly pay? What is his weekly pay?

Lesson 1-6

6. Bettie Ray sells real estate for a 6% straight commission. Her sales totaled $392,400 in September. What is her pay for the month?

Lesson 1-7

7. Eric Erford is an appliance salesman. He earns a graduated commission on sales as shown. His sales for 1 week totaled $5240. What was his total commission?

Commission	Level of Sales
8%	First $1500
10%	Next $3000
12%	Over $4500

A SPREADSHEET APPLICATION

Gross Income

To complete this spreadsheet application, you will need the template diskette for *Mathematics with Business Applications*. Follow the directions in the User's Guide portion of the TRB to complete this activity.

Input the information in the following problems to find the straight-time pay, the overtime pay, and the total or gross pay.

1. Employee: D. Wyse.
   Rate per hour: $9.60.
   Regular hours: 40.
   Overtime hours: 6.
   What is the straight-time pay?
   What is the overtime pay?
   What is the total pay?

2. Employee: K. Yackee.
   Rate per hour: $12.50.
   Regular hours: 36.
   Overtime hours: 8.
   What is the straight-time pay?
   What is the overtime pay?
   What is the total pay?

3. Employee B. Sullivan.
   Rate per hour: $4.50.
   Regular hours: 25.
   Overtime hours: 4.
   What is the straight-time pay?
   What is the overtime pay?
   What is the total pay?

4. Employee: G. Stambaugh.
   Rate per hour: $22.00.
   Regular hours: 40.
   Overtime hours: 12.
   What is the straight-time pay?
   What is the overtime pay?
   What is the total pay?

5. Employee: A. Roth.
   Rate per hour: $27.22.
   Regular hours: 26.
   Overtime hours: 0.
   What is the straight-time pay?
   What is the overtime pay?
   What is the total pay?

6. Employee: W. Pawlowicz.
   Rate per hour: $8.55.
   Regular hours: 36.
   Overtime hours: 9.
   What is the straight-time pay?
   What is the overtime pay?
   What is the total pay?

7. Marc Pempertolie earns $5.50 per hour plus time and a half for overtime. Last week he worked his regular 40 hours plus 6 hours overtime. What is his straight-time pay? What is his overtime pay? What is his total pay?

8. Carey Nofziger earns $6.95 per hour plus time and a half for overtime. Last week she worked her regular 40 hours plus 12 hours overtime. What is her straight-time pay? What is her overtime pay? What is her total pay?

9. Hilarie Mapes earns $16.50 per hour plus time and a half for over-time. Last week she worked her regular 36 hours plus 7.5 hours overtime. What is her straight-time pay? What is her overtime pay? What is her total pay?

10. Judy Dohm earns $19.25 per hour plus time and a half for overtime. She worked her regular 42 hours plus 9.25 hours overtime. What is her straight time pay? What is her overtime pay? What is her total pay?

11. Vernon Cymbola earns $8.63 per hour plus time and a half for over-time on weekends. Last week he worked 35 hours Monday through Friday, 8 hours on Saturday, and 8 hours on Sunday. What is his total pay for the week?

12. Sherri Cordesa earns $18.875 per hour plus double time for overtime on weekends. Last week she worked 28 hours Monday through Friday, 7.5 hours on Saturday, and 9.5 hours on Sunday. What is her total pay for the week?

13. Angela Perez earns $11.36 an hour plus time and a half for overtime. Last week she worked her regular 39 hours and 3 hours overtime. What is her total pay for the week?

14. Robert Lewis earns $12.42 an hour plus double time for overtime. Last week he worked his regular 40 hours plus 6.5 hours overtime. What is his total pay for the week?

15. Stephanie Martin earns $8.50 an hour plus time and a half for over-time. Last week she worked her regular 37 hours plus 10 hours overtime. What is Stephanie's total pay for the week?

16. Dwight Madsen earns $7.25 an hour plus time and a half for over-time. Last week he worked his regular 39 hours plus 7.25 hours on Saturday. What is his total pay for the week?

Salesperson

Denise Hobson is a salesperson in the Mountaintop Ski Shop. She enjoys meeting new people and helping them purchase ski equipment.

When Denise began work at the shop about four years ago, she entered a training program. In it, she learned how to relate to customers and how to advise them, and she learned the company's product line inside and out. Due to her enthusiasm and hard work, Denise now earns $850 a month as a salary and 6% of each sale she makes in commission.

Last year, Denise decided to make a chart of her sales performance. She constructed a bar graph in which her sales total for each month of the year was represented by a bar or column. The bar graph below shows, for example, that in January (the first J on the horizontal axis) of last year, she sold $12,500 worth of ski equipment.

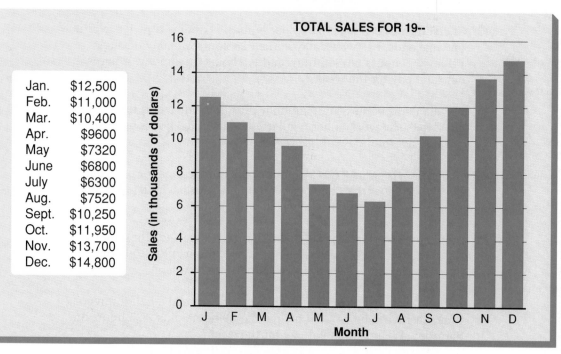

Jan.	$12,500
Feb.	$11,000
Mar.	$10,400
Apr.	$9600
May	$7320
June	$6800
July	$6300
Aug.	$7520
Sept.	$10,250
Oct.	$11,950
Nov.	$13,700
Dec.	$14,800

TOTAL SALES FOR 19--

Sales (in thousands of dollars) vs Month (J F M A M J J A S O N D)

1. From the graph, in which month was Denise's income the least? In which month was her income the greatest?

2. Compute Denise's income for last February.

3. Compute Denise's income for last July.

4. To offset the sag in summer sales, Denise might suggest to the owner that Mountaintop offer a product line in addition to its ski line. What product or products might she suggest? Discuss your answers.

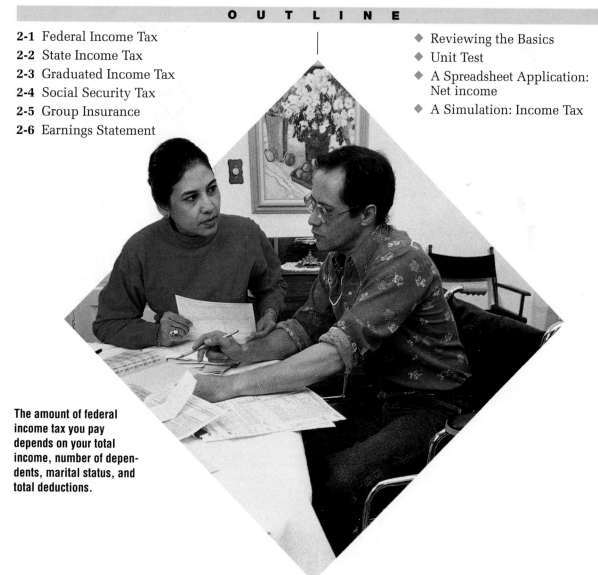
Net Income

Out of your *gross pay,* the total amount of money you earn, employers withhold money for federal, state, and social security taxes. These *deductions* from your gross pay result in your *net pay,* the money you actually receive each pay period. *Federal income taxes* depend on your salary and the number of people you support, or your *dependents* (also called withholding allowances or personal exemptions). Deductions from your gross pay may include the cost of medical insurance, unemployment insurance, disability insurance, savings, stocks, union dues, and charitable contributions. These deductions are shown on your *earnings statement,* which is attached to your paycheck.

O U T L I N E

The amount of federal income tax you pay depends on your total income, number of dependents, marital status, and total deductions.

Federal Income Tax

OBJECTIVE

Use tables to find the amount withheld for FIT.

Employers are required by law to withhold a certain amount of your pay for federal income tax (FIT). The Internal Revenue Service provides employers with tables that show how much money to withhold. The amount withheld depends on your income, marital status, and withholding allowances. You may claim one allowance for yourself and one allowance for your spouse if you are married. You may claim additional allowances for any others you support.

WEEKLY Payroll MARRIED				
Wages		Allowances		
At least	But less than	0	1	2
280	290	33	26	20
290	300	34	28	22
300	310	36	29	23
310	320	37	31	25
320	330	39	32	26
330	340	40	34	28
340	350	42	35	29
350	360	43	37	31
360	370	45	38	32
370	380	46	40	34

EXAMPLE *Skill* 1 *Application* C *Term* Income tax

Velma Miller's gross pay for this week is $321.85. She is married and claims 2 allowances, herself and her husband. What amount will be withheld from Velma's pay for federal income tax?

SOLUTION

A. Find the **income range** from the table.

B. Find the column for 2 **allowances**.

At least	But less than
320	330

C. Amount of income to be withheld is $26.00.

✔ SELF-CHECK Complete the problems, then check your answers in the back of the book.

1. Weekly income: $285.40.
1 allowance.

2. Weekly income: $333.83.
1 allowance.

PROBLEMS

Use the MARRIED Persons—WEEKLY Payroll Period table above to find the amount withheld for federal income tax.

	3.	4.	5.	6.	7.
Employee	G. Chesson	B. Brown	J. R. Foos	J. Isch	F. Restivo
Weekly Income	$283.00	$348.00	$289.50	$334.20	$320.00
Allowances	2	1	0	2	0
Amount Withheld	?	?	?	?	?

SINGLE Persons—WEEKLY Payroll Period

And the wages are—		And the number of withholding allowances claimed is—										
Wages		0	1	2	3	4	5	6	7	8	9	10
At least	But less than	The amount of income tax to be withheld shall be—										
170	175	22	16	10	4	0	0	0	0	0	0	0
175	180	23	17	11	4	0	0	0	0	0	0	0
180	185	24	18	11	5	0	0	0	0	0	0	0
185	190	25	18	12	6	0	0	0	0	0	0	0
190	195	25	19	13	7	0	0	0	0	0	0	0
195	200	26	20	14	7	1	0	0	0	0	0	0
200	210	27	21	15	9	2	0	0	0	0	0	0
210	220	29	22	16	10	4	0	0	0	0	0	0
220	230	30	24	18	12	5	0	0	0	0	0	0
230	240	32	25	19	13	7	1	0	0	0	0	0
240	250	33	27	21	15	8	8	0	0	0	0	0
250	260	35	28	22	16	10	10	0	0	0	0	0
260	270	36	30	24	18	11	11	0	0	0	0	0
270	280	38	31	25	19	13	13	0	0	0	0	0
280	290	39	33	27	21	14	14	2	0	0	0	0
290	300	41	34	28	22	16	16	3	0	0	0	0
300	310	42	36	30	24	17	17	5	0	0	0	0
310	320	44	37	31	25	19	19	6	0	0	0	0
320	330	45	39	33	27	20	20	8	2	0	0	0
330	340	47	40	34	28	22	22	9	3	0	0	0

F.Y.I.

In 1988, the federal, state, and local income taxes collected amounted to $490 billion.

Use the table above to find the amount withheld.

8. Jeannie Cooley, single.
Earns $330.15 weekly.
Claims 2 allowances.
What is the FIT withheld?

9. Matt Osborn, single.
Earns $170 weekly.
Claims 3 allowances.
What is the FIT withheld?

10. Hubert Pacetta, single.
Earns $232.40 weekly.
Claims 3 allowances.
What is the FIT withheld?

11. Emily Zasada, single.
Earns $300 weekly.
Claims 2 allowances.
What is the FIT withheld?

12. Shirley Delaney earns $325 a week. She is single and claims 2 allowances. What amount is withheld weekly for federal income tax?

13. Hank Michalat earns $321 a week. He is single and claims no allowances. What amount is withheld weekly for federal income tax?

14. Harold Leu is single and earns $300 a week. He claims 1 allowance. What amount is withheld weekly for federal income tax?

15. Tina Ford is single, earns $252 a week, and claims 1 allowance. What amount is withheld weekly for federal income tax?

Use the MARRIED Persons—WEEKLY Payroll Period table on page 643 to find the amount withheld for federal income tax.

16. Carol Meinka is married, earns $615 a week, and claims 3 allowances. What amount is withheld weekly for federal income tax?

17. Luther Myers earns $455 a week. He is married and claims 2 allowances. What amount is withheld weekly for federal income tax?

18. Georgette Young is paid a weekly wage of $491.60. She is married and claims 2 allowances. Beginning next year, she will claim another allowance for supporting her mother. How much less will be withheld for federal income tax for the year?

19. Lamoin Kelb earns $272.44 a week. He is married and claimed 2 allowances last year. He hopes to receive a refund on his next tax return by claiming no allowances this year. How much more in withholdings will be deducted weekly if he claims no allowances?

20. A Brief Case Some companies use a percentage method instead of the tax tables to compute the income tax withheld. Use the table at the right to find the amount withheld for the SINGLE employees below. Each weekly allowance is $41.35.

WEEKLY PAYROLL PERIOD, SINGLE PERSON

If the amount of wages (after subtracting withholding allowances) is:		The amount of income tax to be withheld shall be:
Not over $24 0		
Over—	But not over—	of excess over—
$24	—$415 . . . . .15%	—$24
$415	—$972 . . . .$58.65 plus 28%	—$415
$972	. . . .$214.61 plus 31%	—$972

a. Find the **allowance amount.**
Number of Allowances × $41.35

b. Find the **taxable wages.**
Weekly Wage − Allowance Amount

c. Find the **amount withheld** for the employees below.

Employee	Weekly Wage	Number of Allowances	Allowance Amount	Taxable Wages	Amount Withheld
D. Faust	$ 184.00	1	$ 41.35	$142.65	$17.80
J. Wray	$ 490.00	3	$124.05	$365.95	?
E. Kohn	$ 690.00	2	?	?	?
K. Bird	$1150.00	4	?	?	?

MAINTAINING YOUR SKILLS   Look up the skills in parentheses if you need help or more practice.

Give the place and value of the underlined digit. **(Skill 1)**

21. 8<u>4</u>   **22.** 9<u>6</u>5   **23.** <u>6</u>532   **24.** 36.9<u>4</u>   **25.** <u>6</u>72   **26.** 14.78<u>3</u>

Find the percentage. **(Skill 30)**

27. 22.5% × $120

28. 17.50% × $430

29. $\frac{1}{2}$% × $298

30. 128% × $82

Subtract. **(Skill 6)**

31. $344.50 − $68.40   **32.** $432.60 − $83.70   **33.** $1298.65 − $423.76

◇ **2-2** ◇

State Income Tax

OBJECTIVE
Compute the state tax on a straight percent basis.

Most states require employers to withhold a certain amount of your pay for state income tax. In some states, the tax withheld is a percent of your taxable wages. Your taxable wages depend on personal exemptions, or withholding allowances, allowed for supporting yourself and others in your family.

Taxable Wages = Annual Gross Pay − Personal Exemptions

Annual Tax Withheld = Taxable Wages × Tax Rate

EXAMPLE   *Skills* 6, 30   *Application* A   *Term* Personal exemptions

James Bowler's gross pay is $18,800 a year. The state income tax rate is 3% of taxable wages. James takes a married exemption for himself and his wife and 2 personal exemptions for his 2 children. How much is withheld a year from his gross earnings for state income tax?

PERSONAL EXEMPTIONS
Single—$1500
Married—$3000
Each Dependent—$700

SOLUTION

A. Find the **taxable wages.**
  Annual Gross Pay − Personal Exemptions
    $18,800.00   − ($3000.00 + $700.00 + $700.00) = $14,400.00

B. Find the **annual tax withheld.**
  Taxable Wages × Tax Rate
    $14,400.00   ×   3%   = $432.00 annual tax withheld

  3000 + 700 + 700 = 4400 M+ 18800 − RM 4400 = 14400 × 3 % 432

✔ SELF-CHECK   Complete the problems, then check your answers in the back of the book.

Find the annual tax withheld.

1. Gross pay: $18,900.
  Single, 1 dependent.
  State income tax rate: 4%.

2. Gross pay: $34,000.
  Married, 3 dependents.
  State income tax rate: 5%.

PROBLEMS

(Annual Gross Pay	− Personal Exemptions	= Taxable Wages)	× State Tax Rate	= Tax Withheld
3. ($17,000	− $3000	= $14,000) ×	2%	= ?
4. ($24,000	− $2200	= ?) ×	5%	= ?
5. ($ 9,500	− $1500	= ?) ×	3.5%	= ?

Use the table on page 103 for personal exemptions and find the amount withheld.

6. Linda Raabe.
Earns $15,900 annually.
Single, no dependents.
What are her personal exemptions?

7. Harold Gibbons.
Earns $13,840 annually.
Married, no dependents.
What are his personal exemptions?

8. Roger Hoblet.
Earns $79,500 annually.
Married, 3 dependents.
State tax rate is 4.6%.
What are his personal exemptions?
What is withheld for state tax?

9. Sarah Krouse.
Earns $17,300 annually.
Single, 1 dependent.
State tax rate is 2.5%.
What are her personal exemptions?
What is withheld for state tax?

10. Henry Altman earns $24,200 annually as a traffic analyst. He is married and supports 2 children. The state tax rate in his state is 2% of taxable income. What amount is withheld yearly for state income tax?

11. Kristi Maher earns $39,940 per year. Her personal exemptions include herself and her husband. The state tax rate in her state is 4.5% of taxable income. What amount is withheld yearly for state income tax?

12. Heidi Harse is a registered nurse. She earns $29,830 a year and is single. The state income tax rate is 5% of taxable income. What amount is withheld yearly for state income tax?

Use the tables on pages 642–643 for federal withholdings.

13. Joseph Ryczke earns $32,000 a year as a city planner. He is paid on a weekly basis. He is married, has no dependents, and claims 2 withholding allowances for federal income tax purposes. The state tax rate is 2% of taxable income. How much is withheld annually from Joseph's gross pay for state and federal income taxes?

14. Wendy Chou earns $26,320 a year as a nurse's aide. She is paid on a weekly basis. She is single, has 1 dependent, and claims 2 withholding allowances. The state tax rate is 2.5% of taxable income. How much is withheld each year from her gross pay for state and federal income taxes?

MAINTAINING YOUR SKILLS   Look up the skills in parentheses if you need help or more practice.

Subtract. **(Skill 6)**

15. $82.19 - 16.32$

16. $39 - 16.2$

17. $46.2 - 14.297$

18. $900.12 - 612.89$

Find the percentage. **(Skill 30)**

19. 120% of 160

20. 8% of 122

21. $2\frac{1}{2}$% of 500

22. $14\frac{3}{5}$% of 200

2-3

Graduated State Income Tax

OBJECTIVE
Compute the state tax on a graduated income basis.

Some states have a graduated income tax. Graduated income tax involves a different tax rate for each of several levels of income. The tax rate increases as income increases. The tax rate on low incomes is usually 1% to 3%. The tax rate on high incomes may be as much as 20%.

$$\text{Tax Withheld per Pay Period} = \frac{\text{Annual Tax Withheld}}{\text{Number of Pay Periods per Year}}$$

EXAMPLE   *Skills* 5, 11   *Application* K   *Term* Graduated income tax

Lois Ryan's annual salary is $24,800. She is paid semimonthly. Her personal exemptions total $1500. How much does her employer deduct from each of Lois's semimonthly paychecks for state income tax?

STATE TAX	
Annual Gross Pay	**Tax Rate**
First $1000	1.5%
Next $2000	3.0%
Next $2000	4.5%
Over $5000	5.0%

SOLUTION

A. Find the **taxable wages.**
Annual Gross Pay − Personal Exemptions
   $24,800.00   − $1500.00   = $23,300.00

B. Find the **annual tax withheld.**
  (1) First $1000: 1.5% of $1000.00 ....$ 15.00
  (2) Next $2000: 3.0% of $2000.00 ....$ 60.00
  (3) Next $2000: 4.5% of $2000.00 ....$ 90.00
  (4) Over $5000: 5.0% of ($23,300.00 − $5000.00)
           5.0% of $18,300.00   = $ 915.00
                            Total $1080.00

C. Find the **tax withheld per pay period.**
Annual Tax ÷   Number of Pay
Withheld   Periods per Year
 $1080.00   ÷   24   = $45.00 tax withheld semimonthly

24800 − 1500 = 23300 1000 × 1.5 % 15 M+ 2000 × 3 % 60 M+ 2000 ×
4.5 % 90 M+ 23300 − 5000 = 18300 × 5 % 915 M+ RM 1080 ÷ 24 = 45

✔ SELF-CHECK   Complete the problems, then check your answers in the back of the book.

Find the tax withheld per pay period.

1. Annual salary: $19,400.
Personal exemptions: $1500.
24 pay periods.

2. Annual salary: $21,350.
Personal exemptions: $3000.
26 pay periods.

3. Anna Vail.
Annual gross pay of $6200.
Personal exemptions of $3000.
1.5% state tax on first $2000.
3% tax on amount over $2000.
What is her taxable income?

4. Clyde Browning.
Annual gross pay of $7500.
Personal exemptions of $2000.
1.5% state tax on first $2000.
3% tax on amount over $2000.
What is his taxable income?

5. Melissa Brossia.
Annual gross pay of $22,000.
Personal exemptions of $4400.
1% state tax on first $2000.
3% tax on next $3000.
4.5% tax on amount over $5000.
How much state tax is withheld?

6. Curtis Spiess.
Annual gross pay of $12,350.
Personal exemptions of $1000.
1% state tax on first $2000.
3% tax on next $3000.
4.5% tax on amount over $5000.
How much state tax is withheld?

Use the tables below to compute the state tax for problems 7–10.

7. Milton Chandler's annual gross pay is $14,400. He is single and is paid on a monthly basis. How much is withheld monthly for state tax?

PERSONAL EXEMPTIONS	
Single	$1500
Married	3000
Each dependent	700

STATE TAX	
Annual Gross Pay	**Tax Rate**
First $3500	3%
Next $3500	4.5%
Over $7000	7%

8. Nancy Palino's annual gross pay is $34,460. She is married with no dependents. How much is withheld from her biweekly paycheck for state income tax?

9. August Daily earns $24,600 a year as a parks manager. He is married and has 2 children. How much is withheld from his weekly paycheck for state income tax?

10. Suzanne Pollitz earns $18,235 a year as a legal secretary. She is single and is paid on a weekly basis. How much is withheld each week for state income tax?

Add. **(Skill 5)**

11.
```
   0.005
   0.319
+  0.223
```

12.
```
   465.6
     1.627
+   15.11
```

13.
```
$  42.60
   187.90
+    5.42
```

14.
```
$2041.42
  106.90
+  22.84
```

Divide. Round answers to the nearest hundredth. **(Skill 11)**

15. 397 ÷ 24   **16.** 28.618 ÷ 1.3   **17.** 0.00891 ÷ 0.31   **18.** 27 ÷ 15

Find the number of occurrences. **(Application K)**

19. Quarterly for 3 years   **20.** Monthly for $1\frac{1}{2}$ years   **21.** Weekly for 4 years

2-4

Social Security Tax

OBJECTIVE

Compute the amount of income withheld for social security tax.

The Federal Insurance Contributions Act (FICA) requires employers to deduct 7.65% of the first $55,500 of your annual income for social security taxes. There are no deductions on income over $55,500. The employer must contribute an amount that equals your contribution. The federal government uses social security taxes to provide hospitalization insurance for people over 65, retirement income, survivor's benefits, and disability benefits.

Social Security Tax Withheld = Gross Pay × Tax Rate

EXAMPLE   *Skills* 30, 2   *Application* A   *Term* Social security

Carl Nichol's gross weekly pay is $232.00. His earnings to date for the year total $11,136. What amount is deducted from his pay this week for social security taxes?

SOLUTION

A. Are earnings to date less than $55,500?
Yes. $11,136 is less than $55,500.

B. Find the **social security tax withheld.**
Gross Pay × Tax Rate
$232.00 × 7.65% = $17.748 = $17.75 tax withheld

232 × 7.65 % 17.748

✔ SELF-CHECK   Complete the problems, then check your answers in the back of the book.

Find the social security tax withheld for this pay period.

1. Monthly salary: $2400.
Earnings to date: $26,400.

2. Weekly salary: $134.
Earnings to date: $4690.

PROBLEMS

	Gross Pay	×	Tax Rate	=	Social Security Tax Withheld
3.	$ 67.00	×	7.65%	=	?
4.	$ 345.00	×	7.65%	=	?
5.	$ 139.50	×	7.65%	=	?
6.	$1500.00	×	7.65%	=	?
7.	$4820.00	×	7.65%	=	?

Use the FICA tax rate of 7.65% of the first $55,500 in solving.

8. Matt Chester, fire fighter.
$9450 earned this year to date.
$450 gross pay this week.
How much deducted this week
for social security?

9. Alicia Tsongas, pilot.
$3440 gross pay this check.
$56,040 earned this year to date.
How much deducted this pay
period for social security?

10. Cassius Mortier, a farm advisor for Delta Grain Co-Op, is paid on a monthly basis. His gross pay this month is $2174. His earnings to date for this year are $15,218. How much is deducted from his paycheck this month for social security tax?

11. Ann Brewer is an appraiser for a real estate firm. She earns $27,140 a year and is paid on a semimonthly basis. How much is deducted per pay period for social security tax?

12. Beth Deiger was hired for the position of receptionist. She is to be paid on a biweekly basis. How much will be deducted from each of her paychecks for social security?

F.Y.I.

Maximum Wage Subject to FICA Tax	
1988	$45,600
1989	48,000
1990	51,300
1991	54,600
1992	55,500

RECEPTIONIST
$16,900 salary. Meet, greet clients. Park free. Great company. Call Charles Guidry 555-1325 for appointment.

Use the tables on pages 642–643 for federal withholding taxes.

13. Halle Poole is employed by Madison Toy Manufacturing as a designer. She is married, earns $472 weekly, and claims no allowances. Her gross pay to date this year is $9912. How much is deducted from her paycheck this week for federal income and social security taxes?

14. Jasin Pawloski earns $550 each week as a chief designer for Everett Plastics. He is married and claims 2 allowances. How much is withheld from his paycheck for the last week of December for federal income and social security taxes?

15. Nick Peralta was hired on January 2 for the supervisory position shown. He will be paid on a monthly basis. If Pete's social security deductions are 7.65% of the first $55,500, in which month will the last deduction be made?

MECHANICAL ENGINEER
Supervise four engineers in special machine design. Good communicator. Downtown location. $64,000 salary.

MAINTAINING YOUR SKILLS   Look up the skills in parentheses if you need help or more practice.

Find the percentage. **(Skill 30)**

16. 7.15% of $480   **17.** $2\frac{1}{2}$% of 500   **18.** $15\frac{3}{5}$% of 80   **19.** $\frac{3}{4}$% of 20

Round to the place value indicated. **(Skill 2)**

Nearest ten   **20.** 219   **21.** 322.46   **22.** 3954.97   **23.** 23,981.98

Nearest hundred 24. 7928   **25.** 3764   **26.** 63,782   **27.** 147,366.20

2-5

Group Insurance

OBJECTIVE

Compute the deduction for group insurance.

Many businesses offer group insurance plans to their employees. You can purchase group insurance for a lower cost than individual insurance. Businesses often pay part of the cost of the insurance. The remaining cost is deducted from your pay.

$$\text{Deduction per Pay Period} = \frac{\text{Total Annual Amount Paid by Employee}}{\text{Number of Pay Periods per Year}}$$

EXAMPLE   *Skills* 6, 11, 30   *Application* A   *Term* Group insurance

Nikki Bort is a carpenter for Houck Construction Co. She has family medical coverage through the group medical plan that Houck provides for its employees. The annual cost of Nikki's family membership is $4200. The company pays 80% of the cost. How much is deducted from her weekly paycheck for medical insurance?

S O L U T I O N

A. Find the **percent paid by employee.**
100% − 80% = 20%

B. Find the **total amount paid by employee.**
$4200.00 × 20% = $840.00

C. Find the **deduction per pay period.**

Total Amount Paid by Employee	÷	Number of Pay Periods per Year		
$840.00	÷	52	= $16.153	= $16.15 deducted per pay period

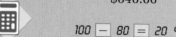 100 ⊟ 80 ▭ 20 4200 ⊠ 20 ▭% 840 ÷ 52 ▭ 16.1538

✔ SELF-CHECK   Complete the problems, then check your answers in the back of the book.

Find the deduction per pay period.

1. Annual cost of insurance: $2400.
Employer pays 80%.
12 pay periods.

2. Annual cost of insurance: $3272.
Employer pays 70%.
52 pay periods.

PROBLEMS

3. Rebbekah Roots, geologist.
Annual group insurance costs $3800.
Company pays 40% of the cost. How much does Rebbekah pay semimonthly?

4. Ron Stover, hydrologist.
Annual group insurance costs $3400.
Employer pays 75% of the cost. How much does Ron pay yearly?

5. Janet Ingel's group medical insurance coverage costs $3580 a year. The company pays 85% of the cost. How much is deducted each month from her paycheck for medical insurance?

6. Bev Katz, a clerk at La Mirage Motel, earns $245.22 weekly. Her group medical insurance costs $2350 a year, of which the company pays 60% of the costs. How much is deducted weekly from her paycheck for medical insurance?

7. Veronica McMame earns $312.48 a week as a security guard for the city. Because she is an employee of the city, the city pays 90% of the cost of any insurance coverage. Her family medical insurance costs $3228 a year. How much is deducted each week for medical insurance?

8. Joel LaVita is a painter for the city. He earns an annual salary of $26,000. His annual medical insurance costs $4312 a year. The city pays 90% of the costs. How much is deducted each week from his paycheck for medical insurance?

9. Lee Munger is employed at McDermott International as a TV technician. Her annual salary is $28,600 and she is paid on a semimonthly basis. Her annual group medical coverage costs $3360, of which the company pays 75%. How much is deducted each pay period from Lee's paycheck for medical coverage?

10. Phil Akers is a surveyor for Globe Land Co. He is covered by travel insurance in addition to the basic medical coverage. Medical insurance costs $3844 a year and travel insurance costs another $347 a year. The company pays 80% of all insurance costs. How much is deducted monthly for travel insurance? How much is deducted monthly for basic medical coverage?

11. Helen Poling has medical, dental, and term life insurance coverages through the company for which she works. Medical coverage costs $2444 a year, dental coverage costs $298 a year, and term life insurance costs $96.72 a year. The company pays 75% of the medical and 50% of the dental coverage. Term life insurance is entirely paid for by the company. What is the total amount deducted weekly for these coverages?

MAINTAINING YOUR SKILLS   Look up the skills in parentheses if you need help or more practice.

Divide. Round answers to the nearest hundredth. **(Skill 11)**

12. 600.48 ÷ 24   **13.** 16.97 ÷ 1.3

14. 0.00392 ÷ 0.30   **15.** 0.692 ÷ 0.008

Find the percentage. **(Skill 30)**

16. 8% of 240   **17.** 9.6% of 80   **18.** $5\frac{1}{4}$% of 1600   **19.** $8\frac{3}{10}$% of 95

2-6

Earnings Statement

OBJECTIVE

Compute the net pay per pay period.

You may have additional deductions taken from your gross pay for union dues, contributions to community funds, savings plans, and so on. The earnings statement attached to your paycheck lists all your deductions, your gross pay, and your **net pay** for the pay period.

Net Pay = Gross Pay − Total Deductions

EXAMPLE   *Skills* 5, 6   *Application* A   *Term* Net pay

Juan Teijeiro's gross weekly salary is $400. He is married and claims 2 allowances. The social security tax is 7.65% of the first $55,500. The state tax is 1.5% of gross pay. Each week he pays $10.40 for medical insurance and $2.50 for charity. Is Juan's earnings statement correct?

DEPT.	EMPLOYEE	CHECK #	WEEK ENDING	GROSS PAY	NET PAY
04	Teijeiro, J.	20566	9/17/—	400.00	312.50

TAX DEDUCTIONS				PERSONAL DEDUCTIONS		
FIT	FICA	STATE	LOCAL	MEDICAL	UNION DUES	OTHERS
38.00	30.60	6.00	—	10.40	—	2.50

SOLUTION

A. Find the **total deductions**.

(1) Federal withholding:
(from table on page 643) $38.00
(2) Social security: 7.65% of $400.00 .... 30.60
(3) State tax: 1.5% of $400.00 .... 6.00
(4) Medical insurance 10.40
(5) Charity 2.50

Total $87.50

B. Find the **net pay**.
Gross Pay − Total Deductions
$400.00 − $87.50 = $312.50 net pay; his statement is correct.

38 M+ 400 × 7.65 % 30.60 M+ 400 × 1.5 % 6 M+ 10.40 M+ 2.50
M+ 400 − RM 87.5 = 312.5

✔ SELF-CHECK   Complete the problem, then check your answer in the back of the book.

1. Ron Rice is single and claims 1 allowance. His gross weekly salary is $320. Each week he pays federal and social security taxes, $16.20 for medical insurance, and $25 for the credit union. What is his net pay?

Find the deductions and the net pay. The FICA is 7.65% of the first $55,500. Use the tax tables on pages 642–643 for federal tax. For problems 2–4, the state tax is 2% of gross pay and the local tax is 1.5% of gross pay.

2. Terry Medley is single and claims 1 allowance.

DEPT.	EMPLOYEE	CHECK #	WEEK ENDING	GROSS PAY	NET PAY
07	MEDLEY, T.	54601	8/24/—	250.50	?

TAX DEDUCTIONS				PERSONAL DEDUCTIONS		
FIT	FICA	STATE	LOCAL	MEDICAL	UNION DUES	OTHERS
?	?	?	?	12.75	—	—

3. Brad Herzig is married and claims 3 allowances.

DEPT.	EMPLOYEE	CHECK #	WEEK ENDING	GROSS PAY	NET PAY
12	HERZIG, B.	11352	5/03/—	304.20	?

TAX DEDUCTIONS				PERSONAL DEDUCTIONS		
FIT	FICA	STATE	LOCAL	MEDICAL	UNION DUES	OTHERS
?	?	?	?	16.60	10.00	2.00

4. Sue Shore is married and claims 1 allowance.

DEPT.	EMPLOYEE	CHECK #	WEEK ENDING	GROSS PAY	NET PAY
A	SHORE, S.	8002349	1/16/—	455.00	?

TAX DEDUCTIONS				PERSONAL DEDUCTIONS		
FIT	FICA	STATE	LOCAL	MEDICAL	UNION DUES	OTHERS
?	?	?	?	17.25	12.00	—

5. Veronica Bell, an interior decorator for Crown Interior Design, is married and claims 1 allowance. Her state personal exemption is $35.46 a week. The state tax rate is 2.5% of taxable wages. Local tax is 1.75% of gross pay.

DEPT.	EMPLOYEE	CHECK #	WEEK ENDING	GROSS PAY	NET PAY
M	BELL, V.	347528	11/02/—	421.25	?

TAX DEDUCTIONS				PERSONAL DEDUCTIONS		
FIT	FICA	STATE	LOCAL	MEDICAL	UNION DUES	OTHERS
?	?	?	?	22.00	—	—

6. Glen Graybar is a painter and earns $24,700 a year. He is single and claims 3 allowances. Weekly deductions include FIT, FICA, state tax of 3.0% on gross earnings, local tax of 1.75% on gross earnings, and $14.50 for medical insurance. Union dues are $12 a week.

DEPT.	EMPLOYEE	CHECK #	WEEK ENDING	GROSS PAY	NET PAY
15	GRAYBAR, G.	91666	3/31/—	?	?

TAX DEDUCTIONS				PERSONAL DEDUCTIONS		
FIT	FICA	STATE	LOCAL	MEDICAL	UNION DUES	OTHERS
?	?	?	?	?	?	—

7. Donna Ovsky is employed by Hiram Associates and works in the marketing department. She is single and claims 2 allowances. Weekly deductions include FIT, FICA, state tax of 2.5% on gross earnings, local tax of 1.75% on gross earnings, and medical insurance. The company pays 70% of the $3448 annual medical cost.

DEPT.	EMPLOYEE	CHECK #	WEEK ENDING	GROSS PAY	NET PAY
14	OVSKY, D.	10-142	10/31/—	460.50	?

TAX DEDUCTIONS				PERSONAL DEDUCTIONS		
FIT	FICA	STATE	LOCAL	MEDICAL	UNION DUES	OTHERS
?	?	?	?	?	—	—

8. Todd Damask is employed as a payroll supervisor and earns $9.60 per hour for a 40-hour week. He is married and claims 6 allowances. The state tax is 3.5% of gross earnings and the local tax is 0.5% of gross earnings. He pays $27.40 for medical insurance and $15 in union dues. During the week ending 4/12, he worked 40 hours.

DEPT.	EMPLOYEE	CHECK #	WEEK ENDING	GROSS PAY	NET PAY
PY	DAMASK, T.	10-942	4/12/—	?	?

TAX DEDUCTIONS				PERSONAL DEDUCTIONS		
FIT	FICA	STATE	LOCAL	MEDICAL	UNION DUES	OTHERS
?	?	?	?	?	?	—

9. A Brief Case   John Clark, a purchasing agent at the Barnett Co., earns $8.00 an hour with time and a half for working over 40 hours a week. He is married and claims 4 allowances. The state tax rate is 25% of the federal tax. The local tax is 2.5% of gross earnings. Medical insurance costs $2432 a year, of which the company pays 80% of the cost. Credit union deductions are $35 a week. What is his net pay for a week in which he worked 48 hours?

MAINTAINING YOUR SKILLS   Look up the skills in parentheses if you need help or more practice.

Add. **(Skill 5)**

10.
```
   16.002
  161.320
+ 342.117
```

11.
```
    16
     9.3
+ 42.016
```

12.
```
  429.9
  76.107
+  3.05
```

13.
```
  62.90
   3.52
+ 186.81
```

14.
```
  0.005
  0.319
+ 0.223
```

15.
```
  465.6
    1.627
+  15.11
```

16.
```
  42.60
 187.90
+  5.42
```

17.
```
 2041.42
  106.90
+  22.84
```

Subtract. **(Skill 6)**

18.
```
  571.21
-  16.96
```

19.
```
  69.3
-  6.24
```

20.
```
  62.39
- 86.14
```

21.
```
  88.19
- 22.81
```

Reviewing the Basics

Skills

(Skill 1)

Write the number that is greater.

1. $239 or $230.98   **2.** $194.45 or $200.45

Round to the nearest cent.

(Skill 2)

3. $154.807   **4.** $25.209   **5.** $40.0953   **6.** $1.523

Solve. Round answers to the nearest cent.

(Skill 5)   **7.** $48.72 + $54.59   **8.** $10.41 + $11.03 + $6.88

(Skill 6)   **9.** $84.72 − $60.71   **10.** $8.89 − $0.62   **11.** $143.32 − $7.89

(Skill 11)   **12.** $892.50 ÷ 12   **13.** $988 ÷ 52   **14.** $1147.27 ÷ 24

(Skill 30)   **15.** 4.5% of $105.45   **16.** 3.5% of $404.10   **17.** 3.2% of $148.89

Applications

Use the following formula to solve problems 18–19.

Tax Withheld = Tax Rate × Gross Pay

(Application A)

18. Tax rate: 5%.
Weekly wage: $320.19.
Find the tax withheld.

19. Tax rate: 6.5%.
Monthly wage: $979.67.
Find the tax withheld.

(Application C)

20. Marge Fuller is single, earns $253.86 weekly, and claims 2 allowances. Use the table to find the weekly federal income tax withheld.

WEEKLY Payroll SINGLE Persons—				
Wages		Allowances		
		Tax Withheld		
At least	But less than	0	1	2
240	250	33	27	21
250	260	35	28	22
260	270	36	30	24
270	280	38	31	25
280	290	39	33	27

Find the number of pay periods.

(Application K)

21. Semimonthly for 1 year

22. Biweekly for 2 years

23. Monthly for 6 years

Terms

Write your own definition for each item.

24. Group insurance   **25.** Income tax   **26.** Net pay

27. Social security   **28.** Graduated income tax   **29.** Personal exemptions

Refer to your reference files at the back of the book if you need help.

Unit Test

Lesson 2-1

1. Julie Miles, a project manager, earns $445.20 a week. She is married and claims 2 allowances. How much is withheld from her weekly paycheck for federal income tax?

WEEKLY Payroll MARRIED Persons—				
Wages		Allowances		
		Tax Withheld		
At least	But less than	0	1	2
430	440	55	49	36
440	450	57	50	38
450	460	58	52	39
460	470	60	53	41
470	480	61	55	42

Lesson 2-2

2. Melvia Hoskins earns $18,000 a year as a librarian. The state income tax rate is 3.6% of taxable income. Her personal exemptions total $3700. How much is withheld each week from Melvia's gross pay for state income tax?

Lesson 2-3

3. Waylon Lewis, a meteorologist, earns an annual salary of $37,420. He is paid biweekly. His personal exemptions total $3000. How much is deducted each pay period from his paycheck for state income tax?

STATE TAX	
Taxable Wages	Tax Rate
First $2000	2%
Next $4000	3%
Next $4000	4.5%
Over $10,000	6%

Lesson 2-4

4. Chris Huen, an oceanographer, is paid $576.20 a week. His earnings to date this year total $21,895.60. The social security tax rate is 7.65% of the first $55,500 earned. How much is deducted from his paycheck this week for social security tax?

Lesson 2-5

5. Sherry Reese, a technical writer for Ace Electronics, earns $423.08 a week. Her medical insurance costs $3219 a year, of which her company pays 75% of the costs. How much is deducted each week from her paycheck for medical insurance?

Lesson 2-6

6. John Jacobson, a title insurance officer, is married and claims 2 allowances. He earns $432.75 a week. The social security tax rate is 7.65% of the first $55,500 earned. The state tax is $6.95 a week. He has weekly deductions of $21 for medical insurance and $30 for payroll savings. Use the FIT table above to find John's federal tax withheld. What is his net pay for a week?

Lesson 2-6

7. You are a travel agent earning $27,400 annually, single, and claim 1 allowance. The social security rate is 7.65%. The state tax is 1% of your gross pay and the local tax is 1.25% of gross pay. You pay $25 a week to the credit union and $10.50 a week for medical insurance. What is your net pay for a week? You will need to use the table on page 642.

A SPREADSHEET APPLICATION

Net Income

To complete this spreadsheet application, you will need the template diskette for *Mathematics with Business Applications*. Follow the directions in the *User's Guide* to complete this activity.

The Kreo Ice Cream Store employs high school students after school and in the summer. Kreo pays a standard hourly rate of $4.85. Deductions are taken for federal withholding (FIT), social security (FICA), and city income tax (CIT).Input the information in the following problems to determine the net income.

1. Week of: June 15

Employee	Employee Number	Hours Worked	Income Tax Information
a. Cole, Dean	1001	32	Single, 1 allowance
b. Drake, Ann	1002	36	Single, 0 allowances
c. Lusetti, Marie	1003	25	Single, 1 allowance
d. Pappas, Mike	1004	30	Single, 0 allowances
e. Smith, Luellen	1005	32	Single, 1 allowance
f. Trotter, Robert	1006	38	Single, 1 allowance
g. Young, Ann	1007	40	Single, 0 allowances

2. Week of: June 22

Employee	Employee Number	Hours Worked	Income Tax Information
a. Cole, Dean	1001	34	Single, 1 allowance
b. Drake, Ann	1002	35	Single, 0 allowances
c. Lusetti, Marie	1003	27	Single, 1 allowance
d. Pappas, Mike	1004	30	Single, 0 allowances
e. Smith, Luellen	1005	20	Single, 1 allowance
f. Trotter, Robert	1006	28	Single, 1 allowance
g. Young, Ann	1007	39	Single, 0 allowances

3. Week of: June 29

Employee	Employee Number	Hours Worked	Income Tax Information
a. Cole, Dean	1001	39	Single, 1 allowance
b. Drake, Ann	1002	40	Single, 0 allowances
c. Lusetti, Marie	1003	33	Single, 1 allowance
d. Pappas, Mike	1004	38	Single, 0 allowances
e. Smith, Luellen	1005	29	Single, 1 allowance
f. Trotter, Robert	1006	27	Single, 1 allowance
g. Young, Ann	1007	38	Single, 0 allowances

If you are employed, some money is probably withheld by your employer from each of your paychecks for federal income tax. You must, by law, prepare an income tax return by April 15 of each year and send it to the Internal Revenue Service (IRS).

On an income tax return, you report your adjusted gross income, which is the total of your wages, salaries, tips, interest, and other income. Based on your adjusted gross income, you use tax tables to figure out your tax liability, the amount of income tax you must pay. Some of your tax liability is already paid by your withholdings. If your tax liability is greater than your withholdings, you must pay the IRS an amount called the "amount you owe." If your tax liability is less than your withholdings, the IRS will return the extra money to you as a tax refund.

Each year your employer must send you a Wage and Tax Statement form, called a W-2 form. This form tells how much money you earned, how much was withheld for social security tax (FICA), and how much was withheld for federal, state, and local income taxes. You will receive copies of your W-2 form to send with your federal, state, and local income tax returns, as well as a copy to keep for your records. If you have earned interest on a bank account, the bank will send you a form showing the amount of interest.

Suppose that last year you had a part-time job as a clerk at a sporting goods store. Your W-2 form might look like this.

1 Control number		OMB No. 1545-0008							
2 Employer's name, address, and ZIP code			6 Statutory employee ☐ Deceased ☐ Pension plan ☐ Legal rep. ☐ 942 emp. ☐ Subtotal ☐ Deferred compensation ☐ Void ☐						
			7 Allocated tips			8 Advance EIC payment			
			9 Federal income tax withheld **$396**			10 Wages, tips, other compensation **$5775**			
3 Employer's identification number	4 Employer's state I.D. number		11 Social security tax withheld			12 Social security wages			
5 Employee's social security number			13 Social security tips			14 Medicare wages and tips			
19 Employee's name, address, and ZIP code			15 Medicare tax withheld			16 Nonqualified plans			
			17 See Instrs. for Box 17			18 Other			
20	21		22 Dependent care benefits			23 Benefits included in Box 10			
24 State income tax	25 State wages, tips, etc.	26 Name of state	27 Local income tax	28 Local wages, tips, etc.	29 Name of locality				

Copy B To Be Filed With Employee's FEDERAL Tax Return   Department of the Treasury—Internal Revenue Service

Form **W-2 Wage and Tax Statement 1991**

This information is being furnished to the Internal Revenue Service.

A SIMULATION

Preparing a 1040EZ Income Tax Return

You are allowed to use the 1040EZ form in preparing your income tax because (1) your income was all from wages, salaries, tips, and taxable scholarships or fellowships; (2) you did not have more than $400 in taxable interest and dividends; (3) your filing status is single; (4) you do not claim any dependents; (5) you are under 65 and not blind and; (6) your income is less $50,000. This return is shown on page 120.

Use the information in the W-2 form on page 118 and assume that you received $10.66 in interest on your savings account.

1. What amount would you write on line 1?

2. What amount would you write on line 2?

3. Add the amounts on lines 1 and 2 to find your adjusted gross income. Write this sum on line 3.

4. Assume you will be claimed as a dependent on another person's return. Then check "Yes" on line 4 and complete the following:

Standard deduction worksheet for dependents

A. Enter the amount from line 1.   A. _____

B. Minimum amount.   B. __550__

C. **Compare** the amounts on lines A and B above. Enter the larger of the two amounts here.   C. _____

D. Maximum amount.   D. __3,400.00__

E. **Compare** the amounts on lines C and D above. Enter the smaller of the two amounts here and on line 4.   E. _____

5. Subtract line 4 from line 3 and write this difference on line 5. This is your taxable income.

6. What amount would you write on line 6? (See your W-2 form.)

7. Refer to the tax table on page 122. What amount would you write as your tax on line 7?

8. Is line 6 larger than line 7? If yes, subtract line 7 from line 6. This is your refund.

9. Is line 7 larger than line 6? If yes, subtract line 6 from line 7. This is the amount you owe.

To complete the tax return, you would sign your name, write the date, and attach a copy of your W-2 form. If you owe a balance due, you would also attach a check or money order for the amount you owe. If you owe less than $1, you do not have to pay. You would make a copy of your completed return for your records. Then you would send the return to the address listed in the instruction booklet that came with the return.

Department of the Treasury—Internal Revenue Service

Form
1040EZ

Income Tax Return for
Single Filers With No Dependents ⊺ **1991**

OMB No. 1545-0675

Name & address

Use the IRS label (see page 10). If you don't have one, please print.

Please print your numbers like this:

9 8 7 6 5 4 3 2 1 0

L A B E L	Print your name (first, initial, last)
H E R E	Home address (number and street). (If you have a P.O. box, see page 11.)	Apt. no.
	City, town or post office, state, and ZIP code. (If you have a foreign address, see page 11.)

Your social security number

Please see instructions on the back. Also, see the Form 1040EZ booklet.

Presidential Election Campaign (see page 11)
Do you want $1 to go to this fund?

Note: *Checking "Yes" will not change your tax or reduce your refund.* ▶

Yes No

Dollars   Cents

Report your income

1 Total wages, salaries, and tips. This should be shown in Box 10 of your W-2 form(s). (Attach your W-2 form(s).) **1**

Attach Copy B of Form(s) W-2 here. Attach tax payment on top of Form(s) W-2.

2 Taxable interest income of $400 or less. If the total is more than $400, you cannot use Form 1040EZ. **2**

Note: *You **must** check Yes or No.*

3 Add line 1 and line 2. This is your **adjusted gross income.** **3**

4 Can your parents (or someone else) claim you on their return?
☐ **Yes.** Do worksheet on back; enter amount from line E here.
☐ **No.** Enter 5,550.00. This is the total of your standard deduction and personal exemption. **4**

5 Subtract line 4 from line 3. If line 4 is larger than line 3, enter 0. This is your **taxable income.** **5**

Figure your tax

6 Enter your Federal income tax withheld from Box 9 of your W-2 form(s). **6**

7 **Tax.** Use the amount on **line 5** to find your tax in the tax table on pages 16-18 of the booklet. Enter the tax from the table on this line. **7**

Refund or amount you owe

8 If line 6 is larger than line 7, subtract line 7 from line 6. This is your **refund.** **8**

9 If line 7 is larger than line 6, subtract line 6 from line 7. This is the **amount you owe.** Attach your payment for full amount payable to the "Internal Revenue Service." Write your name, address, social security number, daytime phone number, and "1991 Form 1040EZ" on it. **9**

Sign your return

Keep a copy of this form for your records.

I have read this return. Under penalties of perjury, I declare that to the best of my knowledge and belief, the return is true, correct, and complete.

| Your signature | Date |
| X | Your occupation |

For IRS Use Only — Please do not write in boxes below.

For Privacy Act and Paperwork Reduction Act Notice, see page 4 in the booklet.

Cat. No. 11329W   Form 1040EZ (1991)

1991   **Instructions for Form 1040EZ**

Use this form if

- Your filing status is single.
- You do not claim any dependents.
- You had **only** wages, salaries, tips, and taxable scholarship or fellowship grants, and your taxable interest income was $400 or less. **Caution:** *If you earned tips (including allocated tips) that are not included in Box 13 and Box 14 of your W-2, you may not be able to use Form 1040EZ. See page 12 in the booklet.*
- You were under 65 and not blind at the end of 1991.
- Your taxable income (line 5) is less than $50,000.
- You did not receive any advance earned income credit payments.

If you are not sure about your filing status, see page 6 in the booklet. If you have questions about dependents, see Tele-Tax (topic no. 155) on page 25 in the booklet.

If you can't use this form, see Tele-Tax (topic no. 152) on page 25 in the booklet.

Completing your return

Please print your numbers inside the boxes. Do not type your numbers. Do not use dollar signs.

Most people can fill out the form by following the instructions on the front. But you will have to use the booklet if you received a scholarship or fellowship grant or tax-exempt interest income (such as on municipal bonds). Also use the booklet if you received a 1099-INT showing income tax withheld (backup withholding) or if you had two or more employers and your total wages were more than $53,400.

Remember, you must report your wages, salaries, and tips even if you don't get a W-2 form from your employer. You must also report all your taxable interest income, including interest from savings accounts at banks, savings and loans, credit unions, etc., even if you don't get a Form 1099-INT.

If you paid someone to prepare your return, that person must also sign it and show other information. See page 15 in the booklet.

Standard deduction worksheet for dependents who checked "Yes" on line 4

Fill in this worksheet to figure the amount to enter on line 4 if someone can claim you as a dependent (even if that person chooses not to claim you).

A. Enter the amount from line 1 on front.   **A.** _____

B. Minimum amount.   **B.** _____ 550.00

C. Compare the amounts on lines A and B above. Enter the LARGER of the two amounts here.   **C.** _____

D. Maximum amount.   **D.** _____ 3,400.00

E. Compare the amounts on lines C and D above. Enter the SMALLER of the two amounts here and on line 4 on front.   **E.** _____

If you checked "No" because no one can claim you as a dependent, enter 5,550.00 on line 4. This is the total of your standard deduction (3,400.00) and personal exemption (2,150.00).

Avoid common mistakes

This checklist is to help you make sure that your form is filled out correctly.

1. Are your name, address, and social security number on the label correct? If not, did you correct the label?

2. If you didn't get a label, did you enter your name, address (including ZIP code), and social security number in the spaces provided on page 1 of Form 1040EZ?

3. Did you check the "Yes" box on line 4 if your parents (or someone else) can claim you as a dependent on their 1991 return (even if they choose not to claim you)? If no one can claim you as a dependent, did you check the "No" box?

4. Did you enter an amount on line 4? If you checked the "Yes" box on line 4, did you fill out the worksheet above to figure the amount to enter? If you checked the "No" box, did you enter 5,550.00?

5. Did you check your computations (additions, subtractions, etc.) especially when figuring your taxable income, Federal income tax withheld, and your refund or amount you owe?

6. Did you use the amount from **line 5** to find your tax in the tax table? Did you enter the correct tax on line 7?

7. Did you attach your W-2 form(s) to the left margin of your return? And, did you sign and date Form 1040EZ and enter your occupation?

Mailing your return

Mail your return by **April 15, 1992.** Use the envelope that came with your booklet. If you don't have that envelope, see page 19 in the booklet for the address to use.

A SIMULATION
(CONTINUED)

Tax Tables

For line 7 of a 1040EZ return, you need to find your tax liability. To do this, you use tax tables like the one below. Tax tables are included in the instruction booklet that comes with the return. Your tax liability depends on your filing status, the number of exemptions you claim, and your taxable income (line 5). For example, suppose your taxable income is $2012, you are single, and a 1040EZ filer. Notice that your taxable income is at least $2000 but less than $2025, so your tax liability is $302.

Use the tax table to find the tax liabilities for taxpayers with the following taxable incomes. Assume that all are 1040EZ filers.

	10.	11.	12.	13.
Taxable income	$2310	$11,180	$21,390	$2380
Tax	?	?	?	?

1991 1040EZ Tax Table

If line 5 is at least—	But less than—	Your tax is—	If line 5 is at least—	But less than—	Your tax is—	If line 5 is at least—	But less than—	Your tax is—
2,000			**11,000**			**21,000**		
2,000	2,025	302	11,000	11,025	1,654	21,000	21,025	3,359
2,025	2,050	306	11,025	11,050	1,661	21,025	21,050	3,373
2,050	2,075	309	11,050	11,075	1,669	21,050	21,075	3,387
2,075	2,100	313	11,075	11,100	1,676	21,075	21,100	3,401
2,100	2,125	317	11,100	11,125	1,684	21,100	21,125	3,415
2,125	2,150	321	11,125	11,150	1,691	21,125	21,150	3,429
2,150	2,175	324	11,150	11,175	1,699	21,150	21,175	3,443
2,175	2,200	328	11,175	11,200	1,706	21,175	21,200	3,457
2,200	2,225	332	11,200	11,225	1,714	21,200	21,225	3,471
2,225	2,250	336	11,225	11,250	1,721	21,225	21,250	3,485
2,250	2,275	339	11,250	11,275	1,729	21,250	21,275	3,499
2,275	2,300	343	11,275	11,300	1,736	21,275	21,300	3,513
2,300	2,325	347	11,300	11,325	1,744	21,300	21,325	3,527
2,325	2,350	351	11,325	11,350	1,751	21,325	21,350	3,541
2,350	2,375	354	11,350	11,375	1,759	21,350	21,375	3,555
2,375	2,400	358	11,375	11,400	1,766	21,375	21,400	3,569
2,400	2,425	362	11,400	11,425	1,774	21,400	21,425	3,583
2,425	2,450	366	11,425	11,450	1,781	21,425	21,450	3,597
2,450	2,475	369	11,450	11,475	1,789	21,450	21,475	3,611
2,475	2,500	373	11,475	11,500	1,796	21,475	21,500	3,625

A SIMULATION
(CONTINUED)

Calculating Your Tax

A few years later, you are working full-time. You are single and have no dependents. Your only sources of income are your salary and the interest on your savings account. Here are your W-2 form and your statement of earnings from your savings bank. You are not claimed as a dependent on another's return.

1 Control number		
2 Employer's name, address, and ZIP code	OMB No. 1545-0008	

2 Employer's name, address, and ZIP code		**6** Statutory employee / Deceased / Pension plan / Legal rep. / 942 emp. / Subtotal / Deferred compensation / Void
		7 Allocated tips	**8** Advance EIC payment

	9 Federal income tax withheld $3900	**10** Wages, tips, other compensation $26,800
3 Employer's identification number	**4** Employer's state I.D. number	**11** Social security tax withheld	**12** Social security wages
5 Employee's social security number		**13** Social security tips	**14** Medicare wages and tips
19 Employee's name, address, and ZIP code		**15** Medicare tax withheld	**16** Nonqualified plans
		17 See Instrs. for Box 17	**18** Other
20	**21**	**22** Dependent care benefits	**23** Benefits included in Box 10
24 State income tax	**25** State wages, tips, etc.	**26** Name of state	**27** Local income tax	**28** Local wages, tips, etc.	**29** Name of locality

Copy B To Be Filed With Employee's FEDERAL Tax Return   Department of the Treasury—Internal Revenue Service

Form **W-2 Wage and Tax Statement 1991**

Use the formulas above, the return on page 120, and the tax table on page 122 to answer these questions.

14. What is your salary? your interest income? your taxable income?

15. How much was withheld for federal income tax? What is your tax liability?

16. Will you have a tax refund or a balance due? How much?

HAMILTON SAVINGS BANK
HAMILTON, INDIANA

PLEASE RETAIN FOR YOUR RECORDS
STATEMENT OF EARNINGS FOR   19____

For your protection, we ask that you compare this information below with your record and notify our auditing department of any discrepancy.

IDENTIFYING NO.:

ACCOUNT NUMBER	INTEREST
13579	41.25

AS OF THE CLOSE OF BUSINESS 12/31/—
YOUR CURRENT BALANCE WAS 824.36
YOUR PASSBOOK BALANCE WAS 824.36

IN ACCORDANCE WITH FEDERAL REGULATIONS, WE HAVE REPORTED INTEREST EARNED ON YOUR ACCOUNT(S) TO THE INTERNAL REVENUE SERVICE. THESE EARNINGS ARE CONSIDERED AS INTEREST FOR TAX REPORTING PURPOSES AND SHOULD BE USED DURING PREPARATION OF YOUR INCOME TAX RETURN.

A SIMULATION
(CONTINUED)

Other Forms and Services

There are other tax return forms besides the 1040EZ form. In some cases, you should use a 1040A or 1040 form, called the long form. For example, you must use a 1040 form if you itemize deductions. By itemizing deductions, people who had large medical expenses, certain interest payments, or certain other expenses may be able to reduce their tax liability. If you itemize deductions, you must file another form with the 1040 Form that lists the expenses you are deducting.

You must file additional forms if, for example:

- You had more than $400 in interest or dividends.
- You had business expenses that were not paid by your employer.
- You had capital gains or losses.
- You had supplemental income from rents, royalties, etc.
- You had farm income and expenses.
- You claim credit for elderly or permanently disabled people.

Check your return carefully after you complete it. The sooner you send it in, the more quickly you will receive your refund.

It is important to be aware that income tax rules change and to consult the Internal Revenue Service when you feel it is necessary. The IRS offers free tax help in preparing tax returns in most areas to older, handicapped, and non-English-speaking individuals. Some institutions and public-spirited groups often make themselves available to the community prior to April 15 to offer help and advice in preparing returns. Federal tax information is available toll free by telephone throughout the United States. The government also offers a telephone service that provides tax refund informtion. See the "Tele-Tax" listing under "IRS."

The special toll-free telephone listing "Problem Resolution" is for those taxpayers who have been unable to resolve their problems with the IRS. Of course, it is advisable to write directly to the IRS District Director with any tax problem that you cannot resolve in the ordinary manner. It is also a good idea to contact your local IRS office and ask for Problem Resolution assistance. Tax laws or technical decisions cannot be changed through these procedures, but you can be helped to obtain a resolution to problems resulting from previous contacts.

You can obtain tax forms from your local IRS office in person or by telephone. All the IRS numbers are listed in your telephone directory under "United States Government, Internal Revenue Service."

UNIT

3

Checking Accounts

A bank *checking account* allows you to pay for goods and services by *check* instead of cash. A check is a form of payment to another person. The bank deducts the sum from your account and pays it to the person named on the check. You keep a record of your *deposits* and the

checks you have written in a *check register*. A monthly *bank statement* sent by the bank also records your deposits, withdrawals, and checks written. The amount of money in your check register and bank statement should agree, or *balance.* Many checking accounts earn interest.

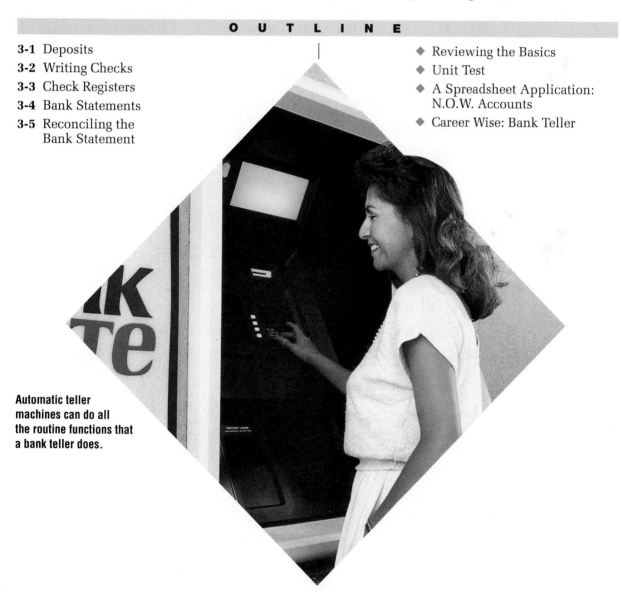

Automatic teller machines can do all the routine functions that a bank teller does.

Deposits

A **deposit** is an amount of money that you put into a bank account. You use a deposit slip to record the amounts of currency, coins, and checks you deposit. To open a checking account, you must make a deposit.

Total Deposit = (Currency + Coins + Checks) − Cash Received

EXAMPLE | *Skills* 5, 6 *Application* A *Term* Deposit

Margaret Miller has a check for $235.42 and a check for $55.47. She would like to receive $40 in cash and deposit the rest of the money in her checking account. What is Margaret's total deposit?

SOLUTION

		DOLLARS	CENTS
CASH	CURRENCY		
	COINS		
CHECKS	LIST SEPARATELY 76-39	235	42
	117-52	55	47
	SUBTOTAL	290	89
	LESS CASH RECEIVED	40	00
	TOTAL DEPOSIT	250	89

In the account of
Margaret C. Miller
Check's bank number
DATE March 11 19 —

FIRST CITY BANK
Currency + Coins + Checks

5 3 ⊩ ꞁꞁ 8 7 6 5 2 ← Margaret's account number

CHECKING ACCOUNT DEPOSIT

✓ SELF-CHECK Complete the problems, then check your answers in the back of the book.

1. ($60.00 + $0.90 + $14.00) − $0.00 = total deposit of how much?

2. ($45.00 + $80.00) − $20.00 = total deposit of how much?

PROBLEMS

Find the subtotal and total deposit.

	3.	**4.**	**5.**	**6.**	**7.**	**8.**
Currency	$30.00	$74.00	—	—	$400.00	$975.00
Coins	$11.80	$ 9.65	—	—	—	$ 40.00
Checks	—	—	$84.50	$124.26	$734.40	$986.53
	—	—	$93.70	$ 48.79	$141.55	$ 91.11
Subtotal	?	?	?	?	?	?
Less Cash Received	—	—	$10.00	$ 20.00	$ 35.00	$ 80.00
Total Deposit	?	?	?	?	?	?

Find the subtotal and total deposit.

		DOLLARS	CENTS
CASH	CURRENCY		
	COINS	5	85
CHECKS	LIST SEPARATELY 14-2	87	18
	7-43	342	41
9.	SUBTOTAL		?
	◊ LESS CASH RECEIVED	75	00
10.	TOTAL DEPOSIT		?

		DOLLARS	CENTS
CASH	CURRENCY	74	00
	COINS	3	89
CHECKS	LIST SEPARATELY	121	74
	120-18	399	59
11.	SUBTOTAL		?
	◊ LESS CASH RECEIVED		
12.	TOTAL DEPOSIT		?

13. Olive Baker has a paycheck for $173.45 and a refund check for $3. She would like to receive $25 in cash and deposit the remaining amount. What is her total deposit?

14. Jess Norton deposited a check for $474.85 and a check for $321.15. He received $50 in cash. What was his total deposit?

15. Bill and Mary Randall deposited their paychecks for $611.33 and $701.45 and a check from their insurance company for $75.25. They received $150 in cash. What was their total deposit?

16. Morgan Meers deposited his paycheck for $201.20, a refund check from a store for $19.78, and $34.23 in cash. What was his total deposit?

17. Carole Winer deposits the following in her checking account: 7 five-dollar bills, 3 two-dollar bills, 18 one-dollar bills, 9 half dollars, 15 quarters, 99 dimes, 48 nickels, 16 pennies, and a check for $28.32. What is her total deposit?

18. Duane Coldren has a check for $343 and a check for $88.91. He would like to deposit the checks and receive 7 ten-dollar bills, 4 one-dollar bills, 9 quarters, and 15 dimes. What is his total deposit?

MAINTAINING YOUR SKILLS   Look up the skills in parentheses if you need help or more practice.

Add. **(Skill 5)**

19. $321.00 + $9.30 + $65.75   **20.** $400.00 + $0.95 + $374.65

21. $350.00 + $8.00 + $15.93   **22.** $694.75 + $321.57

Subtract. **(Skill 6)**

23. $540.00 − $33.00   **24.** $920.00 − $630.00   **25.** $734.40 − $75.00

26. $619.20 − $40.00   **27.** $718.32 − $65.00   **28.** $316.37 − $55.00

Writing Checks

OBJECTIVE
Write a check.

After you have opened a checking account and made a deposit, you can write checks. A check directs a bank to deduct money from your checking account to make a payment. Your account must contain as much money as the amount of the check you are writing so that you do not overdraw your account.

EXAMPLE   *Skill* 1   *Term* Check

Margaret Miller is buying a gift at Hud's Department Store. The cost of the gift is $45.78. Margaret is paying by check. How should Margaret write the check?

SOLUTION
A. Write the date.
B. Write the name of the person or organization to whom payment will be made.
C. Write the amount of the check as a numeral.
D. Write the amount of the check in words with cents expressed as a fraction of a dollar.
E. Make a notation on the check to indicate its purpose.
F. Sign the check.

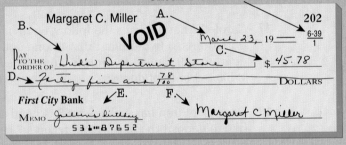

✔ SELF-CHECK   Complete the problems, then check your answers in the back of the book.

1. Write two hundred fifteen and $\frac{32}{100}$ dollars as a numeral.

2. Write $143.32 in words with cents expressed as a fraction of a dollar.

PROBLEMS

Write each amount in words as it would appear on a check.

3. $40.40	**4.** $703.00	**5.** $63.74	**6.** $7.94
7. $34.06	**8.** $66.00	**9.** $1917.00	**10.** $17,200.00
11. $201.09	**12.** $172.61	**13.** $5327.17	**14.** $47,983.39

15. Alice Chino wrote check number 311 to Pugh Health Clinic for $98.72. Did she write the amount in words correctly? If not, write the amount in words correctly.

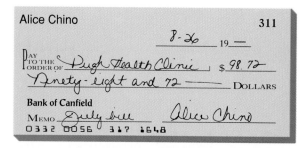

16. For what purpose did Alice Chino write the check to the Pugh Health Clinic?

17. Jim Liebert is paying for an auto repair bill with a check in the amount of $247.25. Did Jim write the amount of the check correctly, both as a numeral and in words?

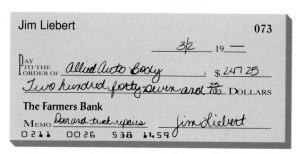

18. What is the number of the check that Jim Liebert used to pay for his auto repair bill?

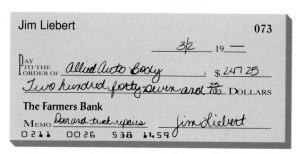

19. Amy Lepon wrote check number 055 to Downtown Electric Company to pay her bill for the month of June. The bill totaled $29.78. Is the check made out in the correct amount?

20. Is the check Amy Lepon wrote made out to the correct company? If not, write the correct company name.

F.Y.I.
The amount written in words takes precedence over the amount written as a numeral.

MAINTAINING YOUR SKILLS   Look up the skill in parentheses if you need help or more practice.

Write the number. **(Skill 1)**

21. Thirty-five and $\frac{15}{100}$ dollars

22. Seventy-four and $\frac{31}{100}$ dollars

Write in word form with cents expressed as a fraction of a dollar. **(Skill 1)**

23. $19.25

24. $50.32

25. $435.00

26. $345.42

27. $5274.19

28. $11,871.63

29. $6110.50

30. $39,974.12

Check Registers

You use a check register to keep a record of your automatic teller deposits, regular deposits, electronic transfers, and the checks you have written. The balance is the amount of money in your account. When you make a deposit, add the amount of the deposit to the balance. When you write a check, subtract the amount of the check from the balance.

New Balance = Previous Balance − Check Amount

New Balance = Previous Balance + Deposit Amount

EXAMPLE *Skills* 5, 6 *Application* A *Term* Balance

Margaret Miller's checking account had a balance of $313.54. She wrote a check for $45.78 on March 23. On March 25, she made a deposit of $240.32. What is the new balance in Margaret's account?

SOLUTION

CHECK NO.	DATE	CHECKS ISSUED TO OR DESCRIPTION OF DEPOSIT	AMOUNT OF CHECK	✓	AMOUNT OF DEPOSIT	BALANCE	
		Previous Balance − Check Amount	BALANCE BROUGHT FORWARD →			313	54
202	³/23	Hud's Dept. Store	45	78		267	76
	³/25	Deposit			240 32	508	08
		Previous Balance + Deposit Amount					

✓ SELF-CHECK Complete the problems, then check your answers in the back of the book.

1.	Balance	$1236.29	**2.**	Balance	$992.71
	Amount of check	− 35.28		Deposit	+ 138.84
	Balance	?		Balance	?

PROBLEMS

Find the new balance after each check or deposit.

	AMOUNT OF CHECK	✓	AMOUNT OF DEPOSIT	BALANCE	
	BALANCE BROUGHT FORWARD →			448	35
3.	46 92				?
4.			216 84		?
5.	251 55				?

	AMOUNT OF CHECK	✓	AMOUNT OF DEPOSIT	BALANCE	
	BALANCE BROUGHT FORWARD →			475	19
6.	75 99				?
7.	31 87				?
8.			108 39		?

9. Your balance is $89.75 on May 23.
Deposit $156.90 on May 30.
Write a $34.79 check on April 2.
What is your new balance?

10. Your balance is $131.02 on April 4.
Write a $31.28 check on April 9.
Write a $45.92 check on April 14.
What is your new balance?

11. Mac Valent opened a new checking account by depositing his paycheck for $209.81. The check register shows his transactions since opening his account. What is his balance after each transaction?

CHECK NO.	DATE	CHECKS ISSUED TO OR DESCRIPTION OF DEPOSIT	AMOUNT OF CHECK		✓	AMOUNT OF DEPOSIT		BALANCE	
		BALANCE BROUGHT FORWARD →						209	81
101	8/12	Acme Light & Power	47	15					?
102	8/18	Foodtown	53	03					?
103	8/20	Shireen Sportswear	107	30					?

12. Paula Melinte's checkbook balance was $149.21 on October 5. Her check register shows her transactions since. What is Paula's balance after each transaction?

CHECK NO.	DATE	CHECKS ISSUED TO OR DESCRIPTION OF DEPOSIT	AMOUNT OF CHECK		✓	AMOUNT OF DEPOSIT		BALANCE	
		BALANCE BROUGHT FORWARD →						149	21
571	10/6	Pettisville Flowers	45	79					?
	10/9	Deposit				213	80		?
572	10/10	Grisier's Music Inc.	16	94					?
573	10/19	Cellular Phone Co.	75	25					?

13. Ralph Snow's latest transactions are shown on the check register. Find his balance after each transaction.

CHECK NO.	DATE	CHECKS ISSUED TO OR DESCRIPTION OF DEPOSIT	AMOUNT OF CHECK		✓	AMOUNT OF DEPOSIT		BALANCE	
		BALANCE BROUGHT FORWARD →						314	30
472	11/17	Oak Park Garden Ctr.	46	00					?
473	11/18	Cash	50	00					?
	11/20	Deposit				286	70		?
474	11/21	Swanton Health Care	126	35					?
	12/4	Deposit				291	00		?

14. Brad Adams opened a new checking account by depositing his IRS tax refund of $145.75. The check register shows his transactions to date. What is his balance after each transaction?

CHECK NO.	DATE	CHECKS ISSUED TO OR DESCRIPTION OF DEPOSIT	AMOUNT OF CHECK	✓	AMOUNT OF DEPOSIT	BALANCE
		BALANCE BROUGHT FORWARD →				145 75
1	4/20	Delta High School	12 00			?
2	4/25	Ames Department Store	31 43			?
3	4/26	VOID	0			?
	5/1	Deposit			120 00	?
4	5/3	Wyse Office Supply	22 59			?
5	5/3	Kayling Fabric Store	34 67			?
6	5/4	Georgio's Restaurant	98 50			?
7	5/15	Crown Bookstore	57 32			?

15. Leah Ahren's latest transactions are shown on the check register. Find her balance after each transaction.

CHECK NO.	DATE	CHECKS ISSUED TO OR DESCRIPTION OF DEPOSIT	AMOUNT OF CHECK	✓	AMOUNT OF DEPOSIT	BALANCE
		BALANCE BROUGHT FORWARD →				397 16
916	9/20	Baggett Auto Store	83 28			?
	9/22	Deposit			45 10	?
917	10/3	Rolland Buchele	134 67			?
918	10/4	First Federal S&L	201 99			?
	10/5	Deposit			139 40	?
919	10/19	Vollmars Ceramics	111 15			?
920	10/20	Cash	30 00			?
921	10/22	Rapids Pharmacy	48 92			?

MAINTAINING YOUR SKILLS   Look up the skills in parentheses if you need help or more practice.

Add. **(Skill 5)**

16. $414.85 + $265.50

17. $845.96 + $400.00

18. $192.78 + $112.50

19. $72.85 + $393.36

20. $2371.81 + $491.48

21. $371.80 + $444.75

Subtract. **(Skill 6)**

22. $579.23 − $212.60

23. $347.89 − $99.92

24. $261.85 − $8.47

25. $141.82 − $64.73

26. $3427.80 − $635.60

27. $671.82 − $314.91

Bank Statements

OBJECTIVE
Compute the present balance on a checking account bank statement.

When you have a checking account, you receive a statement and canceled checks from the bank each month. Canceled checks are the checks that the bank has paid by deducting money from your account. Your statement lists all your checks that the bank has paid and your deposits that the bank has recorded since your last statement. The statement may include a service charge for handling the account, and it may also show an interest credit if you have an interest bearing account.

$$\text{Present Balance} = \text{Previous Balance} + \text{Deposits Recorded} - \text{Checks Paid} - \text{Service Charge} + \text{Interest}$$

EXAMPLE   *Skills* 5, 6   *Application* A   *Term* Statement

Margaret Miller received her bank statement and canceled checks for March. She checks the statement. What is her present balance?

SOLUTION

FIRST CITY BANK

Margaret C. Miller

| DATE LAST STATEMENT |
| 3/1/— |
| BALANCE LAST STATEMENT |
| 124.17 |

CHECKING ACCOUNT NUMBER 531-87652

| CLOSING DATE |
| 4/1/— |

CHECKS AND OTHER CHARGES			DEPOSITS AND CREDITS		BALANCE
DATE	NUMBER	AMOUNT	DATE	AMOUNT	
3/23	202	45.78	3/11	250.89	375.06
3/27	203	124.35	3/25	240.32	329.28
					569.60
SERVICE CHARGE		0.30			445.25
					444.95

Total of all deposits processed

Total of all checks processed

PREVIOUS BALANCE	DEPOSITS RECORDED	CHECKS PAID	SER. CHG. OR INT.	PRESENT BALANCE
124.17 +	491.21 −	170.13 −	0.30 =	444.95

124.17 [+] 491.21 [−] 170.13 [−] .3 [=] 444.95

✔ SELF-CHECK   Complete the problems, then check your answers in the back of the book.

	Previous Balance	+	Deposits Recorded	−	Checks Paid	−	Service Charge	+	Interest	=	Present Balance
1.	$280.00	+	$120.00	−	$140.00	−	$2.50	+	$1.20	=	?
2.	$275.50	+	$105.00	−	$312.60	−	$4.00	+	0	=	?

	3.	4.	5.	6.	7.	8.
Previous Balance	$ 40.10	$487.67	$949.07	$500.00	$9421.99	$5513.11
Total Deposits	$200.00	$430.75	$401.00	$373.96	$7509.23	$1412.46
Total Checks	$190.10	$598.17	$319.80	$289.34	$1397.86	$1209.32
Service Charge	$ 2.34	0	$ 4.15	0	0	$ 5.60
Interest	0	$ 1.38	0	$ 2.49	0	$ 27.41
Present Balance	?	?	?	?	?	?

9. A portion of Susan Dixon's bank statement is shown. Her previous balance was $271.31. What is her present balance?

CHECKS AND OTHER CHARGES			DEPOSITS AND CREDITS		BALANCE
DATE	NUMBER	AMOUNT	DATE	AMOUNT	
6/11	304	19.45	6/12	115.90	
6/15	305	21.02	6/19	115.90	
6/30	307	95.98	6/26	345.85	
SERVICE CHARGE		2.85			

10. A portion of Chester Weis's bank statement is shown below. His previous balance was $341.72. What is his present balance?

CHECKS AND OTHER CHARGES			DEPOSITS AND CREDITS		BALANCE
DATE	NUMBER	AMOUNT	DATE	AMOUNT	
11/2	1031	312.00	11/9	215.55	
11/5	1033	52.38	11/19	112.48	
11/20	1032	47.98	11/29	106.80	
SERVICE CHARGE		5.25			

11. A portion of Kim Hudik's bank statement is shown. Her previous balance was $39.37. What is her present balance?

CHECKS AND OTHER CHARGES			DEPOSITS AND CREDITS		BALANCE
DATE	NUMBER	AMOUNT	DATE	AMOUNT	
1/18	386	27.50	1/20	504.60	
1/22	389	35.00	1/31	121.37	
1/30	387	317.78	2/12	164.22	
2/5	AUTO TELLER	50.00	INTEREST	2.46	
SERVICE CHARGE		.50			

12. Dale Holt received his checking account statement and canceled checks for September. His previous balance was $372.48. What is his present balance?

CHECKS AND OTHER CHARGES			DEPOSITS AND CREDITS		BALANCE
DATE	NUMBER	AMOUNT	DATE	AMOUNT	
9/3	451	37.15	9/7	100.00	335.33
9/9	452	268.75	9/21	200.00	435.33
					166.58
9/21	453	41.89			366.58
9/21	454	45.00			324.69
SERVICE CHARGE		3.50			279.69
					276.19
PREVIOUS BALANCE	DEPOSITS RECORDED		CHECKS PAID	SER. CHG. OR INT.	PRESENT BALANCE
?	?		?	?	?

13. Kay Hoover received her checking account statement and canceled checks for June. Her previous balance was $581.63. What is her present balance?

CHECKS AND OTHER CHARGES			DEPOSITS AND CREDITS		BALANCE
DATE	NUMBER	AMOUNT	DATE	AMOUNT	
6/4	916	515.25	6/10	1314.50	66.38
6/12	917	185.00			1380.88
					1195.88
6/20	918	217.85			978.03
6/21	919	112.35	6/24	1015.40	865.68
			INTEREST	7.17	1881.08
					1888.25
PREVIOUS BALANCE	DEPOSITS RECORDED		CHECKS PAID	SER. CHG. OR INT.	PRESENT BALANCE
?	?		?	?	?

MAINTAINING YOUR SKILLS   Look up the skills in parentheses if you need help or more practice.

Add. **(Skill 5)**

14. $346.50 + $215.50 + $35.97

15. $543.07 + $172.40 + $351.23

16. $917.35 + $448.55 + $627.25

17. $43,906.54 + $3172 + $4.20

Subtract. **(Skill 6)**

18. $915.87
− 748.42

19. $684.31
− 183.49

20. $342.18
− 191.84

21. $2346.39
− 983.42

22. 462.28
− 128.59

23. 1329.84
− 483.49

24. 363.790
− 44.461

25. 1191.863
− 326.399

Reconciling the Bank Statement

OBJECTIVE

Reconcile a check register and a bank statement.

When you receive your bank statement, you compare the canceled checks, the bank statement, and your check register to be sure they agree. You may find some outstanding checks and deposits that appear in your register but did not reach the bank in time to be processed and listed on your statement. You reconcile the statement to make sure that it agrees with your check register.

$$\begin{array}{c} \text{Adjusted} \\ \text{Balance} \end{array} = \begin{array}{c} \text{Statement} \\ \text{Balance} \end{array} - \begin{array}{c} \text{Outstanding} \\ \text{Checks} \end{array} + \begin{array}{c} \text{Outstanding} \\ \text{Deposits} \end{array}$$

EXAMPLE *Skills* 5, 6 *Application* A *Term* Reconcile

Margaret Miller's check register balance is $447.38. She compared her statement, canceled checks, and check register. For each check and deposit listed on her statement, she placed a check mark next to the information in her register. Margaret found these outstanding checks and deposits:

check #201 $81.32   check #204 $39.00   deposit $122.45

How does Margaret reconcile her statement?

SOLUTION

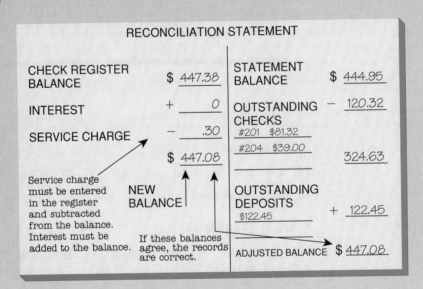

RECONCILIATION STATEMENT

CHECK REGISTER BALANCE	$ 447.38	STATEMENT BALANCE	$ 444.95
INTEREST	+ 0	OUTSTANDING CHECKS	− 120.32
SERVICE CHARGE	− .30	#201 $81.32	
	$ 447.08	#204 $39.00	324.63

Service charge must be entered in the register and subtracted from the balance. Interest must be added to the balance.

NEW BALANCE

If these balances agree, the records are correct.

OUTSTANDING DEPOSITS
$122.45   + 122.45

ADJUSTED BALANCE $ 447.08

447.38 − .3 = 447.08 81.32 + 39 = 120.32 M+ 444.95 − RM 120.32 =
324.63 + 122.45 = 447.08

✔ SELF-CHECK   Complete the problems, then check your answers in the back of the book.

1. Statement balance $374.47
  Outstanding checks   − 238.98
  Outstanding deposits + 140.00
  Adjusted balance   ?

2. Statement balance $772.33
  Outstanding checks   − 283.75
  Outstanding deposits + 427.75
  Adjusted balance   ?

PROBLEMS

Complete the table.

Do the register and statement balances agree?

	3.	4.	5.	6.
Check Register Balance	$147.60	$505.85	$1439.76	$4581.62
Interest	0	0	$ 7.13	$ 22.91
Service Charge	$ 4.70	$ 9.80	0	0
NEW BALANCE	?	?	?	?
Statement Balance	$388.29	$507.21	$1360.44	$1328.66
Outstanding Checks	$345.39	$132.90	$ 432.81	$ 421.99
Outstanding Deposits	$100.00	$121.74	$ 519.26	$3697.86
ADJUSTED BALANCE	?	?	?	?

7. After comparing her bank statement, canceled checks, and checkbook register, Jane Scott completes the reconciliation statement shown below. What is the adjusted balance? Do the register and statement balances agree?

RECONCILIATION STATEMENT

CHECK REGISTER BALANCE	$ 321.04	STATEMENT BALANCE	$ 398.70
INTEREST	+ 0	OUTSTANDING CHECKS	
SERVICE CHARGE	− 4.50	#071 $87.00	−
NEW BALANCE	$ 316.54	#080 $91.44	$

OUTSTANDING DEPOSITS
$96.28   +

** RECONCILIATION STATEMENT FOR YOUR CONVENIENCE

ADJUSTED BALANCE $ _____

8. Charlie Tanner completes the reconciliation statement shown. What are the new and adjusted balances? Do they agree?

RECONCILIATION STATEMENT

CHECK REGISTER BALANCE	$ 792.51	STATEMENT BALANCE	$ 621.89
INTEREST	+ 3.63	OUTSTANDING CHECKS	
SERVICE CHARGE	− 0	#609 $123.96	−
NEW BALANCE	$ _____	#610 $36.87	
		#611 $159.28	

OUTSTANDING DEPOSITS
$294.36   +
$200.00

** RECONCILIATION STATEMENT FOR YOUR CONVENIENCE

ADJUSTED BALANCE $ _____

9. Pattie Millberg received her bank statement and canceled checks for the period ending November 20. She compared the check register with the canceled checks and deposits listed on the statement, then placed a check mark next to the items processed. She reconciles the bank statement.

   a. What total amount does she have in outstanding checks?

   b. What total amount does she have in outstanding deposits?

c. What is her adjusted balance?

d. What is her new check register balance?

e. Do the register and adjusted balances agree?

CHECK NO.	DATE	CHECKS ISSUED TO OR DESCRIPTION OF DEPOSIT	AMOUNT OF CHECK	✓	AMOUNT OF DEPOSIT	BALANCE	
		BALANCE BROUGHT FORWARD →				389	49
			47	29	✓		342	20
047	11/3	Country Pride			✓	400	00	742	20
	11/4	Deposit					704	85
048	11/9	General Telephone Co.	37	35	✓		654	85
	11/15	Cash–Auto Teller	50	00	✓			
	11/20	Deposit				191	37	846	22
049	11/20	Consumers Power Co.	44	89			801	33

DATE LAST STATEMENT
10/20/—

BALANCE LAST STATEMENT
389.49

CLOSING DATE
11/20/—

CHECKING ACCOUNT NUMBER   83-30283

CHECKS AND OTHER CHARGES			DEPOSITS AND CREDITS		BALANCE
DATE	NUMBER	AMOUNT	DATE	AMOUNT	
11/3	047	47.29	11/4	400.00	342.20
11/9	048	37.35			742.20
11/15	A.T.	50.00			704.85
SERVICE CHARGE		6.50			654.85
					648.35

PREVIOUS BALANCE	DEPOSITS RECORDED	CHECKS PAID	SER. CHG. OR INT.	PRESENT BALANCE
389.49	400.00	134.64	6.50	648.35

CHECK REGISTER BALANCE	$ 803.33	STATEMENT BALANCE	$ 648.35
INTEREST	+ 0	OUTSTANDING CHECKS #49 $44.89	− 44.89
SERVICE CHARGE	− 6.50		$
NEW BALANCE	$ 794.83		
		OUTSTANDING DEPOSITS $191.37	+ 191.37
** RECONCILIATION STATEMENT FOR YOUR CONVENIENCE		ADJUSTED BALANCE	$ 794.83

10. A Brief Case Jasper Henning received his bank statement and canceled checks for the month of May. He compared them against the items listed in his check register. He found the items shown below outstanding. Reconcile his bank statement.

CHECK NO.	DATE	CHECKS ISSUED TO OR DESCRIPTION OF DEPOSIT	AMOUNT OF CHECK	✓	AMOUNT OF DEPOSIT	BALANCE	
		BALANCE BROUGHT FORWARD →				202	09
						65	07
			137 02		421 17	486	24
						454	24
108	5/29	Computer World					
	6/1	Deposit	32 00			222	60
		Samsels Health Club	231 64			422	60
109	6/4	Hotel Soffetil			200 00	422	60
110	6/6						
	6/11	Deposit					

DATE LAST STATEMENT
5/1/—

BALANCE LAST STATEMENT
1147.62

CLOSING DATE
6/1/—

CHECKING ACCOUNT NUMBER 30-92854

CHECKS AND OTHER CHARGES			DEPOSITS AND CREDITS		BALANCE
DATE	NUMBER	AMOUNT	DATE	AMOUNT	
5/8	103	47.90	5/7	200.00	1347.62
5/11	105	241.03	5/11	100.00	1299.72
5/12	101	50.00	5/15	572.12	1158.69
5/12	102	75.02			1033.67
5/14	106	36.40			997.27
5/25	107	940.00			1569.39
5/27	104	427.30			629.39
			INTEREST	5.26	202.09
					207.35

PREVIOUS BALANCE	DEPOSITS RECORDED	CHECKS PAID	SER. CHG. OR INT.	PRESENT BALANCE
1147.62	872.12	1817.65	5.26	207.35

CHECK REGISTER BALANCE	$ 422.60	STATEMENT BALANCE	$ 207.35
INTEREST	+ 5.26	OUTSTANDING CHECKS	
SERVICE CHARGE	− 0	#108 $137.02	−
NEW BALANCE	$ _____	#109 $32.00	$ _____
		#110 $231.64	
		OUTSTANDING DEPOSITS	
		$421.17	+ _____
** RECONCILIATION STATEMENT FOR YOUR CONVENIENCE		$200.00	
		ADJUSTED BALANCE	$ _____

MAINTAINING YOUR SKILLS Look up the skill in parentheses if you need help or more practice.

Subtract. **(Skill 6)**

11. $412.30 − $1.25

12. $219.63 − $2.50

13. $96.78 − $3.00

14. $349.82 − $116.78

15. $491.60 − $247.83

16. $127.80 − $64.92

17. $421.92 − $250.00

18. $216.91 − $150.00

19. $97.83 − $51.47

Reviewing the Basics

Skills

Write the amount in words with cents expressed as a fraction of a dollar.

(Skill 1)

1. $25.79   **2.** $299.46

3. $1372.35   **4.** $14,551.00

Solve.

(Skill 5)

5. $81.27 + $99.64 + $81.27   **6.** $1408.34 + $57.84 + $2.57

7. $473.93 + $147.85   **8.** $8402.03 + $998.57

(Skill 6)

9. $416.37 − $57.02   **10.** $783.19 − $531.90

11. $527.05 − $315.37   **12.** $172.21 − $132.41

Applications

Use the following formula to solve problems 13–14.

Total Deposit = (Currency + Coins + Checks) − Cash Received

(Application A)

13. Deposit slip for Jill Dohr.
Check for $701.32.
Coins totaling $15.97.
Currency totaling $108.
What is the total deposit?

14. Deposit slip for Bill Shockey.
Check for $1321.17.
Currency totaling $3217.
Received a $50 bill.
What is the total deposit?

Terms

Match each term with its definition on the right.

15. balance

16. reconcile

17. deposit

18. statement

19. check

a. to obtain agreement between two financial records by accounting for outstanding items

b. an amount of money that you put into an account

c. the amount of money in your account

d. a written order directing the bank to deduct money from your checking account to make a payment

e. items that did not reach the bank in time to be processed and listed in your statement

f. a record prepared by the bank listing all transactions the bank has recorded

Refer to your reference files at the back of the book if you need help.

Unit Test

1. A portion of Cheryl Dodge's deposit slip is shown. What is her total deposit?

2. Herman Stone wrote a check to Acton Electronics in the amount of $1394.45 for the purchase of a micro-computer. The check number was 363. Write the amount of the check in words with cents expressed as a fraction of a dollar.

		DOLLARS	CENTS
CASH	CURRENCY		
	COINS	79	25
CHECKS	LIST SEPARATELY 9-14	143	85
	18-4	396	71
	SUBTOTAL		
	LESS CASH RECEIVED	80	00
	TOTAL DEPOSIT		

3. A portion of Ralph Ray's check register is shown listing his transactions since June 5. What is his new balance?

CHECK NO.	DATE	CHECKS ISSUED TO OR DESCRIPTION OF DEPOSIT	AMOUNT OF CHECK	✓	AMOUNT OF DEPOSIT	BALANCE
		BALANCE BROUGHT FORWARD →			827	31
	6/5	Deposit			418	90	
082	6/9	Miller's Hardware	83	97		
083	6/18	Corner Drug	44	96		

4. A portion of Marie Dunn's bank statement is shown. Her balance from last month's statement was $491.34. What is her present balance?

CHECKS AND OTHER CHARGES			DEPOSITS AND CREDITS		BALANCE
DATE	NUMBER	AMOUNT	DATE	AMOUNT	
10/4	902	42.95	10/8	335.35	
10/6	904	31.21	10/22	391.61	
10/15	903	285.00			
SERVICE CHARGE		8.35			

5. Marie Dunn's statement balance is $952.64. She finds that she has check numbers 901 and 905 outstanding in the amounts of $47.27 and $59.23. She also has an outstanding deposit of $200. She reconciles the statement. What should her adjusted balance be?

A SPREADSHEET APPLICATION

N.O.W. Accounts

To complete this spreadsheet application, you will need the template diskette for *Mathematics with Businss Applications*. Follow the directions in the User's Guide to complete this activity.

Input the transactions for a N.O.W. account in the following problems to find the interest earned and the new balance. Assume the statement period begins on the first day and ends on the last day of the month.

1. Here is a portion of Alan Cook's bank statement. The previous balance is $827.80.

CHECKS AND OTHER CHARGES			DEPOSITS AND CREDITS		BALANCE
DATE	NUMBER	AMOUNT	DATE	AMOUNT	
6/9	301	24.50	6/10	150.00	
6/11	302	145.67	6/15	45.00	
6/15	303	9.86			

2. Here is a portion of Kelly Beck's bank statement. The previous balance is $248.68.

CHECKS AND OTHER CHARGES			DEPOSITS AND CREDITS		BALANCE
DATE	NUMBER	AMOUNT	DATE	AMOUNT	
7/12	101	54.50	7/15	250.00	
7/17	103	284.90	7/29	65.00	
7/24	105	38.78			

3. Here is a portion of Diana Roger's bank statement. The previous balance is $949.99.

CHECKS AND OTHER CHARGES			DEPOSITS AND CREDITS		BALANCE
DATE	NUMBER	AMOUNT	DATE	AMOUNT	
11/5	819	983.78	11/1	450.00	
11/23	820	421.99	11/19	1095.00	
11/28	823	18.23	11/28	120.50	

4. Here is a portion of R. J. Perez's bank statement. The previous balance is $1124.53.

CHECKS AND OTHER CHARGES			DEPOSITS AND CREDITS		BALANCE
DATE	NUMBER	AMOUNT	DATE	AMOUNT	
3/1	235	763.91	3/5	321.98	
3/6	238	21.80			
3/15	239	108.77			

5. Here is a portion of Fong Lo's bank statement. The previous balance is $596.89.

CHECKS AND OTHER CHARGES			DEPOSITS AND CREDITS		BALANCE
DATE	NUMBER	AMOUNT	DATE	AMOUNT	
2/7	522	224.50	2/10	355.16	
2/11	524	391.16	2/20	445.00	
2/22	521	31.68	2/25	40.00	
2/27	527	15.19			

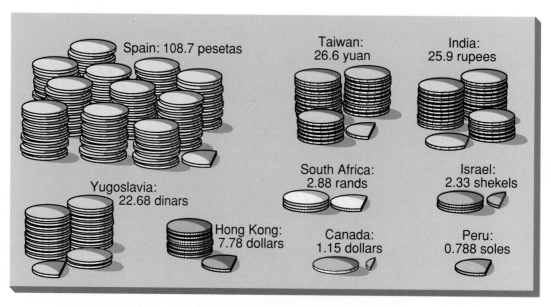

▼

CAREER WISE

$ $

Bank Teller

Colleen Miller is a bank teller at a local savings bank. At the bank, she handles customers' deposits and withdrawals. Some people bring their checks for their mortgage payments to her for processing. Many people come to her to purchase money orders and traveler's checks. Using the bank's computer, Colleen can easily provide customers with accurate statements about their current balances.

Colleen is comfortable with numbers, and so she finds it easy to reconcile the day's transactions at closing time.

Hired for her dependability, knowledge of banking, and courteous attitude, she hopes to become a head teller some day.

Many travelers purchase traveler's checks and convert some cash into foreign currency before going on a trip. The graphic below shows how nine different foreign currencies compared recently to one American dollar.

Spain: 108.7 pesetas
Taiwan: 26.6 yuan
India: 25.9 rupees
Yugoslavia: 22.68 dinars
South Africa: 2.88 rands
Israel: 2.33 shekels
Hong Kong: 7.78 dollars
Canada: 1.15 dollars
Peru: 0.788 soles

1. Complete: 108.7 pesetas has the same value as _____ shekels.

2. Of the monetary units above, which is the smallest relative to the American dollar?

3. Which monetary unit above is the closest in value to the American dollar?

4. How many dinars are equivalent to $150 in American money?

5. How much of an American dollar is 1 rupee?

Savings Accounts

A *savings account* is used for savings only. You cannot write checks on this account. Instead of a check register, you keep track of your deposits, withdrawals, and *interest* with a *passbook* or an *account statement.* Each time you deposit or withdraw money, your bank records the transaction in your passbook. The bank also records the interest earned on the money deposited in your savings account. Interest is money earned on the funds, or *principal,* in your account. The more often the interest is computed, or *compounded,* the more money your account will earn.

O U T L I N E

A savings account may pay a higher interest rate than a checking account.

Deposits

OBJECTIVE

Complete a savings account deposit slip and compute the total deposit.

To open a savings account, you must make a deposit. Each time you make a deposit, it is added to the balance of your account. You fill out on savings account deposit slips the cash and checks that you deposit. If you want to receive cash, you subtract the amount from the subtotal to find the total deposit amount.

Total Deposit = (Currency + Coins + Checks) − Cash Received

EXAMPLE | *Skills* 5, 6 *Application* A *Term* Deposit

Robert Cassidy has 28 one-dollar bills in currency, $7.27 in coins, and a check for $29.34 to deposit in his savings account. He wants to receive a twenty-dollar bill in cash. How much will he deposit?

S O L U T I O N

BRIDGETOWN BANK & TRUST

DATE *November 12, 19—*

Currency + Coins + Checks

ACCOUNT NUMBER

| 0 | 1 | 1 | 3 | 0 | 1 | 4 |

NAME *Robert Cassidy*

ADDRESS *18 Laurel Lane*

Bridgetown, CT 05120

		DOLLARS	CENTS
CASH	CURRENCY	28	00
	COINS	7	27
CHECKS	LIST SEPARATELY 14-6	29	34
	SUBTOTAL	64	61
	LESS CASH RECEIVED	20	00
	TOTAL DEPOSIT	44	61

SAVINGS ACCOUNT DEPOSIT

28 + 7.27 + 29.34 = 64.61 − 20 = 44.61

✔ SELF-CHECK

Complete the problems, then check your answers in the back of the book.

Find the total deposit.

1. Deposited currency of $37. Deposited checks for $40 and $86.

2. Deposited checks for $32 and $94. Received $25 in cash.

PROBLEMS

Find the subtotal and total deposit.

	3.	**4.**	**5.**	**6.**	**7.**
Deposits	$44.00	$76.00	$52.96	$180.81	$ 64.89
	8.35	9.27	39.75	115.35	39.57
	26.80	44.38			928.12
Subtotal	?	?	?	?	?
Less Cash Received	0	0	$30.00	$150.00	$ 20.00
Total Deposit	?	?	?	?	?

8. Sandi Spencer.
Deposits $74 in cash.
Deposits $3.95 in coins.
What is the total deposit?

9. Ernest McMahon.
Deposits $73.23 in currency.
Deposits a check for $124.17.
What is the total deposit?

10. Kenneth Hall.
Deposits a check for $335.28.
Deposits a check for $29.50.
Deposits $90 in cash.
What is the total deposit

11. Susan Allen.
Deposits a check for $823.40.
Deposits a check for $61.88.
Deposits $50 in cash.
What is the total deposit?

12. Hazel Bruot fills out the savings deposit form shown. What is her total deposit?

13. Joe Gryster deposited a check for $475.77 and another check for $94.26 in his savings account. He received $70 in cash. What was his total deposit?

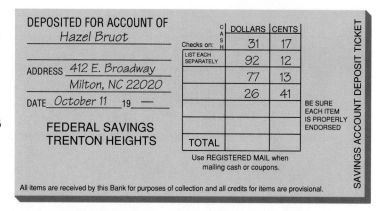

14. Norman Frances completed a savings account deposit slip on which he recorded checks for $327.19 and $52.88 for deposit. He received $38 in cash. What was his total deposit?

15. Barbara Hagedorn deposited 4 twenty-dollar bills, 9 ten-dollar bills, 35 quarters, 8 dimes, 97 pennies, and a check for $75.96 in her savings account. What was her total deposit?

16. David Rodero had 2 checks for $478.80 and $54.29. He would like to deposit the 2 checks and receive 4 twenty-dollar bills, 1 one-dollar bill, and 25 quarters. What would be his total deposit?

17. Loretta Miller would like to deposit 2 checks for $136.40 and $889.11. She would like to receive $75 in cash. What would be her total deposit?

MAINTAINING YOUR SKILLS   Look up the skills in parentheses if you need help or more practice.

Add. **(Skill 5)**

18. $31.50
   + 42.45

19. $40.46
   + 18.32

20. $173.79
   + 45.93

21. $551.16
   + 146.81

22. $89.70
   4.32
   + 26.92

Subtract. **(Skill 6)**

23. $98.93
   − 20.00

24. $692.57
   − 35.40

25. $103.33
   − 60.00

26. $687.28
   − 75.00

27. $68.25
   − 17.65

4-2

Withdrawals

OBJECTIVE
Complete a savings account withdrawal slip.

A **withdrawal** is a sum of money that you take out of your savings account. Your withdrawal is subtracted from the balance of your account. You fill out on withdrawal slips the money you are taking from your account.

EXAMPLE　　　　*Skills* 1　*Term* Withdrawal

Robert Cassidy would like to withdraw $45 from his savings account. His account number is 0113014. How should he fill out the withdrawal slip?

SOLUTION
A. Write the date of withdrawal.
B. Write the savings account number.
C. Write the amount withdrawn in words with cents expressed as a fraction of a dollar.
D. Write the amount withdrawn as a numeral.
E. Sign the withdrawal slip.

BRIDGETOWN BANK & TRUST

A. → DATE *November 14*　19– –

B. → | 0 | 1 | 1 | 3 | 0 | 1 | 4 |

SAVINGS ACCOUNT NUMBER

Forty-five and 00 / 100 ⌒ DOLLARS
C. ↗

D. ↘ $ | 45 | 00
dollars　cents

WITHDRAWAL

E. → *Robert Cassidy*
SIGN HERE

✔ SELF-CHECK　Complete the problems, then check your answers in the back of the book.

1. Write $45.76 in words with cents expressed as a fraction of a dollar.

2. Write two hundred ninety-one and $\frac{42}{100}$ dollars as a numeral.

PROBLEMS

Write each amount in words as it would appear on a withdrawal slip.

3. $17.35　**4.** $21.25　**5.** $44.93　**6.** $68.74

7. $406.00　**8.** $137.51　**9.** $7852.03　**10.** $9182.14

For each of the following exercises, write (a) the account number, (b) the amount as a numeral, and (c) the amount in words.

11. FARMER'S MERCANTILE — SAVINGS ACCOUNT NUMBER 1 7 5 9 4 1 7 9
DATE 5/19/— $ 831 95
PAY TO MYSELF
OR TO Home Finance Co.
_____ DOLLARS
AND CHARGE TO THE ABOVE NUMBERED ACCOUNT
SIGN HERE Calvin Gordon
WITHDRAWAL

12. FARMER'S MERCANTILE — SAVINGS ACCOUNT NUMBER 1 6 0 1 0 3 6 8
DATE Jan. 12, 19 — $ 374 28
PAY TO MYSELF
OR TO Odessa French
_____ DOLLARS
AND CHARGE TO THE ABOVE NUMBERED ACCOUNT
SIGN HERE Odessa French
WITHDRAWAL

13. Avis Bogart's savings account number is 81-0-174927. He fills out a savings withdrawal slip for $318.29 to purchase a gift.

14. Mona Hornack has been saving for a trip abroad. Her travel agent has arranged a trip to Europe that will cost $2460. Mona withdraws the amount from her savings account. Her account number is 13-122-541.

15. Mike Reed has been saving to buy a commemorative stamp for his stamp collection. He fills out the savings withdrawal slip shown.

16. Clyde Murphy has finally saved enough money to buy a video camera. The total purchase price is

NAME Mike Reed   DATE 11/8/19 —
ADDRESS 9 Bellvue Drive
Mission Hill, KA 58050
0 6 0 2 9 1 7 5 $ 76.60
SAVINGS ACCOUNT NUMBER   AMOUNT
RECEIVED THE SUM OF _____
CITIZEN SAVINGS AND TRUST
⑆041206261⑈   SIGNATURE Mike Reed
SAVINGS WITHDRAWAL

$947.31. He fills out a withdrawal slip for the amount. His savings account number is 7086-5.

MAINTAINING YOUR SKILLS Look up the skills in parentheses if you need help or more practice.

Write in words with cents expressed as a fraction of a dollar. **(Skill 1)**

17. $94.78   **18.** $219.34   **19.** $162.05   **20.** $15.71   **21.** $4211.15

Write as a numeral. **(Skill 1)**

22. Thirty-nine and $\frac{41}{100}$ dollars   **23.** Two hundred fifty-one and $\frac{27}{100}$ dollars

24. Six thousand three hundred forty and $\frac{22}{100}$ dollars.

25. Twenty-five thousand six hundred ninety-six and $\frac{29}{100}$ dollars

4-3

Passbooks

OBJECTIVE

Compute the new balance in a savings account passbook.

Your bank may provide you with a savings account passbook. When you make a deposit or withdrawal, a bank teller records in your passbook the transaction, any interest earned, and the new balance.

New Balance = Previous Balance + Interest + Deposits − Withdrawals

---

EXAMPLE   *Skills* 5, 6   *Application* A   *Term* Passbook

Ann Hardy has a passbook savings account. On April 1, she deposited $173.12 in her account. The bank teller recorded the transaction and the interest earned during the past savings period. What is the new balance in her account?

| DEPOSITOR: | Ann Hardy |
| ACCOUNT NO.: | 08-021235-4 |

	DATE	WITH-DRAWAL	DEPOSIT	INTEREST	BALANCE
18	01-01			9.75	793.90
19	01-12		250.00		1043.90
20	02-11		35.50		1079.40
21	03-03	30.00			1049.40
22	04-01			13.72	?
23	04-01		173.12		?
24					

SOLUTION   Find the **new balance.**

Previous Balance + Interest + Deposits − Withdrawals
$1049.40   + $13.72 + $173.12 −   0 = $1236.24 new balance

1049.4 + 13.72 + 173.12 − 0 = 1236.24

✔ SELF-CHECK   Complete the problems, then check your answers in the back of the book.

Find the new balance.

1. Previous balance, $900; interest, $11; deposit, 0; withdrawal, $60.

2. Previous balance, $3074; interest, $42.30; deposit, $75; withdrawal, 0.

PROBLEMS

	3.	**4.**	**5.**	**6.**	**7.**
Previous Balance	$481.32	$737.02	$81.31	$3019.20	$478.97
Interest	$ 6.80	$ 20.83	$ 1.77	$ 170.71	$ 27.08
Deposit	$ 80.00	—	$40.00	—	$110.00
Withdrawal	—	$145.00	$10.00	$ 921.00	—
New Balance	?	?	?	?	?

8. On November 20, Consuela Mendez withdrew $430 from her savings account. What is the new balance in her account?

DEPOSITOR: Consuela Mendez

DATE	WITHDRAWAL	DEPOSIT	INTEREST	BALANCE
10/2				3151.70
11/15			88.56	3240.26
11/20	430.00			?

9. On January 16, Keith Fair deposited $98.25 in his savings account. The teller also recorded $18.60 in interest to his account. What is the new balance in his account?

DEPOSITOR: Keith Fair

DATE	WITHDRAWAL	DEPOSIT	INTEREST	BALANCE
1/01				1000.00
1/05		240.25		1240.25
1/16		98.25	18.60	?

10. Find the balance for each date on this savings account passbook.

PASSBOOK ACCOUNT NUMBER: 07-17615

DATE	WITHDRAWAL	DEPOSIT	INTEREST	BALANCE
2/06				847.51
3/15		241.00		?
4/01			7.98	?
5/15	100.00			?
6/01		301.43		?
7/01			10.12	?
8/20		214.76		?
9/15	100.00			?
10/01			15.44	?
10/05	125.00			?
10/12	100.00			?
10/20		127.45		?
11/1	75.00			?

MAINTAINING YOUR SKILLS Look up the skills in parentheses if you need help or more practice.

Add. **(Skill 5)**

11. $350.00 + $88.00

12. $325.17 + $11.05

13. $858.60 + $314.58

14. $921.73 + $9.15 + $370.00

Subtract. **(Skill 6)**

15. $250 − $75

16. $66.97 − $12.50

17. $517.54 − $120.00

18. $815.35 − $140

19. $6114.98 − $39.73

20. $84,317.21 − $5829.37

Account Statements

When you have a savings account, your bank may mail you a monthly or quarterly account statement. The account statement shows all deposits, withdrawals, and interest credited to your account.

New Balance = Previous Balance + Interest + Deposits − Withdrawals

EXAMPLE Skills 5, 6 *Application* A *Term* Statement

Mary Witkew received this savings account statement. After checking her passbook and transactions to be sure all items have been recorded correctly, she checks the calculations. What is the balance in her account on July 1?

DEPOSITOR: Mary Witkew				
DATE	WITHDRAWAL	DEPOSIT	INTEREST	BALANCE
4-15		250.00		524.50
5-11		125.00		649.50
6-10	100.00			549.50
7-01			6.77	556.27
7-01		80.00		

PREVIOUS STATEMENT		THIS STATEMENT	
DATE	BALANCE	DATE	BALANCE
04-01	$274.50	07-01	?

SOLUTION Find the **new balance.**

Previous Balance + Interest + Deposits − Withdrawals

$274.50 + $6.77 + $455.00 − $100.00 = $636.27 new balance

274.5 + 250 = 524.5 + 125 = 649.5 − 100 = 549.5 + 6.77 = 556.27 + 80 = 636.27

✔ SELF-CHECK Complete the problem, then check your answers in the back of the book.

1. Previous balance, $700; interest, $9.50; deposits of $100 and $250; withdrawals of $80 and $110. What is the new balance?

PROBLEMS

Previous Balance +	Interest +	Deposits −	Withdrawals =	New Balance
2. $ 400.00 +	$19.00 +	$ 50.00 −	$150.00 =	?
3. $ 485.00 +	$19.50 +	$125.00 −	$200.00 =	?
4. $ 674.00 +	$12.19 +	$160.00 −	$190.00 =	?
5. $7381.19 +	$96.41 +	$231.43 −	$180.00 =	?

6. Judi Imhoff.
Previous balance: $717.52.
Interest: $4.36.
Deposits: $125 and $276.95.
Withdrawal: $90.
What is her new balance?

7. Scott Roan.
Previous balance: $2161.41.
Interest: $20.04.
Deposits: $345 and $575.80.
Withdrawals: $210 and $945.
What is his new balance?

8. Ruth Schuler.
Previous balance: $74,561.49.
Interest: $1017.98.
Deposits: $918.37 and $944.56.
Withdrawals: $959.40 and
 $14,391.47.
What is her new balance?

9. Louis Robinson.
Previous balance: $6817.86.
Interest: $60.41.
Deposits: $439.35 and $865.82.
Withdrawals: $235 and $4116.
What is his new balance?

10. Harry Karras checks the calculations on his savings account statement.
What is the balance in his account on September 30?

DEPOSITOR: Harry Karras			ACC. NO. 020-79-98	
DATE	WITHDRAWAL	DEPOSIT	INTEREST	BALANCE
7/09	314.00			421.34
8/30		217.50		638.84
9/29	310.00			328.84
9/30			6.10	

PREVIOUS STATEMENT		THIS STATEMENT	
DATE	BALANCE	DATE	BALANCE
06/30	$735.34	9/30	?

11. Ed Bogin received his savings account statement. What is the balance in his account on July 15?

NAME: Edward Bogin		ACCOUNT NUMBER: 12-36-5000		
DATE	WITHDRAWAL	DEPOSIT	INTEREST	BALANCE
01-28		45.00		548.27
02-03		80.40		628.67
02-15			2.85	631.52
03-15			2.86	634.38
04-10	400.00			234.38
04-15			2.37	?
05-01		335.60		?
05-15			2.28	?
06-15			2.61	?
07-15			2.62	?

PREVIOUS STATEMENT		THIS STATEMENT	
DATE	BALANCE	DATE	BALANCE
12-15	$503.27	07-15	?

12. Sylvia Road received her savings account statement. Find the balance for each date on her statement.

DEPOSITOR: Sylvia Road   ACC. NO. 10-257-190

DATE	WITHDRAWAL	DEPOSIT	INTEREST	BALANCE
7/01		21.43		953.00
7/07		123.88		1076.88
7/10	300.00			776.88
7/15		220.00	5.07	1001.95
7/22		35.15		?
8/01	180.00			?
8/07		220.00		?
8/10	50.00			?
8/15			3.45	?

PREVIOUS STATEMENT		THIS STATEMENT	
DATE	BALANCE	DATE	BALANCE
6/30	$931.57	8/15	?

13. Leona Biniakiewicz received her savings account statement. Find the balance for each date on her statement.

DEPOSITOR: Leona Biniakiewicz   ACC. NO. 16-86-1453

DATE	WITHDRAWAL	DEPOSIT	INTEREST	BALANCE
9/30		500.00	118.08	18,519.75
10/5		100.00		18,619.75
10/15	400.00			?
10/30			120.18	?
11/15	400.00			?
11/30		500.00	119.46	?
12/15	400.00			?
12/20	100.00			?
12/30			119.85	?

PREVIOUS STATEMENT		THIS STATEMENT	
DATE	BALANCE	DATE	BALANCE
9/30	$17,901.67	12/31	?

MAINTAINING YOUR SKILLS   Look up the skills in parentheses if you need help or more practice.

Add. **(Skill 5)**

14. $450.00 + $9.50 + $40.00   **15.** $385 + $7.52 + $875

16. $793.60 + $2.38 + $5.00   **17.** $426.30 + $278.41 + $342.91

18. $7812.39 + $37.90 + $527.84 + $477.91

Subtract. **(Skill 6)**

19. $7942.70 − $3453.80   **20.** $16,865.95 − $14,991.39

21. $338.49 − $299.39   **22.** $41,215.24 − $11,645.91

23. $3647.98 − $1647.08   **24.** $10,451.47 − $9.99

Simple Interest

OBJECTIVE

Compute the simple interest.

When you deposit money in a savings account, you are permitting the bank to use the money. The amount you earn for permitting the bank to use your money is called interest. The principal is the amount of money earning interest. The annual interest rate is the percent of the principal that you earn as interest based on one year. Simple interest is interest paid on the original principal.

Interest = Principal × Rate × Time

EXAMPLE   *Skills* 28, 14, 30   *Application* J   *Term* Simple interest

Joyce Tyler deposited $900 in a savings account at Hamler State Bank. The account pays an annual interest rate of $5\frac{1}{2}\%$. She made no other deposits or withdrawals. After 3 months, the interest was calculated.

A. How much simple interest did her money earn?

B. Suppose the account paid 6%. How much would she have earned?

SOLUTION

A. Find the **interest** at $5\frac{1}{2}\%$.
Principal × Rate × Time
$900.00 \times 5\frac{1}{2}\% \times \frac{3}{12}$
$900.00 × 0.055 × 0.25 =
$12.375 = $12.38 interest

B. Find the **interest** at 6%.
Principal × Rate × Time
$900.00 \times 6\% \times \frac{3}{12}$
$900.00 × 0.06 × 0.25 =
$13.50 interest

900 ☒ 5.5 ☒% 49.5 ☒ 3 ÷ 12 = 12.375 or 900 ☒ .055 ☒ .25 = 12.375

✔ SELF-CHECK   Complete the problems, then check your answers in the back of the book.

1. Principal: $400.
Annual interest rate: $6\frac{1}{2}\%$.
What is the interest after 3 months?

2. Principal: $1500.
Annual interest rate: $7\frac{1}{2}\%$.
What is the interest after 6 months?

PROBLEMS

	3.	**4.**	**5.**	**6.**	**7.**
Principal **Rate**	$720.00 × 0.06	$960.00 × 0.06	$327.00 × 0.12	$4842.00 × 0.10	$3945.37 × 0.065
	?	?	?	?	?
Time (Months)	$\times \frac{3}{12}$	$\times \frac{9}{12}$	$\times \frac{6}{12}$	$\times \frac{12}{12}$	$\times \frac{3}{12}$
Interest	?	?	?	?	?

8. Lillian Hanby's savings account.
$975 on deposit.
$5\frac{1}{2}\%$ annual interest rate.
Time is 3 months.
Find the interest.

9. Trevor Tyler's savings account.
$1342 on deposit.
$6\frac{1}{2}\%$ annual interest rate.
Time is 6 months.
Find the interest.

10. Ameil Trost's savings account.
$8460 on deposit.
$5\frac{3}{4}\%$ annual interest rate.
Time is 9 months.
Find the interest.

11. Norman Robert's savings
account.
$927.15 on deposit.
$8\frac{1}{2}\%$ annual interest rate.
Time is 5 months.
Find the interest.

12. Perry Wells deposited $760 in a new savings account at Maumee Savings and Loan Association. No other deposits or withdrawals were made. After 3 months, the interest was computed at an annual interest rate of $5\frac{1}{2}\%$. How much simple interest did his money earn?

13. Loretta Patterson deposited $429 in a new savings account at Henry County Bank and Trust. She made no other deposits or withdrawals. After 6 months, the interest was computed at an annual rate of $6\frac{3}{4}\%$. How much simple interest did her money earn?

14. On May 1, Harold Bothan opened a savings account at Fulton Savings Bank with a deposit of $1250. No other deposits or withdrawals were made. How much simple interest did his money earn by August 1?

$ FULTON SAVINGS BANK $

GOOD NEWS to our regular savings account customers!

$5\frac{1}{2}\%$

*Interest computed quarterly and paid quarterly.

*Effective May 1.

15. Vernon Taber deposited his $2000 scholarship money in a savings account at State Home Savings Bank on June 1. At the end of July, interest was computed at an annual interest rate of $6\frac{1}{2}\%$. How much simple interest did his money earn?

16. On March 1, Tessa Obee deposited a refund check for $9364.85 in a savings account at Napoleon Trust Co. At the end of August, interest was computed at an annual interest rate of $7\frac{1}{4}\%$. How much simple interest did her money earn?

MAINTAINING YOUR SKILLS Look up the skills in parentheses if you need help or more practice.

Change the fractions and mixed numbers to decimals. **(Skill 14)**

17. $\frac{1}{2}$ 18. $\frac{3}{4}$ 19. $\frac{1}{4}$ 20. $5\frac{1}{2}$ 21. $6\frac{1}{4}$ 22. $7\frac{3}{4}$

Write the percents as decimals. **(Skill 28)**

23. $5\frac{1}{2}\%$ 24. $6\frac{1}{4}\%$ 25. $9\frac{1}{2}\%$ 26. $7\frac{3}{4}\%$ 27. $10\frac{5}{8}\%$ 28. $5\frac{3}{8}\%$

Compound Interest

OBJECTIVE

Compute the compound interest and the amount.

Interest that you earn in a savings account during an interest period is added to your account. The new balance is used to calculate the interest for the next interest period. Compound interest is interest earned not only on the original principal but also on the interest earned during previous interest periods. The amount is the balance in your account at the end of an interest period.

Amount = Principal + Interest

EXAMPLE *Skills* 28, 14, 30 *Application* K *Term* Compound interest

John Heldt deposited $1800 in a savings account that earns 6% interest compounded quarterly. He made no deposits or withdrawals. What was the amount in the account at the end of the second quarter?

SOLUTION Find the **amount** for each quarter.

First Quarter: Interest = Principal × Rate × Time

Interest = $1800.00 × 6% × $\frac{1}{4}$ = $27.00

Amount = Principal + Interest

Amount = $1800.00 + $27.00 = $1827.00

Second Quarter: Interest = $1827.00 × 6% × $\frac{1}{4}$ = $27.41

Amount = $1827.00 + $27.41 = $1854.41

1800 M+ × 6 % 108 × .25 = 27 + RM 1800 = 1827 CM M+ × 6 %
109.62 × .25 = 27.405 + RM 1827 = 1854.405

✔ SELF-CHECK   Complete the problems, then check your answers in the back of the book.

1. $2000 is deposited at 8% compounded semiannually for 1 year. Find the amount after 1 year.

PROBLEMS

	2.	3.	4.	5.	6.
Principal	$900.00	$400.00	$2,360.00	$18,260.00	$27,721.00
Annual Interest Rate	6%	6%	$8\frac{1}{2}\%$	$7\frac{1}{2}\%$	9.513%
Interest Period	quarterly	monthly	semiannually	quarterly	annually
First Period Interest	$ 13.50	$ 2.00	$ 100.30	?	?
Amount	$913.50	$402.00	?	?	?
Second Period Interest	?	?	?	?	?
Amount	?	?	?	?	?

7. Alicia Martin's savings account.
Principal is $800.
6% interest compounded quarterly.
What is the amount in the account at the end of the second quarter?

8. Angelo Larragu's savings account.
Principal is $1200.
6% interest compounded quarterly.
What is the amount in the account at the end of the third quarter?

9. Elmer Pasture deposited $860 in a new regular savings account that earns 5.5% interest compounded semiannually. He made no other deposits or withdrawals. What was the amount in the account at the end of 1 year?

10. Jana Dejute deposited $1860 in a new credit union savings account on the first of a quarter. The principal earns 8% interest compounded quarterly. She made no other deposits or withdrawals. What was the amount in her account at the end of 6 months?

11. Betty and Sam Eassavore's savings account had a balance of $9544 on May 1. The account earns interest at a rate of 5.25% compounded monthly. What is the amount in their account at the end of August if no deposits or withdrawals were made during the period?

12. Ernie Boddy had $3620 on deposit at Suburban Savings Bank on July 1. The money earns interest at a rate of 6.5% compounded quarterly. What is the amount in the account on April 1 of the following year if no deposits or withdrawals were made?

13. The Vassillis opened a savings account with a deposit of $1800 on July 1. The account pays interest at 9% compounded semiannually. What amount will they have in their account on July 1 one year later?

14. Elaine Deck had $10,000 put into a trust fund for her on her 16th birthday. The trust fund pays $9\frac{1}{4}$% interest compounded annually. What will her trust fund be worth on her 21st birthday?

MAINTAINING YOUR SKILLS Look up the skills in parentheses if you need help or more practice.

Write the percents as decimals. **(Skill 28)**

15. $5\frac{1}{4}$%   **16.** $8\frac{1}{2}$%   **17.** $5\frac{3}{4}$%   **18.** 6.9%   **19.** 7.25%

Find the percentage. **(Skill 30)**

20. $950 × 8%   **21.** $760 × $6\frac{1}{2}$%   **22.** $3620 × 9%

23. $12,380 × $9\frac{1}{2}$%   **24.** $23,260 × $4\frac{1}{2}$%   **25.** $36,200 × 1%

Write the fractions as decimals. Round answers to the nearest hundredth. **(Skill 14)**

26. $\frac{3}{10}$ **27.** $\frac{4}{12}$ **28.** $\frac{5}{8}$ **29.** $\frac{3}{5}$

Compound Interest Tables

To compute compound interest quickly, you can use a compound interest table, which shows the amount of $1.00 for many interest rates and interest periods. To use the table, you must know the total number of interest periods and the interest rate per period.

AMOUNT OF $1.00			
Total Interest Periods	Interest Rate per Period		
	1.250%	1.375%	1.500%
1	1.01250	1.01375	1.01500
2	1.02515	1.02768	1.03022
3	10.3797	1.04182	1.04567
4	1.05094	1.05614	1.06136
5	1.06408	1.07066	1.07728
6	1.07738	1.08538	1.09344
7	1.09085	1.10031	1.10984
8	1.10448	1.11544	1.12649

Amount = Original Principal × Amount of $1.00

Compound Interest = Amount − Original Principal

EXAMPLE   *Skills* 11, 8   *Application* C   *Term* Compound interest

Farmers and Merchants Bank pays 6% interest compounded quarterly on regular savings accounts. Carol Whiteside deposited $3000 for 2 years. She made no other deposits or withdrawals. How much interest did Carol earn during the 2 years?

SOLUTION

A. Find the **total interest periods.**
2 years × 4 quarters per year = 8 total interest periods

B. Find the **interest rate per period.**
6% ÷ 4 quarters = 1.5%   interest rate per period

C. Find the **amount of $1.00** from the table for 8 periods (2 years).
1.12649

D. Find the **amount.**
Original Principal × Amount of $1.00
$3000.00   ×   1.12649   = $3379.47

E. Find the **compound interest.**
Amount   −   Original Principal
$3379.47 − $3000   = $379.47 compound interest

✔ SELF-CHECK   Complete the problems, then check your answers in the back of the book.

1. $2000 invested at 5.5% compounded quarterly for 2 years. Find the amount.

2. $4500 invested at 3% compounded semiannually for 2 years. Find the amount.

AMOUNT OF $1.00

Total Interest Periods	Interest Rate per Period						
	1.250%	1.375%	1.500%	2.750%	2.875%	3.000%	3.125%
1	1.01250	1.01375	1.01500	1.02749	1.02875	1.03000	1.03125
2	1.02515	1.02768	1.03022	1.05575	1.05832	1.06090	1.06347
3	1.03797	1.04182	1.04567	1.08478	1.08875	1.09272	1.09671
4	1.05094	1.05614	1.06136	1.11462	1.12005	1.12550	1.13098
5	1.06408	1.07066	1.07728	1.14527	1.15225	1.15927	1.16632
6	1.07738	1.08538	1.09344	1.17676	1.18538	1.19405	1.20277
7	1.09085	1.10031	1.10984	1.20912	1.21946	1.22987	1.24036
8	1.10448	1.11544	1.12649	1.24237	1.25452	1.26677	1.27912
9	1.11829	1.13078	1.14339	1.27654	1.29059	1.30477	1.31909
10	1.13227	1.14632	1.16054	1.31165	1.32769	1.34391	1.36031
11	1.14642	1.16209	1.17795	1.34772	1.36586	1.38423	1.40282
12	1.16075	1.17806	1.19562	1.38478	1.40513	1.42576	1.44666
13	1.17526	1.19426	1.21355	1.42286	1.44553	1.46853	1.49187
14	1.18995	1.21068	1.23175	1.46199	1.48709	1.51258	1.53849
15	1.20482	1.22733	1.25023	1.50219	1.52984	1.55796	1.58657
16	1.21988	1.24421	1.26898	1.54350	1.57382	1.60470	1.63615
17	1.23513	1.26132	1.28802	1.58595	1.61907	1.65284	1.68728
18	1.25057	1.27866	1.30734	1.62956	1.66562	1.70243	1.74000
19	1.26620	1.29624	1.32695	1.67438	1.71351	1.75350	1.79438
20	1.28203	1.31406	1.34685	1.72042	1.76277	1.80611	1.85045
21	1.29806	1.33213	1.36706	1.76773	1.81345	1.86029	1.90828
22	1.31428	1.35045	1.38756	1.81635	1.86559	1.91610	1.96791
23	1.33071	1.36902	1.40838	1.86630	1.91922	1.97358	2.02941
24	1.34735	1.38784	1.42950	1.91762	1.97440	2.03279	2.09283

PROBLEMS

Use the compound interest table above to solve. Round answers to the nearest cent.

	3.	4.	5.	6.	7.
Principal	$900.00	$640.00	$1340.00	$6231.40	$3871.67
Annual Interest Rate	5.5%	6%	5%	5.75%	18%
Interest Periods per Year	4	2	4	2	12
Total Time	2 years	1 year	3 years	5 years	6 months
Amount	?	?	?	?	?
Compound Interest	?	?	?	?	?

8. Principal is $936.
5.5% annual interest rate.
Compounded quarterly for
1 year.
What is the compound
interest?

9. Principal is $1236.35.
6% annual interest rate.
Compounded quarterly for
2 years.
What is the compound
interest?

10. National Savings and Trust pays 5.5% interest compounded semiannually on regular savings accounts. Iva Howe deposited $2800 in a regular savings account for 2 years. She made no other deposits or withdrawals during the period. How much interest did her money earn?

11. Alistair Russel deposited $900 in a savings plan with the employees' credit union where she works. The credit union savings plan pays 6% interest compounded quarterly. If she makes no other deposits or withdrawals, how much interest will her money earn in 1 year?

12. Dime Bank pays 6.25% interest compounded semiannually on special notice savings accounts. Jessie McKenzie deposited $3438.70 in a special notice savings account for $2\frac{1}{2}$ years. No other deposits or withdrawals were made during the period. How much interest did his money earn?

F.Y.I.

The compound interest formula is:

$A = P(1 + i)^n$

where P is the principal, i is the rate per period, n is the number of periods, and A is the amount.

13. Union Savings and Loan pays 5% interest compounded quarterly on regular savings accounts. Rose and Bob Yung had $4000 on deposit for 1 year. At the end of the year, they withdrew all their money and deposited $4000 at Volunteer Co-Operative Bank. How much more did the $4000 earn at Volunteer Co-Operative for 1 year?

VOLUNTEER CO-OPERATIVE BANK
announces new interest rate!

5.75%

on regular savings accounts.
INTEREST COMPOUNDED SEMIANNUALLY. MINIMUM OF $10 DEPOSIT REQUIRED.

14. Wilma Bracken opened a savings account at Dallas Trust Bank on March 1. Dallas Trust pays 15% compounded monthly on money left on deposit for at least 6 months. Wilma opened her account with an initial deposit of $10,000. What is the $10,000 worth on October 1? What did the $10,000 earn by October 1?

15. Maryann Radner opened a savings account at New England Savings and Loan on January 1 with a deposit of $800. New England Savings and Loan pays 11.5% compounded quarterly. What will the $800 be worth on January 1 of the following year? What will the $800 have earned by January 1 of the following year?

16. Martha and Mike Aieta deposited $4000 in a savings account in their child's name when their child was born. The $4000 earns interest at $6\frac{1}{4}$% compounded semiannually. What will the $4000 be worth on the child's 12th birthday?

MAINTAINING YOUR SKILLS   Look up the skill in parentheses if you need help or more practice.

Divide. **(Skill 11)**

17. $8\% \div 4$	**18.** $4\% \div 4$	**19.** $6\% \div 12$	**20.** $7\% \div 4$
21. $8.5\% \div 2$	**22.** $3.5\% \div 4$	**23.** $5.25\% \div 4$	**24.** $16.5\% \div 2$
25. $7\frac{1}{2}\% \div 4$	**26.** $6\frac{1}{4}\% \div 2$	**27.** $19\frac{1}{2}\% \div 12$	**28.** $11\frac{1}{2}\% \div 4$

4-8

Daily Compounding

OBJECTIVE

Find compound interest using tables.

Usually the more frequently interest is compounded, the more interest you will earn. Many banks offer savings accounts with daily compounding. When interest is compounded daily, it is computed each day and added to the account balance. The account will earn interest from the day of deposit to the day of withdrawal. A table can be used to calculate the amount and interest for daily compounding.

AMOUNT OF $1.00 AT 5.5% COMPOUNDED DAILY, 365-DAY YEAR			
Day	Amount	Day	Amount
21	1.00316	31	1.00468
22	1.00331	32	1.00483
23	1.00347	33	1.00498
24	1.00362	34	1.00513
25	1.00377	35	1.00528

Amount = Original Principal × Amount of $1.00

Compound Interest = Amount − Original Principal

EXAMPLE *Skill* 8 *Application* G *Term* Compound interest

On May 30, Deloris Zelms deposited $1000 in a savings account that pays $5\frac{1}{2}\%$ interest compounded daily. On July 1, how much interest had been earned on the principal in her account?

SOLUTION

A. Find the **number of days** from May 30 to July 1. Use the table on page 647.
July 1 is day 182. May 30 is day 151.
182 − 151 = 31 days

B. Find the **amount.**
Original Principal × Amount of $1.00
$1000.00 × 1.00468 = $1004.68

C. Find the **compound interest.**
Amount − Original Principal
$1004.68 − $1000.00 = $4.68 compound interest

✔ SELF-CHECK Complete the problems, then check your answers in the back of the book.

$6000 deposited at 5.5% compounded daily for 25 days.

1. Find the amount.

2. Find the compound interest.

Use the daily compound interest table on page 644 to solve. Round answers to the nearest cent. Interest is 5.5% compounded daily.

	3.	**4.**	**5.**	**6.**	**7.**
Principal	$80,000	$900	$6,500	$3,800	$15,321
Number of Days	25	31	50	90	120
Amount	?	?	?	?	?
Compound Interest	?	?	?	?	?

8. Welton Bower's savings account. $980 on deposit for 90 days. $5\frac{1}{2}$% annual interest rate. Interest compounded daily. What is the compound interest?

9. Donna Wayne's savings account. $4160 on deposit for 50 days. $5\frac{1}{2}$% annual interest rate. Interest compounded daily. What is the compound interest?

10. On June 10, Bertha Scisloski deposited $8241.78 in a savings account that pays $5\frac{1}{2}$% interest compounded daily. How much interest will the money earn in 31 days?

11. Marylin Robertson has a savings account that earns $5\frac{1}{2}$% interest compounded daily. On May 5, the amount in the account was $28,214.35. How much interest will the money earn in 90 days?

12. On April 11, Rob Walthall had $6521.37 in his savings account. The account pays $5\frac{1}{2}$% interest compounded daily. How much interest will the money earn by June 30?

13. On August 23, Richard McCormick had $1432.19 in his savings account at Camden Savings and Trust. The account earns $5\frac{1}{2}$% interest compounded daily. What will be the amount in his savings account when he closes his account on October 1?

14. Sandy Hane's savings account shows a balance of $904.31 on March 1. The same day, she made a deposit of $375 to the account. The bank pays interest at a rate of $5\frac{1}{2}$% compounded daily. What will be the amount in her account on May 10? on June 29?

Multiply. **(Skill 8)**

15. $4000 × 1.02131

16. $9000 × 1.00135

17. $550 × 1.00907

18. $1437 × 1.00392

19. $950 × 1.00392

20. $1416 × 1.01059

21. $7370 × 1.00347

22. $41,520 × 1.00407

23. $94 × 1.00196

24. $389 × 1.00301

25. $7925.14 × 1.01670

26. $327.78 × 1.00015

Reviewing the Basics

Skills

(Skill 1)

Write in words.

1. $479.00   **2.** $37.74   **3.** $3091.47   **4.** $15,263.59

Solve. Round answers to the nearest cent.

(Skill 5)

5. $193.74 + $18.31   **6.** $42,481.91 + $53.40 + $6.19

(Skill 6)

7. $871.95 − $136.99   **8.** $1400.00 − $519.87

(Skill 8)

9. 1.12051 × $6000   **10.** 1.104361 × $961.82

(Skill 10)

11. 8.5 ÷ 4   **12.** 7.5 ÷ 2   **13.** 6.74 ÷ 4   **14.** 9.5 ÷ 12

Write as a decimal.

(Skill 14)

15. $8\frac{1}{4}$ **16.** $4\frac{1}{2}$ **17.** $21\frac{3}{4}$ **18.** $7\frac{5}{8}$

(Skill 28)

19. $6\frac{3}{4}\%$ **20.** $4\frac{7}{8}\%$ **21.** 19.95%   **22.** 12.351%

Applications

(Application C)

Use the table on page 644 to find the amount of $1.00.

23. 9 interest periods.
Rate per period 1.5%.

24. 4 interest periods.
Rate per period 1.25%.

(Application G)

Find the elapsed time in days. Use the table on page 645.

25. From March 4 to May 31   **26.** From June 10 to July 30

Write as a fraction of a year.

(Application J)

27. 3 months   **28.** 6 months   **29.** 9 months   **30.** 18 months

(Application K)

Find the total number of interest periods.

31. Quarterly for 4 years   **32.** Semiannually for $2\frac{1}{2}$ years

Terms

Use each term in a sentence that explains the term.

33. Passbook   **34.** Withdrawal   **35.** Compound interest

36. Simple interest   **37.** Deposit   **38.** Statement

Refer to your reference files in the back of the book if you need help.

Unit Test

Lesson 4-1

1. George Pinewood deposited $54.21 in currency and a check for $384.19 in his savings account. What was his total deposit?

Lesson 4-2

2. Lynn Moyer has to withdraw $218.31 from her savings account. Write the amount as a numeral and in words.

Lesson 4-3

3. On September 13, Calvin Muir deposited $341.75 in his savings account. What is the new balance?

DEPOSITOR: Calvin Muir				
DATE	WITHDRAWAL	DEPOSIT	INTEREST	BALANCE
8-24				$415.80
9-13		341.75		?

Lesson 4-4

4. Susan Allard's savings account statement shows the following: a previous balance of $387.21, a withdrawal of $218.30, total deposits of $522.61, and interest of $1.23. What is the new balance?

Lesson 4-5

5. Rhea Franken deposited $419.31 in a savings account that pays an annual interest rate of $6\frac{3}{4}$ %. She made no other deposits or withdrawals. How much simple interest did her money earn in 3 months?

Lesson 4-6

6. Dan Dorsey deposited $981.40 in a regular savings account that earns 5.5% interest compounded quarterly. He made no other deposits or withdrawals. What is the amount at the end of the second quarter?

Lesson 4-7

7. Boston Savings Bank pays 6% interest compounded quarterly on regular savings accounts. Fred Pilliod deposited $1800 in a regular savings account for $1\frac{1}{2}$ years. He made no other deposits or withdrawals. Use the table to find the interest earned.

AMOUNT OF $1.00		
Total Interest Periods	Rate Per Period	
	1.375%	1.500%
4	1.05614	1.06136
5	1.07067	1.07728
6	1.08539	1.09344

Lesson 4-8

8. Irene Myers deposited $571.31 in a savings account that pays 5.5% interest compounded daily. How much interest will her money earn in 31 days?

AMOUNT OF $1.00 AT 5.5% COMPOUNDED DAILY, 365-DAY YEAR			
Day	Amount	Day	Amount
11	1.00165	31	1.00468
12	1.00180	32	1.00483
13	1.00196	33	1.00498

A SPREADSHEET APPLICATION

Compound Interest

To complete this spreadsheet applicaton, you will need the template diskette for *Mathematics with Business Applications.* Follow the directions in the *User's Guide* to complete this activity. (Then pick up copy pertinent to each unit's problem.)

1. Mike Oyer.
   Principal: $1000.
   Interest rate: 6%
   Periods per year: 4.
   Total time: 2 years.
   What is the amount?
   What is the interest?

2. Max Markley
   Principal: $10,000.
   Interest rate: 6.5%.
   Periods per year: 12.
   Total time: 1 year.
   What is the amount?
   What is the interest?

3. Helen Hallett.
   Principal: $5000.
   Interest rate: 7.5%.
   Periods per year: 4.
   Total time: 0.5 years.
   What is the amount?
   What is the interest?

4. Peggy Delaney
   Principal: $6540.
   Interest rate: 8%
   Periods per year: 2.
   Total time: 4 years.
   What is the amount?
   What is the interest?

5. Jean Deilman
   Principal: $568.65.
   Interest rate: 5.5%.
   Periods per year: 365.
   Total time: 2 years.
   What is the amount?
   What is the interest?

6. Angelo Diaz.
   Principal: $6987.25.
   Interest rate: 9.5%.
   Periods per year: 365.
   Total time: 0.25 years.
   What is the amount?
   What is the interest?

7. Elmer O'Neal deposited $40,000 in a savings plan paying 9.85% interest compounded monthly for 5 years. What is the amount? What is the interest?

8. Cheryle Contikos opened a $40,000 savings account at 9.85% interest compounded quarterly for 5 years. What is the amount? What is the interest?

9. Juan Carrero deposited $2000 in a passbook savings account paying 5.5% interest compounded quarterly for 2 years. What is the amount? What is the interest?

10. Elva and Paul Schrerez deposited $5000 in a 5-year certificate of deposit paying 15% interest compounded daily (365). How much interest did they earn?

11. Laticia Olrich deposited $1308.25 in a passbook savings account earning 5.5% interest compounded quarterly for 1 year. How much interest did she earn?

12. Maria and Randall Zapata deposited $50,000 in a 20-year certificate of deposit paying 6.5% interest compounded monthly. How much interest did they earn in 20 years?

5

Cash Purchases

A *sales receipt* is a proof of purchase showing the cost of your purchase and the *sales tax,* where applicable. To find the best buy based on price, you can use the *unit price,* or the cost per unit of an item. Many stores offer *coupons* and *rebates* as an incentive for consumers to purchase an item. Stores may also sell products at *sale prices,* which are lower than the regular selling price. The *markdown* is the difference between the regular selling price and the sale price.

O U T L I N E

5-1 Sales Tax

5-2 Total Purchase Price

5-3 Unit Pricing

5-4 Finding the Better Buy

5-5 Coupons and Rebates

5-6 Markdown

5-7 Sale Price

◆ Reviewing the Basics

◆ Unit Test

◆ A Spreadsheet Application: Cash Purchases

◆ Career Wise: Cashier

◆ Cumulative Review (Skills/ Applications) Units 1–5

◆ Cumulative Review Test Units 1–5

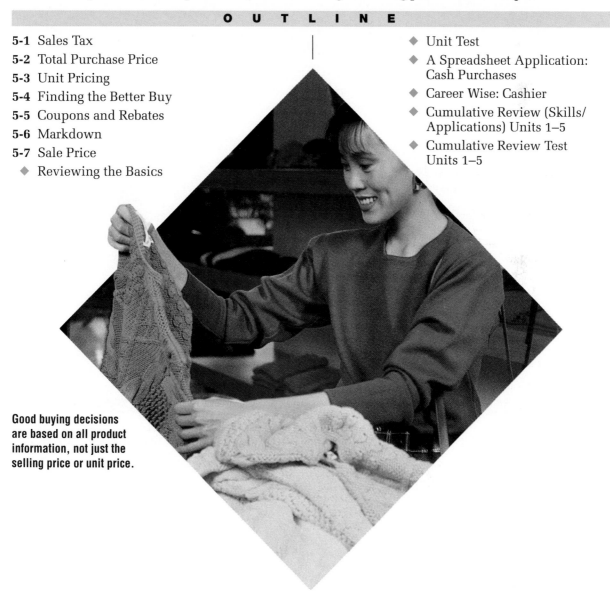

Good buying decisions are based on all product information, not just the selling price or unit price.

Sales Tax

OBJECTIVE
Compute the sales tax.

Many state, county, and city governments charge a sales tax on certain items or services that you buy. The sales tax rate is usually expressed as a percent.

Sales Tax = Total Selling Price × Sales Tax Rate

EXAMPLE   *Skills* 30, 2   *Application* A   *Term* Sales tax

Ron Engle bought a camera that had a selling price of $129.99 and 3 rolls of film that had a selling price of $18.95. What is the sales tax on his purchases if the tax rate is: (a) 4%? (b) 5.5%?

SOLUTION

A. Find the **total selling price.**
$129.99 + $18.95 = $148.94

B. Find the **sales tax.**
Total Selling Price × Sales Tax Rate

(1) $148.94 × 4%
$148.94 × 0.04 = $5.9576 = $5.96 sales tax
(2) $148.94 × 5.5%
$148.94 × 0.055 = $8.1917 = $8.19 sales tax

129.99 ⊞ *18.95* ⊟ *148.94* ⊠ *4* % *5.9576*

✔ SELF-CHECK   Complete the problems, then check your answers in the back of the book.

Find the sales tax.

1. Total selling price: $420.
   Sales tax rate: 5%.

2. Total selling price: $11,400.
   Sales tax rate: 4.5%.

PROBLEMS

	Item	Total Selling Price	×	Sales Tax Rate	=	Sales Tax
3.	Clothing	$ 81.39	×	3%	=	?
4.	Pillowcases	$ 22.99	×	4%	=	?
5.	Calculator	$ 21.75	×	5%	=	?
6.	Compact disc	$ 12.95	×	4.5%	=	?
7.	Computer	$995.00	×	6%	=	?
8.	Breakfast	$ 7.95	×	6.25%	=	?

Find the sales tax.

9.

```
PNAPL        .79
DAIRY        .89
SUGAR       1.39
CHEESE      1.69
SPAGHETTI   1.19
```

Tax rate: 5%

10.

```
13.95   ATLAS
 7.95   MAPS
 8.49   TOTE BG
29.95   BOOK
15.95   BOOK
19.95   BOOK
```

Tax rate: 4%

11.

```
PICTURE FRAME  $49.50
HOOKS           1.29
MATTING        11.45
MATTING        14.55
```

Tax rate: 6%

F.Y.I.
Alaska, Delaware, Montana, New Hampshire, and Oregon have no statewide sales tax.

12. Lloyd DeVoe purchases a $29.95 blanket, a $14.95 mini blind, an adjustable screen for $15.95, and a 12-foot extension cord for $9.49. The sales tax rate is 6%. What is the total selling price? What is the sales tax?

13. Wilber Haddad purchases a weed trimmer for $79, potting soil for $4.95, a 50-foot garden hose for $9.44, and a chaise lounge for $17.95. The sales tax rate is 5.5%. What is the sales tax?

14. Sally Hahn purchases the 13" color television shown. The sales tax rate is 4.5%. What is the sales tax?

13" Color Television
$199.99

15. At Alvin's Book Shop Rick Reimer purchases a record for $9.95, a dictionary for $21.50, a paperback book for $4.25, and a magazine for $4.00. The sales tax rate is 4.5%. What is the sales tax on his purchases?

16. Paula Ducat purchases a $109.95 phone answering machine, a $169.95 cordless phone, and a phone cord for $9.45. The sales tax rate is 6.5%. What is the sales tax on her purchases?

17. Fran Oates purchases a $25.95 leather handbag, a pair of running shoes for $37.95, and four polyester knit tops at $8.75 each. The sales tax rate is 3.75%. What is the sales tax on her purchases?

MAINTAINING YOUR SKILLS   Look up the skills in parentheses if you need help or more practice.

Find the percentage. **(Skill 30)**

18. 8% of $29.9   **19.** $4\frac{1}{2}$% of $13,840   **20.** 120% of $90   **21.** $\frac{1}{2}$% of $240

Round to the place value indicated. **(Skill 2)**

Nearest thousand	**22.** 85,713	**23.** 4139	**24.** $31,234	**25.** 2,436,541
Nearest ten	**26.** 435	**27.** 35,229	**28.** 94	**29.** 1131
Nearest cent	**30.** $34.125	**31.** $279.5429	**32.** $357.346	**33.** $2.8556
Nearest tenth	**34.** 37.614	**35.** 357.982	**36.** 131.025	**37.** 521.681

Total Purchase Price

OBJECTIVE

Compute the total purchase price.

Most stores give you a sales receipt as proof of purchase. The sales receipt may be a handwritten sales slip or a cash register tape. The receipt shows the selling price of each item or service you purchased, the total selling price, any sales tax, and the total purchase price.

Total Purchase Price = Total Selling Price + Sales Tax

EXAMPLE Skills 8, 5 Application A Term Sales recepit

Mike Gustweller purchases the golf items listed on the sales receipt. He checks the information on the receipt. The sales tax rate is 5%. What is the total purchase?

SOLUTION

F.Y.I.
Note that $163.18 is 105% of $155.41.

WILEY'S PRO SHOP: 5th and Main Street

QUANTITY	DESCRIPTION	PRICE	AMOUNT
1	ProFlite 8-iron set	139 95	139 95
2	3-ball packages-golf balls	2 89	5 78
1	Deluxe golf cart cover	9 68	9 68
	PRICE PER ITEM	TOTAL PRICE	
	NUMBER PURCHASED		
	TOTAL SELLING PRICE		
	TOTAL PURCHASE PRICE	Subtotal	155 41
	SALES SLIP	Sales tax	7 77
	Customer's copy	Total	163 18

139.95 M+ 2 × 2.89 = 5.78 M+ 9.68 M+ RM 155.41 × 5 % 7.7705 + RM 155.41 = 163.1805

✔ SELF-CHECK Complete the problem, then check your answer in the back of the book.

1. Kate Farison bought a dome tent priced at $179, 2 coolers priced at $18 each, and a propane stove priced at $40. The sales tax rate is 2%. Find the total purchase price.

PROBLEMS

	Item	Total Selling Price	×	Sales Tax Rate	=	Sales Tax	Total Purchase Price
2.	Rubber cement	$ 1.65	×	5%	=	$0.08	?
3.	Note pad	$ 1.15	×	6%	=	$0.07	?
4.	Pencil/pen set	$23.95	×	4%	=	?	?

The sales tax rate is 6%. Find the total purchase price.

5.

```
2.49    GRAP
1.69    MILK
2.75    PROD
1.49    PROD
 .65    GROC
2.29    DELI
 .40    ONIO
```

6.

```
 .95    LETT
 .79    FLOU
2.15    BRED
 .52    MUSH
 .99    CRAN
1.15    RICE
2.29    ORAN
```

7.

```
BIRDSEED      9.94
THISTLES      8.95
FEEDER       27.45
SUET          1.39
SUNFLOWER    12.79
```

8. Sales slip for Anita Ewing.
Shirt for $11.99.
Tie for $9.95.
Sales tax of $1.21.
What is the total purchase price?

9. Sales slip for Carl Gonzales.
Briefcase for $59.99.
2 notepads at $0.59 each.
Sales tax of $2.45.
What is the total purchase price?

10. Sales slip for Aurelio Esparza.
Computer disk file for $14.95.
Surge protector for $19.95.
2 boxes of labels at $19.99 each.
Sales tax rate of 4%.
What is the total purchase price?

11. Sales slip for Norma Engel.
Microwave for $149.99.
Microwave cart for $119.95.
Microwave cookware for $19.95
Sales tax rate of 5.5%.
What is the total purchase price?

12. Ralph McDounagh purchases a $45.85 power saw and a $12.95 extension
cord. The sales tax rate is 6%. What is the total purchase price?

13. Gloria LeVect purchases the items
shown on the sales slip. The sales
tax rate is 2.5%. What is the total
purchase price?

```
2  SWEATSHIRTS    29.50 EA
2  SWEATPANTS      24.50 EA
3  TURTLENECKS      7.00 EA
```

14. Frank Norris purchases a 35-mm camera body for $299.97, a 50-mm lens
for $69.97, a tripod for $49.97, and 3 rolls of film at $2.95 each. The sales
tax rate is 5.4%. What is the total purchase price?

15. Tanya Burley purchases 6 cans of motor oil at $1.09 a can, 8 spark plugs
at $2.25 per plug, and an oil filter for $4.99. The sales tax rate is 5.5%.
What is the total purchase price?

16. At Big Discount Store, Liz Doyle purchases a 22-lb turkey at 99¢ per
pound, 6 oranges at 4 for $1.00, a $3.99-box of laundry detergent, a
gallon of milk for $1.69, a 99¢ frozen pie, and 6 cans of condensed soup
at 2 cans for 74¢. The sales tax rate is 3%. There is no sales tax charged
on food items. What is the total purchase price?

17. Luther Strik purchases the items shown on the sales slip. The sales tax rate is 5.225%. There is no sales tax charged on items of clothing. What is the total purchase price?

```
1 Calfskin Golves $36.95

1 Sunglasses        34.95

2 Pair Socks         7.95 EA

2 Flashlights        8.95 EA

2 Lawnchairs        14.95 EA
```

18. Nancy Topok purchases a suit for $129.75, a blouse for $29, a pair of shoes for $46, and a matching bracelet and necklace for $33.95 each. She pays a state tax of 4% and a local tax of 2%. What is the total sales tax on her purchases? What is the total purchase price?

19. Jodi Lodyard purchased the camping items listed on the following sales receipt. The sales tax is 6%. Complete the sales receipt.

	SPORTLAND		
QUANTITY	DESCRIPTION	PRICE	AMOUNT
1	Propane cylinder	$ 29.98	?
1	Propane cooker	119.95	?
2	Pie cookers	11.98	?
3	Folding camp stools	8.98	?
2	Insect repellent	3.86	?
		Subtotal	?
		Sales tax	?
		Total	?

MAINTAINING YOUR SKILLS Look up the skills in parentheses if you need help or more practice.

Multiply. (Skill 8)

20. 7.4×9.2 **21.** 15.6×9.142 **22.** 0.91×0.004 **23.** 0.03×1.07

Add. (Skill 5)

24.	**25.**	**26.**	**27.**	**28.**
322.29	19.38	9.65	708.326	542.36
174.00	7.8	0.0024	57.942	400.50
+ 4.52	+ 0.007	+ 342.1	+ 9.888	+ 0.724

29.	**30.**	**31.**	**32.**	**33.**
414.6	37.3	33	475.86	416.015
8.05	84.9	2.45	861.2	513.02
+ 960	+ 260.05	+ 90.8	+ 994.9	+ 322.811

5-3

Unit Pricing

OBJECTIVE
Compute the unit price.

Grocery stores often give **unit price** information for their products. Shoppers can use this information to determine which size of a product is the better buy based solely on price. The unit price of an item is its cost per unit of measure or count, such as dollars per pound or cents per dozen.

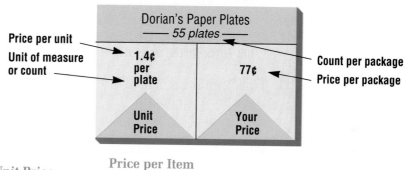

Unit Price $= \dfrac{\text{Price per Item}}{\text{Measure or Count}}$

EXAMPLE   *Skills* 11, 2   *Application* A   *Term* Unit price

Shane Burns purchased a 6.5-oz can of tuna for 89¢ and a dozen oranges for $3.96. What is the unit price of the items to the nearest tenth of a cent?

SOLUTION Find the **unit price.**
Price per Item ÷ Measure or Count
Tuna: $0.89 ÷ 6.5 = $0.1369 = 13.69¢ = 13.7¢ unit price per ounce

Oranges: $3.96 ÷ 12 = $0.33 = 33¢ unit price per orange

✔ SELF-CHECK   Complete the problems, then check your answers in the back of the book.

1. A 32-oz can of spaghetti sauce costs 99¢. Find the unit price per ounce to the nearest tenth of a cent.

2. A package of 8 hot dog buns costs 99¢. Find the price per bun to the nearest hundredth of a cent.

PROBLEMS

Round to the nearest tenth of a cent.

	Price per Item	÷	Measure or Count	=	Unit Price
3.	99¢	÷	12 oz	=	? per oz
4.	$ 7.99	÷	40 oz	=	? per oz
5.	$10.54	÷	16 m	=	? per m
6.	89¢	÷	2 L	=	? per L

Round to the nearest cent.

7. Roasted peanuts.
10 ounces for $1.89.
What is the price per ounce?

8. Soft drinks.
12 cans for $2.69.
What is the price per can?

9. 9-volt batteries.
2 per pack.
Sells for $1.99.
What is the price per battery?

10. Mustard.
10.5-oz jar.
Sells for $1.29.
What is the price per ounce?

11. Laura Zeedy recently purchased 4 tires for her automobile. The total purchase price was $178.44. What was the price per tire?

12. Alan Kinsman sees that Polaroid film is on sale for 2 rolls for $15.99. What is the price per roll?

13. Tania Ramirez purchases a 7.25-oz package of egg rolls for $1.09. What is the price per ounce?

14. A jar of barbecue sauce costs $1.29. There are 346 milliliters in the jar. What is the unit price to the nearest tenth of a cent?

15. Tracy Haserman purchases a package containing 3 video cassettes for $11.25. What is the unit price?

16. A 4-roll pack of bath tissue is priced at 2 packs for $1.75. What is the price per pack? What is the unit price?

17. What is the unit price for each item in the advertisement?
  a. Paper plates
  b. Paper towels
  c. Two cans of peaches
  d. Potatoes

Shopper's Specials

PAPER PLATES, 50 ct.. . . . . $1.49
PAPER TOWELS, 2 rolls $1.49
PEACHES, 59¢ ea. . . . . . 2 / $1.10
POTATOES, 5-lb bag. $0.98

MAINTAINING YOUR SKILLS Look up the skills in parentheses if you need help or more practice.

Divide. Round answers to the nearest hundredth. **(Skill 11)**

18. 190.50 ÷ 76 **19.** 13.4 ÷ 2.9 **20.** 27.218 ÷ 7.6 **21.** 0.91 ÷ 40

Round to the place value indicated. (Skill 2)

Nearest ten	**22.** 62.9	**23.** 19.8	**24.** 411.61	**25.** 1347.9
Nearest tenth	**26.** 21.78	**27.** 741.92	**28.** 2.348	**29.** 1132.562
Nearest hundredth	**30.** 3.218	**31.** 87.161	**32.** 430.144	**33.** 999.999

5-4

Finding the Better Buy

OBJECTIVE

Find the better buy based on unit price.

Compare the unit prices of products when you're deciding which size to buy. The size with the lower unit price is the better buy. If the larger size of a product is the better buy, consider whether you will be able to use up the product before it spoils.

EXAMPLE   *Skills* 11, 1   *Application* A   *Term* Unit price

Delman's raisins are sold in two sizes. The price of a 12-oz package is $1.29. The price of a 24-oz package is $1.99. Based on price alone, which package is the better buy?

SOLUTION

A. Find the **unit price** for each item.
Price per Item ÷ Measure or Count

12-oz package: $1.29 ÷ 12 = $0.1075 = 10.75¢
unit price per ounce

24-oz package: $1.99 ÷ 24 = $0.0829 = 8.29¢
unit price per ounce

B. Find the **better buy.**
Compare: 10.75¢ and 8.29¢ ◄— Better Buy

✔ SELF-CHECK   Complete the problems, then check your answers in the back of the book.

Find the better buy.

1. A 20-oz package costs 62¢.
A 15-oz package costs 54¢.

2. A box of 48 tea bags costs $1.99.
A box of 100 tea bags costs $2.39.

PROBLEMS

Find the unit price to the nearest tenth of a cent. Then determine the better buy based on price alone.

	Small Size	Large Size	Better Buy
3.	59¢ ÷ 16 oz = ?	89¢ ÷ 22 oz = ?	?
4.	65¢ ÷ 10 = ?	$6.70 ÷ 100 = ?	?
5.	$9.25 ÷ 14 oz = ?	$19.95 ÷ 30 oz = ?	?

6. Scouring pads.
  a. Box of 18 for $1.69.
  b. Box of 10 for $1.05.
  Which is the better buy?

7. Tortilla chips.
  a. 16-oz bag for $2.49.
  b. 12-oz bag for $1.75.
  Which is the better buy?

8. Alicia Murphy has a choice of two sizes of spray paint. The 16-oz can sells for $3.29 while a 13-oz can sells for $2.69. Which size is the better buy?

9. Jeff Evans wants to purchase some insect repellent. A 6-oz can sells for $3.86, while a 9-oz can sells for $5.59. Which size can is the better buy?

10. Kathy Kruse is shopping in the meat department of the grocery store and sees the turkey advertised. Which turkey is the better buy?

<table>
<tr><td colspan="2">HOLIDAY SPECIAL
This week only</td></tr>
<tr><td colspan="2">TURKEY</td></tr>
<tr><td>10 – 12lb
$14.25</td></tr>
<tr><td>18 – 20lb
$23.75</td></tr>
</table>

11. Nathan Ayers reads in the newspaper that his favorite cola is on sale at two stores in his neighborhood. The Country Store is selling a 24-can box for $5.99, while Food Mart is selling a 15-can box for $3.49. Which size box is the better buy?

12. Guiseppe Gerken wants to purchase some paint thinner. A 64-oz can costs $2.78, while a 16-oz can costs $1.09. Which is the better buy?

13. Kenny Chape's favorite peanut butter is available in the following sizes: a 12-oz jar for $1.69, an 18-oz jar for $2.19, a 28-oz jar for $3.49, and a 40-oz jar for $4.79. Which size jar is the best buy?

14. Wu Chong is buying mint tea bags. The brand he prefers is sold in boxes of 16 bags for 89¢, 48 bags for $1.99, 100 bags for $2.49, and 150 bags for $3.39. Which size box is the best buy?

15. The Co-Op Food Chain sells 4 sizes of a generic laundry detergent. The family size contains 25 pounds and costs $10.99. The giant size contains 14 pounds and costs $5.89. The regular size contains 6 pounds 15 ounces and costs $3.09. The small size contains 4 pounds and costs $1.99. What is the unit price of each? Which size is the better buy?

16. A Brief Case Heather Baker is shopping for the best buy of a brand of dog food. The dog food comes in the following sizes: a 2.27-kg bag for $2.29, a 4.54-kg bag for $3.19, and an 11.35-kg bag for $5.89. What is the unit price of each? Which size bag is the best buy?

MAINTAINING YOUR SKILLS Look up the skills in parentheses if you need help or more practice.

Divide. Round answers to the nearest hundredth. **(Skill 11)**

17. $2.3\overline{)49.16}$ **18.** $0.250\overline{)18.5}$ **19.** $0.006\overline{)0.8703}$ **20.** $24\overline{)328.20}$

Which number is greater? **(Skill 1)**

21. $1568 or $1586

22. $0.2326 or $0.2349

23. 5.1 or 5.01

24. $96 or $106

5-5

Coupons and Rebates

O B J E C T I V E

Compute the final price after using a coupon or rebate.

Many manufacturers, stores, and service establishments offer customers discounts through **coupons** and **refunds** or **rebates**. These special discounts are an incentive to purchase a particular item. Manufacturer's or store coupons are redeemed at the time of purchase. In order to obtain a manufacturer's rebate, the consumer must mail in a rebate coupon along with the sales slip and the UPC (Universal Product Code) label from the items.

Final Price = Total Selling Price − Total Savings

EXAMPLE   *Skill* 6   *Application* A   *Terms* Coupon, rebate

Gus Senton purchased a 32-ounce bottle of liquid detergent for $2.19. He had a coupon for $1.00 off. What is the final price of the detergent?

$100 **MANUFACTURER'S COUPON** $100
NO EXPIRATION DATE

SAVE
$1⁰⁰

on SAVON Liquid Detergent
Any Size
$100 $100

S O L U T I O N   Find the **final price.**
Total Selling Price − Total Savings
$2.19   − $1.00   = $1.19 final price

✔ SELF-CHECK   Complete the problems, then check your answers in the back of the book.

1. Cheese slices cost $2.09. Coupon for $1.00. Find the final price.

2. Trash bags cost $4.67. Coupon for 40¢. Find the final price.

Find the final price of each item.

	Item	Total Selling Price	Coupon	Final Price
3.	Fabric softener	$2.39	25¢	?
4.	Deodorant	$1.98	35¢	?
5.	Shampoo	$2.89	50¢	?
6.	Crackers	$1.89	50¢	?
7.	Freezer bags	$1.83	25¢	?
8.	Cleanser	$2.79	20¢	?

9. Audrey Bailey purchased a 22-ounce can of cherry pie filling for $1.79. She had a store coupon for 35¢. What is the final price of the pie filling?

10. Jonnie Marker purchased 3 boxes of crackers for $1.79 per box. He had two coupons, one for 30¢ and one for 25¢. What is the final price of the crackers?

11. Shawn Derricotte purchased an AM/FM headset radio (7-1625S) for $28.95. What is the price after the rebate?

12. Tammy Duffey purchased a personal stereo (3-5415) for $127.45. What is the price after the rebate?

13. Joel Jurcevich purchased a clock radio (7-4646) for $57.67. What is the price after the rebate?

14. An AM/FM headset radio (7-1625BLS) sells for $48.95. What is the price after the rebate?

15. A personal stereo (3-5432) sells for $84.60. What is the final price after the rebate if an envelope costs $0.15 and a stamp costs $0.29?

16. Tom Keener does his own service work on his car. He made use of the manufacturer's rebates. He purchased these items: 2 battery cables for $7.95 each, 6 spark plugs at $1.29 each, 1 can of carburetor cleaner for $1.19, 1 can of engine cleaner for $1.29, an air filter for $8.97, an oil filter for $3.99, a spark plug wire set for $14.58, and 6 quarts of oil at $0.99 a quart. Determine the total cost of the service work on the car. How much will Tom have paid after he receives the rebates?

Mail-in certificate (Not payable at the retail store)

CASH BACK BONUS REBATES

Clock Radio/Telephone

7-4712	$5.00
7-4712WH	$5.00

AM/FM Headset Radios

7-1625BLS	$3.00
7-1625S	$3.00
7-1625	$3.00

Personal Stereo

3-5415	$3.00
3-5415S	$3.00
3-5432	$3.00
3-5432S	$3.00

BONUS REBATES

25¢ off on Spark Plugs

50¢ off on Automotive Chemicals:
Gas Treatment
Oil Treatment
Carburetor Cleaner
Power Steering Fluid
Engine Cleaner
Brake Fluid

$1.00 off on Filters
(Oil Filter or Air Filter)

$3.00 off on Spark Plug Wire Sets

MAINTAINING YOUR SKILLS Look up the skill in parentheses if you need help or more practice.

Subtract. **(Skill 6)**

17.	**18.**	**19.**	**20.**	**21.**
$1.65	$11.65	$1.79	$127.98	$1.01
− .35	− 3.00	− .25	− 5.00	− .15

5-6

Markdown

OBJECTIVE
Compute the dollar amount of the markdown.

Stores often sell products at sale prices, which are lower than their regular selling prices. The markdown, or discount, is the amount of money that you save by purchasing a product at the sale price.

The markdown rate, or discount rate, of an item is its markdown expressed as a percent of its regular selling price. Businesses frequently advertise the markdown rate rather than the sale price.

Markdown = Regular Selling Price − Sale Price

Markdown = Regular Selling Price × Markdown Rate

EXAMPLE 1 *Skill* 6 *Application* A *Term* Markdown

Nora Maag purchased a VCR at a sale price of $399.99. The regular selling price is $549.99. What was the markdown?

SOLUTION Find the **markdown.**
Regular Selling Price − Sale Price
　　$549.99　　　　−　　$399.99 = $150.00 markdown

EXAMPLE 2 *Skills* 30, 2 *Application* A *Term* Markdown rate

Furniture World is selling all summer furniture for 35% off the regular selling price. Chad Wolfrum purchased furniture there that regularly sold for $799. How much did he save by buying the furniture during the sale?

> **END OF SEASON SALE!**
> *This month only*
>
> Save 35% on all
> Summer Furniture
>
> **FURNITURE WORLD**

SOLUTION Find the **markdown.**
Regular Selling Price × Markdown Rate
　　$799.00　　　×　　　35%　 = $279.65 markdown

✔ **SELF-CHECK** Complete the problems, then check your answers in the back of the book.

1. The regular selling price of a camcorder is $999.95. The sale price is $419.95. Find the markdown.

2. Shoes are on sale for 40% off the regular selling price of $89.50. Find the markdown.

	Item	Regular Selling Price	−	Sale Price	=	Markdown
3.	Table	$169.99	−	$99.99	=	?
4.	Shirt	$ 16.99	−	$11.99	=	?
5.	Handbag	$ 25.99	−	$15.50	=	?

	Item	Regular Selling Price	−	Sale Price	=	Markdown
6.	Cookware	$19.99	×	40%	=	?
7.	Toaster	$12.99	×	23%	=	?
8.	Lamp	$80.00	×	35%	=	?
9.	Humidifier	$49.99	×	$\frac{1}{3}$ off	=	?

10. Computer software. Regularly sells for $74.95. Sale price is $59.95. What is the markdown?

11. Mini van. Regularly sells for $14,495. Sale price is $11,595. What is the markdown?

12. A pair of dress pants. Regularly sells for $87.99. Markdown rate is 45%. What is the markdown?

13. Plastic cups. Regularly sell for $2.29. Markdown rate is 25%. What is the markdown?

14. Bargains on games usually occur in January. Angie Sigler purchases a backgammon set that regularly sells for $39.95. The set is on sale for $24.95. What is the markdown?

15. April is usually a good month to purchase home improvement supplies. A sheet of cedar paneling at Wetzler Lumber Yard regularly sells for $38.50. Jud Stassi purchases the paneling on sale for $24.95 a sheet. What is the markdown?

16. A 27″ color television that regularly sells for $899.95 is marked down 33% during McSurley's May Sale. What is the markdown?

17. Tori Topps sees a winter coat that regularly sells for $174.95 on sale at 50% off. How much can she save by purchasing the coat during the sale?

18. TBA Auto Parts holds a sale on car care products. All polishing cloths and cellulose sponges are marked down 35%. What is the markdown on polishing cloths that regularly sell for 90¢ each? What is the markdown on cellulose sponges that regularly sell for $1.35 each?

19. Tires are generally on sale during the month of September. Melanie Peters purchased 4 of the glass-belted tires shown on sale. What was the total markdown on her purchase?

Ojai Tire Center

A78-13 Glass-belted
****** RADIALS ******

reg. $49.95 per tire

SALE PRICE $33.95 ea.

20. Typewriter sales often occur in the month of June. Pitre Office Supplies has 3 electronic typewriters Model 118E remaining in stock. The Model 118E regularly sells for $835 but is on sale for $585. What is the markdown?

21. Gift items, such as the ones shown, are frequently discounted after the holidays in December and January. What is the markdown for each item advertised?

CRYSTAL SERVING ACCESSORIES
50% off regular prices

Salt and pepper set, reg. $14.00

Sugar and creamer, reg. $27.50

Handled mugs, set of 4, reg. $29.50

Jam jar with spoon, reg. $12.50

3-pc. salad set, reg. $38.99

22. Jeremy Berkowitz purchases a down jacket during an April coat sale at Westside Ski Shop. All coats are discounted 35%. The down jacket regularly sells for $74.50. What is the markdown?

23. January, May, July, and August are months in which many department stores hold "white sales." Use the advertisement to find the markdown on each item. Then compare the results with a markdown rate of 30% off the regular selling price on each item. Which results in a larger markdown?

August White Sale

Full flat or fitted sheets
Reg. $6.99 . . . . 2 / $9.75
Queen flat or fitted sheets
Reg. $12.99 . . . . 2 / $18.00
King flat or fitted sheets
Reg. $19.99 . . . . 2 / $28.00
Standard cases,
Reg. $5.99 . . . . $7.99 pr.
King cases
Reg. $7.99 . . . . $5.59 pr.

MAINTAINING YOUR SKILLS   Look up the skills in parentheses if you need help or more practice.

Subtract. **(Skill 6)**

24.   401.44
       −   98.25

25.   79.7
       −   9.924

26.   5.6
       − 4.6301

27.   73.291
       − 62.824

Find the percentage. **(Skill 30)**

28. 30% of $120   **29.** 40% of $280   **30.** $\frac{1}{4}$ % of 850   **31.** 230% of 550

Sale Price

OBJECTIVE

Compute the sale price when markdown rate is known.

For items that are on sale, some stores advertise the amount of markdown and the regular selling price. If you know this information, you can calculate the sale price.

Sale Price = Regular Selling Price − Markdown

EXAMPLE   *Skills* 30, 6   *Application* A   *Term* Sale price

F.Y.I.
Note that 38% off implies that you pay 62%. Thus, 62% of $47.95 is $29.73.

Debralee Medford is purchasing this clock radio at Highland's Hi-fi. What will she pay for the radio?

Dual-Alarm Clock Radio

Cut 38% Reg. 47.95

SOLUTION

A. Find the **markdown.**
Regular Selling Price × Markdown Rate
    $47.95   ×   38%   = $18.221 = $18.22 markdown

B. Find the **sale price.**
Regular Selling Price −   Markdown
    $47.95   − $18.22   = $29.73 sale price

47.95 [M+] [×] 38 [%] 18.221 [M−] [RM] 29.729

✔ SELF-CHECK   Complete the problems, then check your answers in the back of the book.

1. The regular selling price of a skirt is $45. The markdown rate is 30%. Find the sale price.

2. The regular selling price of a bicycle is $219. The markdown rate is 20%. Find the sale price.

PROBLEMS

	Regular Selling Price	× Markdown Rate	= Markdown	Sale Price
3.	$120.00	× 40%	= ?	?
4.	$ 79.90	× 30%	= ?	?
5.	$ 10.99	× 25%	= ?	?
6.	$175.00	× 15%	= ?	?
7.	$450.00	× 65%	= ?	?

8. Rechargeable lantern. Regularly sells for $21.95. Markdown rate is 30%. What is the sale price?

9. Compact stereo. Regularly sells for $129.95. Markdown rate is 53%. What is the sale price?

10. Amity's Shoe Store has marked down men's work and outdoor boots 25% during its spring sale. What is the sale price of a pair of boots with a regular price of $49.99?

25% OFF
ALL MEN'S WORK and OUTDOOR BOOTS

Regularly
14.99 – 49.99

11. During an August pre-school sale, Casual Apparel, Inc. has marked down its line of dress shirts 30%. These shirts are regularly priced at $17.95 each. What is the sale price of these shirts?

12. The Wyoming Pizza Shack discounts 15% off the regular selling price on the purchase of 3 or more pizzas. What is the sale price on the purchase of 3 pizzas that regularly sell for $5.75, $7.75, and $9.95?

13. Owen's Discount Store placed this advertisement in the local papers. Verify that you save 29%.

19.79 Save 29%

Our 27.88 **Rechargeable Hand Vac** with crevice tool, comfort-grip handle

14. A Brief Case   Ruiz's Furniture Market carried this ad for recliner rockers. Determine the regular selling price. Determine the percent markdown.

Recliners
Sale ▽ **249**⁹⁹

SAVE $200

MAINTAINING YOUR SKILLS   Look up the skills in parentheses if you need help or more practice.

Find the percentage. Round answers to the nearest hundredth. **(Skill 30)**

15. 25% of $74.80   **16.** 7.5% of 425   **17.** 8% of 322   **18.** 35% of 82.5

Subtract. **(Skill 6)**

19.	**20.**	**21.**	**22.**	**23.**
471.814	37	0.016	$138.26	$22.85
− 389.008	− 26.321	− 0.0019	− 125.18	− 18.36

Reviewing the Basics

Skills

(Skill 1)

Write the number that is smaller.

1. 14.4¢ or 15.2¢   **2.** 50.01¢ or 50.1¢   **3.** 17.25¢ or 17.3¢

(Skill 2)

Round to the nearest cent.

4. $12,2806   **5.** $1.723   **6.** 49.36¢   **7.** 421.71¢

Solve.

(Skill 5)

8. $4.49 + $2.57 + $26.08   **9.** $139.15 + $12.20 + $127.85

(Skill 6)

10. $86.50 − $18.59   **11.** $103.71 − $32.51   **12.** $75 − $1.07

(Skill 8)

13. $4.69 × 8   **14.** $45.60 × 24   **15.** $74.27 × 1.04

(Skill 11)

16. 95¢ ÷ 16   **17.** 85¢ ÷ 12   **18.** $4.32 ÷ 10

(Skill 30)

19. 18% of $18.95   **20.** 3.2% of $60   **21.** 8.5% of $372.80

Applications

Use the following formula to find the markdown.

Markdown = Regular Selling Price × Markdown Rate

(Application A)

22. Sewing machine.
Regular price $479.95.
Markdown rate of 30%.

23. TV set.
Regular price $799.95.
Markdown rate of 35%.

Terms

Match each term with its definition on the right.

24. Sales tax

25. Sales receipt

26. Unit price

27. Markdown

28. Markdown rate

29. Sale price

a. the amount of money that is saved by buying an item at the sale price

b. the purchase price of an item after markdown

c. the total purchase price including sales tax

d. the markdown of an item expressed as a percent of the regular selling price

e. a cash register tape or sales slip showing the selling price of items purchased, any sales tax, and the total purchase price

f. the cost per unit measure or count of an item

g. a tax charged on the selling price of an item or service provided

Refer to your reference files at the back of the book if you need help.

Unit Test

Lesson 5-1

1. Debra Rothchild purchases a garden hose for $12.95, 5 packs of seeds at $0.59 each, and a flat of flowers for $7.25. The sales tax rate is 6%. What is the sales tax on her purchases?

Lesson 5-2

2. Jack Roth purchases 6 cans of oil at 99¢ a can, 8 spark plugs at $2.25 each, a $7.49 air filter, and an oil filter for $4.89. The sales tax rate is 4%. What is the total purchase price of his purchases?

Lesson 5-3

3. A brand of salad dressing that Sandy Beckman prefers is on sale at Foodway. What is the unit price of each package?

Shop and Save at
FOODWAY
This week's specials:
Creamy Dressing
8-oz bottle for $1.29
12-oz bottle for $1.89

Lesson 5-4

4. Martin Gillis compares the prices for paper towels. A 3-roll package is priced at $2.89. A single roll is priced at $0.89. Based on price alone, which is the better buy?

Lesson 5-5

5. Jim Masters purchased 2 boxes of crackers for $1.99 per box. He had two coupons, one for 30¢ and one for 25¢. What is the final price of the crackers?

Lesson 5-5

6. A roll of film sells for $3.97. What is the price after a $1.00 rebate if an envelope costs 15¢ and a stamp costs 29¢?

Lesson 5-6

7. Mary Patterson purchases a CB radio on sale for $79.95. The regular selling price was $139.95. What is the markdown on the radio?

Lesson 5-6

8. Fuller's Bedland regularly sells a posture classic king-sized mattress for $309.95. For an August clearance sale, the mattress is discounted 30% off the regular selling price. How much is saved by purchasing the mattress during the sale?

Lesson 5-7

9. Bob Kenton purchases a jacket that regularly sells for $39.95. The jacket has been discounted 33%. What is the sale price of the jacket?

Lesson 5-7

10. A rechargeable flashlight regularly sells for $15.97. It is on sale for $11.97 and has a rebate of $2.00. What is the final price after the rebate if an envelope costs $0.15 and a stamp costs $0.29? How much is saved altogether?

A SPREADSHEET APPLICATION

Cash Purchases

To complete this spreadsheet application, you will need the template diskette for *Mathematics with Business Applications*. Follow the directions in the User's Guide to complete this activity.

Input the information in the following problems to find the sale price and total purchase price.

1. Item: chair.
   Selling price: $340.
   Percent markdown: 25%.
   What is the sale price?
   Sales tax rate: 6.5%.
   What is the total purchase price?

2. Item: shirt.
   Selling price: $15.25.
   Percent markdown: 10%.
   What is the sale price?
   Sales tax rate: 4%.
   What is the total purchase price?

3. Item: couch.
   Selling price: $568.99.
   Percent markdown: 20%.
   What is the sale price?
   Sales tax rate: 7.55%.
   What is the total purchase price?

4. Item: socks.
   Selling price: $3.65.
   Percent markdown: 33%.
   What is the sale price?
   Sales tax rate: 7.25%.
   What is the total purchase price?

5. Item: wallet.
   Selling price: $15.99.
   Percent markdown: 25%.
   What is the sale price?
   Sales tax rate: 6.50%.
   What is the total purchase price?

6. Item: jeans.
   Selling price: $24.99.
   Percent markdown: 15%.
   What is the sale price?
   Sales tax rate: 7%.
   What is the total purchase price?

7. Mary Zimmerman purchased a microwave oven that was marked down 20%. The regular price is $156 and the sales tax rate is 8.5%. What is the sale price? What is the total purchase price?

8. Cleo Voss purchased a computer that was marked down 30%. The regular price is $2345 and the sales tax rate is 6.5%. What is the sale price? What is the total purchase price?

9. A lawn mower regularly sells for $239.99. If it is currently marked down 17%, what is the sale price? If the sales tax rate is 6%, what is the total purchase price?

10. A washing machine regularly sells for $519.99. If it is currently marked down 23%, what is the sale price? If the sales tax rate is 6.50%, what is the total purchase price?

11. Carter Auto Sales advertised 10% off the list price of all automobiles. A station wagon has a list price of $15,550. The sales tax rate is 8%. What is the total purchase price?

12. Perkins Clothing Store advertised 45% off all winter clothing. Diane Peebles picked out several items with a total selling price of $435.42. The sales tax rate is 5.25%. What is the total purchase price?

13. An oak desk regularly sells for $589.99. If it is marked down 24%, what is the sale price? If the sales tax rate is 6.5%, what is the total price?

14. A curio cabinet regularly sells for $319.99. If it is marked down 14%, what is the sale price? The sales tax rate is 5.75%. What is the total purchase price?

15. Everhard's Appliance Store advertised 25% off the list price of all washing machines in stock. The most popular model has a list price of $639.99. The sales tax rate is 7%. What is the total purchase price?

16. Cassie's Health Foods advertised 10% off the list price of Choice Herbs. Sue Trundell selected items that totaled $122.43. The sales tax rate is 6.25%. What is the total selling price?

CAREER WISE

Cashier

In 1991 there were 47,489 new books and new editions published in the United States. This is no surprise to Tomas Ortiz who works as a cashier at a downtown bookstore. In a typical day, he will sell books for students, children's books, computer and technology books, cookbooks, craft books, and many other types of books.

Recently, he set aside a collection of books that had been marked down for clearance. Markdowns ranged from 25% to 40%. The paperback biographies were marked down the most and sold quickly.

As a cashier, Tomas uses a computerized cash register that automatically records the sale price, computes the tax, and gives the total purchase. The register also indicates the correct change. When the day is done, he reconciles the day's transactions with the starting and ending cash amounts.

The pie chart at the left below shows a percentage breakdown by subject area of the 47,489 new books and new editions for 1991. The chart at the right below shows how the academic books comprise the "academic slice" of the pie.

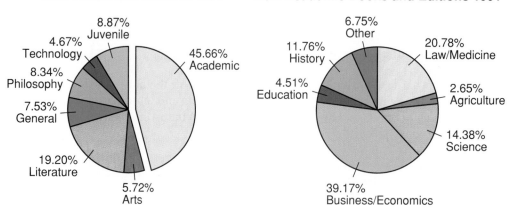

New Books and Editions 1991

8.87% Juvenile
4.67% Technology
8.34% Philosophy
7.53% General
19.20% Literature
5.72% Arts
45.66% Academic

New Academic Books and Editions 1991

6.75% Other
11.76% History
4.51% Education
20.78% Law/Medicine
2.65% Agriculture
14.38% Science
39.17% Business/Economics

1. Refer to the pie chart at the left above. In which category might you find "how-to" books?

2. Which are the two largest slices of the pie chart for academic books?

3. How do the two answers to question 3 compare?

4. Of the 47,489 new books and new editions in 1991, how many were academic? How many were science books?

Cumulative Review
(Skills/Applications) Units 1–5

Skills

(Skill 1)

Write the amount in words with cents expressed as a fraction of a dollar.

1. $47.59   **2.** $385.63   **3.** $2035.26

Write the number that is less.

4. 7.3¢ or 6.9¢   **5.** $321 or $298   **6.** $0.013 or $0.0129

Round to the nearest cent.

(Skill 2)

7. $33.175   **8.** $0.1456   **9.** $137.997   **10.** $4716.3042

Round to the nearest tenth of a cent.

11. 15.415¢   **12.** 8.573¢   **13.** 33.0192¢   **14.** 95.1775¢

Solve.

(Skill 4)

15. $96 − $74   **16.** $619 − $479   **17.** $8914 − $7105

18. $3734 − $1573   **19.** $10,913 − $315   **20.** $1117 − $927

(Skill 5)

21. $421.00 + $785.00   **22.** $47.69 + $13.87

23. $629.78 + $26.05 + $9.00   **24.** $3.67 + $12.19 + $0.89

(Skill 6)

25. $900.24 − $300.29   **26.** $321.83 − $200.44

27. $136.75 − $23.84   **28.** $461.90 − $322.21

29. $4325.60 − $190.25

(Skill 8)

30. $7.25 × 40   **31.** $9.50 × 36   **32.** $4.25 × 27.5

33. $0.99 × 0.207   **34.** $0.25 × 0.004   **35.** $3.70 × 0.06

(Skill 11)

36. $9.66 ÷ 24   **37.** $4.56 ÷ 12   **38.** $352.96 ÷ 52

39. $43.02 ÷ 0.32   **40.** $9.75 ÷ 0.25   **41.** $149.28 ÷ 0.75

(Skill 30)

42. 35% of $375.95   **43.** 3.5% of $9290

44. 33.5% of $2240   **45.** 8.13% of $11,800.50

46. $7\frac{3}{4}$% of $9600   **47.** $8\frac{1}{2}$% of $4968.15

(Skill 16)

48. $\frac{1}{2} + \frac{3}{4}$ **49.** $3\frac{1}{2} + 2\frac{1}{2}$ **50.** $8\frac{3}{4} + 6\frac{1}{2}$

Write as a decimal.

(Skill 14)

51. $\frac{1}{2}$ **52.** $\frac{1}{10}$ **53.** $25\frac{1}{4}$ **54.** $42\frac{3}{8}$

(Skill 28)

55. 7.25%   **56.** 37.30%   **57.** $12\frac{1}{2}$%   **58.** $45\frac{3}{4}$%

Use the following formula to find the markdown.

Markdown = Regular Selling Price × Markdown Rate

59. Luggage bag.
Regular price is $39.95.
Markdown rate is 25%.

60. Barbecue grill.
Regular price is $79.95.
Markdown rate is 40%.

61. Sports coat.
Regular price is $125.00.
Markdown rate is 20%.

62. Lawn mower.
Regular price is $219.95.
Markdown rate is 33%.

Use the table on page 644 to find the amount of $1.

63. 4 interest periods.
Rate per period is 2.875%.

64. 9 interest periods.
Rate per period is 3.000%.

65. 7 interest periods.
Rate per period is 2.875%.

66. 20 interest periods.
Rate per period is 1.500%.

Find the elapsed time.

67. From 9:15 a.m. to 11:55 a.m.

68. From 1:10 p.m. to 6:30 p.m.

69. From 7:52 a.m. to 5:20 p.m.

70. From 8:17 a.m. to 5:30 p.m.

Use the table on page 645 to find the elapsed time in days.

71. From May 7 to May 26

72. From June 2 to June 30

73. From April 15 to July 23

74. From March 10 to May 21

Write as a fraction of a year.

75. 3 months

76. 6 months

77. 10 months

78. 16 months

Find the number of occurrences.

79. Weekly for 3 years

80. Biweekly for 2 years

81. Monthly for $1\frac{1}{2}$ years

82. Semiannually for 4 years

Write you own definition for each term.

83. Straight-time pay

84. Overtime pay

85. Commission

86. Income tax

87. Social security

88. Group insurance

89. Sales tax

90. Unit price

91. Markdown

92. Sale price

93. Deposit

94. Reconcile

95. Withdrawal

96. Simple interest

97. Compound interest

98. Time card

99. Passbook

100. Piecework

Refer to your reference files in the back of the book if you need help.

Cumulative Review Test
Units 1–5

Lesson 1-2

1. Julius Harrison works as a truck driver and earns $9.40 an hour for a regular 40-hour week. His overtime rate is $1\frac{1}{2}$ times his regular hourly rate. This week he worked his regular 40 hours plus $7\frac{3}{4}$ hours of overtime. What is his total pay?

Lesson 1-5

2. Bonnie Webster is an accountant for Radio News Corp. She earns an annual salary of $42,312 paid on a semimonthly basis. What is her semimonthly pay?

Lesson 1-6

3. R. T. Sloan sells hospital equipment for a 9% straight commission. Her sales totaled $94,321 in September. What is her pay for the month?

Lesson 2-1

4. Royce Adill, a shipping supervisor, earns $448.50 a week. He is married and claims 2 allowances. How much is withheld from his weekly paycheck for federal income tax?

WEEKLY Payroll Married					
Wages		Allowances			
At least	But less than	0	1	2	3
420	430	55	49	43	37
430	440	56	50	44	39
440	450	58	52	46	40
450	460	59	53	47	42
460	470	61	55	49	43
470	480	62	56	50	45
480	490	64	58	52	46
490	500	65	59	53	48
500	510	67	61	55	49
510	520	68	62	56	51

Lesson 2-2

5. Julia Capper earns $32,564 a year as a systems analyst. The state income tax rate is 3.6% of taxable income. Her personal exemptions total $4692. How much is withheld each week from her gross pay for state income tax?

Lesson 2-4

6. Wilbur Clark, a medical technician, is paid $418.25 a week. His earnings to date this year total $15,057. The social security tax rate is 7.65% of the first $55,500 earned. How much is deducted from Wilbur's paycheck this week for social security tax?

Lesson 2-6

7. Cletus Carr, a research assistant for Winzler Chemicals, is married and claims no allowances. He earns $505.20 a week. The social security tax rate is 7.65% of the first $55,500 earned. The state tax is $10.15 a week. He has weekly deductions of $12 for medical insurance and $60 for payroll savings. Use the table above to find his federal tax withheld. What is his net pay for a week?

Lesson 3-2

8. Gary Webster wrote a check to Seaton Oil Co. in the amount of $337.19. The check number was 363. Write the amount of the check in words with cents expressed as a fraction of a dollar.

9. A portion of Betty Bigg's bank statement is shown. Her balance from last month's statement was $829.32. What is her present balance?

CHECKS AND OTHER CHARGES			DEPOSITS AND CREDITS		BALANCE
DATE	NUMBER	AMOUNT	DATE	AMOUNT	
10/4	902	41.94	10/8	347.90	
10/4	904	29.63	10/22	271.40	
10/15	903	237.00			
SERVICE CHARGE		4.50			

10. Betty Biggs compares her checkbook register, canceled checks, and bank statement. She finds that she has check numbers 901 and 905 outstanding in the amounts of $41.83 and $95.92. She also has an outstanding deposit of $400. She reconciles the statement. What should her adjusted balance be?

11. Viola Eyster's savings account statement shows the following: a previous balance of $913.82, a withdrawal of $241.93, total deposits of $518.25, and interest of $5.12. What is the new balance?

12. Esther Marshal deposited $3472.50 in an account that pays an annual interest rate of $6\frac{1}{2}\%$. She made no other deposits or withdrawals. How much simple interest did Beverly's money earn in 6 months?

13. Swanton Savings Bank pays $5\frac{1}{2}\%$ interest compounded quarterly on regular savings accounts. Willard Boehler deposited $4400 in a regular savings account for $1\frac{1}{2}$ years. He made no other deposits or withdrawals. Use the table to find the interest earned.

Total Interest Periods	AMOUNT OF $1.00 Rate Per Period	
	1.375%	1.500%
4	1.05614	1.06136
5	1.07067	1.07728
6	1.08539	1.09344

14. Michael Noll purchases 6 cans of soup at 77¢ a can, 3 pounds of green beans at 89¢ a pound, a $3.49-box of laundry detergent, and a package of chicken for $6.74. The sales tax rate is 6.5%. What is the total purchase price?

15. Sue Duffey compares the prices for fabric softener dryer sheets. A box of 20 is priced at $0.89. A box of 40 is priced at $1.59 and a box of 60 is priced at $2.29. Based on price alone, which is the better buy?

16. Conner Appliance regularly sells a Model K2504E 25″ console color television for $849.99. For an August clearance sale, the Model K2504E is discounted 40% off the regular selling price. How much is saved by purchasing the television during the sale?

Charge Accounts and Credit Cards

A *charge account* allows you to "buy now and pay later." You pay for goods or services with a *credit card*. At the end of a month, the credit company bills you for all the items you purchased on credit. This account *statement* shows any *finance charges* you must pay if your previous bill was not paid in full. A finance charge is interest charged on the amount you owe. Finance charges are figured by the *previous-balance method,* the *unpaid-balance method*, or the *average-daily-balance method.*

O U T L I N E

Credit cards can be used to purchase consumer goods.

Sales Receipts

OBJECTIVE
Compute the total purchase price.

When you make a purchase with a credit or charge card, the salesclerk prepares a **sales receipt.** The receipt shows your name and account number, the price of each item you purchased, the sales tax, and the total purchase price. You sign the sales receipt and receive a copy for your records.

ECONO-CHARGE

1212 430 235 170 19

GOOD THRU
LAST DAY OF   10/93

X002250
MARY M. KEY

Total Purchase Price = Total Selling Price + Sales Tax

EXAMPLE   *Skill* 30 *Application* A   *Term* Sales receipt

Martha Palmer purchased the items listed on the sales receipt. The sales tax rate is 4%. Is the total purchase price correct?

SOLUTION

101 04076

Martha's account number
Martha Palmer

D & L DEPARTMENT STORE

The issuer of the card identified on this item is authorized to pay the amount shown as TOTAL upon proper presentation. I promise to pay such TOTAL (together with any other charges due thereon) subject to and in accordance with the agreement governing the use of such card.

**CARDHOLDER
SIGN HERE ×**

Martha must sign the receipt.

Date of purchase

DATE	AUTHORIZATION NO.	SALESCLERK	DEPT.	IDENTIFICATION	TAKE
9/10					SEND

QUAN.	CLASS	DESCRIPTION	PRICE	AMOUNT	
2		Wool sweaters	28.00	56 00	
1		Denim jeans		29 95	
1		Cotton shirt		18 50	
		Total selling price			
		CURRENCY CONVERSION DATE	SUB-TOTAL	104 45	
		DATE	SALES TAX	4 18	
		AMOUNT	TOTAL	108 63	

SALES SLIP CUSTOMER COPY

Total purchase price

A. Find the **total selling price.**
(2 × $28.00) + $29.95 + $18.50
$56.00   + $29.95 + $18.50 = $104.45 total selling price

B. Find the **sales tax.** Total Selling Price + Sales Tax Rate
$104.45 × 4% = $4.178 = $4.18

C. Find the **total purchase price.** Total Selling Price + Sales Tax
$104.45 + $4.18 = $108.63 Total purchase price is correct.

2 ⊠ 28 ▭ 56 ⊞ 29.95 ⊞ 18.5 ▭ 104.45 M+ ⊠ .04 ▭ 4.178 M+
RM 108.628

✔ SELF-CHECK   Complete the problems, then check your answers in the back of the book.

Find the sales tax and the total purchase price.

1. Cost of coat: $116.
Sales tax rate: 6%.

2. Price of boots: $99.95.
Sales tax rate: 5%.

	Total Selling Price	×	Sales Tax Rate	=	Sales Tax	Total Purchase Price
3.	$144.00	×	7%	=	?	?
4.	$ 63.00	×	4%	=	?	?
5.	$129.55	×	6.5%	=	?	?

Complete the sales receipt.

DATE 3/14/–	AUTH NO 42	IDENTIFICATION	CLERK J.R.	REG/DEPT	☑TAKE ☐SEND
QTY	CLASS	DESCRIPTION	PRICE	AMOUNT	
1		Lamp		49 ¦ 95	
1		Lock set		29 ¦ 95	

SUBTOTAL ?
CUSTOMER SIGNATURE × Darcia Adams   TAX   4 ¦ 19
SALES SLIP   TOTAL ?

6.
7.

DATE 6/1/–	AUTH NO 86430	IDENTIFICATION	CLERK D.L.	REG/DEPT	☑TAKE ☐SEND
QTY	CLASS	DESCRIPTION	PRICE	AMOUNT	
1		Knit Shirt		15 ¦ 98	
2		Pairs Socks	2.59 ea	5 ¦ 18	
1		Golves		5 ¦ 75	

SUBTOTAL ?
CUSTOMER SIGNATURE × Steve Rizzo   TAX   1 ¦ 35
SALES SLIP   TOTAL ?

8.
9.

50057 394856 009
LEWIS PETERRO
BARNEY'S SERVICE
DATE: 10 / 6 / –

	QTY	Price	Amount
Supreme	Regular	Unleaded	12.8	1 ¦ 15	14 ¦ 72
Motor oil		1 Qt	2 ¦ 25	2 ¦ 25
Windshield washer fluid				1 ¦ 35
		SALES TAX	¦ 19
SIGN HERE × Lewis Peterro		TOTAL	?

10.

DATE 9/2	AUTH NO —	IDENTIFICATION	CLERK R.H.	REG/DEPT A	☑TAKE ☐SEND
QTY	CLASS	DESCRIPTION	PRICE	AMOUNT	
1		Fan		39 ¦ 95	
3		Drill Bits	2.09	?	
1		Storage Box		13 ¦ 39	

SUBTOTAL ?
CUSTOMER SIGNATURE × Janet Rye   TAX   3 ¦ 13
SALES SLIP   TOTAL ?

11.
12.
13.

14. Sales slip for Jim Baker.
Tune-up for $44.75.
2 tires at $45.99 each.
Headlight for $8.79.
4 quarts of oil at $1.85 each.
Sales tax is $11.09.
What is the total purchase price?

15. Sales slip for Lisa Chong.
2 bags of lawn seed at $29.95 each.
1 bag of potting soil for $7.95.
12 packs of seeds at $1.09 each.
Lawn sprinkler for $24.95.
Sales tax is $7.41.
What is the total purchase price?

16. Darlene Zink charges 12.5 gallons of gasoline. The gas costs $1.299 per gallon. What is the total purchase price?

17. Allen Zintec purchases a ring at Tillman's Jewelers. The ring costs $42.95 plus a 4% sales tax. He charges the purchase on his bank charge card. What is the total purchase price?

18. Roy Horst purchases a typewriter at Hall Office Supplies. The typewriter costs $199.99 plus a sales tax of 5.5%. What is the total purchase price?

19. Dorothy Mathias purchases a stereo at Electronics, Inc. The stereo costs $379.99 plus a sales tax of 6%. What is the total purchase price?

20. Quentin Cassidy charges the following items on his bank charge card: a suit for $259.99, 3 shirts at $24.99 each, 2 ties at $18.50 each, and a pair of shoes for $89.99. There is a sales tax of 5%. What is the total purchase price?

21. Mary Jane Drolshagen charges the following items on her department store charge card: a suit for $179.99, 2 skirts for $65.99 each, 2 sweaters at $25.99 each, and 2 pair of shoes for $49.99 each. There is a sales tax of 6%. What is the total purchase price?

22. At Bargain City, Emily Williams purchases a gallon of paint thinner for $3.98, 3 gallons of paint at $17.95 a gallon, 2 tubes of caulking at $1.89 each, 2 paint rollers at $2.79 each, and 2 paintbrushes at $5.99 each. There is a sales tax of 6%. What is the total purchase price?

23. At Budget Mart, Leonard Milum purchases a baby stroller for $34.97, 3 plush animals at $9.79 each, a wastebasket for $9.95, and a box of detergent for $4.99. There is a sales tax of 3.5%. He charges the purchases on his bank charge card. What is the total purchase price?

24. At the Harvest Inn, Theora and Frankie Watson dined on clams and lobster. A portion of the check for the dinner is shown. There is an 8% meal tax. Frankie charged the dinner on his bank charge card and added a tip of 15% on the cost of the meal before tax. What was the total cost of dining out?

HARVEST INN

1	Steamed clams	4	95
2	Lobster dinners	43	90
	@ $21.95 each		
2	Cups coffee	1	80

MAINTAINING YOUR SKILLS Look up the skills in parentheses if you need help or more practice.

Write the percents as decimals. **(Skill 28)**

25. 8%   **26.** 7%   **27.** 4.5%   **28.** 8.75%   **29.** 9.2%

Add. **(Skill 5)**

30. $34.65 + $22.55 + $45.88   **31.** $223.43 + $64.89 + $7.95

32. $46.75 + $8.65 + $324.45   **33.** $8.25 + $35.76 + $42.03 + $1.23

Find the percentage. **(Skill 30)**

34. 9% of 120   **35.** 7% of 42   **36.** 6.5% of 20   **37.** 8.2% of 86

38. 9.2% of $120   **39.** 8.4% of $425   **40.** 6.75% of $93.60

Account Statements

OBJECTIVE

Compute the new balance in a charge account.

With a credit card or charge account, you receive a monthly statement listing all transactions processed by the closing date for that month. If your previous bill was not paid in full, a finance charge is added. A finance charge is interest that is charged for delaying payment.

New Balance = Previous Balance + Finance Charge + New Purchases − (Payments + Credits)

EXAMPLE   *Skills* 6, 5   *Application* A   *Term* Finance charge

Martha Palmer received this charge account statement on October 2. What is her new balance?

MAIL THIS PORTION WITH PAYMENT TO P.O. BOX 1027

D & L DEPARTMENT STORE   Martha's account number

☐ IF ADDRESS IS INCORRECT, PLEASE CHECK
BOX AND CORRECT ON REVERSE SIDE

Martha Palmer
1234 MAIN ST.
TOLEDO, OH 43605

ACCOUNT NUMBER					101-04076	
NEW BALANCE					?	
TOTAL MINIMUM AMOUNT NOW DUE					43.00	
SEND PAYMENT TO REACH US BY					10/30/--	

DATE	DEPT. NO.	DESCRIPTION	PURCHASES	CREDITS	PAYMENTS	ITEM NO.
09/07	162	SPORTSWEAR		23.42		0-06-18005
09/10	363	SPORTSWEAR	108.63			0-07-20015
09/17		PAYMENT-THANK YOU			50.00	0-06-47023
09/19	214	JR SPORTSWEAR		8.52		0-07-45005
09/19	53	HOUSEWARES	19.53			0-07-45006
09/27	34	COSMETICS	25.62			0-07-64004
09/29	1	LINENS-TOWELS	19.25			0-07-70012

Last date transactions were processed ⌐
Amount Martha owed last month
$50 payment toward last month's bill
Amount Martha now owes

BILL CLOSING DATE	PREVIOUS BALANCE	FINANCE CHARGE	TOTAL PURCHASES THIS MONTH	CREDITS	PAYMENTS	NEW BALANCE
9/30	332.64	5.64	173.03	31.94	50.00	?

Pay new balance in full within 25 days of bill closing date to avoid finance charge next month. Finance charge, if any, is computed on the average daily balance of $375.72 by applying monthly periodic rates as follows:

—— Credit for items returned

MONTHLY PERIODIC RATE: 1.50 % 1.00 %	BALANCE FIRST $600 OVER $600	ANNUAL PERCENTAGE RATE: 18% 12%

SUBJECT TO A MINIMUM FINANCE CHARGE OF $.50.
NOTICE: SEE REVERSE SIDE FOR IMPORTANT INFORMATION

ACCOUNT NUMBER	101-04076
NEW BALANCE	
TOTAL MINIMUM AMOUNT NOW DUE	
SEND PAYMENT TO REACH US BY	

SOLUTION   Find the **new balance.**

Previous Balance	+	Finance Charge	+	New Purchases	−	(Payments + Credits)
$332.64	+	$5.64	+	$173.03	−	($50.00 + $31.94)
		$511.31			−	$81.94 = $429.37

new balance

332.64 + 5.64 + 173.03 − 81.94 = 429.37

Find the new balance.

	Previous Balance	+	Finance Charge	+	New Purchases	−	Payments & Credits		New Balance
1.	$600.00	+	$7.50	+	$90.00	−	$100.00	=	?
2.	$278.75	+	$4.18	+	$35.85	−	$48.00	=	?

PROBLEMS

What is the new balance for the credit statements shown?

3.

BILLING DATE	PREVIOUS BALANCE	FINANCE CHARGE	NEW PURCHASES	PAYMENTS & CREDITS	NEW BALANCE
8/15/--	$600.00	$9.00	$140.00	$100.00	?

4.

BILLING DATE	PREVIOUS BALANCE	FINANCE CHARGE	NEW PURCHASES	PAYMENTS & CREDITS	NEW BALANCE
1/22/--	$410.75	$7.20	$175.00	$150.00	?

5.

BILLING DATE	PREVIOUS BALANCE	FINANCE CHARGE	NEW PURCHASES	PAYMENTS & CREDITS	NEW BALANCE
3/1/--	$450.95	$6.76	$39.95	$160.00	?

6.

BILLING DATE	PREVIOUS BALANCE	FINANCE CHARGE	NEW PURCHASES	PAYMENTS & CREDITS	NEW BALANCE
9/4/--	$233.23	$2.33	$40.36	$133.23	?

7.

BILLING DATE	PREVIOUS BALANCE	FINANCE CHARGE	NEW PURCHASES	PAYMENTS & CREDITS	NEW BALANCE
6/15/--	$675.19	0	$416.34	$675.19	?

8.

BILLING DATE	PREVIOUS BALANCE	FINANCE CHARGE	NEW PURCHASES	PAYMENTS & CREDITS	NEW BALANCE
8/1/--	$2494.21	$43.65	$137.25	$450.00	?

9. What is the new balance for the credit statement shown?

PAYMENTS (P) & CREDIT (C) DEDUCTED-DATES	
20.00 P 04-07	
20.00 P 03-28	
PREVIOUS BALANCE	PURCHASES ADDED
265.69	145.99
FINANCE CHARGE ADDED 3.98	NEW BALANCE ?
MINIMUM PAYMENT 25.00	MUST BE RECEIVED BY 05-15
	TO AVOID ADDITIONAL FINANCE CHARGE
ANNUAL PERCENTAGE RATE	PERIODIC MONTHLY RATE
TO $500 18%	1-1/2 %
OVER $500 12%	1 %

10. What is the new balance for the credit statement shown?

ACCOUNT NUMBER 064-173-388-62		BILLING DATE 03-10
PREVIOUS BALANCE	− PAYMENTS	RECEIVED ON
366 | 72	366 | 72	02-26
− CREDITS	+ FINANCE CHARGE	+ PURCHASES
24 | 59	5 | 50	45 | 23
+ INS. PREMIUMS	= NEW BALANCE	AMOUNT DUE
	?	40 | 00
	NEXT BILLING DATE 04-12	ANNUAL PERCENTAGE RATE(S)
	PERIODIC RATE(S)	
UP TO $500	1.50%	18.0%
OVER $500	1.00%	12.0%
MINIMUM CHARGE		.50

11. Monthly statement for
Mae Keyes.
Credit Master bankcard.
Previous balance of $307.85.
Payment of $40.
New purchases: $9.50, $41.75.
Finance charge of $4.62.
What is the new balance?

12. Monthly statement for
Bob Ross.
J-MART Store charge card.
Previous balance of $144.79.
Payment and credit totaling $144.79.
Finance charge of $2.53.
No new purchases.
What is the new balance?

13. Monthly statement for
Walt Klin.
Department store charge card.
Previous balance of $787.29.
Payment of $100.
New purchases: $47.97, $49.28.
Finance charge of $11.81.
What is the new balance?

14. Monthly statement for
Jane Cook.
All-Charge bankcard.
Previous balance of $529.78.
Payment of $85.
New purchases: $277.32, $38.20.
Finance charge of $7.95.
What is the new balance?

15. Marci Cassidy received this charge account statement. Find her
new balance.

DEPT.	DESCRIPTION	CHARGES	PAYMENT/CREDIT	DATE	REF. #
109	Garden Shop	42.75		1/25	6004
85	Menswear	145.98		1/25	7018
	PAYMENT		74.40	2/1	8014
71	Appliances		35.50	2/2	3113
BILLING DATE: 2/16					

PREVIOUS BALANCE	PAYMENTS & CREDITS	UNPAID BALANCE	FINANCE CHARGE	NEW PURCHASES	NEW BALANCE
$285.92	?	— —	$4.29	?	?

MAINTAINING YOUR SKILLS   Look up the skills in parentheses if you need help or more practice.

Add. **(Skill 5)**

16. $532.75 + $45.90 + $38.90 + $16.55

17. $44.29 + $324.60 + $8.65 + $27.50

18. $44.52 + $923.17 + $9.20 + $337.63

19. $8.16 + $24.50 + $39.25 + $234.78

Subtract. **(Skill 6)**

20. $41.50 − $9.50   **21.** $321.65 − $12.35   **22.** $427.16 − $45.12

23. $337.42 − $49.58 **24.** $798.87 − $389.68 **25.** $4296.83 − $999.49

Finance Charge—
Previous-Balance Method

OBJECTIVE

Compute the finance charge by the previous-balance method.

Some credit card companies use the previous-balance method to compute finance charges. They compute the finance charge based on the amount you owed on the closing date of your last statement. The periodic rate is the monthly finance charge rate.

Finance Charge = Previous Balance × Periodic Rate

New Balance = Previous Balance + Finance Charge + New Purchases − (Payments + Credits)

EXAMPLE   *Skills* 30, 2   *Application* A   *Term* Previous balance

Carl Byers has a charge account with a store that charges 1.5% of the previous balance for the finance charge. A portion of Carl's statement is shown. He checks the computations. What is the new balance of his account on October 5?

PREVIOUS BALANCE	CLOSING DATE THIS MONTH		CLOSING DATE LAST MONTH	
$125.60	October 5, 19--		September 6, 19--	
TOTAL PURCHASES	PAYMENTS & CREDITS	FINANCE CHARGE	NEW BALANCE	MINIMUM PAYMENT
$122.15	$48.75	?	?	$20.00

SOLUTION

A. Find the **finance charge.**

Previous Balance × Periodic Rate

$125.60 × 1.5% = $1.884 = $1.88 finance charge

B. Find the **new balance.**

$$\text{Previous Balance} + \text{Finance Charge} + \text{New Purchases} - (\text{Payments} + \text{Credits})$$

$125.60 + $1.88 + $122.15 − $48.75

$249.63 − $48.75 = $200.88 new balance

✔ SELF-CHECK   Complete the problems, then check your answers in the back of the book.

The periodic rate is 1.5%. Find the finance charge.

1. Previous balance of $180.

2. Previous balance of $87.90.

Use the previous-balance method to find the finance charge. Round answers to the nearest cent.

	3.	4.	5.	6	7.
Previous Balance	$60.00	$150.00	$148.00	$197.30	$287.42
Periodic Rate	× 1.5%	× 1.5%	× 1.6%	× 1.25%	× 1.75%
Finance Charge	?	?	?	?	?

The finance charge is 1.5% of the previous balance. Find the finance charge and the new balance for each statement shown.

8.

BILLING DATE	PREVIOUS BALANCE	FINANCE CHARGE	NEW PURCHASES	PAYMENTS & CREDITS	NEW BALANCE
4/11/--	$400.00	?	$200.00	$160.00	?

9.

BILLING DATE	PREVIOUS BALANCE	FINANCE CHARGE	NEW PURCHASES	PAYMENTS & CREDITS	NEW BALANCE
5/7/--	$170.00	?	$64.00	$45.00	?

10.

BILLING DATE	PREVIOUS BALANCE	FINANCE CHARGE	NEW PURCHASES	PAYMENTS & CREDITS	NEW BALANCE
3/1/--	$32.50	?	$15.50	$20.00	?

11.

BILLING DATE	PREVIOUS BALANCE	FINANCE CHARGE	NEW PURCHASES	PAYMENTS & CREDITS	NEW BALANCE
9/15/--	$497.00	?	$35.95	$80.00	?

12.

BILLING DATE	PREVIOUS BALANCE	FINANCE CHARGE	NEW PURCHASES	PAYMENTS & CREDITS	NEW BALANCE
12/12/--	$564.28	?	$221.82	$125.00	?

13.

BILLING DATE	PREVIOUS BALANCE	FINANCE CHARGE	NEW PURCHASES	PAYMENTS & CREDITS	NEW BALANCE
10/30/--	$1282.29	?	0	$225.00	?

Use the previous-balance method of computing finance charges to solve.

14. Rick Demski's charge account.
Previous balance of $188.
Periodic rate is 2.0%.
   What is the finance charge?
New purchases totaling $42.50.
Payment and credit totaling $25.
   What is the new balance?

15. Marie Burch's charge card.
Previous balance of $144.30.
Periodic rate is 1.25%.
   What is the finance charge?
New purchases totaling $97.32.
Payments totaling $120.
   What is the new balance?

16. Kay Maxwell has a charge account at Simon's Department Store where the finance charge is 1.5% of the previous balance. A portion of her account statement is shown. Find the finance charge and the new balance.

PREVIOUS BALANCE	CLOSING DATE THIS MONTH		CLOSING DATE LAST MONTH	
$157.53	July 20, 19--		June 21, 19--	
TOTAL PURCHASES	PAYMENTS & CREDITS	FINANCE CHARGE	NEW BALANCE	MINIMUM PAYMENT
$42.91	$27.18	?	?	$30.00

17. Dale Watson has a charge account where the finance charge is 1.25% of the previous balance. A portion of his statement for May is shown. What is the total of his payments and credits? What is the finance charge? What is the new balance of his account?

DATE	REFERENCE #	DEPT.	DESCRIPTION	PURCHASES	PAYMENTS	CREDITS
5-04	34029	03	Sport coat	135.29		
5-12	40085	06	Robe			49.79
5-23	29450	10	Payment		50.00	
5-23	37047	03	Men's slacks	47.35		
5-25	77460	04	Hardware	27.15		

PREVIOUS BALANCE	CLOSING DATE THIS MONTH		CLOSING DATE LAST MONTH	
$141.58	May 31, 19--		April 30, 19--	
TOTAL PURCHASES	PAYMENTS & CREDITS	FINANCE CHARGE	NEW BALANCE	MINIMUM PAYMENT
$209.79	?	?	?	$20.00

18. A Brief Case Many credit account companies charge periodic rates that vary depending on the amount of the previous balance. Use the periodic rates on the statement shown to calculate the finance charge and the new balance according to the previous-balance method.

PREVIOUS BALANCE	– PAYMENTS	RECEIVED ON
849 | 74	120 | 00	12-09
– CREDITS	+ FINANCE CHARGE	+ PURCHASES
35 | 93	?	221 | 75
+ INS. PREMIUMS	= NEW BALANCE	AMOUNT DUE
	?	90 | 00

PAY NEW BALANCE BY TO AVOID FINANCE CHARGE ▶	NEXT BILLING DATE 01-27	
BALANCES	PERIODIC RATE(S)	ANNUAL PERCENTAGE RATE(S)
UP TO $500	1.50%	18.0%
OVER $500	1.00%	12.0%
MINIMUM CHARGE	.50	

MAINTAINING YOUR SKILLS Look up the skills in parentheses if you need help or more practice.

Round answers to the nearest hundredth. **(Skill 2)**

19. 43.155   **20.** 8.241   **21.** 9.5371   **22.** 14.8981   **23.** 2.9971

Round answers to the nearest cent. **(Skill 2)**

24. $2.385   **25.** $3.7521   **26.** $9.5145   **27.** $0.4354   **28.** $18.015

Find the percentage. **(Skill 30)**

29. 8% of 438   **30.** 3.75% of 988   **31.** 1.25% of 42.24   **32.** 1.75% of 365.5

6-4

Finance Charge—
Unpaid-Balance Method

OBJECTIVE

Compute the finance charge by the unpaid-balance method.

Some companies use the unpaid-balance method of computing finance charges. They compute the finance charge based on that portion of the previous balance that you have not paid.

Unpaid Balance = Previous Balance − (Payments + Credits)

Finance Charge = Unpaid Balance × Periodic Rate

New Balance = Unpaid Balance + Finance Charge + New Purchases

EXAMPLE   *Skill* 30   *Application* A   *Term* Unpaid balance

A portion of Lucille Sherman's charge account statement is shown. The monthly finance charge is 1.5% of the unpaid balance. What is the new balance of her account?

88	PAYMENT/Thank you		40.00		

BILLING DATE: 2/16

PREVIOUS BALANCE	PAYMENTS & CREDITS	UNPAID BALANCE	FINANCE CHARGE	NEW PURCHASES	NEW BALANCE
$132.40	$40.00	?	?	$79.55	?

SOLUTION

A. Find the **unpaid balance.**
Previous Balance − (Payments + Credits)
$132.40   − $40.00 = $92.40 unpaid balance

B. Find the **finance charge.**
Unpaid Balance × Periodic Rate
$92.40   × 1.5% = $1.386 = $1.39 finance charge

C. Find the **new balance.**
Unpaid Balance + Finance Charge + New Purchases
$92.40   + $1.39   + $79.55 = $173.34 new balance

✔ SELF-CHECK   Complete the problems, then check your answers in the back of the book.

The periodic rate is 1.5%. Find the unpaid balance, the finance charge, and the new balance.

	Previous Balance	Payments and Credits	New Purchases
1.	$400.00	$100.00	$70.00
2.	$220.00	$150.00	$95.00

For problems 3 through 10, use a periodic rate of 1.5% and the unpaid-balance method of computing the finance charge.

	(Previous Balance	−	Payments + Credits	=	Unpaid Balance)	+	Finance Charge	+	New Purchases	=	New Balance
3.	($500.00	−	$100.00	=	?	) +	?	+	$ 80.00	=	?
4.	($300.00	−	$150.00	=	?	) +	?	+	$ 45.00	=	?
5.	($350.00	−	$ 75.00	=	?	) +	?	+	$ 90.00	=	?
6.	($125.50	−	$ 45.50	=	?	) +	?	+	$ 42.50	=	?
7.	($473.50	−	$ 57.50	=	?	) +	?	+	$222.50	=	?
8.	($173.43	−	$100.00	=	?	) +	?	+	$127.91	=	?
9.	($491.87	−	$119.00	=	?	) +	?	+	$147.94	=	?
10.	($738.27	−	$145.00	=	?	) +	?	+	$199.95	=	?

11. Ruth Kean's account statement.
Unpaid balance of $88.
Periodic rate is 1.5%.
　What is the finance charge?
New purchases of $40.
　What is the new balance?

12. Don Vester's account statement.
Unpaid balance of $19.70.
Periodic rate is 1.25%.
　What is the finance charge?
New purchases of $431.85.
　What is the new balance?

13. Midge Duez's account statement.
Unpaid balance of $121.60.
Periodic rate is 1.6%.
　What is the finance charge?
New purchases of $72.19.
　What is the new balance?

14. Liz Cole's account statement.
Unpaid balance of $921.35.
Periodic rate is 1.75%.
　What is the finance charge?
New purchases of $75.43.
　What is the new balance?

15. A portion of Alvin Sujkowski's charge account statement is shown. The finance charge is 2% of the unpaid balance. What is the new balance?

PREVIOUS BALANCE	PAYMENTS & CREDITS	UNPAID BALANCE	FINANCE CHARGE	NEW PURCHASES	NEW BALANCE
$419.29	$45.00	?	?	$79.31	?

16. A portion of Verda Buell's charge account statement is shown. The finance charge is 1.25% of the unpaid balance. What is the new balance?

PREVIOUS BALANCE	PAYMENTS & CREDITS	UNPAID BALANCE	FINANCE CHARGE	NEW PURCHASES	NEW BALANCE
$556.71	$147.55	?	?	$21.64	?

17. June Ray has a charge account at Parker's Discount Store where the finance charge is 2% of the unpaid balance. Find the indicated amounts on the statement shown.

DEPT.	DESCRIPTION	CHARGES	PAYMENT/CREDIT	DATE	REF. #
55	Battery	65.67		3/2	6982
44	Smoke alarm	9.85		3/2	8640
32	Muffler	54.96		3/2	9000
98	Daywear		21.45	3/5	7640
88	PAYMENT		50.00	3/9	500

BILLING DATE: 3/10

PREVIOUS BALANCE	PAYMENTS & CREDITS	UNPAID BALANCE	FINANCE CHARGE	NEW PURCHASES	NEW BALANCE
$379.13	?	?	?	?	?

18. Gary Green has a charge account at Knapp's Department Store where the finance charge is 1.75% of the unpaid balance. Find the indicated amounts on the statement shown.

DEPT.	DESCRIPTION	CHARGES	PAYMENT/CREDIT	DATE	REF. #
09	Linens	84.39		9/12	9064
14	PAYMENT		50.00	9/13	A345
03	Sporting goods	27.83		9/26	11309
08	Sporting goods		27.83	10/02	3960
15	Electronics	239.95		10/02	11714

BILLING DATE: 10/10

PREVIOUS BALANCE	PAYMENTS & CREDITS	UNPAID BALANCE	FINANCE CHARGE	NEW PURCHASES	NEW BALANCE
$338.65	?	?	?	?	?

MAINTAINING YOUR SKILLS   Look up the skills in parentheses if you need help or more practice.

Add. **(Skill 5)**

19. $425.10 + $38.75 + $29.51 + $4.22

20. $5.95 + $38.75 + $71.19 + $314.75

Subtract. **(Skill 6)**

21. $499.24 − $88.31   **22.** $391.37 − $79.43   **23.** $523.89 − $154.79

24. $310.01 − $58.75   **25.** $87.01 − $9.54   **26.** $808.76 − $39.41

Find the percentage. Round answers to the nearest hundredth. **(Skill 30)**

27. 4% of 220   **28.** 8% of 60   **29.** 3.9% of 500   **30.** 2.8% of 460

31. 0.7% of 500   **32.** 3.15% of 181.2   **33.** 1.75% of 851

Finance Charge—Average Daily Balance (No new purchases included)

OBJECTIVE

Compute the finance charge based on the average daily balance, no new purchases included.

Many companies calculate the finance charge using the average-daily-balance method where no new purchases are included. The average daily balance is the average of the account balance at the end of each day of the billing period. For this method of computing finance charges, new purchases posted during the billing period are not included when figuring the balance at the end of the day.

$$\text{Average Daily Balance} = \frac{\text{Sum of Daily Balances}}{\text{Number of Days}}$$

EXAMPLE   *Skills* 8, 11   *Application* G   *Term* Average daily balance

A portion of Dewey Napp's credit card statement is shown.

REFERENCE	POSTING DATE	TRANSACTION DATE	DESCRIPTION	PURCHASES & ADVANCES	PAYMENTS & CREDITS
131809	9/05	8/24	Health Club	48.75	
265118	9/18		PAYMENT		44.85
407372	9/20	9/01	Wilson's	37.85	
329416	10/01	8/30	Ed's Discount	20.99	

BILLING PERIOD		PREVIOUS BALANCE	PERIODIC RATE	AVERAGE DAILY BALANCE	FINANCE CHARGE
9/04 - 10/03		$194.85	2%	?	?
PAYMENTS & CREDITS		PURCHASES & ADVANCES	NEW BALANCE	MINIMUM PAYMENT	PAYMENT DUE
$44.85		$107.59	?	$20.00	10/25

A finance charge was added to Dewey's account balance because he did not pay his last bill in full. The finance charge was computed using the average daily balance where new purchases were not included. Only the payment of $44.85 affected the average daily balance. What is the average daily balance?

A. Find the **sum of daily balances.**

Dates	Payment	End-of-Day Balance		Number of Days	Sum of Balances
9/4–9/17		$194.85	×	14	$2727.90
9/18	$44.85	150.00	×	1	150.00
9/19–10/3		150.00	×	15	2250.00
			Total	30	$5127.90

B. Find the **average daily balance.**
Sum of Daily Balances ÷ Number of Days
$5127.90   ÷ 30 = $170.93 average daily balance

194.85 ⊠ 14 ⊟ 2727.9 M+ 194.85 ⊟ 44.85 ⊟ 150 M+ 150 ⊠ 15 ⊟ 2250 M+
RM 5127.9 ÷ 30 ⊟ 170.9

✔ SELF-CHECK   Complete the problems, then check your answers in the back of the book.

Find the average daily balance, excluding new purchases.

Dates	Payment	End-of-Day Balance	×	Number of Days	=	Sum of Balances
9/9–9/18		$500.00	×	10	=	**1.** ?
9/19	$100.00	$400.00	×	1	=	**2.** ?
9/20–10/8		**3.** ?	×	**4.** ?	=	**5.** ?
			Total	**6.** ?		**7.** ?

and **8.** ? ÷ **9.** ? = **10.** average daily balance

The finance charge is calculated by multiplying the average daily balance times the periodic rate.

Finance Charge = Average Daily Balance × Periodic Rate

New Balance = Unpaid Balance + Finance Charge + New Purchases

EXAMPLE   *Skills* 8, 11   *Application* G   *Term* Average daily balance

Dewey checks the finance charge and the new balance. The finance charge is 2% of the average daily balance. What is the new balance?

A. Find the **unpaid balance.**
Previous Balance − (Payments + Credits)
    $194.85   − $44.85   = $150.00 unpaid balance

B. Find the **finance charge.**
Average Daily Balance × Periodic Rate
    $170.93   ×   2% = $3.418 = $3.42 finance charge

C. Find the **new purchases.** $107.59

D. Find the **new balance.**
Unpaid Balance + Finance Charge + New Purchases
    $150.00   + $3.42   + $107.59   = $261.01
                                                  new balance

The finance charge is 1.5% of the daily balance.

11. Find the finance charge.   **12.** Find the new balance.

PREVIOUS BALANCE	TOTAL CHARGES	PAYMENTS AND CREDITS	AVERAGE DAILY BALANCE	FINANCE CHARGE	NEW BALANCE
$180.00	$40.00	$30.00	$165.00	?	?

PROBLEMS

	Billing Periods	Payment	End-of-Day Balance	Number of Days	Sum of Balances	
13.	6/01–6/15		$75.00	15	$1125.00	What is the average daily balance?
	6/16	$50.00	25.00	1	25.00	
	6/17–6/30		25.00	14	?	
	TOTALS			?	?	
14.	7/15–8/2		$400.00?	?		What is the average daily balance?
	8/3	$100.00	300.00	?	?	
	8/4–8/14		300.00	?	?	
	TOTALS			?	?	

F.Y.I.
On December 1, 1991, the average monthly credit card annual percentage rate was 18.87%.

15. Lee Hoshino has a bank charge card. Use the portion of the account statement shown to find the average daily balance, excluding new purchases, and finance charge.

REFERENCE	POSTING DATE	TRANSACTION DATE	DESCRIPTION	PURCHASES & ADVANCES	PAYMENTS & CREDITS
6646598	8/25	8/20	Al's Hardware	44.80	
7000507	8/28		PAYMENT RECEIVED		50.00

BILLING PERIOD	PREVIOUS BALANCE	PERIODIC RATE	AVERAGE DAILY BALANCE	FINANCE CHARGE
8/01 - 8/31	$250.00	1.5%	?	— —

The finance charge is computed using the average-daily-balance method where no new purchases are included. Find the finance charge and the new balance for the following statements.

16.

BILLING PERIOD	PREVIOUS BALANCE	PERIODIC RATE	AVERAGE DAILY BALANCE	FINANCE CHARGE
2/3 - 3/2	$196.00	2%	$156.00	?

PAYMENTS & CREDITS	PURCHASES & ADVANCES	NEW BALANCE	MINIMUM PAYMENT	PAYMEMT DUE
$60.00	0	?	$15.00	3/20

17.

BILLING PERIOD	PREVIOUS BALANCE	PERIODIC RATE	AVERAGE DAILY BALANCE	FINANCE CHARGE
11/15 - 12/14	$322.49	1.75%	$277.21	?

PAYMENTS & CREDITS	PURCHASES & ADVANCES	NEW BALANCE	MINIMUM PAYMENT	PAYMEMT DUE
$123.49	$49.51	?	$20.00	1/2

18.

BILLING PERIOD	PREVIOUS BALANCE	PERIODIC RATE	AVERAGE DAILY BALANCE	FINANCE CHARGE
7/1 - 8/1	$74.50	1.25%	$45.66	?

PAYMENTS & CREDITS	PURCHASES & ADVANCES	NEW BALANCE	MINIMUM PAYMENT	PAYMEMT DUE
$44.50	$23.95	?	$10.00	8/20

19. A portion of Helena Strege's account statement for March from Inbank Charge Company is shown. The finance charge is computed using the average-daily-balance method where new purchases are excluded. Find the average daily balance, the finance charge, and the new balance.

REFERENCE	POSTING DATE	TRANSACTION DATE	DESCRIPTION	PURCHASES & ADVANCES	PAYMENTS & CREDITS
450345	3/20		PAYMENT		24.66
458343	3/27	3/14	Aston Oil Co.	81.30	

BILLING PERIOD	PREVIOUS BALANCE	PERIODIC RATE	AVERAGE DAILY BALANCE	FINANCE CHARGE
3/4 - 4/3	$94.66	2%	?	?

PAYMENTS & CREDITS	PURCHASES & ADVANCES	NEW BALANCE	MINIMUM PAYMENT	PAYMENT DUE
$24.66	$81.30	?	$10.00	4/21

20. Edith Bertelli received this statement from Bertrand's. Find the average daily balance, the finance charge, and the new balance.

REFERENCE	POSTING DATE	TRANSACTION DATE	DESCRIPTION	PURCHASES & ADVANCES	PAYMENTS & CREDITS
1027485	4/11		PAYMENT		40.00
4500298	4/15	4/01	Menswear	39.95	
5473390	4/23	4/21	Housewares	15.99	
1374655	4/25		PAYMENT		50.00

BILLING PERIOD	PREVIOUS BALANCE	PERIODIC RATE	AVERAGE DAILY BALANCE	FINANCE CHARGE
4/1 - 5/1	$175.00	1.2%	?	?

PAYMENTS & CREDITS	PURCHASES & ADVANCES	NEW BALANCE	MINIMUM PAYMENT	PAYMENT DUE
$90.00	$55.94	?	$25.00	5/25

21. Louis Minier received this statement from Crowley Bank. Find the average daily balance, the finance charge, and the new balance.

REFERENCE	POSTING DATE	TRANSACTION DATE	DESCRIPTION	PURCHASES & ADVANCES	PAYMENTS & CREDITS
1616787	4/25	4/19	A-1 Plumbing	61.45	
3945557	4/30		PAYMENT		79.60
3957475	5/5	5/1	Ace Hardware	32.45	
4000076	5/20		PAYMENT		50.00

BILLING PERIOD		PREVIOUS BALANCE	PERIODIC RATE	AVERAGE DAILY BALANCE	FINANCE CHARGE
4/21 - 5/20		$179.60	2%	?	?
PAYMENTS & CREDITS		PURCHASES & ADVANCES	NEW BALANCE	MINIMUM PAYMENT	PAYMENT DUE
$129.60		$93.90	?	$30.00	6/12

22. **A Brief Case** Alice Kruse received this statement from Garrison's. Find the indicated amounts on the statement shown.

REFERENCE	POSTING DATE	TRANSACTION DATE	DESCRIPTION	PURCHASES & ADVANCES	PAYMENTS & CREDITS
31784	7/25		PAYMENT		30.00
103645	7/25	7/12	Men's shoes	79.48	
116748	7/30	7/28	Electronics		19.48
345803	8/8		PAYMENT		40.00
57845	8/8	8/7	Bakery	12.45	

BILLING PERIOD		PREVIOUS BALANCE	PERIODIC RATE	AVERAGE DAILY BALANCE	FINANCE CHARGE
7/22 - 8/21		$379.46	1.75%	?	?
PAYMENTS & CREDITS		PURCHASES & ADVANCES	NEW BALANCE	MINIMUM PAYMENT	PAYMENT DUE
?		?	?	$30.00	9/11

MAINTAINING YOUR SKILLS Look up the skills in parentheses if you need help or more practice.

Multiply. (**Skill 8**)

23. 12×200 **24.** 18×150 **25.** 22×37.5 **26.** 9×34.56

27. 7×225.5 **28.** 14×42.50 **29.** 35×61.8 **30.** 37×341.5

Divide. **Round answers to the nearest hundredth. (Skill 10)**

31. $3750 \div 30$ **32.** $3360 \div 28$ **33.** $1129.95 \div 31$ **34.** $10,852.5 \div 30$

Find the average. (**Application Q**)

35. 44, 73, 92, 88, 63

36. 324, 406, 958, 285, 785, 374

37. 135, 415, 364, 419, 84

38. 52.8, 63.41, 14.66, 34.89, 9.35

6-6

Finance Charge—Average Daily Balance (New purchases included)

OBJECTIVE

Compute the finance charge based on the average daily balance, new purchases included.

Some companies compute the finance charge using the average-daily-balance method where new purchases are included when figuring the daily balances during the posting period.

$$\text{Average Daily Balance} = \frac{\text{Sum of Daily Balances}}{\text{Number of Days}}$$

EXAMPLE   *Skills* 8, 11   *Application* G   *Term* Average daily balance

Scott McTique has a charge account where the finance charge is computed using the average-daily-balance method that includes new purchases. He checks to be sure the average daily balance is correct.

REFERENCE	POSTING DATE	TRANSACTION DATE	DESCRIPTION	PURCHASES & ADVANCES	PAYMENTS & CREDITS
1-32734	12/10	12/8	Housewares	25.85	
2-44998	12/20		PAYMENT		70.00

BILLING PERIOD	PREVIOUS BALANCE	PERIODIC RATE	AVERAGE DAILY BALANCE	FINANCE CHARGE
12/1 - 12/31	$125.80	2%	?	?

PAYMENTS & CREDITS	PURCHASES & ADVANCES	NEW BALANCE	MINIMUM PAYMENT	PAYMENT DUE
$70.00	$25.85	?	$20.00	1/21

SOLUTION   **A.** Find the **sum of daily balances**.

Dates	Payment	Purchase	End-of-Day Balance		Number of Days	Sum of Balances
12/1–12/9			$125.80	×	9	$1132.20
12/10		$25.85	151.65	×	1	151.65
12/11–12/19			151.65	×	9	1364.85
12/20	$70.00		81.65	×	1	81.65
12/21–12/31			81.65	×	11	898.15
				Total	31	$3628.50

B. Find the **average daily balance**.

Sum of Daily Balances ÷ Number of Days

$3628.50   ÷ 31 = $117.048 = $117.05 average daily balance

125.8 × 9 = 1132.2 M+ 125.8 + 25.85 = 151.65 M+ × 9 = 1364.85 M+
151.65 − 70 = 81.65 M+ × 11 = 898.15 M+ MR 3628.5 ÷ 31 = 117.048

Lesson 6-6 Finance Charge—Average Daily Balance (New purchases included) ◆ **211**

Find the average daily balance, with new purchases.

Dates	Payment	Purchase	End-of-Day Balance	×	Number of Days	=	Sum of Balances
9/9–9/15			$500.00	×	7	=	**1.** ?
9/16		$100.00	$600.00	×	1	=	**2.** ?
9/17–9/21			**3.** ?	×	5	=	**4.** ?
9/22	$150.00		**5.** ?	×	1	=	**6.** ?
9/23–10/8			**7.** ?	×	16	=	**8.** ?
					Total **9.** ?		**10.** ?

The finance charge is calculated by multiplying the average daily balance times the periodic rate.

Finance Charge = Average Daily Balance × Periodic Rate

New Balance = Unpaid Balance + Finance Charge + New Purchases

EXAMPLE   *Skills* 8, 11   *Application* G   *Term* Average daily balance

Scott checks the finance charge and the new balance. The finance charge is 2% of the average daily balance. What is the new balance?

SOLUTION

A. Find the **unpaid balance.**
Previous Balance − (Payments + Credits)
$125.80   − $70.00   = $55.80 unpaid balance

B. Find the **finance charge.**
Average Daily Balance × Periodic Rate
$117.05   ×   2% = $2.3418 = $2.34 finance charge

C. Find the **new purchases.** $25.85

D. Find the **new balance.**
Unpaid Balance + Finance Charge + New Purchases
$55.80   + $2.34   + $25.85 = $83.99 new balance

PROBLEMS

Find the average daily balance, new purchases included.

11. Dates	Payment	Purchase	End-of-Day Balance	Number of Days	Sum of Balances	
9/6–9/17			$600.00	?	?	What is the average daily balance?
9/18		$140.00	?	?	?	
9/19–9/24		?	?	?		
9/25	$120.00	?	?	?		
9/26–10/5		?	?	?		
			TOTAL	?	?	

12. Edith Bertelli received this statement from Bertrand's. Find the average daily balance (new purchases included), the finance charge, and the new balance.

REFERENCE	POSTING DATE	TRANSACTION DATE	DESCRIPTION	PURCHASES & ADVANCES	PAYMENTS & CREDITS
1027485	4/11		PAYMENT		40.00
4500298	4/15	4/01	Menswear	39.95	
5473390	4/23	4/21	Housewares	15.99	
1374655	4/25		PAYMENT		50.00
BILLING PERIOD		PREVIOUS BALANCE	PERIODIC RATE	AVERAGE DAILY BALANCE	FINANCE CHARGE
4/1 - 5/1		$175.00	1.2%	?	?
PAYMENTS & CREDITS	PURCHASES & ADVANCES	NEW BALANCE	MINIMUM PAYMENT	PAYMENT DUE	
$90.00	$55.94	?	$25.00	5/25	

13. Louis Minier received this statement from Crowley Bank. Find the average daily balance (new purchases included), the finance charge, and the new balance.

REFERENCE	POSTING DATE	TRANSACTION DATE	DESCRIPTION	PURCHASES & ADVANCES	PAYMENTS & CREDITS
1616787	4/25	4/19	A-1 Plumbing	61.45	
3945557	4/30		PAYMENT		79.60
3957475	5/5	5/1	Ace Hardware	32.45	
4000076	5/20		PAYMENT		50.00
BILLING PERIOD		PREVIOUS BALANCE	PERIODIC RATE	AVERAGE DAILY BALANCE	FINANCE CHARGE
4/21 - 5/20		$179.60	2%	?	?
PAYMENTS & CREDITS	PURCHASES & ADVANCES	NEW BALANCE	MINIMUM PAYMENT	PAYMENT DUE	
$129.60	$93.90	?	$30.00	6/12	

F.Y.I.

On December 1, 1991, the average monthly credit card balance was $1626.00.

MAINTAINING YOUR SKILLS Look up the skills in parentheses if you need help or more practice.

Multiply. **(Skills 7, 8)**

14. 7 × 145 **15.** 5 × 360 **16.** 31 × 56.23 **17.** 8 × 385.21

18. 12 × 14.5 **19.** 11 × 99.78 **20.** 18 × 1455 **21.** 2 × 891.52

Divide. Round answers to the nearest hundredth. **(Skill 10)**

22. 1608.75 ÷ 30 **23.** 1329.89 ÷ 31

24. 1029 ÷ 28 **25.** 9505.08 ÷ 37

Find the average. Round answers to the nearest hundredth. **(Application Q)**

26. 36, 45, 58, 62, 48 **27.** 456.2, 364.8, 471.5, 392.18

28. 36.25, 56.87, 33.10, 93.51 **29.** 365.23, 577.10, 782.20

Reviewing the Basics

Skills

(Skill 2)

Round to the nearest cent.

1. $173.1426   **2.** $172.183   **3.** $37.998   **4.** $5.692

Solve.

(Skill 5)

5. $485.43 + $41.97 + $9.98

6. $538.26 + $6.19 + $16.71 + $157.22

(Skill 6)

7. $478.51 − $358.92   **8.** $624.84 − $37.89   **9.** $38.10 − $5.99

(Skill 8)

10. $460 × 8   **11.** $520 × 16   **12.** $1420.85 × 30

13. $22.15 × 9   **14.** $35.75 × 15   **15.** $66.95 × 19

(Skill 11)

16. $150.60 ÷ 30   **17.** $1923.75 ÷ 28   **18.** $7859.60 ÷ 30

19. $582.86 ÷ 31   **20.** $7245.79 ÷ 31   **21.** $22,812.59 ÷ 31

(Skill 30)

22. 1.5% of $720   **23.** 1% of 543.21   **24.** 2% of 267.15

Applications

(Application G)

Find the elapsed time in days.

25. From June 4 to June 30   **26.** From April 4 to April 25

27. From May 21 to June 9   **28.** From July 15 to August 28

Find the average.

(Application Q)

29. 48, 42, 74, 91, 51, 65   **30.** 12.8, 14.1, 21.2, 18.3, 20.6

31. $45, $94, $65, $83, $66   **32.** 4.54, 8.23, 9.10, 7.54

Terms

Match each term with its definition on the right.

33. Sales receipt

34. Finance charge

35. Previous balance

36. Unpaid balance

37. Average daily balance

a. the average of the balances in a charge account at the end of each day in the billing period

b. the amount of money that remains unpaid in a charge account

c. a record of the total purchase that a charge card user signs each time the account is used

d. interest that a charge or credit card user pays for delaying payment

e. the monthly finance charge rate

f. the amount that a charge or credit card user owed at the close of the last billing period

Refer to your reference files in the back of the book if you need help.

Unit Test

Lesson 6-1

1. Alma Ying used her bank charge card to purchase a stereo. The stereo cost $995.99 plus a 6% sales tax. What was the total purchase price on the sales receipt?

Lesson 6-2

2. Find the new balance on the charge account statement.

PREVIOUS BALANCE	CLOSING DATE THIS MONTH		CLOSING DATE LAST MONTH	
$175.41	November 13, 19--		October 14, 19--	
TOTAL PURCHASES	PAYMENTS & CREDITS	FINANCE CHARGE	NEW BALANCE	MINIMUM PAYMENT
$72.59	$50.00	$1.90	?	— —

Lesson 6-3

3. Ray Etzel has a charge account that charges 1.5% of the previous balance for a finance charge. His statement shows purchases totaling $98.15, a previous balance of $670, and a payment of $95. What is the new balance?

Lesson 6-4

4. A portion of Ethel Lewis's account statement is shown. The finance charge is 1.5% of the unpaid balance. Find the unpaid balance, the finance charge, and the new balance.

BILLING DATE	PREVIOUS BALANCE	PAYMENTS & CREDITS	UNPAID BALANCE	FINANCE CHARGE	NEW PURCHASES	NEW BALANCE
10/13	$413.65	$137.35	?	?	$37.10	?

Lesson 6-5

5. Find the average daily balance where new purchases are not included, the finance charge, and the new balance.

REFERENCE	POSTING DATE	TRANSACTION DATE	DESCRIPTION	PURCHASES & ADVANCES	PAYMENTS & CREDITS
13100	9/14	8/23	Westford Theatre	65.00	
20046	9/25	8/27	Record Mart	14.50	
54545	9/28		PAYMENT		40.00
BILLING PERIOD	**PREVIOUS BALANCE**	**PERIODIC RATE**	**AVERAGE DAILY BALANCE**	**FINANCE CHARGE**	
9/1 - 9/30	$150.00	1.5%	?	?	
PAYMENTS & CREDITS	**PURCHASES & ADVANCES**	**NEW BALANCE**	**MINIMUM PAYMENT**	**PAYMENT DUE**	
$40.00	?	?	$20.00	10/21	

Lesson 6-6

6. In problem 5, find the average daily balance, finance charge, and the new balance using the average-daily-balance method where new purchases are included.

A SPREADSHEET APPLICATION

Charge Accounts and Credit Cards

To complete this spreadsheet application, you will need the diskette *Spreadsheet Applications for Business Mathematics,* which accompanies this textbook. Follow the directions in the *User's Guide* to complete this activity.

Input the transactions indicated in the following problems to find the average daily balance including new purchases, finance charge, and new balance.

Note that for a payment you must enter a negative number. You do this by entering a negative or a minus sign in front of the number. Thus, a payment of $30.00 would be entered as -30 (do not use a dollar sign). Enter the periodic rate as a decimal. Thus, a periodic rate of 2% is entered as 0.02. The billing period coincides with the calendar month.

1. Balance on January 1: $675.00.
Purchase on January 5: $100.00.
Purchase on January 23: $25.00.
Payment on January 28: $200.00.
Periodic rate: 2%.
What is the average daily balance?
What is the finance charge?
What is the new balance?

2. Balance on January 1: $233.55.
Purchase on January 9: $23.55.
Purchase on January 18: $45.25.
Payment on January 25: $45.00.
Periodic rate: 1.50%.
What is the average daily balance?
What is the finance charge?
What is the new balance?

3. Balance on March 1: $379.46.
Purchase on March 11: $59.97.
Purchase on March 19: $23.56.
Purchase on March 25: $156.20.
Payment on March 30: $180.00.
Period rate: 1.75%.
What is the average daily balance?
What is the finance charge?
What is the new balance?

4. Balance on March 1: $785.23.
Purchase on March 3: $23.21.
Purchase on March 5: $156.32.
Purchase on March 10: $97.58.
Payment on March 20: $450.00.
Periodic rate: 1.75%.
What is the average daily balance?
What is the finance charge?
What is the new balance?

5. Balance on July 1: $563.25.
Payment on July 5: $350.00.
Purchase on July 12: $123.20.
Purchase on July 19: $98.54.
Purchase on July 26: $31.25.
Periodic rate: 1.65%.
What is the average daily balance?
What is the finance charge?
What is the new balance?

6. Balance on July 1: $563.25.
Purchase on July 12: $123.20.
Purchase on July 19: $98.54.
Purchase on July 26: $31.25.
Payment on July 30: $350.00.
Periodic rate: 1.65%.
What is the average daily balance?
What is the finance charge?
What is the new balance?

Loans

A *loan* is money that you have borrowed and must repay with *interest*. Interest is the cost of borrowing money. A *single-payment loan* (sometimes called a promissory note) is repaid in one payment after a specified period of time. An *installment loan* is a loan for which you pay a portion of the loan and a portion of the interest in several installments. Because all banks do not charge the same amount of interest for a sum of money, you should shop around for the best buy. Using the *annual percentage rate* is one way to compare the cost of borrowing money. If you repay a loan early, you may receive a *refund* of part of the finance charge.

O U T L I N E

Before you can get a loan, you must fill out a credit application form. A credit investigation is made from the information about your income, bank accounts, and any credit that has been extended to you.

Single-Payment Loans

OBJECTIVE

Compute the maturity value and interest rate of a single-payment loan.

A single-payment loan is a loan that you repay with one payment after a specified period of time. A promissory note is a type of single-payment loan. It is a written promise to pay a certain sum of money on a certain date in the future. The maturity value of the loan is the total amount you repay. It includes both the principal and the interest owed. The principal is the amount borrowed.

The term of a loan is the amount of time for which the loan is granted. A single-payment loan may be granted for a stated number of years, months, or days. When the term is a certain number of days, the lending agency may calculate interest in one of two ways. Ordinary interest is calculated by basing the time of the loan on a 360-day year. Exact interest is calculated by basing the time on a 365-day year.

Maturity Value = Principal + Interest Owed

EXAMPLE   *Skills* 28, 14   *Application* J   *Term* Single-payment loan

Anita Sloane's bank granted her a single-payment loan of $7200 for 91 days at an interest rate of 12%. What is the maturity value of the loan if: (1) her bank charges ordinary interest? (2) her bank charges exact interest?

SOLUTION

A. (1) Find the **ordinary interest owed.**
Principal × Rate × Time
$7200.00 × 12% × $\frac{91}{360}$ = $218.40 ordinary interest

(2) Find the **exact interest owed.**
Principal × Rate × Time
$7200.00 × 12% × $\frac{91}{365}$ = $215.408 = $215.41 exact interest

B. (1) Find the **maturity value.**
Principal + Interest Owed
$7200.00 + $218.40 = $7418.40 maturity value

(2) Find the **maturity value.**
Principal + Interest Owed
$7200.00 + $215.41 = $7415.41 maturity value

7200 [M+] [×] 12 [%] [×] 91 [÷] 360 [=] 218.4 [M+] [RM] 7418.4

✔ SELF-CHECK   Complete the problems, then check your answers in the back of the book.

1. Compute the ordinary interest and the maturity value.
$600 × 10% × $\frac{90}{360}$ = ?

2. Compute the exact interest and the maturity value.
$800 × 12% × $\frac{75}{365}$ = ?

		3.	4.	5.	6.	7.
Principal		$900	$9600	$850	$7500	$9435
Annual Interest Rate		12%	15%	11%	12.5%	18.72%
Term		45 days	30 days	72 days	123 days	275 days
Ordinary Interest		$13.50	?	?	?	?
Maturity Value		?	?	?	?	?

8. Maria Rodriquez.
Single-payment loan of $1000.
Interest rate of 15%.
108 days, ordinary interest.
What is the interest owed?
What is the maturity value?

9. Manuel Bruins.
Single-payment loan of $8400.
Interest rate of 12%.
146 days, exact interest.
What is the interest owed?
What is the maturity value?

10. Joseph Henning's bank granted him a single-payment loan of $2400 for 144 days at an interest rate of 18%. His bank charges ordinary interest. What is the maturity value of his loan?

11. Vanessa Tackett's bank granted her a single-payment loan of $21,000 for 45 days at an interest rate of 15%. Her bank charges ordinary interest. What is the maturity value of her loan?

12. Jessie Ardella obtained a single-payment loan of $225 to pay a repair bill. He agreed to repay the loan in 31 days at an interest rate of 14.75% exact interest. What is the maturity value of his loan?

13. Gordon Hamen obtained a single-payment loan of $44,000 to pay for tree spray for his commercial orchard. He agreed to repay the loan in 270 days at an interest rate of 14.35% exact interest. What is the maturity value of his loan?

14. Mark Norwalk would like to borrow $8000 for 90 days. Advance Finance Company charges an interest rate of 18% at ordinary interest. AAA Finance Company also charges an interest rate of 18% but at exact interest. What would be the maturity value of the loan at Advance Finance Company? What would be the maturity value of the loan at AAA Finance Company?

MAINTAINING YOUR SKILLS   Look up the skills in parentheses if you need help or more practice.

Write the percents as decimals. **(Skill 28)**

15. 40%   **16.** 90.5%   **17.** 7%   **18.** 8.25%   **19.** 15.64%

Reduce the fractions to lowest terms. **(Skill 12)**

20. $\frac{40}{60}$ **21.** $\frac{180}{360}$ **22.** $\frac{90}{360}$ **23.** $\frac{120}{360}$ **24.** $\frac{73}{365}$

Change the fractions to decimals. **(Skill 14)**

25. $\frac{40}{60}$ **26.** $\frac{45}{360}$ **27.** $\frac{126}{360}$ **28.** $\frac{146}{365}$ **29.** $\frac{219}{365}$

Installment Loans

OBJECTIVE
Compute the amount financed on an installment loan.

An installment loan is a loan that you repay in several equal payments over a specified period of time. Usually when you purchase an item with an installment loan, you must make a down payment. The down payment is a portion of the cash price of the item you are purchasing. The amount financed is the portion of the cash price that you owe after making the down payment.

Amount Financed = Cash Price − Down Payment

EXAMPLE   *Skills* 30, 28, 2   *Application* A   *Term* Installment loan

Rebecca Clay purchased a washer and a dryer for $1140. She used the store's installment credit plan to pay for the items. She made a down payment and financed the remaining amount. What amount did Rebecca finance if she made: (1) a 20% down payment? (2) a 25% down payment?

SOLUTION

(1) Find the 20% **down payment.**
$1140.00 × 20% = $228.00

(2) Find the 25% **down payment.**
$1140.00 × 25% = $285.00

(1) Find the **amount financed.**
Cash Price − Down Payment
$1140.00 − $228.00
= $912.00 amount financed

(2) Find the **amount financed.**
Cash Price − Down Payment
$1140.00 − $285.00
= $855.00 amount financed

1140 [M+] [×] 20 [%] 228 [M−] [RM] 912

✔ SELF-CHECK   Complete the problems, then check your answers in the back of the book.

Find the down payment and the amount financed.

1. Waterbed.
Cash price of $1360.
20% down payment

2. Television set.
Cash price of $725.
30% down payment

PROBLEMS

	3.	4.	5.	6.	7.	8.
Cash Price	$640	$4860	$1774	$3600	$9480	$5364
Percent Down Payment	20%	30%	25%	40%	15%	25%
Down Payment	$128	?	?	?	?	?
Amount Financed	?	?	?	?	?	?

9. Cash price of $1265.
20% down payment.
What amount is financed?

10. Cash price of $14,470.
25% down payment.
What amount is financed?

11. Cash price of $8371.39.
15% down payment.
What amount is financed?

12. Cash price of $18,936.50.
30% down payment.
What amount is financed?

13. Milt Gibson purchased computer equipment for $4020. He used the store's credit plan. He made a 20% down payment. What amount did he finance?

14. Linda Chevez purchased a stereo for her car. The stereo cost $279.50. Using the store's credit plan, she made a 30% down payment. What amount did she finance?

15. Ardella Haubert purchased living room furniture for $987.95. She made a down payment of 20% and financed the remaining amount using the store's installment plan. What amount did she finance?

16. Bev and Tom Hoffman went on a two-week vacation at a total cost of $1876. They financed the trip through Sentinel Bank. They made a 20% down payment and financed the remaining amount on the installment plan. What amount did they finance?

> **SENTINEL BANK offers**
> *TRAVEL and ADVENTURE*
>
> Go NOW . . . . Vacation loans
> Pay   ONLY
> Later . . . . 20% down

17. Amy and Cliff Martin want to remodel their kitchen. They would like to finance part of the cost but do not want the amount financed to be more than $9000. The total cost of remodeling the kitchen is $12,000. What percent of the total cost should their down payment be?

18. Mack Casey wants to purchase a car costing $14,590. He will finance the car with an installment loan from the bank but would like to finance no more than $10,000. What percent of the total cost of the car should his down payment be?

MAINTAINING YOUR SKILLS   Look up the skills in parentheses if you need help or more practice.

Write the percents as decimals. **(Skill 28)**

19. 32%   **20.** 45%   **21.** 25%   **22.** 30%   **23.** 20%

Find the percentage. **(Skill 30)**

24. 440 × 30%   **25.** 325 × 20%   **26.** 1240 × 25%   **27.** 950 × 15%

Round to the nearest cent. **(Skill 2)**

28. $49.9638   **29.** $178.3813   **30.** $413.995   **31.** $17.6309

7-3

Simple Interest
Installment Loans

OBJECTIVE

Compute the monthly payment, total amount repaid, and finance charge on an installment loan.

When you obtain a simple interest installment loan, you must pay finance charges for the use of the money. Usually you repay the amount financed plus the finance charge in equal monthly payments. The amount of each monthly payment depends on the amount financed, the number of payments, and the annual percentage rate (APR). The annual percentage rate is an index showing the relative cost of borrowing money.

MONTHLY PAYMENT ON A $100 LOAN

Term in Months	10%	12%	15%	18%
6	$17.16	$17.25	$17.40	$17.55
12	8.79	8.88	9.03	9.17
18	6.01	6.10	6.24	6.38
24	4.61	4.71	4.85	4.99
30	3.78	3.87	4.02	4.16
36	3.23	3.32	3.47	3.62
42	2.83	2.93	3.07	3.23
48	2.54	2.63	2.78	2.94

$$\text{Monthly Payment} = \frac{\text{Amount of Loan}}{\$100} \times \text{Monthly Payment for a \$100 loan}$$

$$\text{Total Amount Repaid} = \text{Number of Payments} \times \text{Monthly Payment}$$

$$\text{Finance Charge} = \text{Total Amount Repaid} - \text{Amount Financed}$$

EXAMPLE   *Skills* 8, 6   *Application* C   *Term* Annual percentage rate

Clara Hart obtained an installment loan of $1850 to purchase new furniture. The annual percentage rate is 15%. She must repay the loan in 18 months. What is the finance charge?

SOLUTION

A. Find the **monthly payment.** (Refer to the table above.)

$$\frac{\text{Amount of Loan}}{\$100} \times \text{Monthly Payment for a \$100 loan}$$

$$\frac{\$1850}{\$100} \times \$6.24 = \$115.44 \text{ monthly payment}$$

B. Find the **total amount repaid.**

Number of Payments × Monthly Payment

18 × $115.44 = $2077.92 total amount repaid

C. Find the **finance charge.**

Total Amount Repaid − Amount Financed

$2077.92 − $1850.00 = $227.92 finance charge

1850 [M+] [÷] 100 [×] 6.24 [=] 115.44 [×] 18 [=] 2077.92 [−] [RM] 1850 [=] 227.92

1. Find the monthly payment, total amount repaid, and the finance charge for a $1650 installment loan at 18% for 24 months.

PROBLEMS

Use the table on page 222 to solve the following.

	2.	**3.**	**4.**	**5.**	**6.**
APR	10%	15%	18%	12%	15%
Term (Months)	6	18	24	6	36
Amount Financed	$1000	$1780	$600	$900	$4350
Monthly Payment	?	?	?	?	?
Total Repaid	?	?	?	?	?
Finance Charge	?	?	?	?	?

7. Hazel Basnett.
   Installment loan of $2000.
   12 monthly payments.
   APR is 18%.
   What are the monthly payments?
   What is the finance charge?

8. Brian Anderson.
   Installment loan of $750.
   24 monthly payments.
   APR is 10%.
   What are the monthly payments?
   What is the finance charge?

9. Bob Wozniak obtained an installment loan of $2400 to put a roof on his house. The APR is 12%. The loan is to be repaid in 36 monthly payments. What is the finance charge?

10. Jim Wilson obtained an installment loan of $1450 to pay for some new furniture. He agreed to repay the loan in 18 monthly payments at an APR of 15%. What is the finance charge?

11. Mark and Pam Voss obtained an installment loan of $2460. They obtained the loan at an APR of 10% for 12 months. What is the finance charge?

12. Herb and Marci Rahla are purchasing a dishwasher with an installment loan that has an APR of 18%. The dishwasher sells for $699.95. They agree to make a down payment of 20% and to make 12 monthly payments. What is the finance charge?

13. Andrew and Ruth Bacon would like to obtain an installment loan of $1850 to repair the gutters on their home. They can get the loan at an APR of 15% for 24 months or at an APR of 18% for 18 months. Which loan costs less? How much do the Bacons save by taking the loan that costs less?

14. Lola Samaria would like an installment loan of $950. Walton Savings and Loan will loan her the money at 15% for 12 months. Horton Finance Company will loan her the money at 18% for 24 months. Which loan costs less? How much will she save by taking the loan that costs less?

15. Adolfo Ramirez obtained an installment loan of $2800 for a used car. He financed the purchase with a finance company and agreed to repay the loan in 24 monthly payments at an APR of 22%. What is the finance charge?

MONTHLY PAYMENT ON A $100 LOAN				
Term in Months	Annual Percentage Rate			
	20%	22%	24%	26%
6	$17.65	$17.75	$17.85	$17.95
12	9.26	9.36	9.45	9.55
18	6.48	6.57	6.67	6.77
24	5.09	5.19	5.29	5.39
30	4.26	4.36	4.46	4.57
36	3.72	3.82	3.92	4.03
42	3.33	3.43	3.54	3.65
48	3.04	3.15	3.26	3.37

16. Aurora Kaylow obtained an installment loan of $6000 on a used sailboat. She financed the purchase through the boat dealer and agreed to repay the loan in 48 monthly payments at an APR of 24%. What is the finance charge?

17. Pauline and Eldon Kharche would like to obtain an installment loan of $1800. They can get the loan at an APR of 22% for 24 months or at an APR of 20% for 18 months. Which loan costs less? How much do the Kharches save by taking the loan that costs less?

18. Lucretia and Don Protsman would like an installment loan of $3280. City Loan will loan the money at 24% for 24 months. Economy Line Finance Company will loan the money at 22% for 30 months. Which loan costs less? How much will be saved by taking the loan that costs less?

19. Sue and Tom Weber plan to buy a home computer that costs $2999.95. They have a down payment of $499.95 and will finance the remainder at an APR of 22% for 24 months. What is the finance charge?

20. Harold and Nora O'Desky obtained an installment loan to purchase the garage advertised for $3295. They made a down payment of $295. The bank has agreed to loan them the remainder of the money at an APR of 20% for 48 months. What is the finance charge?

MAINTAINING YOUR SKILLS   Look up the skills in parentheses if you need help or more practice.

Multiply. **(Skills 30, 20)**

21. $4000 \times 6\% \times \frac{1}{12}$

22. $1240 \times 12\% \times \frac{1}{12}$

23. $1800 \times 9\% \times \frac{1}{12}$

24. $\$1500 \times 12\% \times \frac{1}{12}$

25. $\$4250 \times 15\% \times \frac{1}{12}$

26. $\$3600 \times 15\% \times \frac{1}{12}$

Subtract. **(Skill 6)**

27. 4200
    − 42

28. 1240
    − 15.50

29. 1224.50
    − 15.31

30. 7321.65
    − 29.86

Installment Loans—Allocation of Monthly Payment

A simple interest installment loan is repaid in equal monthly payments. Part of each payment is used to pay the interest on the unpaid balance of the loan, and the remaining part is used to reduce the balance. The interest is calculated each month using the simple interest formula. The amount of principal that you owe decreases with each monthly payment. A repayment schedule shows the distribution of interest and principal over the life of the loan. The repayment schedule here shows the interest and principal on an installment loan of $1800 for 6 months at 15%.

REPAYMENT SCHEDULE FOR AN $1800 LOAN AT 15% FOR 6 MONTHS

Payment Number	Monthly Payment	Amount for Interest	Amount for Principal	New Principal	
1	$313.20	$22.50	$290.70	$1509.30	
2	313.20	18.87	294.33	1214.97	Note that the last
3	313.20	15.19	298.01	916.96	payment would be
4	313.20	11.46	301.74	615.22	increased by $0.38
5	313.20	7.69	305.51	309.71	in order to zero
6	313.20	3.87	309.33	0.38	← out the loan.

Payment to Principal = Monthly Payment − Interest

New Principal = Previous Principal − Payment to Principal

EXAMPLE *Skills* 6, 8, 30 *Application* A *Term* Repayment schedule

Stephanie and Donald Cole obtained the loan of $1800 at 15% for 6 months shown in the repayment schedule. Show the calculation for the first payment. What is the interest? What is the payment to principal? What is the new principal?

SOLUTION

A. Find the **interest.**
Principal × Rate × Time
$1800.00 × 15% × $\frac{1}{12}$ = $22.50 interest

B. Find the **payment to principal.**
Monthly Payment − Interest
$313.20 − $22.50 = $290.70 payment to principal

C. Find the **new principal.**
Previous Principal − Payment to Principal
$1800.00 − $290.70 = $1509.30 new principal

✔ SELF-CHECK Complete the problems, then check your answers in the back of the book.

1. Interest the second month is: $1509.300 × 15% × 1/12 = ?

2. Payment to principal is: $313.20 − ? = ?

3. The new balance is: $1509.30 − ? = ?

	Amount Financed	Interest Rate	Monthly Payment	Amount for Interest	Amount for Principal	New Principal
4.	$1200	12%	$106.56	$12.00	$94.56	?
5.	$2400	15%	$116.40	$30.00	?	?
6.	$3460	10%	$207.95	?	?	?
7.	$1680	20%	$ 85.51	?	?	?
8.	$ 860	24%	$ 45.49	?	?	?
9.	$ 975	26%	$ 66.01	?	?	?

10. Joan and Bill Kelly.
   Furniture loan of $2400.
   Interest rate is 12%.
   Monthly payment is $113.04.
   How much of the first monthly
      payment is for interest?

11. Annie and Fred Petroff.
   Appliance loan of $2000.
   Interest rate of 18%.
   Monthly payment is $99.80.
   How much of the first monthly
      payment is for interest?

12. Wilma and Glen Barnes.
   Used car loan of $3600.
   Interest rate is 20%.
   Monthly payment is $133.92.
   How much of the first monthly
      payment is for interest?
   How much of the first monthly
      payment is for principal?
   What is the new balance?

13. Cora and Jessie Vineyard.
   Personal loan of $1500.
   Interest rate of 24%.
   Monthly payment is $79.35.
   How much of the first monthly
      payment is for interest?
   How much of the first monthly
      payment is for principal?
   What is the new balance?

14. Maxine Berlen obtained a 12-month loan of $1800 for tuition from First Federal Savings and Loan of Delta. The interest rate is 10%. Her monthly payment is $158.22. For the first payment, what is the interest? What is the payment to principal? What is the new principal?

15. Arthur Greenler obtained a 24-month loan of $6000 for a used car from Economy Loan Inc. The interest rate is 20%. His monthly payment is $305.40. For the first payment, what is the interest? What is the payment to principal? What is the new principal?

16. Orvil Marshall obtained a 42-month loan of $5550 for new windows for his home. The interest rate is 15%. His monthly payment is $170.39. For the first payment, what is the interest? What is the payment to principal? What is the new principal?

17. Phyllis Peterson obtained an 18-month loan of $3200 for a used sailboat from the boat dealer. The interest rate is 22%. Her monthly payment is $213.44. For the first payment, what is the interest? What is the payment to principal? What is the new principal?

Refer to the table of page 646–647 for problems 18 and 19.

18. Anna McGee obtained an 18-month loan of $3980 for a used car from the car dealer. The interest rate is 14%. What is the monthly payment? For the first payment, what is the interest? What is the payment to principal? What is the new principal?

19. Tyrone Murdoch obtained a 48-month loan of $8400 for a used truck from the dealer. The interest rate is 16%. What is the monthly payment? For the first payment, what is the interest? What is the payment to principal? What is the new principal?

For problems 20 to 26, complete the repayment schedule for a loan of $2400 at 12% for 12 months.

Payment Number	Monthly Payment	Amount for Interest	Amount for Principal	New Principal
1	$213.12	$24.00	$189.12	$2210.88
2	213.12	22.11	191.01	2019.87
3	213.12	20.20	192.92	1826.95
4	213.12	18.27	194.85	1632.10
5	213.12	16.32	196.80	1435.30
20. 6	213.12	14.35	198.77	?
21. 7	213.12	12.37	?	?
22. 8	213.12	?	?	?
23. 9	213.12	?	?	?
24. 10	213.12	?	?	?
25. 11	213.12	?	?	?
26. 12	?	?	?	?

REPAYMENT SCHEDULE FOR A $2400 LOAN AT 12% FOR 12 MONTHS

MAINTAINING YOUR SKILLS Look up the skill in parentheses if you need help or more practice.

Find the percentage. (Skill 30)

27. 12% of $5000
28. 15% of $6000
29. 8% of $8400
30. 22% of $1282.15
31. 26% of $2348.90
32. 20% of $456.21
33. 6% of $340.80
34. 15% of $9845.20
35. 7% of $12,346.97

Paying Off Simple Interest Installment Loans

OBJECTIVE

To compute the final payment when paying off a simple interest installment loan.

The Truth-in-Lending Law specifies that if a loan is paid off early, the lender must disclose the method for paying off the loan. Because interest is always paid on the unpaid balance, you just pay the previous balance plus the current month's interest if you pay off a simple interest installment loan before the end of the term. The final payment is the previous balance plus the current month's interest.

Final Payment = Previous Balance + Current Month's Interest

EXAMPLE   *Skills* 5, 6, 8, 30   *Application* A   *Term* Final payment

The first three months of the repayment schedule for Doug and Donna Collins's loan of $1800 at 12% for 6 months is shown. What is the final payment if they pay the loan off with payment number 4?

REPAYMENT SCHEDULE FOR AN $1800 LOAN AT 12% FOR 6 MONTHS

Payment Number	Monthly Payment	Amount for Interest	Amount for Principal	New Principal
1	$310.50	$18.00	$292.50	$1507.50
2	310.50	15.08	295.42	1212.08
3	310.50	12.12	298.38	913.70

SOLUTION

A. Find the **previous balance.** It is $913.70.

B. Find the **interest** for the fourth month.

Principal	×	Rate	×	Time		
$913.70	×	12%	×	$\frac{1}{12}$	= $9.137	= $9.14 interest

C. Find the **final payment.**

Previous Balance	+	Current Month's Interest		
$913.70	+	$9.14	=	$922.84 final payment

✔ SELF-CHECK   Complete the problems, then check your answers in the back of the book.

Find the interest for a month and then the final payment.

1. Previous balance of $800 at 12%.

2. Previous balance of $1280 at 15%.

PROBLEMS

Find the interest and the final payment.

	3.	**4.**	**5.**	**6.**	**7.**
Interest Rate	12%	15%	10%	18%	22%
Previous Balance	$4800.00	$3000.00	$1460.80	$3987.60	$3265.87
Interest	$ 48.00	?	?	?	?
Final Payment	?	?	?	?	?

8. Chris Worthington.
   Previous balance of $2460.
   Interest rate is 20%.
   What is the interest?
   What is the final payment?

9. Walter Tavinier.
   Previous balance of $8258.
   Interest rate is 15%.
   What is the interest?
   What is the final payment?

10. Willard Hudson took out a simple interest loan of $6000 at 10% for 24 months. After 4 payments, the balance is $5081.23. He pays off the loan when the next payment is due. What is the interest? What is the final payment?

11. Lillian Hartwick took out a simple interest loan of $3600 at 18% for 12 months. After 6 payments, the balance is $1879.90. She pays off the loan when the next payment is due. What is the interest? What is the final payment?

12. Scott DuBois took out a simple interest loan of $1800 for home repairs. The loan is for 12 months at 10% interest. After 8 months, the balance is $620.26. He pays off the loan when the next payment is due. What is the final payment?

13. Carolyn Frincke took out a simple interest loan of $6000 for a used car. The loan is for 24 months at 20% interest. After 17 payments, the balance is $2001.46. She pays off the loan when the next payment is due. What is the final payment?

14. Nicholas and Dorothea Schrodt were looking over the repayment schedule for their boat loan of $5550 at 15% for 42 months. They note the following:

    a. Balance after payment 18 is $3525.03.
    b. Balance after payment 24 is $2742.99.
    c. Balance after payment 30 is $2045.25.

    Determine the final payment if they pay the loan off when the 19th payment is due. What if they wait until the 25th payment? The 31st payment?

15. **A Brief Case**   Stanley Huston purchased a Rototiller for $1318.45. He made a down payment of 20%. The dealer financed the remainder at 12% simple interest for 1 year. Stan's custom gardening prospered, and he paid the loan off at the time the fifth payment was due. How much was his final payment?

Multiply. (Skill 8)

16. 5489×0.15

17. 2729×0.22

18. 9032×0.18

19. 983.54×0.12

20. 657.80×0.06

21. 6118.53×0.09

Find the percentage. (Skill 30)

22. $430 \times 18\%$

23. $3561.90 \times 9\%$

24. $\$10,907.45 \times 15\%$

7-6

Determining the APR

OBJECTIVE

Use a table to find the annual percentage rate of a loan.

If you know the number of monthly payments and the finance charge per $100 of the amount financed, you can use a table to find the annual percentage rate of the loan. You can use the APR of loans to compare the relative cost of borrowing money.

APR	10.00%	10.25%	10.50%	10.75%	11.00%	11.25%	11.50%	11.75%	12.00%	12.25%	12.50%
Term	Finance Charge Per $100 of Amount Financed										
6	$ 2.94	$ 3.01	$ 3.08	$ 3.16	$ 3.23	$ 3.31	$ 3.38	$ 3.45	$ 3.53	$ 3.60	$ 3.68
12	5.50	5.64	5.78	5.92	6.06	6.20	6.34	6.48	6.62	6.76	6.90
18	8.10	8.31	8.52	8.73	8.93	9.14	9.35	9.56	9.77	9.98	10.19
24	10.75	11.02	11.30	11.58	11.86	12.14	12.42	12.70	12.98	13.26	13.54

$$\text{Finance Charge per \$100} = \$100 \times \frac{\text{Finance Charge}}{\text{Amount Financed}}$$

EXAMPLE   *Skills* 11, 2   *Application* C   *Term* Annual percentage rate

Paul Norris obtained an installment loan of $1500 to pay for ham radio equipment. The finance charge is $146.25. He agreed to repay the loan in 18 monthly payments. What is the annual percentage rate?

SOLUTION

A. Find the **finance charge per $100.**
$100 ×(Finance Charge ÷ Amount Financed)
$100 ×   ($146.25 ÷ $1500.00)
$100 ×   0.0975   = $9.75 finance charge per $100

B. Find the **APR.** (Refer to the table above.)
In the row for 18 payments, find the number closest to $9.75. It is $9.77. Read the APR at the top of the column. APR is 12.00%.

✔ SELF-CHECK   Complete the problems, then check your answers in the back of the book.

1. 6-month loan.
$100 × ($24.64 ÷ $800) = ?
APR = ?

2. 24-month loan.
$100 × ($96.22 ÷ $850) = ?
APR = ?

Complete the table. Use the APR table on page 646–647.

	3.	4.	5.	6.	7.
Finance Charge	$33.10	$434.16	$421.50	$ 1652	$597.66
Amount Financed	$1,000	$ 2,400	$ 3,000	$5,400	$ 4,200
Finance Charge per $100	$3.31	?	?	?	?
Number of Payments	6	24	18	36	30
Annual Percentage Rate	?	?	?	?	?

8. Ed Naiman.
Installment loan of $2500.
Finance charge of $430.50.
24 monthly payments.
What is the APR?

9. Betty Arca.
Installment loan of $800.
Finance charge of $170.40.
36 monthly payments.
What is the APR?

10. Webster Larkin.
Installment loan of $300.
Finance charge of $9.96.
6 monthly payments.
What is the APR?

11. Kenneth Bryant.
Installment loan of $9365.
Finance charge of $2823.
36 monthly payments.
What is the APR?

12. Marie Brenson obtained an installment loan of $460 to purchase computer software. The finance charge is $19.32. She agreed to repay the loan in 6 monthly payments. What is the annual percentage rate?

13. Herb Stanley obtained an installment loan of $6800 to pay his daughter's college tuition. The finance charge is $731. He agreed to repay the loan in 24 monthly payments. What is the APR?

14. Jeff Stapleton obtained an installment loan of $395 to pay for car repairs. The finance charge is $36.34. He agreed to repay the loan in 12 monthly payments. What is the APR?

15. Julia Bourne obtained an installment loan of $3800 to purchase a lawn and garden tractor. The finance charge is $722.38. She agreed to repay the loan in 24 monthly payments. What is the APR?

F.Y.I.
The national average APR on simple interest installment loans for new cars was 10.95% on December 1, 1991.

16. Oneta Correy wants to obtain an installment loan of $1960 to purchase the used car advertised. The bank has agreed to loan her the money. She must repay the loan in 24 months. The finance charge is $337.51. What is the APR of her loan?

SALE $1960.00
1983 4-dr sedan, a/c, power equipment, one owner. No rust. Easy financing.

17. Helen Olson needs an installment loan of $950 to purchase a VHS recorder. The store has agreed to loan her the money. She must repay the loan in 24 months. The finance charge is $150.10. What is the APR of her loan?

18. Brent and Lola Miller are buying a washer that costs $399.95 and a dryer that costs $249.99. To use the store's installment plan, they need a down payment of $49.94. They must make 18 monthly payments of $37.08 each. What is the APR on their installment loan?

19. Jorge Holland is buying a new furnace that costs $2600. The bank requires a down payment of 20% and 36 monthly payments of $66.95 each. What is the APR on his loan?

20. Ty Chin is buying a new color television set that costs $659.38. To use the installment plan available at the department store, he must make a down payment of 25% and make 30 monthly payments of $19.46 each. What is the APR on his loan?

21. Andrew Stachowick would like an installment loan of $5000 to be repaid in 36 months. Nathaniel Loan Company will grant the loan with a finance charge of $1215.60. City Finance Company will grant the loan with a finance charge of $1532.50. What is the APR on each loan?

22. Wayne Charles would like an installment loan of $8000 to be repaid in 24 months. ABC Finance Company will grant the loan with a finance charge of $1984. Atco Financial Service will grant the loan with a finance charge of $2010. What is the APR on each loan?

23. Eleanor Penny purchased a $1987 dining room set. She made a down payment of $87 and through the store's installment plan agreed to pay $169 a month for 12 months. What is the finance charge? What is the APR? What is the total amount paid for the set?

24. Kathleen Dunn purchased a home computer for $3249.29. She made a down payment of $649.29 and financed the remainder by agreeing to pay $129.23 per month for 24 months. What is the finance charge? What is the APR? What is the total amount paid for the computer?

MAINTAINING YOUR SKILLS   Look up the skills in parentheses if you need help or more practice.

Round to the nearest cent. **(Skill 2)**

25. $19.439   **26.** $12.4162   **27.** $40.3072   **28.** $19.3019

Divide. Round answers to the nearest hundredth. **(Skills 10, 11)**

29. $17\overline{)510}$ **30.** $24\overline{)1060}$ **31.** $47\overline{)2642}$ **32.** $984\overline{)67,182}$

33. $1.4\overline{)98.22}$ **34.** $8.2\overline{)219.6}$ **35.** $76.1\overline{)94.88}$ **36.** $64.2\overline{)104.76}$

Multiply. **(Skill 8)**

37. 100×4.21 **38.** 10×0.3174

39. 100×0.0192 **40.** 100×0.71442

7-7

Refund of Finance Charge

OBJECTIVE

Compute the finance charge refunded for early repayment of a loan.

If you repay an installment loan that is not a simple interest installment loan before the final due date, you may be entitled to a refund of part of the finance charge. The refund is usually stated as a percent of the finance charge. The lender may determine the percent refund using a rebate schedule, which is a table of percent refunds.

Refund = Finance Charge × Percent Refund

F.Y.I.

The rebate schedule is based on the "Rule-of-78."

Term of Loan	Number of Months Loan Has Run													
	1	2	3	4	5	6	7	8	9	10	11	12	13	14
3	50.00	16.67	0											
6	71.43	47.62	28.57	14.29	4.76	0								
9	80.00	62.22	46.67	33.33	22.22	13.33	6.67	2.22	0					
12	84.62	70.51	57.69	46.15	35.90	26.92	19.23	12.82	7.69	3.85	1.28	0		
15	87.50	75.83	65.00	55.00	45.83	37.50	30.00	23.33	17.50	12.50	8.33	5.00	2.50	0.83
18	89.47	79.53	70.18	61.40	53.22	45.61	38.60	32.16	26.32	21.05	16.37	12.28	8.77	5.85

EXAMPLE *Skills* 30, 2 *Application* C *Term* Rebate schedule

Viola Stambaugh had a 12-month installment loan. The total finance charge on the loan was $113.80. How much was Viola's refund if she paid off the loan with the:

A. 7th payment?   **B.** 10th payment?

SOLUTION

A. (1) Find the **percent refund.**
Refer to the table above.
Term of loan in months   12
Number of months loan
has run 7
Value from table is 19.23%

(2) Find the **refund.**
Finance × Percent
Charge   Refund
$113.80 × 19.23% = $21.883
= $21.88 refund

B. (1) Find the **percent refund.**
Refer to the table above.
Term of loan in months 12
Number of months loan
has run 10
Value from table is 3.85%

(2) Find the **refund.**
Finance × Percent
Charge   Refund
$113.80 ×3.85% = $4.3813
= $4.38 refund

✔ SELF-CHECK Complete the problems, then check your answers in the back of the book.

A 15-month loan has run for 8 months. The finance charge was $252.80.
  1. Find the percent refund.   **2.** Find the refund.

Use the table on page 233 to find the percent refund.

	3.	**4.**	**5.**	**6.**	**7.**	**8.**
Term of Loan (Months)	15	15	9	18	6	12
Finance Charge	$60.00	$140.00	$350.00	$417.75	$281.60	$727.12
Months Loan Has Run	7	4	5	13	4	6
Percent Refund	30.00%	55.00%	?	?	?	?
Amount of Refund	?	?	?	?	?	?

9. Chloe Carter.
15-month installment loan.
$224.50 total finance charge.
Repaid loan in 6 months.
What is the refund?

10. Kevin Taylor.
9-month installment loan.
$82.50 total finance charge.
Repaid loan in 6 months.
What is the refund?

11. Carlo Blanco had a 15-month installment loan that he repaid in 6 months. The total finance charge was $224. How much was his refund?

12. Ross Wyman had an 18-month installment loan that he repaid in 7 months. The total finance charge was $439.44. How much was his refund?

13. Ruth Royer took a 12-month installment loan to help pay for her new washer. The total finance charge was $49.96. Ruth repaid the loan in 4 months. How much was her refund?

14. Eileen Carl obtained an installment loan of $2400 from the bank. She agreed to pay $218.50 a month for 12 months. The finance charge totaled $222. How much of the finance charge can Eileen save by repaying the loan in 10 months rather than 12 months?

15. Teshona Ku had an installment loan of $8000 in which he agreed to pay $487.87 a month for 18 months. How much is the finance charge? How much can he save by repaying the loan in 12 months?

MAINTAINING YOUR SKILLS   Look up the skills in parentheses if you need help or more practice.

Round to the nearest hundredth. **(Skill 2)**

16. 87.334   **17.** 421.065   **18.** 492.596   **19.** 339.996

Find the percentage. **(Skill 30)**

20. 80 × 18% **21.** 340 × 12% **22.** 68 × 7.82% **23.** 39.4 × 17.50%

Reviewing the Basics

Skills

(Skill 2)

Round to the nearest cent.

1. $732.513   **2.** $48.266   **3.** $84.599   **4.** $24.683

Solve.

(Skill 6)

5. $839.44 − $721.68   **6.** $35.29 − $7.20   **7.** $4113.97 − $873.98

(Skill 8)

8. $19 × 18   **9.** $13.52 × 12   **10.** $123.24 × 48

(Skill 11)

11. $48.00 ÷ $380   **12.** $32.25 ÷ $250   **13.** $714.44 ÷ $1850

(Skill 30)

14. 26% of $500   **15.** 45% of $440   **16.** 32.5% of $221.98

Write as a decimal.

(Skill 14)

17. $\frac{3}{4}$ **18.** $\frac{36}{360}$ **19.** $\frac{7}{12}$ **20.** $\frac{250}{360}$ **21.** $\frac{73}{365}$

(Skill 28)

22. 73%   **23.** 12%   **24.** 9.25%   **25.** 15.3%   **26.** 10.125%

Write as a percent.

(Skill 26)

27. 0.15   **28.** 0.45   **29.** 0.137   **30.** 0.356   **31.** 0.2235

32. 0.125   **33.** 0.454   **34.** 1.571   **35.** 20.512   **36.** 0.0051

Applications

(Application C)

Use the table to find the monthly payment.

37. $700 financed for 12 months at 15% APR.

38. $3140 financed for 6 months at 12% APR.

MONTHLY PAYMENT ON A $100 LOAN			
Term in Months	Annual Percentage Rate		
	10%	12%	15%	18%
6	$17.16	$17.25	$17.40	$17.55
12	8.79	8.88	9.03	9.17
18	6.01	6.10	6.24	6.38
24	4.61	4.71	4.85	4.99

Write as a fraction of a year.

(Application J)

39. 10 months   **40.** 6 months   **41.** 7 months   **42.** 3 months

43. 185 days (exact year)   **44.** 170 days (ordinary year)

45. 90 days (ordinary year)   **46.** 90 days (exact year)

Terms

Use each term in a sentence that explains the term.

47. Single-payment loan   **48.** Rebate schedule   **49.** Down payment

50. Annual percentage rate   **51.** Installment loan   **52.** Interest

Refer to your reference files in the back of the book if you need help.

Unit Test

Lesson 7-1

1. Charles Quick's bank granted him a single-payment loan of $3240 for 100 days at an annual interest rate of 14%. His bank charges ordinary interest. What is the maturity value of his loan?

Lesson 7-2

2. Ed Wallace purchased a TV for $629 using the store's installment credit plan. Ed made a 25% down payment and financed the remaining amount. What amount did he finance?

Lesson 7-3

3. Lisa Snow obtained an installment loan of $2300. The annual percentage rate is 18%. She plans to repay the loan in 24 months. Use the table to find the finance charge.

Term in Months	MONTHLY PAYMENT ON A $100 LOAN			
	Annual Percentage Rate			
	10%	12%	15%	18%
6	$17.16	$17.25	$17.40	$17.55
12	8.79	8.88	9.03	9.17
18	6.01	6.10	6.24	6.38
24	4.61	4.71	4.85	4.99

Lesson 7-4

4. Tom DuVall obtained a 36-month loan of $4350 for a used car. The interest rate is 15%. His monthly payment is $150.95. For the first payment, what is the interest? What is the payment to principal? What is the new principal?

Lesson 7-5

5. Linda Hartman took out a simple interest loan of $3600 at 18% for 12 months. After 9 payments, the balance is $960.48. She pays off the loan when the next payment is due. What is the interest? What is the final payment?

Lesson 7-6

6. Juan Corvez obtained an installment loan of $625 to pay for a new refrigerator. The finance charge is $102.44. He agreed to make 24 payments of $30.31 each. Use the table to find the annual percentage rate.

Number of Payments	FINANCE CHARGE PER $100 FINANCED		
	APR		
	14.50%	14.75%	15.00%
6	$ 4.27	$ 4.35	$ 4.42
12	8.03	8.17	8.31
18	11.87	12.08	12.29
24	15.80	16.08	16.37

Lesson 7-7

7. Jane Tripp repaid a 12-month installment loan after 3 months. The total finance charge was $135.17. Use the rebate table to find Jane's refund.

Term of Loan	Number of Months Loan Has Run			
	1	2	3	4
3	50.00	16.67	0	0
6	71.43	47.62	28.57	14.29
9	80.00	62.22	46.67	33.33
12	84.62	70.51	57.69	46.15

A SPREADSHEET APPLICATION

Loans

To complete this spreadsheet application, you will need the template-diskette *Mathematics with Business Applications,* which accompanies this textbook. Follow the directions in the *User's Guide* to complete this activity.

Input the information in the following problems to determine the monthly payment, total interest, and the monthly payment allocation.

1. Auto loan of $4000.
12 monthly payments.
APR is 12%.
What is the payment?
What is the total interest?
What is the loan balance
after payment 6?

2. Stereo loan of $1500.
12 monthly payments.
APR is 15%.
What is the payment?
What is the total interest?
What is the loan balance
after payment 9?

3. Furniture loan of $2360.
24 monthly payments.
APR is 16.5%.
What is the payment?
What is the total interest?
What is the loan balance
after payment 16?

4. Auto loan of $12,000.
48 monthly payments.
APR is 13.25%.
What is the payment?
What is the total interest?
What is the loan balance
after payment 42?

5. Auto loan of $16,000.
60 monthly payments.
APR is 9%.
What is the payment?
What is the total interest?
What is the loan balance
after payment 36?

6. Home repairs of $3425.
60 monthly payments.
APR is 16.25%.
What is the payment?
What is the total interest?
What is the loan balance
after payment 30?

7. Boat loan of $19,960.
72 monthly payments.
APR is 14.5%.
What is the payment?
What is the total interest?
What is the loan balance
after payment 71?

8. Personal loan of $8700.
48 monthly payments.
APR is 12.25%.
What is the payment?
What is the total interest?
What is the loan balance
after payment 48?

CAREER WISE

Loan Officer

Keisha Themba handles applications for credit at a major credit card company. Typically, an application involves questions about the applicant, employment status, savings, other credit cards, and so on.

Keisha has met many college students through her job. Because many of them work part-time in the afternoons or evenings, they want to get their first credit card. After the information is verified, a credit card is issued. Their credit limit is usually lower than that of people with long and healthy credit histories. In recent years, she has helped thousands of people begin to establish credit.

Five stores that accept credit cards were surveyed. Sales figures for TVs and stereos are recorded in the pictograph below. A pictograph uses a picture to represent a quantity.

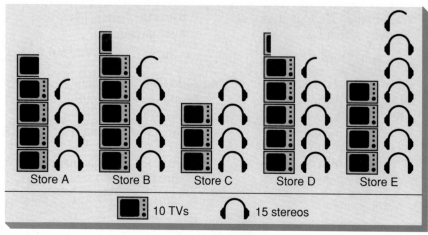

1. How many TVs are represented by [TV icon]?

2. Estimate the number of stereos sold at Store E.

3. If the average price of a TV at Store B is $249, estimate the sales from the TVs at Store B.

4. The company's finance rate is 1.65% per month on any purchase not paid within the first 25 days after the purchase and any previous balance. What is the finance charge for a month on an outstanding balance of $875?

5. According to the survey, 82% of the stereos sold were purchased with credit cards. How many stereos is this?

Automobile Transportation

A *sticker price* on a new automobile shows the *base price,* the cost of *options,* and the *destination charge* for delivering the car from the factory to the dealer. The auto dealer may take less than the sticker price if you make an offer that is higher than the *dealer's cost.* If you buy a used car, there are *used-car guides* that show the approximate price you can expect to pay for a particular model. In addition to the cost of purchasing an automobile, your transportation costs include *insurance, maintenance,* fuel, and other operating costs.

O U T L I N E

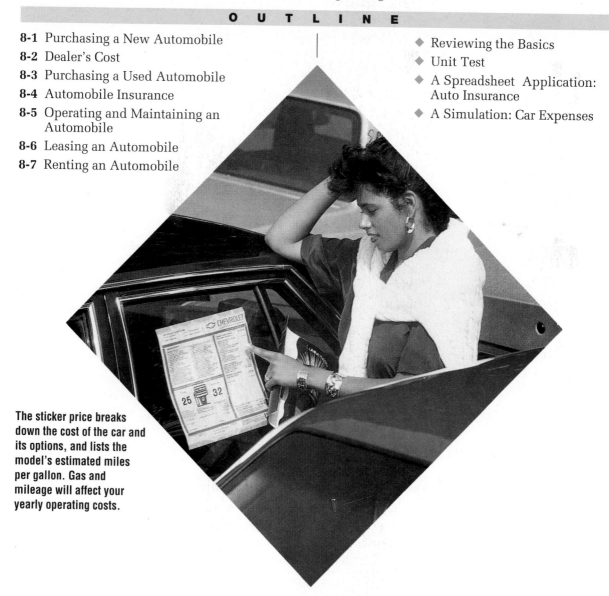

The sticker price breaks down the cost of the car and its options, and lists the model's estimated miles per gallon. Gas and mileage will affect your yearly operating costs.

Purchasing a New Automobile

Automobile manufacturers are required by law to place a sticker on the window of each new car to show all charges for the car. The **base price** is the price of the engine, chassis, and any other piece of standard equipment for a particular model. **Options** are extras for convenience, safety, or appearance, such as a radio, air-conditioning, and tinted glass. The **destination charge** is the cost of shipping the car from the factory to the dealer. The **sticker price** is the total of the base price, options, and destination charge.

Sticker Price = Base Price + Options + Destination Charge

EXAMPLE Skill 5   Application A   Term Sticker price

Scott Huber is shopping for a sports car. A portion of the sticker for one that he is interested in is shown below. What is the sticker price for this sports coupe?

LAMONT 2-DOOR SPORT COUPE		BASE PRICE $11,495
C/C	OPTIONAL EQUIPMENT DESCRIPTION	LIST PRICE
WP1	Rear Window Defogger	145
AT0	Automatic Transmission W/overdrive	490
TG1	Tinted Glass	120
AC3	Air-Conditioning	610
PDL	Power Door Locks	195
D84	Custom Two-Tone Paint	93
	DESTINATION CHARGE	450

SOLUTION

A. Find the **options.**

$145.00 + $490.00 + $120.00 + $610.00 + $195.00 + $93.00 = $1653.00

B. Find the **sticker price.**

Base Price + Options + Destination Charge
$11,495.00 + $1653.00 + $450.00 = $13,598.00
sticker price

145 + 490 + 120 + 610 + 195 + 93 = 1653 M+ 11495 + RM 1653 + 450 = 13598

✓ SELF-CHECK   Complete the problems, then check your answers in the back of the book.

Find the sticker price.

	Base Price	Options	Destination Charge
1.	$8975	$450; $610; $126	$390
2.	$10,400	$195; $675; $810; $92	$325

Find the sticker price.

	3.	4.	5.	6.	7.
Base Price	$9000	$9900	$11,540	$12,654	$19,842
Options	$1600	$2400	$ 1260	$ 2865	$ 3861
Destination Charge	$ 400	$ 350	$ 345	$ 338	$ 425
Sticker Price	?	?	?	?	?

8. Custom Ciera.
Base price is $9851.
Options: 2.8 Liter V6 for $660.
 Auto. trans. for $495.
 A/C for $725.
 Tinted glass for $195.
Destination charge is $345.
What is the sticker price?

9. Salon Classic.
Base price is $15,495.
Options: GT package for $1295.
 AM/FM stereo for $610.
 Bucket seats for $345.
 Power door locks for $190.
Destination charge is $400.
What is the sticker price?

10. Marlo Barsotti is interested in a 4-door sedan that has a base price of $11,975. Factory-installed options total $1576. The destination charge from its assembly plant in St. Louis is $352. What is the sticker price for this car?

11. Andy Dorfman sees a Fleetwood mini van that he is interested in buying. There is a 5% state sales tax on the purchase of the van. If Andy pays the sticker price, how much tax must he pay?

C/C	OPTIONAL EQUIPMENT DESCRIPTION	LIST PRICE
UN2	FLEETWOOD MINI VAN	BASE PRICE 14,895
PS1	Six-Way Driver Power Seat	240
WRD	Electric Rear Window Defogger	145
WW4	Tinted Glass	310
AT2	Automatic Transmission	475
PDL	SB Radial Tires W/ws	88
ITC	Custom Interior	388
D84	Custom Two-Tone Paint	153
RA2	AM/FM Stereo W/disc player	690
	DESTINATION CHARGE	440

MAINTAINING YOUR SKILLS Look up the skill in parentheses if you need help or more practice.

Add. **(Skill 5)**

12. 8850 + 995 + 660 + 242

13. 6770 + 1217 + 648 + 344 + 85

14. 19,453 + 2540 + 199 + 60

15. 9851 + 2889 + 401 + 75 + 90

16. 15,237 + 78.50 + 421.60 + 92

17. 7575.20 + 1243.26 + 791.24

8-2

Dealer's Cost

OBJECTIVE

Compute the dealer's cost of a new automobile.

Automobile dealers pay less than the prices on the sticker for both the basic automobile and the options. Consumer magazines often report the **dealer's cost** as a percent of the sticker price. You may save money when purchasing a new automobile by making an offer that is higher than the estimated dealer's cost but lower than the sticker price.

$$\text{Dealer's Cost} = \frac{\text{Percent of}}{\text{Base Price}} + \frac{\text{Percent of}}{\text{Options Price}} + \frac{\text{Destination}}{\text{Charge}}$$

EXAMPLE Skills 30, 2 Application A Term Dealer's cost

Lisa and Tom Marker want to purchase a new Winsor Sedan. The car has a base price of $12,438, options totaling $2240, and a destination charge of $360. They read in a consumer magazine that the dealer's cost for a Winsor is about 80% of the base price and 77% of the options price. What should they estimate as the dealer's cost?

SOLUTION

A. Find the **percent of base price.**
$12,438.00 × 80% = $9950.40

B. Find the **percent of options price.**
$2240.00 × 77% = $1724.80

C. Find the **dealer's cost.**

Percent of		Percent of		Destination	
Base Price	+	Options Price	+	Charge	
$9950.40	+	$1724.80	+	$360.00	= $12,035.20 dealer's cost

12438 ⊠ 80 % 9950.4 M+ 2240 ⊠ 77 % 1724.8 + RM 9950.4 + 360 =
12035.2

✔ **SELF-CHECK**   Complete the problem, then check your answer in the back of the book.

1. The dealer's cost is 80% of the base price of $9000 and 75% of the options price of $3000 plus a destination charge of $340. Find the dealer's cost.

PROBLEMS

$\left(\begin{array}{c}\text{Base} \\ \text{Price}\end{array} \times \begin{array}{c}\text{Dealer's} \\ \text{Percent}\end{array}\right)$	$+ \left(\begin{array}{c}\text{Options} \\ \text{Price}\end{array} \times \begin{array}{c}\text{Dealer's} \\ \text{Percent}\end{array}\right)$	$+ \begin{array}{c}\text{Destination} \\ \text{Charge}\end{array}$	$= \begin{array}{c}\text{Dealer's} \\ \text{Cost}\end{array}$
2. ($ 8000 × 80%)	+ ($2200 × 75%)	+ $360	= ?
3. ($17,000 × 85%)	+ ($3400 × 80%)	+ $342	= ?
4. ($14,300 × 82%)	+ ($1200 × 78%)	+ $465	= ?

5. MX-7 Diesel pickup.
   Base price is $10,800.
   Options total $1260.
   Destination charge is $370.
   Dealer pays 80% of base price
   and 80% of options price.
   What is the dealer's cost?

6. Rowley SXT.
   Base price is $8800.
   Options total $1900.
   Destination charge is $350.
   Dealer pays 84% of base price
   and 75% of options price.
   What is the dealer's cost?

7. Royal Sport, 2-door.
   Base price is $12,795.
   Options total $2845.
   Destination charge is $290.
   Dealer pays 78% of base price
   and 77% of options price.
   What is the dealer's cost?

8. LeBou Series 720 SE.
   Base price is $15,980.
   Options total $1292.
   Destination charge is $498.
   Dealer pays 84% of base price
   and 78% of options price.
   What is the dealer's cost?

9. Paul Dempsey is looking at a new Cameo GT-20 that has a base price of $8940, options totaling $1440, and a destination charge of $380. The dealer's cost is about 85% of the base price and 77% of the price of the options. What is the sticker price of the car? What should Paul estimate as the dealer's cost?

10. Pauletta Kreger is considering the purchase of a recreational van. She sees one that has a base price of $16,950 and options totaling $2280. The destination charge is $695. She estimates that the dealer's cost is 79% of the base price and 80% of the price of the options. If Pauletta offers the dealer $250 over the estimated dealer's cost, what is her offer?

11. Julia Brown offered an automobile dealer $150 over the estimated dealer's cost on a car that has a base price of $12,900 and options totaling $2480. The dealer's cost is about 85% of the base price and 81% of the price of the options. The destination charge is $520. What was her offer?

12. **A Brief Case** Judith Estes is ordering a new car built to her specifications direct from the factory. The base price is $16,940 and options total $1180. There is a destination charge of $390. The dealer's cost is about 81% of the base price and 88% of the price of the options. The dealer will sell her the car for $200 over the estimated dealer's cost plus a 6% sales tax. What is the total cost of the car?

MAINTAINING YOUR SKILLS   Look up the skill in parentheses if you need help or more practice.

Find the percentage. Round answers to the nearest hundredth. (Skill 30)

13. 8% of 700

14. 15% of 980

15. 22% of 756

16. 24% of 1520

17. 45% of 9800

18. 78% of 3440

19. 88% of 7780

20. 92% of 2146

21. 75% of 8971

22. 30% of 1217

23. 87.25% of 6163

Purchasing a Used Automobile

OBJECTIVE

Compute the average retail price of a used automobile.

Automobile dealers usually advertise used cars for prices that are higher than what they expect you to pay. Used-car guides, published monthly, give the average prices for cars that were purchased from dealers during the previous month. The information can help you make decisions about how much to pay for a used automobile.

$$\frac{\text{Average}}{\text{Retail Price}} = \frac{\text{Average}}{\text{Retail Value}} + \frac{\text{Additional}}{\text{Options}} - \frac{\text{Options}}{\text{Deductions}} - \frac{\text{Mileage}}{\text{Deduction}}$$

EXAMPLE Skills 5, 6 Application C Term Used-car guide

Av'g. Trd.-in	BODY TYPE	Model	Av'g. Loan	Av'g. Ret'l
PALAMINO — V8-AT-PS-VC				
5375 4	H'dtp. 2Dr.	I59	4850	6175
5325 4	H'dtp. "S"	I59	4800	6125
5625 4	H'dtp. Sprt.	I59	5075	6450
	25 Add Rear Window Defroster		25	25
	100 Add AM/FM Stereo.		100	100
	125 Add AM/FM Stereo/Tp.		125	125
	100 Add Cruise Control.		100	100
	100 Add Power Door Locks		100	100
	100 Add Power Windows.		100	100
	50 Add Digital Dash		50	50
	75 Add 2-Tone Paint.		75	75
	75 Add Aluminum Wheels		75	75
	75 Add Tilt Strg. Wheel		75	75
	125 Add GT Pkg.		125	125
	425 Deduct Manual Transmission		425	425
	500 Deduct W/out Air-Cond.		500	500

Jackie Morris would like to purchase a Palamino Spirit, model I59, 2-door hardtop, that is advertised for $5195. It has an AM/FM stereo radio and no air-conditioning. It has been driven 78,000 miles. The used-car guide indicates that $275 should be subtracted if the mileage is between 75,000 and 80,000 miles. What average retail price should Jackie keep in mind when she makes an offer for the car?

SOLUTION Find the **average retail price.**

$$\frac{\text{Average Retail}}{\text{Value}} + \frac{\text{Additional}}{\text{Options}} - \frac{\text{Options}}{\text{Deductions}} - \frac{\text{Mileage}}{\text{Deduction}}$$

$$\$6175.00 \ + \ \$100.00 \ - \ \$500.00 \ - \ \$275.00 \ = \ \$5500.00$$
average retail price

6175 [+] 100 [−] 500 [−] 275 [=] 5500

✔ SELF-CHECK Complete the problem, then check your answer in the back of the book.

	Retail Value	Additional Options	Options Deductions	Mileage Deduction	Average Retail Price
1.	$4795	$325	$425	$250	?

PROBLEMS

Use the table above to find the average retail price.

	Model	Retail Value	AM/FM Stereo	Power Locks	Power Windows	Digital Dash	Man. Trans.	A/C	Avg. Retail Price
2.	2 Dr.	$6175	No	No	No	No	Yes	Yes	?
3.	H'dtp."S"	$6125	Yes	No	No	Yes	Yes	No	?
4.	H'dtp. Sprt.	$6450	Yes	No	Yes	No	No	Yes	?

5. 5-year-old Elite compact. Average retail value is $4725. Add $50 for tilt steering. Add $325 for air-conditioning. Deduct $100 for manual transmission. What is the average retail price?

6. 2-year-old Cramer sedan. Average retail value is $8650. Add $175 for AM/FM stereo/tape. Deduct $450 for no air-conditioning. Deduct $325 for excessive mileage. What is the average retail price?

7. Sue Soto owns a 4-year-old Spector 4-door sedan that she wants to sell in order to purchase a new car. One used-car guide shows the average retail value of her car is $5225. She adds $50 for having a vinyl top, $125 for cassette player, $100 for power windows, and $75 for power locks. She deducts $475 for having no air conditioner. What is the average retail price she can ask as a selling price for her car?

Use the used-car guide shown to solve problems 8–10.

8. Jan Pohn owns a Mystic Sport Coupe 2D. The car has AM/FM stereo, air conditioning, power steering, power door locks, power windows, cruise control, and a manual transmission. What is the average retail price of the car?

9. Ray Sanchez owns a used Mystic Sport Coupe Z28 with air-conditioning, an automatic transmission, AM/FM stereo/tape deck, rear window defroster, and tilt steering wheel. Recently he was asked if he wanted to sell his car. What is the average retail price he should keep in mind when pricing his car?

Av'g. Trd.-in	BODY TYPE	Av'g. Loan	Av'g. Ret'l.
	MYSTIC V8-PS-PE		
8550	11 Spt. Cpe. 2D	7700	9575
9450	12 Berlinetta Cpe.	8525	10575
10350	Spt. Cpe. Z28	9325	11525
175	**Add** Sunroof	175	175
125	**Add** AM/FM Stereo	125	125
150	**Add** AM/FM Stereo/Tp.	150	150
125	**Add** Power Door Locks	125	125
125	**Add** Power Windows	125	125
125	**Add** Cruise Control	125	125
100	**Add** Tilt Strg. Wheel	100	100
75	**Add** Luggage Rack (S/W) . . .	75	75
100	**Add** Aluminum Wheels	100	100
50	**Add** Rear Window Defroster. . .	50	50
125	**Add** Digital Dash	125	125
250	**Add** 6 Cyl.	250	250
475	**Deduct** Manual Transmission. .	475	475
150	**Deduct** Convent Steer.	150	150
575	**Deduct** W/out Air-Cond.	575	575

10. Joan Barryhill wishes to trade in her Mystic Berlinetta Coupe on the purchase of a new automobile. It has AM/FM stereo, air-conditioning, automatic transmission, power steering, power windows, and rear window defroster. She must deduct $375 from the trade-in value for high mileage. What is the average trade-in value of her car?

MAINTAINING YOUR SKILLS   Look up the skills in parentheses if you need help or more practice.

Add. (Skill 3)

11. 4225 + 1200 + 375 + 245

12. 4060 + 225 + 3950 + 325 + 75

13. 4675 + 15 + 85 + 25 + 325

14. 175 + 260 + 3470 + 25 + 25 + 10

Subtract. (Skill 4)

15. 8450 − 475 **16.** 3890 − 2530 **17.** 2205 − 225 **18.** 3625 − 425

Automobile Insurance

Liability insurance, which includes bodily injury insurance and property damage insurance, protects you against financial losses if your car is involved in an accident. Bodily injury limits of 25/100 mean that the insurance company will pay up to $25,000 to any one person injured and up to $100,000 if more than one person is injured. Property damage insurance protects you against financial loss if your automobile damages the property of others.

Comprehensive insurance on your automobile protects you from losses due to fire, vandalism, theft, and so on. Collision insurance will pay to repair the damage to your automobile if it is involved in an accident. Either policy may be written with a deductible clause. A $50-deductible clause means that you pay the first $50 of the repair bill. The annual base premium is determined by the amount of insurance you want, the age group of your car, and the insurance-rating group. The insurance-rating group depends on the size and value of your car.

The annual premium is the amount you pay each year for insurance coverage. Your annual premium depends on the base premium and your driver-rating factor. The base premium depends on the amount of coverage you want. The driver-rating factor depends on your age, marital status, the amount you drive each week, and so on. If several people drive your car, the highest driver-rating factor among those who drive your car is used to determine the annual premium.

Insurance companies use tables to determine your base premium.

BASE PREMIUM FOR A PRIVATE PASSENGER AUTOMOBILE						
Property Damage Limits	Bodily Injury Limits					
	25/50	25/100	50/100	100/200	100/300	300/300
$ 25,000	$206.40	$218.80	$213.20	$252.00	$258.00	$286.80
50,000	212.40	224.80	237.20	258.00	264.00	293.20
100,000	220.80	233.20	245.60	266.40	272.40	301.20

PHYSICAL DAMAGE PREMIUM							
Coverage	Age Group	Insurance-Rating Group					
		10	11	12	13	14	15
Comprehensive $50-Deductible	A	$76.80	$81.60	$95.20	$108.00	$122.00	$135.60
	B	65.20	77.60	90.40	102.40	115.60	128.40
	C	62.00	74.00	86.00	98.00	110.40	122.80
	D	59.20	70.40	82.00	93.20	105.20	116.80
Collision $50-Deductible	A	$225.60	$246.00	$266.80	$287.20	$307.60	$328.00
	B	214.00	233.20	253.20	272.40	291.60	311.20
	C	204.00	222.80	241.60	260.00	278.40	296.80
	D	194.40	212.00	230.00	247.60	265.20	282.80

$$\text{Annual Base Premium} = \frac{\text{Liability}}{\text{Premium}} + \frac{\text{Comprehensive}}{\text{Premium}} + \frac{\text{Collision}}{\text{Premium}}$$

$$\text{Annual Premium} = \text{Annual Base Premium} \times \text{Driver-Rating Factor}$$

EXAMPLE   *Skills* 5, 8   *Application* C   *Term* Annual premium

Della Welch is the principal operator of her car. Her driver-rating factor is 2.20. Her insurance includes 50/100 bodily injury and $50,000 property damage. Her car is in age group A and insurance-rating group 13 (A, 13). She has $50-deductible comprehensive and $50-deductible collision insurance. What is her annual base premium? What is her annual premium?

SOLUTION

A. Find the **annual base premium.**

Liability Premium	+	Comprehensive Premium	+	Collision Premium		
$237.20	+	$108.00	+	$287.20	=	$632.40 annual base premium

B. Find the **annual premium.**

Annual Base Premium × Driver-Rating Factor
   $632.40   ×   2.20 = $1391.28 annual premium

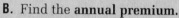 237.2 + 108 + 287.2 = 632.4 × 2.2 = 1391.28

✔ SELF-CHECK   Complete the problem, then check your answer in the back of the book.

Find the annual base premium and the annual premium.

1. 25/100 bodily injury and $100,000 property damage.
Car is in age group C and insurance-rating group 10 (C, 10).
$50-deductible comprehensive and $50-deductible collision.
Driver-rating factor is 1.50.

PROBLEMS

2. Driver-rating factor is 1.30.
Age, rating group is A, 14.
Coverage: 50/100 bodily injury.
  $25,000 property damage.
  $50-deductible comprehensive.
  $50-deductible collision.
What is the annual base premium?
What is the annual premium?

3. Driver-rating factor is 1.60.
Age, rating group is D, 12.
Coverage: 100/300 bodily injury.
  $50,000 property damage.
  $50-deductible comprehensive.
  $50-deductible collision.
What is the annual base premium?
What is the annual premium?

4. Driver-rating factor is 1.15.
Age, rating group is A, 15.
Coverage: 50/100 bodily injury.
  $50,000 property damage.
  $50-deductible comprehensive.
  $50-deductible collision.
What is the annual base premium?
What is the annual premium?

5. Driver-rating factor 3.90.
Age, rating group is B, 14.
Coverage: 300/300 bodily injury.
  $100,000 property damage.
  $50-deductible comprehensive.
  $50-deductible collision.
What is the annual base premium?
What is the annual premium?

6. Paula Williams uses her car primarily to run errands. She has $50-deductible comprehensive, $50-deductible collision, 100/200 bodily injury, and $25,000 property damage coverage. Her driver-rating factor is 1.00 and her car is classified D, 14. What is her annual base premium? What is her annual premium?

7. Scott Hanson uses his car primarily for business. He has $50-deductible comprehensive, $50-deductible collision, 100/300 bodily injury, and $100,000 property damage coverage. His driver-rating factor is 1.35 and his car is classified B, 15. What is his annual base premium? What is his annual premium?

8. Cheryl Obritter uses her car mainly for pleasure. She has $50-deductible comprehensive, $50-deductible collision, 100/300 bodily injury, and $50,000 property damage coverage. Her driver-rating factor is 2.15 and her car is classified C, 12. What is her annual base premium? What is her annual premium?

9. Carl McPeak uses his car to deliver office supplies. He has $50-deductible comprehensive, $50-deductible collision, 100/200 bodily injury, and $50,000 property damage coverage. His driver-rating factor is 3.10 and his car is classified D, 15. What is his annual base premium? What is his annual premium?

10. Henry Klump delivers magazines to retail outlets. His driver-rating factor is 2.15. His insurance coverage includes 25/100 bodily injury and $25,000 property damage. He has $50-deductible comprehensive and $50-deductible collision. His car is in age group C and insurance-rating group 10. What is his annual base premium? What is his annual premium?

11. Esther Miller-Kralik drives to and from work. Her driver-rating factor is 1.85. Her insurance coverage includes 25/50 bodily injury and $25,000 property damage. She has $50-deductible comprehensive and $50-deductible collision. Her car is in age group B and insurance-rating group 11. What is her annual base premium? What is her annual premium?

12. Elijah Kiboda owns and operates Kiboda's Imports, Inc. His driver-rating factor is 3.90. His insurance coverage includes 25/50 bodily injury and $25,000 property damage. He has $50-deductible comprehensive and $50-deductible collision. His convertible is in age group B and insurance-rating group 15. What is his annual base premium? What is his annual premium?

13. Tom Nome uses his car to drive to and from work. His driver-rating factor is 2.85. His insurance coverage includes 25/50 bodily injury and $50,000 property damage. He has $50-deductible comprehensive and $50-deductible collision. His car is in age group D and insurance-rating group 12. What is his annual base premium? What is his annual premium?

14. Lori Peterson uses her car for delivering newspapers. Her insurance coverage includes 25/50 bodily injury and $25,000 property damage. She has no comprehensive and no collision. Her rating factor is 1.90. What is her annual premium?

15. Gordie Meeks drives a farm automobile. His rating factor is 0.90. His insurance coverage only includes 100/200 bodily injury and $100,000 property damage. What is his annual premium?

16. Nathaniel Steiner delivers eggs to retail stores. His 12-year-old car is not worth very much so he doesn't carry comprehensive or collision insurance. He carries 50/100 bodily injury and $50,000 property damage insurance. His rating factor is 3.10. What is his annual premium?

17. Glenn O'Malley owns a car that is classified D, 11. His driver-rating factor is 4.10. He has 25/100 bodily injury, $25,000 property damage, $50-deductible comprehensive, and $50-deductible collision coverage. What is his annual premium? How much more would he pay for 100/300 bodily injury?

18. Sandra Jabara owns and operates Jabara's Bicycle Shop. She purchased a van with which to make deliveries. The van is classified A, 15. She has the following coverage: $100,000 property damage, 100/300 bodily injury, $50-deductible comprehensive, and $50-deductible collision. Her driver-rating factor is 1.35. What is her annual premium? What would be her annual premium if the van were classified A, 12?

19. Arnold Barbuto uses his car mainly for pleasure driving. His car is classified A, 12. His driver-rating factor is now 2.65. He learns that his rating factor will be 2.15 after he is married next year. How much less will his annual premium be if he has $25,000 property damage, 50/100 bodily injury, $50-deductible comprehensive, and $50-deductible collision?

MAINTAINING YOUR SKILLS Look up the skills in parentheses if you need help or more practice.

Add. **(Skill 5)**

20. 429.45 + 87.92 + 36.48 + 73.35

21. 49.55 + 2.82 + 34.59 + 733.20

22. 87.08 + 114.46 + 3.94 + 78.82

23. 480.88 + 65.44 + 749.45 + 9.80

24. 2.50 + 0.89 + 31.30 + 7.97

25. 40.93 + 73.30 + 67.83 + 3.9

Multiply. **(Skill 8)**

26. 1.25×79.90 27. 2.40×360 28. 3.90×67.70

29. 34.42×73.20 30. 654×2.6 31. 2121×3.50

32. 46.58×4.15 33. 371.19×3.20 34. 431.50×2.25

Operating and Maintaining an Automobile

OBJECTIVE

Compute the total cost per mile of operating and maintaining an automobile.

Many costs are involved in operating and maintaining an automobile. Variable costs, like gasoline and tires, increase as the number of miles you drive increases. Fixed costs, like automobile insurance, registration fees, and depreciation, remain about the same regardless of how many miles you drive. Depreciation is a decrease in the value of your car because of its age and condition.

$$\text{Cost per Mile} = \frac{\text{Annual Variable Cost} + \text{Annual Fixed Cost}}{\text{Number of Miles Driven}}$$

EXAMPLE   *Skills* 11, 2   *Application* A   *Term* Depreciation

Ann Kory purchased a used automobile for $4000 one year ago. She drove 9000 miles during the year and kept a record of all her expenses. She estimates the car's present value at $3200. What was the cost per mile for Ann to operate her car last year?

Variable Costs		Fixed Costs	
Gasoline	$345.24	Insurance	$ 385.40
Oil changes	71.85	License/registration	76.25
Maintenance	114.36	Depreciation	
New tire	41.75	($4000 – $3200)	800.00
	$573.20		$1261.65

SOLUTION   Find the **cost per mile**.

$$\left(\begin{array}{c} \text{Annual} \\ \text{Variable Cost} \end{array} + \begin{array}{c} \text{Annual} \\ \text{Fixed Cost} \end{array} \right) \div \begin{array}{c} \text{Number of} \\ \text{Miles Driven} \end{array}$$

($573.20 + $1261.65) ÷ 9000
$1834.85   ÷ 9000 = $0.203 = $0.20
                                             cost per mile

573.2 ➕ 1261.65 🟰 1834.85 ➗ 9000 🟰 0.20387

✔ SELF-CHECK   Complete the problems, then check your answers in the back of the book.

Find the total cost and the cost per mile.

	Variable Cost	Fixed Cost	Miles Driven
1.	$900	$1700	10,000
2.	$1137.26	$2491.24	12,000

Solve. Round answers to the nearest cent.

(Annual Variable Cost	+	Annual Fixed Cost	=	Total Annual Cost)	÷	Miles Driven	=	Cost per Mile
3. ($1000.00	+	$1250.00	=	?)	÷	9000	=	?
4. ($1530.00	+	$1275.00	=	?)	÷	11,000	=	?
5. ($2114.00	+	$3786.00	=	?)	÷	14,700	=	?
6. ($1584.00	+	$934.35	=	?)	÷	6800	=	?
7. ($2312.50	+	$4321.90	=	?)	÷	20,415	=	?

8. Mabel Kite, student. Drove 9500 miles last year. Fixed costs totaled $1215. Variable costs totaled $985. What was the total annual cost? What was the cost per mile?

9. Karim Yakobian, salesperson. Drove 24,500 miles last year. Fixed costs totaled $2460. Variable costs totaled $3940. What was the total annual cost? What was the cost per mile?

10. John Davidson, accountant. Drove 15,460 miles last year. Fixed costs totaled $1127.40. Variable costs totaled $2076.30. What was the cost per mile?

11. Elska Rashidi, sales manager. Drove 26,350 miles last year. Fixed costs totaled $1527.32. Variable costs totaled $2981.65. What was the cost per mile?

12. Hope Kocinski drove her car 12,200 miles last year. Her variable costs totaled $780.35. Her fixed costs totaled $2439. What was the cost per mile for her to operate her car?

13. Alice Powers drove her car 13,550 miles last year. Her variable costs totaled $1776.90. Her fixed costs totaled $2457.15. What was the cost per mile for her to operate her car?

14. J. J. Olmstead drove his van 11,400 miles last year. His variable costs totaled $1965.89. His fixed costs totaled $1884.26. What was the cost per mile for him to operate his car?

15. Carl Collins drove his subcompact car 24,200 miles last year. His variable costs totaled $4059.33. His fixed costs totaled $1973.27. What was the cost per mile for him to operate his car?

16. Pinckney Keil purchased an automobile for $18,350 one year ago. He drove it 11,500 miles during the first year and kept a record of all his expenses. His variable costs were: gasoline, $533.60; oil changes, $95.84; parking, $115.71; and repairs, $91.35. His fixed costs were: insurance, $418; license, $76.75; and depreciation. He estimates the car's present value at $15,350. What is his cost per mile?

17. Clara Nowiejski purchased an automobile for $16,940 one year ago. She drove it 14,500 miles during the first year and kept a record of all her expenses. Her variable costs were: gasoline, $1039; oil changes, $114.96; parking, $109.75; and repairs, $41.20. Her fixed costs were: insurance, $673; license, $46.75; and depreciation. She estimates the car's present value at $12,340. What was her cost per mile?

18. Nora Moskowitz and three friends agreed that they would share equally in the costs of operating and maintaining a car. Last year, the car was driven 11,342 miles. Fixed costs totaled $1399.56, while variable costs totaled $1626.39. How much did it cost per mile to operate the car? How much did it cost per person?

19. Gary Halston and two neighbors carpool to and from work in Gary's subcompact. The three agreed that they would share equally in the cost of operating and maintaining the car. Last year, the car was driven 18,500 miles. Fixed costs totaled $2686.87, while variable costs totaled $3275.27. How much did it cost per mile to operate the car for the year? How much did it cost each person to carpool for the year?

Oil changes	$ 71.55
Tune - up	87.95
Alignment	27.95
Insurance	415.00
Parking	42.20
Registration	68.50
Loan interest	459.70
Depreciation	1520.00
Gasoline	366.24

20. Allison Pappas kept records on the operation and maintenance of her car for the previous year. In the summary shown, which of the items listed are fixed costs? Which are variable costs?

21. Last year, Allison Pappas drove her car 8400 miles. What was the total cost to operate and maintain her car for the year? What was her cost per mile?

22. Earl and Bella Ridder each own the same model car. Each kept records of the operating and maintenance costs. Last year, Earl drove his car 13,700 miles and Bella drove hers 9700 miles. Both had fixed costs totaling $1368.27. Earl's variable costs totaled $2056.95 and Bella's totaled $1597.30. Who spent more per mile to operate and maintain the car? How much more per mile?

MAINTAINING YOUR SKILLS   Look up the skills in parentheses if you need help or more practice.

Round to the nearest hundredth. **(Skill 2)**

23. 21.751   **24.** 15.352   **25.** 4.3981   **26.** 15.9061

27. 0.04126   **28.** 0.3179   **29.** 1.0711   **30.** 0.0617

Divide. Round answers to the nearest hundredth. **(Skill 10)**

31. 641 ÷ 200   **32.** 1500 ÷ 500

33. 850 ÷ 9000   **34.** 3241 ÷ 15,000

35. 1875 ÷ 6800   **36.** 3199 ÷ 23,400

37. 7876 ÷ 16,520   **38.** 4135 ÷ 16,792

Leasing an Automobile

OBJECTIVE

Compute the total cost of leasing an automobile.

Rather than purchasing an automobile, you may want to lease one. When you lease an automobile, you make monthly payments to the leasing company, dealer, or bank for 2 to 5 years. At the end of the lease, you return the automobile to the dealer, or you may purchase it depending on the type of lease. The most common lease is a closed-end lease. With a closed-end lease, you make a specified number of payments, return the car, and owe nothing unless you damaged the car or exceeded the mileage limit. Another type of lease is an open-end lease. At the end of an open-end lease, you can buy the automobile for its residual value. The residual value is the expected value of the car at the end of the lease period. The residual value is established at the signing of the lease. With either lease, you must pay all the monthly payments, a security deposit, title fee, and license fee.

$$\begin{array}{c}\text{Total Lease}\\\text{Cost}\end{array} = \left(\begin{array}{c}\text{Number of}\\\text{Payments}\end{array} \times \begin{array}{c}\text{Amount of}\\\text{Payment}\end{array}\right) + \text{Deposit} + \begin{array}{c}\text{Title}\\\text{Fee}\end{array} + \begin{array}{c}\text{License}\\\text{Fee}\end{array}$$

EXAMPLE

Skills 2, 5, 11 *Application* A *Term* Lease

Ralph Dunn leased an S-10 pickup truck for use in his lawn care business. He pays $168.97 per month for 48 months. His deposit was $200. He paid a $40 title fee and a $15 license fee. What is his total lease cost?

SOLUTION

Find the **total cost.**

Total of payments: 48 × $168.97 = ...	$8110.56
Deposit..................................	200.00
Title fee	40.00
License fee..............................	+ 15.00
Total lease cost	$8365.56

48 ⊠ 168.97 ▢ 8110.56 ⊞ 200 ⊞ 40 ⊞ 15 ▢ 8365.56

✔ SELF-CHECK   Complete the problems, then check your answers in the back of the book.

Find the total lease cost.

1. 54 payments of $139.
Deposit of $500.
Title fee of $35.
License fee of $60.

2. 60 payments of $249.
Deposit of $1200.
Title fee of $40.
License fee of $35.

Find the total lease cost.

(	Number of Payments	×	Amount of Payment	=	Total of Payments	)	+ Deposit	+	Title Fee	+	License Fee	=	Total Lease Cost
3.	(24	×	$219	=	?)		+ $419	+	$8	+	$36	=	?
4.	(48	×	$199	=	?)		+ $749	+	$15	+	$15	=	?
5.	(48	×	$119	=	?)		+ $1200	+	$60	+	$75	=	?
6.	(54	×	$180	=	?)		+ $1200	+	$35	+	$96	=	?
7.	(60	×	$374	=	?)		+ $1500	+	$66	+	$55	=	?

8. Eagle Sport Coupe.
48 payments of $199.
Deposit of $1200.
Title fee of $35.
License fee of $25.
What is the total lease cost?

9. Jeep Wrangler.
48 payments of $149.
Deposit of $200.
Title fee of $60.
License fee of $40.
What is the total lease cost?

10. Rhoda Solberg leased a Jeep Cherokee for personal use. She pays $239 a month for 48 months. She also paid a deposit of $1100, a title fee of $90, and a license fee of $125. What is the total lease cost?

11. Nadine Daniels leased an Accord LX Wagon for $249.50 a month for 60 months. She paid a deposit of $350, a title fee of $25, and a license fee of $215. What is the total lease cost?

12. Doris Boyer has an open-end lease for a Chevy Lumina for her fabric store. The lease costs $219 a month for 48 months. She paid a deposit of $250, a title fee of $25, and a license fee of $135. At the end of the lease, she can buy the car for its residual value of $5446. What is the total lease cost? What is the total cost if she buys the car?

13. Homer Gilwee leased a Chevy Beretta for $199 a month for 48 months. He paid a deposit of $225, a title fee of $15, and a license fee of $60. The lease carried a stipulation that there would be a 10¢-per-mile charge for all miles over 60,000. He drove the car 68,515 miles. What is the total cost of leasing the automobile?

14. **A Brief Case** Alicia Harper can lease a Ford Escort Sport Coupe for $254.95 a month for 48 months. She must pay a deposit of $250, a title fee of $75, and a license fee of $120. At the end of the 48 months, the car is expected to be worth $4117. Instead of leasing, she can purchase the car for $278.96 a month for 48 months plus a $978 down payment and the same title and license fees. Is it less expensive to lease or purchase the car?

MAINTAINING YOUR SKILLS Look up the skill in parentheses if you need help or more practice.

Multiply. **(Skill 8)**

15. 33.90 × 5 **16.** 29.95 × 4 **17.** 7 × 54.65 **18.** 11 × 19.99

Renting an Automobile

OBJECTIVE

Compute the cost per mile of renting an automobile.

From time to time, you may need to **rent** a car. Some automobile rental agencies charge a daily rate plus a rate per mile, while others charge a daily rate with no mileage charge. In either case, you pay for the gasoline used. If you are under 21 years of age, your parents may be required to pay for the insurance on the automobile. If you are 21 or over, the rental agency usually pays for the insurance. The insurance generally has a collision deductible clause that states that you will pay for a portion of any damage to the car if it is in an accident. You can obtain complete insurance coverage with a **collision waiver** by paying an additional charge per day.

$$\text{Cost per Mile} = \frac{\text{Total Cost}}{\text{Number of Miles Driven}}$$

EXAMPLE   *Skills* 11, 2, 5   *Application* A   *Term* Rent

Joe Wozniak rented a compact car for 3 days at $27.95 per day plus 20¢ per mile. He purchased the collision waiver for $10.00 per day. Joe drove 468 miles and paid $21.70 for gasoline. What was the total cost of renting the car? What was the total cost per mile to rent the car?

SOLUTION   **A.** Find the **total cost.**

Daily cost: $27.95 × 3 = ...	$83.85
Mileage cost: $0.20 × 468 = ...	93.60
Gasoline cost:	21.70
Collision waiver: $10.00 × 3 = ...	$30.00
Total cost	$229.15

B. Find the **cost per mile.**

Total Cost   ÷   Number of Miles Driven
$229.15   ÷   468   = $0.489 = $0.49 cost per mile

27.95 ☒ 3 ☐ 83.85 M+ .2 ☒ 468 ☐ 93.6 M+ 21.7 M+ 10 ☒ 3 ☐ 30 M+
RM 229.15 ÷ 468 ☐ .4896

✔ SELF-CHECK   Complete the problems, then check your answers in the back of the book.

Winona Simms rented a compact car for $30.00 for 4 days plus 22¢ per mile. Winona drove 430 miles and spent $18.90 on gasoline.

1. Find the total cost.   **2.** Find the cost per mile.

Solve. Find the cost per mile to the nearest cent.

(Total Daily Cost	+	Total Mileage Cost	+	Gasoline Cost	=	Total Cost)	÷	Miles Driven	=	Cost per Mile
3. ($ 60.00	+	$45.00	+	$11.88	=	$116.88)	÷	300	=	?
4. ($160.00	+	$94.00	+	$23.89	=	?)	÷	500	=	?
5. ($119.95	+	$74.40	+	$30.53	=	?)	÷	620	=	?
6. ($159.95	+	$63.55	+	$41.67	=	?)	÷	420	=	?

Solve. Find the cost per mile to the nearest cent. (No mileage charge)

(Rental Cost	+	Gasoline Cost	=	Total Cost)	÷	Miles Driven	=	Cost per Mile
7. ($ 66.50	+	$10.39	=	?)	÷	240	=	?
8. ($ 96.00	+	$25.98	=	?)	÷	300	=	?
9. ($154.75	+	$28.27	=	?)	÷	476	=	?
10. ($ 89.95	+	$41.38	=	?)	÷	518	=	?
11. ($199.95	+	$74.39	=	?)	÷	646	=	?

12. Wo Chen rented a subcompact.
Cost $28 a day for 4 days.
Drove 520 miles at 33¢ a mile.
Gasoline cost $20.78.
What was the total cost?
What was the cost per mile?

13. Fred Bardi rented a sedan.
Cost $32 a day for 5 days.
Drove 1100 miles at 37¢ a mile.
Gasoline cost $54.95.
What was the total cost?
What was the cost per mile?

14. A. H. Wise rented a standard car.
Cost $159.95 for a week.
Gasoline cost $29.95.
Drove 415 miles.
What was the total cost?
What was the cost per mile?

15. Abrah Bahta rented a luxury car.
Cost $129.97 for a week.
Gasoline cost $17.90.
Drove 270 miles.
What was the total cost?
What was the cost per mile?

16. Sanchez Corado rented a sedan for 4 days at $38.00 a day plus 30¢ a mile. He purchased the collision waiver for $8.50 per day. He drove 420 miles and paid $19.64 for gasoline. What was the total cost of renting the car? What was the cost per mile?

17. Freda Rochetti rented a midsize car for 2 days at $32.95 a day plus 33¢ a mile. She purchased the collision waiver for $7.50 per day. She drove 140 miles and paid $8.39 for gasoline. What was the total cost of renting the car? What was the cost per mile?

18. Ace Car Rentals has a weekly rate of $129.95 for economy cars and no mileage charge. An economy car is driven 450 miles and uses $22.90 in gasoline. What is the cost per mile?

19. Riser Car Rentals has a weekly rate of $159.95 for station wagons and a mileage charge of 30¢ a mile with 100 free miles. A station wagon is driven 432 miles and uses $28.09 in gasoline. What is the cost per mile?

20. Jessica Vick is renting a subcompact car while her own car is being repaired. The subcompact rents for $99.95 a week with a mileage charge of 34¢ a mile with 75 free miles per week. She is going to rent the car for 2 weeks and expects to drive it about 460 miles. She estimates that the gasoline will cost about $20. What is the total cost of renting the car? What is the cost per mile?

21. Wilma Wallace decides to rent a van for her vacation. The van rents for $62 a day or $249 a week with no mileage charge. She will use the van for 8 days and will drive it about 1500 miles. She estimates that gasoline will cost about $180. What will be the total cost to rent the van? How much will it cost per mile?

22. Marty Collins and Paula Green plan to rent a luxury sedan for the prom. They will need the car for 1 day and will drive it about 60 miles. The car rents for $49.95 a day plus 30¢ a mile. They estimate that gasoline will cost about $10. If they share equally in the costs, how much will it cost each of them to rent the sedan?

23. The LaGuardias plan to fly to their vacation spot and then rent a car to tour the island. The vacation package they signed specifies that a sedan will be available for $49.90 a day with no charge for mileage. What will it cost the LaGuardias to rent the car for 5 days if they spend $28.35 for gasoline and $8.50 a day for the collision waiver? What will it cost per mile if they drive about 420 miles?

24. **A Brief Case** Alice Coopersmith is moving to a new home. She can rent a 14-foot panel truck for $40 a day plus 26¢ a mile, or an 18-foot truck for $55 a day plus 29¢ a mile. It would take 4 trips to make the move in the 14-foot truck, but only 3 trips in the larger truck. She estimates that gasoline would cost about $45 for either truck. A round-trip to her new home is 60 miles. Regardless of the number of trips, Alice would need the truck for only 1 day. To save on expenses, which size truck should Alice rent? How much would it cost per mile to rent each truck?

MAINTAINING YOUR SKILLS Look up the skills in parentheses if you need help or more practice.

Add. **(Skill 5)**

25. 7.94 + 34.67 + 86.75 + 378.99

26. 86.03 + 8.75 + 94.01 + 378.19

27. 44.64 + 29.899 + 35.080 + 4.219

28. 39.37 + 58.35 + 19.70 + 88.01

Divide. Round answers to the nearest hundredth. **(Skill 11)**

29. 762.20 ÷ 32   **30.** 684.26 ÷ 42.2   **31.** 502.00 ÷ 361.9

32. 801.83 ÷ 15.5   **33.** 582.19 ÷ 395   **34.** 207.89 ÷ 385

Reviewing the Basics

Skills

(Skill 2)

Round to the nearest cent.

1. $312.7539   **2.** $0.4462   **3.** $223.1091   **4.** $0.2981

Solve. Round answers to the nearest cent.

(Skill 5)

5. $14.36 + $89.09   **6.** $253.41 + $12.29

7. $14 + $15.75 + $39.79

(Skill 6)

8. $56.33 − $39.59   **9.** $426.08 − $229.78   **10.** $19 − $9.88

(Skill 8)

11. $41.60 × 3.15   **12.** $54.42 × 6.5   **13.** $236.55 × 4.10

14. $312.25 × 3.10   **15.** $124.10 × 0.95   **16.** $120.50 × 2.50

(Skill 11)

17. $1280 ÷ 9500   **18.** $3766 ÷ 7740   **19.** $1532.35 ÷ 14,000

(Skill 30)

20. 75% of $3600   **21.** 81% of $5180   **22.** 92% of $4575.24

Applications

(Application C)

Use the table to answer the following.

23. What is the average trade-in value of a G30 custom pickup, model A90, with a manual transmission?

24. What is the average retail value of a G30 deluxe pickup, model A95, with power steering, air-conditioning, and 4-wheel drive?

Av'g. Trd.-in	BODY TYPE Model	Av'g. Loan	Av'g. Ret'l.
	G30 Series E —V8 1-Tn AT		
4325	Customed Wdbed A90	3900	5300
5150	Deluxe Wdbed A95	4650	6175
500	Add 4-wheel drive	450	600
150	Add air-cond.	150	200
150	Add SS equip.	150	200
850	Add camper sp.	600	1000
50	Add power steering	50	75
125	Deduct man. trans.	125	150

Terms

Use each term in one of the sentences.

25. Sticker price

26. Rent

27. Used-car guide

28. Deductible clause

29. Depreciation

a. You can find price information on used cars that were purchased from dealers during the previous month in a _____?_____.

b. A decrease in the value of your car because of its age or condition is called _____?_____.

c. If you do not own an automobile but need to use one on occasion, you can _____?_____. from a rental agency.

d. The _____?_____ in your insurance policy states that you must pay a portion of any repair bill.

e. The total of the base price, options price, and destination charge is the _____?_____.

Unit Test

Lesson 8-1

1. Duwayn Archer sees a mini van that has a base price of $13,785. Factory-installed options total $1732. There is a destination charge of $440. What is the sticker price of the mini van?

Lesson 8-2

2. Nancy Clark is interested in a 4-wheel drive pickup that has a base price of $11,050, options totaling $1349, and a destination charge of $390. She has read that the dealer's cost is about 82% of the base price and 80% of the price of the options. What is the estimated dealer's cost that Nancy should keep in mind when making an offer?

Lesson 8-3

3. August LaVoy owns a used Sport Van model WC21 with automatic transmission, a V8 engine, AM/FM stereo, and no air-conditioning. What is the average retail price that August should keep in mind when pricing her car for sale?

Av'g. Trd.-in	BODY TYPE Model	Av'g. Loan	Av'g. Ret'l.
	STARCEST VAN C10-V8-AT		
11025	Sport Van WC11	9925	12550
12000	Panel Van WC21	10800	13575
12200	Sport Van WC21	11000	13875
100	Add AM/FM Stereo	100	150
75	Add Cruise Control	75	100
550	Deduct W/o Air-Cond.	500	550

Lesson 8-4

4. Beth Houghton uses her car to drive to and from work. Her insurance coverage includes 50/100 bodily injury and $25,000 property damage. Her driver-rating factor is 1.15. She has $50-deductible comprehensive and $50-deductible collision insurance coverage on her car. Her car is in age group B and rating group 12. What is her annual base premium? What is her annual premium?

Property Damage Limits	Bodily Injury Limits		
	25/50	25/100	50/100
$ 25,000	$206.40	$218.80	$213.20
50,000	212.40	224.80	237.20
100,000	220.80	233.20	245.60

Age Group	Insurance-Rating Group		
	12	13	14
	Comprehensive $50-Deductible		
A	$95.20	$108.00	$122.00
B	90.40	102.40	115.60
	Collision $50-Deductible		
A	$266.80	$287.20	$307.60
B	253.20	272.40	291.60

Lesson 8-5

5. Nora Hovey drove her car 13,220 miles last year. Her records show that fixed costs totaled $1290.60 and variable costs totaled $1940.40. How much did it cost per mile for Nora to operate her car last year?

Lesson 8-6

6. Nate Ruoff leased a GMC truck for his flower shop. He pays $299 per month for 48 months. His deposit was $300. He paid a $60 title fee and a $44 license fee. What is the total lease cost?

Lesson 8-7

7. Jen-Shiang Hong flew to Toronto on vacation. While there, he rented a car for 3 days at $35.95 a day plus 22¢ a kilometer. He drove the car 630 kilometers. Gasoline cost $32.60. How much did it cost per kilometer for him to rent the car?

A SPREADSHEET APPLICATION

Auto Insurance

To complete this spreadsheet application, you will need the diskette *Spreadsheet Applications for Business Mathematics,* which accompanies this textbook. Follow the directions in the *User's Guide* to complete this activity.

Input the information in the following problems to find the insurance costs and to compare various amounts of coverage.

You will need to use the premium table on page 246 to determine the base premium. All problems are $50-deductible comprehensive and collision.

1. Driver-rating factor: 1.50.
   Age, rating group: A, 12.
   Property damage: $25,000.
   Bodily injury: $25,000/$50,000.
   What is the annual base premium?
   What is the annual premium?

2. Driver-rating factor: 2.20.
   Age, rating group: D, 12.
   Property damage: $50,000.
   Bodily injury: $25,000/$100,000.
   What is the annual base premium?
   What is the annual premium?

3. Driver-rating factor: 1.65.
   Age, rating group: B, 15.
   Property damage: $50,000.
   Bodily injury: $100,000/$200,000.
   What is the annual base premium?
   What is the annual premium?

4. Driver-rating factor: 2.65.
   Age, rating group: C, 15.
   Property damage: $100,000.
   Bodily injury: $100,000/$300,000.
   What is the annual base premium?
   What is the annual premium?

5. Michelle Ross is an occasional operator of the family car and drives mainly for pleasure. She has $25,000 property damage and 50/100 bodily injury coverage. Her driver-rating factor is 1.70, and the car is classified C, 14. What is her annual base premium? What is her annual premium?

6. Suppose Michelle in problem 5 becomes the principal operator and her rating factor is now 2.20. What is her annual premium? How much more does she have to pay?

7. Jeremy Gobans is an occasional operator of the family car and drives mainly for pleasure. He has $50,000 property damage and 100/200 bodily injury coverage. His driver-rating factor is 2.65, and the car is classified C, 10. What is his annual base premium? What is his annual premium?

8. Suppose Jeremy in problem 7 becomes the principal operator and his rating factor is now 4.10. What is his annual premium? How much more does he have to pay?

9. An unmarried female operator, Jenny Shockey, age 19, is the occasional operator of her father's farm Jeep. Her rating factor is 2.0. She has $25,000 property damage and 25/50 bodily injury coverage. The 1989 Jeep is classified C, 13. What is her annual base premium? What is her annual premium?

A SIMULATION

Car Expenses

According to the American Automobile Association, the average car owner spends approximately $3500 a year in operating and maintaining an automobile. The costs of owning a car include fixed costs, such as finance charges, depreciation, insurance, and license fees. You also have variable costs, such as gasoline, maintenance, repairs, parking fees, and tolls. As a car owner, you may want to calculate the cost per mile. If you drive to work, you may want to know the cost per day.

1. Suppose you drove your 4-year-old car 12,600 miles and the car gets 26 miles per gallon. How many gallons of gasoline did you use? At $1.229 per gallon, how much will it cost?

2. In addition to gasoline, your variable costs include: 5 oil changes at $29.50 each, $89.00 for a tune-up, $225 for brake repairs, $185 for motor repairs, $49.40 for a tire, and $117 for parking and tolls. Your fixed costs include depreciation of $900 and an $85.50 registration fee. What is the total of these expenses?

3. The base premium for your automobile insurance is $570. Assume you are unmarried, under 21, the principal operator, and drive for pleasure. Use the table on page 246 to find your driver-rating factor. What will your annual insurance premium be?

4. What is the total annual cost for gasoline, maintenance, repairs, tires, parking, depreciation, registration, and insurance?

5. What is your cost per mile?

6. Suppose you drive 12 miles round-trip to work each day, 5 days a week, 50 weeks per year. How much does it cost you to drive to work each day? each week? each year?

Find the cost per mile and the cost to drive to work for each of these cases. Use your driver-rating factor from question 3.

	7.	8.	9.	10.
Annual Miles	12,150	11,950	13,700	21,300
Miles per Gallon	40	25	27	15
Cost per Gallon	$0.999	$1.059	$1.189	$1.299
Variable Cost	$690	$460	$330	$850
Fixed Cost	$1200	$600	$1400	$1200
Base Insurance Premium	$520	$660	$790	$672
Cost per Mile	?	?	?	?
Daily Miles	45	30	26	63
Cost per Day to Go to Work	?	?	?	?
	?	?	?	?

Purchase Price and Financing

You have decided to buy a car. One choice you have to make is whether to buy a new car or a used car. A new car has a higher purchase price and greater depreciation. Comprehensive and collision insurance are more expensive for a new car. A used car can be expected to have higher maintenance and repair costs.

There are two cars you are considering: a new Volney and a used Jess. To help choose which one to buy, you decide to calculate how much it would cost per month to run each car for the next three years.

CODE	DESCRIPTION	LIST PRICE
PH2	VOLNEY 2-DOOR HATCHBACK	8975.00
P36	SUNSET RED METALLIC	145.00
R58	AM/FM STEREO TAPE DECK	475.00
SA4	ADJUSTABLE SEATS	260.00
AT2	AUTOMATIC TRANSMISSION	490.00
V31	AIR-CONDITIONING	750.00
829	DESTINATION CHARGE	425.00

11. A portion of the sticker for the Volney is shown. What is the total sticker price?

12. The dealer's cost for the Volney is about 90% of the base price and 85% of the price of the options. If the dealer accepts your offer of $200 over the dealer's cost, what would the Volney cost you, including destination charge and 6% sales tax?

13. You would make a $2000 down payment for the Volney and finance the rest of the cost with a simple interest installment loan at 12% for 48 months.
  a. What is the amount financed?
  b. Use the table on page 222 to find the monthly payment. What is the total amount repaid? What is the finance charge?

14. The used-car guide entry for the Jess is shown. You are interested in a 2-door hardtop with power windows, AM/FM stereo with tape deck, and cruise control. If you bought the Jess for its average retail value, what would it cost, including 6% sales tax?

Av'g. Trd.-in	BODY TYPE Model	Av'g. Loan	Av'g. Ret'l.
	JESS — Series C JB LT		
3875	H'dtp 2 Dr	3500	4800
4275	H'dtp "S"	3850	5225
125	Add AM/FM Stereo/Tp	125	125
75	Add Power Door Locks	75	75
100	Add Power Windows	100	100
100	Add Power Seats	100	100
75	Add Cruise Control	75	75

15. You would make a $2000 down payment for the Jess and finance the rest of the cost at 18% for 36 months. Using the table on page 222, find the monthly payment and the finance charge.

A SIMULATION
(CONTINUED)

Insurance

In order to choose an automobile insurance company, you have read articles in consumer and automobile magazines, talked with relatives and friends, and asked insurance companies about their coverage and premiums. The tables below show the base premiums of the company you have chosen.

Property Damage Limits	Bodily Injury Limits		
	25/100	50/100	100/200
$ 25,000	$218.80	$213.20	$252.00
50,000	224.80	237.20	258.00
100,000	233.20	245.60	266.40

Age Group	Insurance-Rating Group		
	10	11	12
	Comprehensive $50-Deductible		
A	$76.80	$81.60	$95.20
B	65.20	77.60	90.40
	Collision $50-Deductible		
A	$225.60	$246.00	$266.80
B	214.00	233.20	253.20

16. For either car you choose, you plan to have 50/100 bodily injury and $50,000 property damage coverage. What is the annual base premium?

17. You plan to have $50-deductible comprehensive and $50-deductible collision insurance coverage. The Volney is in age group A and rating group 12. What is the annual base premium?

18. The Jess is in age group B and rating group 10. What is the annual base premium for comprehensive and collision insurance?

19. Use the table below to find your driver-rating factor. Assume you are unmarried, under 21, the principal operator, and drive for pleasure. For each car, what is the total cost per month for insurance?

DRIVER-RATING FACTORS FOR AUTOMOBILE INSURANCE					
Driver Classification	Multiple of Base Premium				
	Under 21 Years Occasional Operator	Under 21 Years Principal Operator	21 – 24 Years Occasional Operator	21 – 24 Years Principal Operator	25 – 29 Years Principal Operator
Unmarried youthful operators Females					
Pleasure	1.70	2.20	1.30	1.65	Classify
Farm auto	1.30	1.65	0.95	1.25	as Adult
Males					
Pleasure	2.65	4.10	1.80	2.75	1.65
Farm auto	2.00	3.05	1.35	2.05	1.30

A SIMULATION
(CONTINUED)

Maintenance, Depreciation, and Other Costs

A car depreciates in value most rapidly when it is new. You can estimate how a car will depreciate by looking at the values of older cars of the same model. Depreciation is often expressed as a percent of the purchase price.

Maintenance and repair costs tend to increase as a car ages. You can estimate these costs by talking with mechanics and with people who own the same model. Consumer and automobile magazines sometimes have articles comparing these costs for various models.

20. The Volney should lose 18% of its purchase price (less tax) the first year, 15% of its purchase price (less tax) the second year, and 12% the third year. What is the total percent? How much less will the Volney be worth in three years? What is the average depreciation per month?

21. The Jess can be expected to have a value of $2094 in three years. What is the depreciation per month?

22. For the Volney, you expect to spend $100 the first year, $160 the second year, and $480 the third year for maintenance and repairs. What is the average cost per month?

23. You expect the Jess to have maintenance and repair costs of $440 the first year, $560 the second year, and $750 the third year. It will also need tires for $250 plus 6% sales tax. What is the average cost per month?

24. You plan to drive 12,000 miles per year. Gasoline costs $1.259 per gallon. The Volney gets 30 miles per gallon. What is the cost of gasoline per month?

25. The Jess gets 25 miles per gallon. What is the cost of gasoline per month?

26. Make a table like this one to compare the monthly costs of the cars.

	Monthly cost	Volney	Jess
Fixed Costs	Finance charge	?	?
	Insurance	?	?
	Depreciation	?	?
	License and registration	$3.90	$3.90
Variable Costs	Gasoline	?	?
	Maintenance, repairs, tires	?	?
	TOTAL	?	?
	Cost per Mile	?	?

Housing Costs

Most people finance the cost of a home by taking out a *mortgage* loan. You pay a portion of the selling price and then finance the remaining amount. Most mortgages must be repaid in equal monthly installments that include the amount financed (principal) and the interest. Some monthly payments include a monthly payment for *real estate taxes* and *homeowner's insurance.* Real estate taxes are paid to the city or county for schools, roads, and other expenses. Homeowner's insurance pays for loss by fire, theft of contents, and property damage. Other housing costs include costs for *utilities,* such as heating fuel, water, and electricity.

O U T L I N E

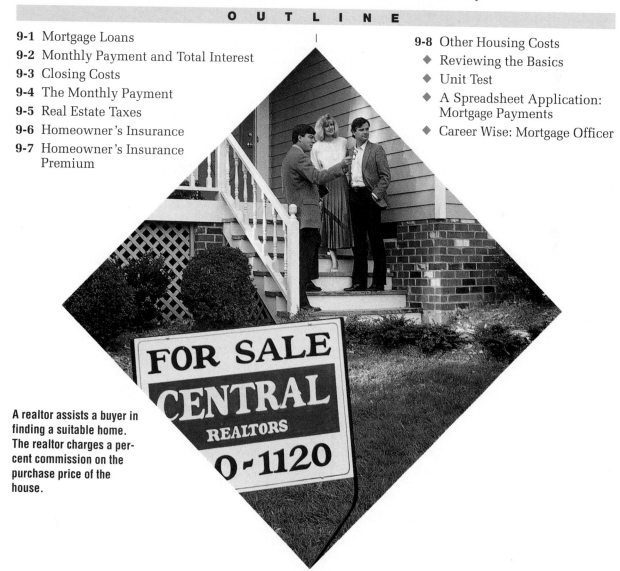

A realtor assists a buyer in finding a suitable home. The realtor charges a percent commission on the purchase price of the house.

Mortgage Loans

OBJECTIVE

Compute the mortgage loan amount.

When you purchase a home, you will probably make a down payment and finance the remaining portion of the selling price with a mortgage loan from a bank or savings and loan association. Generally, the down payment is between 10% and 40% of the selling price. The mortgage loan is usually repaid with interest in equal monthly payments. The mortgage gives the bank the right to sell the property if you fail to make the payments.

Mortgage Loan Amount = Selling Price − Down Payment

EXAMPLE *Skills* 30, 4, 6 *Application* A *Term* Mortgage loan

Jessica and Kirk Cramer are considering the purchase of a new home for $140,000. A 15% down payment is required. What is the amount of the mortgage loan needed to finance the purchase?

SOLUTION

A. Find the **down payment.**
$140,000 × 15% = $21,000 down payment

B. Find the **mortgage loan amount.**
Selling Price − Down Payment
$140,000 − $21,000 = $119,000 mortgage loan amount

140000 ⨯ 15 % 21000 140000 − 21000 = 119000

✔ SELF-CHECK Complete the problems, then check your answers in the back of the book.

Find the down payment and the amount of the mortgage.

1. $80,000 selling price, 25% down.

2. $200,000 selling price, 30% down.

PROBLEMS

	Buyer	Selling Price	Percent Down Payment	Down Payment	Mortgage Loan Amount
3.	Jeff Wilson	$ 87,000	20%	$17,400	?
4.	Beth Ivanoski	$ 62,500	15%	?	?
5.	David Ellison	$ 98,800	25%	?	?
6.	Sherrie DeQuine	$156,000	18%	?	?

7. Marleen Belton.
   Home priced at $48,500.
   20% down payment required.
   What is the down payment?
   What is the mortgage loan amount?

8. Cliff and Megan Sammour.
   Condominium priced at $75,000.
   30% down payment required.
   What is the down payment?
   What is the mortgage loan amount?

9. David and Peggy Chin.
   Mobile home priced at $27,400.
   Made a 20% down payment.
   What is the mortgage loan amount?

10. Alvira and Berry Fukunaga.
    Home priced at $280,000.
    Made a 40% down payment.
    What is the mortgage loan amount?

11. John and Maria Leivowitz decide that the home advertised is ideally suited to the needs of their family. They wish to make a 15% down payment and finance the remaining amount through their bank. What is the amount of their mortgage loan?

> A home with character, charm, and personality, 4 bedrooms, 3 full baths, beautiful vaulted living room ceiling, deluxe kitchen, large master bedroom suite, big family room.
> **At $229,900—it's a steal!**

12. Rita and Alfred Johnson offered $85,000 for a home that had been priced at $90,000. After some negotiating, the Johnsons and the seller agreed on a selling price of $86,500. What is the amount of the mortgage loan if they make a 20% down payment?

13. Richard Darman offered $112,500 for a home that had been priced at $125,000. The seller agreed to the offer. A bank is willing to finance the purchase if he can make a down payment of 25%. What is the amount of his mortgage loan?

14. The Hasbros have decided to purchase a duplex. They would use the rental income from one part of the duplex to help meet the mortgage payments. A selling price of $100,800 was agreed upon. What is the amount of the mortgage loan if a down payment of $22\frac{1}{2}$% is required?

15. **A Brief Case**   Dan and Sue Willingham have saved $14,000 for a down payment on their future home. Their bank has informed them that the minimum down payment required to obtain a mortgage loan is 20%. What is the most that they can spend for a home and expect to receive bank approval for their loan?

MAINTAINING YOUR SKILLS   Look up the skills in parentheses if you need help or more practice.

Find the percentage. **(Skill 30)**

16. 15% of 40,000
17. 25% of 84,000
18. 20% of 90,000
19. 40% of 175,000
20. 22% of 190,000
21. 5% of 425,500

Subtract. **(Skill 4)**

22. 40,000 − 8400
23. 94,000 − 18,000
24. 180,000 − 36,000
25. 98,500 − 12,250
26. 31,000 − 3100
27. 125,500 − 25,100

Monthly Payment and Total Interest

OBJECTIVE

Compute the monthly payment, total amount paid, and total interest charged.

Banks that make mortgage loans charge interest for the use of their money. The interest rate will vary from bank to bank, so it pays to shop around. If you know the annual interest rate, the amount of the loan, and the length of the loan, you can use a table to find the monthly payment, the total amount paid, and the total interest charged.

MONTHLY PAYMENT FOR A $1000 LOAN			
Annual Interest Rate	Length of Loan (Years)		
	20	25	30
10.00%	$ 9.66	$ 9.09	$ 8.78
10.50%	9.99	9.45	9.15
11.00%	10.33	9.81	9.53
11.50%	10.67	10.17	9.91
12.00%	11.02	10.54	10.29
12.50%	11.37	10.91	10.68
13.00%	11.72	11.28	11.07
13.50%	12.08	11.66	11.46

$$\text{Monthly Payment} = \frac{\text{Amount of Mortgage}}{\$1000} \times \frac{\text{Monthly Payment}}{\text{for a \$1000 Loan}}$$

$$\text{Amount Paid} = \text{Monthly Payment} \times \text{Number of Payments}$$

$$\text{Total Interest Charged} = \text{Amount Paid} - \text{Amount of Mortgage}$$

EXAMPLE   *Skills* 8, 6   *Application* C   *Term* Interest

Carol and Adam Burke have applied for an $80,000 mortgage loan at an annual interest rate of 12.00%. The loan is for a period of 30 years and will be paid in equal monthly payments that include interest. What is the total amount of interest charged?

SOLUTION

A. Find the **monthly payment.** (Refer to the table above.)

$$\frac{\text{Amount of Mortgage}}{\$1000} \times \text{Monthly Payment for a \$1000 Loan}$$

$$\frac{\$80,000.00}{\$1000.00} \times \$10.29 = \$823.20 \text{ monthly payment}$$

B. Find the **amount paid.**

Monthly payment × Number of Months
$823.20   × (12 months × 30 years)
$823.20   ×   360   = $296,352.00 amount paid

C. Find the **total interest charged.**

Amount Paid − Amount of Mortgage
$296,352.00 − $80,000.00 = $216,352.00 total interest

80000 ÷ 1000 ✕ 10.29 = 823.2 ✕ 12 ✕ 30 = 296352 − 80000 = 216352

✔ SELF-CHECK   Complete the problem, then check your answer in the back of the book.

1. Find the monthly payment, amount paid, and interest charged for a $90,000 mortgage loan at an annual interest rate of 10% for 20 years.

Use the table on page 268 to solve.

	2.	3.	4.	5.	6.
Mortgage	$50,000	$70,000	$95,000	$225,000	$395,000
Years	20	25	30	20	30
Rate	10.00%	12.00%	11.50%	13.00%	13.50%
Payment	?	?	?	?	?
Amount Paid	?	?	?	?	?
Total Interest	?	?	?	?	?

Use the table on page 645 to solve.

7. Charles and Sandy Compton.
$80,000 mortgage.
Terms: 11% for 25 years.
What is the monthly payment?
What is the total amount paid?
What is the total interest charged?

8. Abigail and Karlis Krisjanis.
$70,000 mortgage.
Terms: 14% for 20 years.
What is the monthly payment?
What is the total amount paid?
What is the total interest charged?

9. Julie Hardy.
$150,000 mortgage.
Terms: 10.5% for 20 years.
What is the total interest charged?

10. Diane Novak.
$50,000 mortgage.
Terms: 10% for 30 years.
What is the total interest charged?

F.Y.I.
The national average mortgage rate dropped to a 15-year low of 9% in December of 1991. It was 10.18% in December of 1990.

11. Ivan and Vicki Egan have obtained a $60,000 mortgage loan at an annual interest rate of 12% for 15 years. What is the monthly payment? What is the total amount to be paid?

12. Ellen and Clyde Perez reached an agreed upon price of $124,000 with the owner for the purchase of a house. They made a down payment of $14,000 and could finance the remaining amount in one of two ways: at 11.5% for 25 years or at 12% for 20 years. Which mortgage results in a larger amount of interest paid? How much greater?

13. How much can be saved in total interest by financing $90,000 at 10% for 15 years rather than 20 years?

MAINTAINING YOUR SKILLS Look up the skills in parentheses if you need help or more practice.

Multiply. (Skill 8)

14. 24 × 120.50 **15.** 36 × 431.2 **16.** 12 × 832.40 **17.** 15 × 342.20

Subtract. (Skill 6)

18. 75,500 − 22,200 **19.** 92,461 − 12,420 **20.** 453,821.50 − 100,000

Closing Costs

OBJECTIVE
Compute the total closing cost.

At the time you sign the documents transferring ownership of your new home, you must pay any closing costs that the bank charges. The closing costs may include fees for lawyers, credit checks and title searches, taxes, and the preparation of the documents. Some banks charge a flat fee regardless of the amount of the loan. Some banks charge a percent of the amount of the loan. Other banks charge itemized fees at the closing.

Closing Costs = Sum of Bank Fees

EXAMPLE   *Skills* 30, 3   *Application* A   *Term* Closing costs

Marla and Glen Carleone have been granted an $80,000 mortgage loan. When they sign the papers to purchase their new home, they will have to pay the closing costs shown. What is the total of the closing costs?

```
Credit report:                          $45
Loan origination fee:  2% of mortgage loan
Abstract of title:                     $120
Attorney fee:                          $250
Documentation stamp:  0.3% of mortgage
Processing fee:       1.10% of mortgage
```

SOLUTION   Find the **sum of the bank fees.**

a. Credit report .... $ 45
b. Loan origination $80,000 x 2% 1600
c. Abstract of title 120
d. Attorney fee 250
e. Documentation stamp: $80,000 × 0.3% 240
f. Processing fee: $80,000 × 1.10% 880

Total closing cost $3135

45 [M+] 80000 × 2 [%] 1600 [M+] 120 [M+] 250 [M+] 80000 [×] .3 [%] 240 [M+]
80000 [×] 1.1 [%] 880 [M+] [RM] 3135

✔ SELF-CHECK   Complete the problem, then check your answer in the back of the book.

1. Use the list of closing costs in the example above to find the total closing cost on a $60,000 mortgage.

PROBLEMS

Use the list of closing costs in the example above to solve.

2. Jeremy Roberts.
   Mortgage loan of $50,000.
   What is the total closing cost?

3. Vincent and Sue Hemsley.
   Mortgage loan of $95,000.
   What is the total closing cost?

4. Ralph and Cristi Sheen.
Mortgage loan of $271,000.
What is the total closing cost?

5. Jack and Dina King.
Mortgage loan of $420,000.
What is the total closing cost?

6. Jude and Rose McDermott are
financing $39,700 at 11.5% for
30 years. They must pay these
closing costs. What is the total
of the closing costs?

```
Appraisal fee.....................$250
Credit report....................$45
Title search.....................$380
Service fee........1.5% of mortgage
Legal fees ......................$250
```

7. Barry and Ella Ellerbee have agreed to purchase a house for $96,500.
Kenmore Savings and Loan Association is willing to lend the money at
12.25% for 25 years provided they can make a $10,000 down payment.
The total closing cost is 3.25% of the amount of the mortgage loan.
What is their closing cost?

8. Rene and Jefferson Franklin are interested in purchasing a $60,000
home. They plan to make a 20% down payment and finance the
remaining amount through Peabody Savings Association. Peabody
Savings has these closing costs: credit report, $30; appraisal report,
$255; title insurance, $190; survey and photographs, $225; recording
fee, $45; legal fees, $280; and property taxes, $389. If the loan is
approved, how much cash will the Franklins need to secure the loan,
including the down payment?

9. A Brief Case   Many states have usury laws that set a legal maximum
interest rate that lending agencies can charge. When the trend in interest rates is above the maximum allowed, many lending agencies will
add points to their basic mortgage charges. A **point** is 1% of the mortgage loan and is included in the closing costs. Five points is a one-time
charge of 5% of the mortgage loan.

Jean and Jim Couric have agreed to purchase a $92,800 home. They plan to make a
25% down payment. Beacon Hill Bank is
willing to finance the mortgage at 10.25%
for 30 years plus 2 points. How much cash
will they need to secure the loan, including
the down payment?

```
Title search:   $155
Appraisal fee:  $200
Credit report:   $55
Legal fee:      $375
Property tax:   $750
Fire insurance: $280
```

MAINTAINING YOUR SKILLS   Look up the skill in parentheses if you need help or more practice.

Find the percentage. (Skill 30)

10. 15% of 9000

11. 4% of 86,000

12. 7% of 252,000

13. 2.4% of 78,000

14. 3.3% of 83,000

15. 1.2% of 30,000

16. 0.3% of 92,000

17. 0.15% of 85,100

18. 0.04% of 22,300

19. 1% of 81,500

20. 1.4% of 150,000

21. 0.61% of 283,300

9-4

The Monthly Payment

OBJECTIVE

Compute the allocation of monthly payment toward principal, interest, and the new principal.

Most mortgage loans are repaid in equal monthly payments. Each payment includes an amount for payment of interest and an amount for payment of the principal of the loan. The amount of interest is calculated using the simple interest formula ($I = P \times R \times T$). The amount of principal that you owe decreases with each payment that you make. The chart shows the interest and principal paid in the first 4 months of an $80,000 mortgage loan.

$80,000 MORTGAGE LOAN AT 12.00% FOR 25 YEARS				
Payment Number	Monthly Payment	Amount for Interest	Amount for Principal	New Balance
1	$843.20	$800.00	$43.20	$79,956.80
2	843.20	799.57	43.63	79,913.17
3	843.20	799.13	44.07	79,869.10
4	843.20	798.69	44.51	79,824.59

Payment to Principal = Monthly Payment − Interest

New Principal = Previous Balance − Payment to Principal

EXAMPLE   *Skills* 6, 8, 30   *Application* A   *Term* Principal

Rod and Carey Finn obtained a 30-year, $80,000 mortgage loan from First Bank and Trust. The interest rate is 12%. Their monthly payment is $823.20. For the first payment, what is the interest? What is the payment to principal? What is the new principal?

SOLUTION

A. Find the **interest.**

Principal × Rate × Time

$80,000.00 × 12% × $\frac{1}{12}$ = $800.00 interest

B. Find the **payment to principal.**

Monthly Payment − Interest

$823.20 − $800.00 = $23.20 payment to principal

C. Find the **new principal.**

Previous Balance − Payment to Principal

$80,000.00 − $23.20 = $79,976.80 new principal

80000 × 12 % 9600 × 1 ÷ 12 = 800 M+ 823.2 − RM 800 = 23.2

80000 − 23.2 = 79976.8

✓ SELF-CHECK   Complete the problems, then check your answers in the back of the book.

In the example above, the new principal is $79,976.80. For the second payment, find the:

1. Interest on $79,976.80.   **2.** Payment to principal.   **3.** New balance.

	Mortgage Amount	Interest Rate	First Monthly Payment	Amount for Interest	Amount for Principal	New Principal
4.	$ 50,000	10%	$ 483.00	$416.67	$66.33	?
5.	$ 70,000	12%	$ 737.80	?	?	?
6.	$ 60,000	13%	$ 676.80	?	?	?
7.	$120,000	14%	$1492.80	?	?	?
8.	$225,000	13%	$2637.00	?	?	?

9. Lois Larczyk.
Mortgage loan of $46,000.
Interest rate is 12%.
Monthly payment is $506.92.
How much of the first monthly
   payment is for interest?

10. Patrick Yunker.
Mortgage loan of $84,000.
Interest rate is 10%.
Monthly payment is $714.
How much of the first monthly
   payment is for interest?

11. Matthew Roberts.
Mortgage loan of $38,600.
Interest rate is 12.5%.
Monthly payment is $438.88.
How much of the first monthly
   payment is for interest?
How much of the first payment
   is for principal?
What is the new principal?

12. Dee Pollom.
Mortgage loan of $98,000.
Interest rate is 14.5%.
Monthly payment is $1254.40.
How much of the first monthly
   payment is for interest?
How much of the first payment
   is for principal?
What is the new principal?

13. Jill Beyley obtained a 25-year, $60,000 mortgage loan from Peoples Savings and Loan Association. The interest rate is 12%. The monthly payment is $632.40. For the first payment, what is the interest? What is the payment to principal? What is the new balance?

14. Norman Foster obtained a 30-year, $180,000 mortgage loan from First Federal Savings and Loan Association. The interest rate is 15%. His monthly payment is $2277. For the first payment, what is the interest? What is the payment to principal? What is the new balance?

15. Amelia McGuire obtained a 20-year, $36,000 mortgage loan from Society Trust Company. The interest rate is 11.5%. Her monthly payment is $384.12. For the first payment, what is the interest? What is the pay272 ment to principal? What is the new balance?

16. Ken Burris obtained a 15-year, $88,500 mortgage loan from Swancreek Trust Company. The interest rate is 10.5%. His monthly payment is $978.81. For the first payment, what is the interest? What is the payment to principal? What is the new balance?

Refer to the table on page 645 for problems 17-21.

17. Emily Johnson obtained a 25-year, $82,000 mortgage loan from the Miners and Merchants Deposit Company. The interest rate is 11.5%. What is her monthly payment? For the first payment, what is the interest? What is the payment to principal? What is the new balance?

18. Hollis McMahn obtained a 15-year, $90,500 mortgage loan from Pacific Trust Company. The interest rate is 10.5%. What is the monthly payment? For the first payment, what is the interest? What is the payment to principal? What is the new balance?

19. Bill and Julie Johnson purchased a home for $672,400. They made a 20% down payment and financed the remaining amount at 11% for 30 years. What is the monthly payment? How much of the first monthly payment is used to reduce the principal?

Rated X! X-tra nice, X-tra large master bedroom with dressing room, X-tra sized screen porch, X-tra space for a garden. Split plan, 3 bedroom, 2 bth, dbl. garage. X-cellent value too. $164,900.

20. Ed and Cathy Brehm purchased the home advertised. They made a down payment of $24,900 and financed the remaining amount at 12.5% for 30 years. What is the monthly payment? What is the new principal after the first monthly payment?

21. A Brief Case Jeff and Rita Contreras purchased a condominium for $67,600. They made a 15% down payment and financed the remaining amount at 11.5% for 20 years. They have made 167 payments to date.

a. Use this portion of the repayment schedule to find the remaining debt after the 170th payment.

F.Y.I.
It takes 22+ years for payment to principal to equal payment to interest on a 30-year mortgage.

Payment Number	Monthly Payment	Amount for Interest	Amount for Principal	New Balance
168	613.10	$306.04	$307.06	$31,627.99
169	613.10	?	?	?
170	613.10	?	?	?

b. Use this portion of the repayment schedule to find the last payment.

Payment Number	Monthly Payment	Amount for Interest	Amount for Principal	New Balance
238	613.10	$14.46	$593.64	$909.71
239	613.10	?	?	?
240	?	?	?	?

MAINTAINING YOUR SKILLS   Look up the skills in parentheses if you need help or more practice.

Subtract. **(Skill 6)**

22. 48,000 − 29.46   **23.** 91,800 − 39.55   **24.** 24,400 − 23.12

25. 78,902 − 22.98   **26.** 18,185 − 45.11   **27.** 14,915 − 107.7

9-5

Real Estate Taxes

OBJECTIVE
Compute the assessed value and taxes of real estate.

When you own a home, you will have to pay city or county real estate taxes. The money collected is used to operate and maintain roads, parks, schools, government offices, and so on. The amount of real estate tax that you pay in one year depends on the assessed value of your property and the tax rate. The assessed value is found by multiplying the market value of your property by the rate of assessment. The market value is the price at which a house can be bought or sold. It is determined by an assessor hired by the municipality. The rate of assessment is a percent.

The tax rate is sometimes expressed in mills per dollar of valuation. A mill is $0.001. A tax rate of 80 mills is a tax rate of $80 per $1000 of assessed value. When working with mills, it is often convenient to express mills in dollars by dividing by 1000.

Assessed Value = Market Value × Rate of Assessment

Real Estate Tax = Tax Rate × Assessed Value

EXAMPLE   *Skills* 8, 30 *Application* A   *Term* Real estate taxes

The Ottawa County tax assessor stated that the market value of the Courtland Farm is $340,000. The rate of assessment in Ottawa County is 40% of market value. The tax rate is 84.32 mills. What is the real estate tax on the Courtland Farm?

SOLUTION   **A.** Find the **assessed value.**
Market Value × Rate of Assessment
    $340,000   ×   40%   = $136,000 assessed value

B. Express the **tax rate** as a decimal.
84.32 mills ÷ 1000 = 0.08432 tax rate

C. Find the **real estate tax.**
Tax Rate × Assessed Value
    0.08432   × $136,000.00 = $11,467.52 real estate tax

340000 ☒ *40* ☒% *136000* M+ *84.32* ÷ *1000* = *.08432* ☒ RM *136000* =
11467.52

✔ SELF-CHECK   Complete the problems, then check your answers in the back of the book.

The tax rate is 65.5 mills, market value is $70,000, and rate of assessment is 40%. Find the following:

1. The tax rate as a decimal.

2. The assessed value.

3. The real estate tax.

4. The M. L. D'Limas' home.
   Market value is $72,000.
   Rate of assessment is 30%.
   What is the assessed value?

5. The Simms' condominium.
   Market value is $59,800.
   Rate of assessment is 40%.
   What is the assessed value?

6. The Posadny Farm.
   Market value is $142,000.
   Rate of assessment is 35%.
   Tax rate is $86 per $1000.
   What is the real estate tax?

7. The Masons' home.
   Market value is $36,000.
   Rate of assessment is 50%.
   Tax rate is 72.35 mills.
   What is the real estate tax?

8. Brenda Roth's home is located in a community where the rate of assessment is 50% of market value. The tax rate is $85 per $1000 of assessed value. Her home has a market value of $92,000. What is its assessed value? What is the property tax?

9. The rate of assessment in Fulton County is 35%. The tax rate is $72.25 per $1000 of assessed value. What is the real estate tax on a piece of property that has a market value of $236,000?

10. Ali and Jackie Erwin live in a locality where the tax rate is 71.385 mills. The rate of assessment is 60%. The property that the Erwins own has a market value of $98,400. What is their real estate tax for a year?

11. Jose and Trudy Engstrom own a home that has a market value of $675,000. They live in an area where the rate of assessment is 45% and the tax rate is 48.535 mills. What is the annual real estate tax?

12. Timothy Oakland owns a mobile home and the property on which it sits. His mobile home has a market value of $24,000 and his property has a market value of $13,000. Timothy lives in a county where the rate of assessment is 100% and the tax rate is $32.85 per $1000 of assessed value. What is his yearly real estate tax?

13. **A Brief Case**   Gina and Tony Jasinski received a tax statement showing that their land has an assessed value of $7500 and their buildings have an assessed value of $42,300. The rate of assessment in their locality is 40%. What is the market value of their property?

MAINTAINING YOUR SKILLS   Look up the skills in parentheses if you need help or more practice.

Multiply. **(Skill 8)**

14. 37.3 × 78.4

15. 13.18 × 9.42

16. 13.3 × 4.37

17. 0.856 × 4.38

18. 83.92 × 44.9

19. 0.274 × 8356

Find the percentage. **(Skill 30)**

20. 90,000 × 20%

21. 140,000 × 45%

22. 45,500 × 33%

23. 328,800 × 92%

24. 78,710 × 38%

25. 49,500 × 70%

Homeowner's Insurance

OBJECTIVE

Compute the amount of coverage.

When you own a home, you will probably purchase homeowner's insurance to protect yourself and your home from losses due to fire, theft of contents, and personal liability. Homeowner's insurance also includes loss-of-use coverage to pay for some of the expenses of living away from home while damage to your home is being repaired. Personal liability and medical coverage protect you from financial losses if someone is injured on your property.

To receive full payment for any loss up to the amount of the policy, you must insure your home for at least 80% of its replacement value. Some companies may require you to insure your home for 90% to 100% of its replacement value. The replacement value is the amount required to reconstruct your home if it is destroyed. Insurance companies use the amount of coverage on your home to calculate the amount of coverage you receive on your garage, personal property, and for loss of use. Many companies use these percents of the amount of coverage on the home for each type of protection.

Coverage	Percent of Coverage
Personal property	50%
Loss of use	20%
Garage and other structures	10%

Amount of Coverage = Amount of Coverage on Home × Percent

EXAMPLE *Skill* 30 *Application* A *Term* Homeowner's insurance

The replacement value of Joy and Ron Amodeo's home is estimated at $94,000. They have insured their home for 80% of its replacement value. According to the guidelines above, what is the amount of coverage on the Amodeos' personal property?

SOLUTION

A. Find the **amount of coverage on home.**
$94,000 × 80% = $75,200 coverage on home

B. Find the **amount of coverage on personal property.**
Amount of Coverage on Home × Percent
$75,200   × 50% = $37,600
coverage on personal property

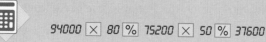

✔ SELF-CHECK Complete the problems, then check your answers in the back of the book.

A home is insured for 90% of its replacement value of $120,000, or $108,000. Using the percents in the table above, find the coverage for:

1. Personal property.   **2.** Loss of use.   **3.** Garage.

Use the table on page 277 to find the percent of coverage.

4. The Bahrs' home.
   Replacement value is $70,000.
   80% coverage on home.
   What is the amount of insurance?

5. The Loos' duplex.
   Replacement value is $95,000.
   100% coverage on home.
   What is the amount of insurance?

6. The Callahans' home.
   Replacement value is $38,500.
   90% coverage on home.
   What is the amount of insurance?
   What is the amount of coverage
       for personal property?

7. The Cloyds' home.
   Replacement value is $144,000.
   80% coverage on home.
   What is the amount of insurance?
   What is the amount of coverage
       for loss of use?

8. The Courtney family is purchasing a home that has a replacement
   value of $324,000. They wish to insure their new home for 90% of its
   replacement value. What is the amount of insurance on their home?
   What is the amount of coverage on their garage?

9. The Frankels recently purchased a home that has a replacement value
   of $324,000. They insured their new home for 80% of its replacement
   value. What is the amount of insurance on their home? What is the
   amount of coverage for personal property? loss of use? garage?

The Jenson Insurance Company uses these guidelines for homeowner's
insurance coverages. The policies must be written for 80% of a home's
replacement value. Use the guidelines to solve the following.

10. The McCauleys own a home that
    has a replacement value of $62,000.
    They insure it for 80% of its re-
    placement value. What is the
    amount of coverage on their
    personal property?

JENSON INSURANCE COMPANY	
Coverage—Home 80% Insured	
Garage/structures	10%
Personal property	50%
Loss of use	20%
Personal liability	50%
Medical payments	1%

11. Hugh O'Neill has a wood-frame home that has a replacement value of
    $470,000. He has insured it for 80% of its replacement value. What is
    the amount of coverage on his home? What is the amount of coverage
    for each type of protection listed in the guidelines?

Look up the skill in parentheses if you need help or more practice.

Find the percentage. (Skill 30)

12. 10% of $90,000

13. 80% of $30,000

14. 50% of 140,000

15. 15% of 420,000

16. 15% of 95,500

17. 90% of 544,000

Homeowner's Insurance Premium

OBJECTIVE

Compute the annual homeowner's insurance premium.

The amount of your homeowner's policy **premium** depends on the amount of insurance, the location of your property, and the type of construction of your home. Your **fire protection class** is a number that reflects the quality of fire protection available in your area.

Amount of Insurance Coverage	Brick/Masonry Veneer					Wood Frame				
	Fire Protection Class					Fire Protection Class				
	1–6	7–8	9	10	11	1–6	7–8	9	10	11
$ 40,000	$ 128	$ 131	$ 173	$ 182	$ 208	$ 137	$ 141	$ 182	$ 191	$ 219
45,000	133	137	173	182	208	144	147	191	200	229
50,000	137	141	185	195	223	146	150	195	204	234
60,000	147	151	199	210	241	158	162	210	221	252
70,000	164	166	219	230	264	173	178	230	242	277
80,000	185	191	252	264	303	198	204	264	279	319
90,000	206	212	281	295	339	222	228	295	310	357
100,000	229	236	313	328	377	246	253	328	345	396
120,000	272	280	372	391	449	293	301	391	411	472
150,000	353	362	481	505	581	379	389	505	532	611
200,000	474	487	647	680	782	509	523	680	716	823
250,000	567	580	739	785	898	600	614	785	835	956
300,000	676	693	882	937	1072	716	733	937	996	1141
400,000	785	804	1024	1087	1244	831	850	1087	1157	1325
500,000	1007	1031	1313	1394	1595	1065	1091	1394	1484	1699

ANNUAL PREMIUMS FOR A TYPICAL HOMEOWNER'S POLICY

EXAMPLE   *Skill* 30   *Application* C   *Term* Premium

The replacement value of Marcia Syke's home is $150,000. She has insured her home for 80% of its replacement value. The home is of wood-frame construction and has been rated in fire protection class 4. What is the annual premium?

SOLUTION

A. Find the **amount of coverage.**
$150,000 × 80% = $120,000 coverage on home

B. Find the **annual premium.** (Refer to the table above.)
$120,000 coverage, wood frame,
fire protection class 4.....................$293 annual premium

150000 ⊠ 80 % 120000

✔ SELF-CHECK   Complete the problem, then check your answer in the back of the book.

1. A home is insured for $150,000. It has a brick/masonry veneer and is in fire protection class 9. Find the premium.

Use the table on page 279 to find the annual premium.

2. The Campbells' wood-frame house. $40,000 homeowner's policy. Fire protection class 8. What is the annual premium?

3. Kuen Yee Ngs's brick home. $80,000 homeowner's policy. Fire protection class 11. What is the annual premium?

4. The Norths' brick home. Replacement value of $100,000. Insured for 90%. Fire protection class 9. What is the annual premium?

5. The Wordeys' wood-frame home. Replacement value of $125,000. Insured for 96%. Fire protection class 5. What is the annual premium?

6. The Smiths own a wood-frame home in an area rated fire protection class 1. Their two-family home has a replacement value of $112,500 and is insured for 80%. What is their annual premium?

7. The Quicks own a brick home that has a replacement value of $375,000. They purchased a homeowner's policy for 80% of its replacement value. They live in an area rated fire protection class 9. What is their annual policy premium?

8. Carla Campodonico has insured her home for $300,000. The two-story, wood-frame house is located in an area rated fire protection class 11. What is the annual policy premium?

9. A Brief Case When you obtain a mortgage loan, the bank may require you to include with your monthly principal and interest payment an amount equal to $\frac{1}{12}$ of the amounts needed to pay your real estate taxes and fire insurance premium. The bank places the money in an **escrow account**. When the real estate taxes and insurance premium are due, the bank uses the money from the escrow account to make the payments.

   Nelia and Gary Penn own a brick home with a market value and replacement value of $150,000. They insured their home for 100% of its replacement value. The Penns live in an area where the rate of assessment is 35%, the tax rate is 51.58 mills, and the fire protection is rated class 6. Their monthly principal and interest payment is $1581. How much is the monthly payment for principal, interest, real estate taxes, and insurance?

Find the percentage. **(Skill 30)**

10. 70% of 90,000

11. 80% of 140,000

12. 96% of 148,000

13. 50% of 70,000

14. 40% of 84,500

15. 15% of 71,400

16. 40% of 925,000

17. 20% of 345,500

18. 33.3% of 124,300

9-8

Other Housing Costs

OBJECTIVE
Compute the total housing cost and compare it with suggested guidelines.

In addition to your monthly mortgage payment, real estate taxes, and insurance payment, you will have expenses for utilities, maintenance, and home improvements. Utilities costs may include charges for electricity, gas, water, telephone, and heating fuel. The Federal Housing Administration (FHA) recommends that your total monthly housing cost be less than 35% of your monthly net pay.

EXAMPLE *Skills* 1, 5, 30 *Term* Utilities cost

Sue and Paul Kwan have a combined monthly take-home pay of $3320. They keep a record of their monthly housing expenses. The list of expenses for May is shown. Were their housing costs for May within the FHA guidelines?

Housing Expenses for May	
Mortgage payment	$698.24
Insurance ($303 ÷ 12)	25.25
Real estate taxes ($1885 ÷ 12)	157.08
Electricity	65.90
Heating fuel	54.20
Telephone	36.18
Water	26.20
Loan payment on oven	50.00
Repair storm door	38.68

SOLUTION

A. Find the **total monthly cost.**
Sum of expenses
above = $1151.73

B. Find the **recommended maximum.**
$3320.00 × 35% = $1162.00

C. Compare. Is **total monthly cost** less than **recommended maximum**?
Is $1151.73 less than $1162.00? . . . . . Yes, the Kwans are within the guidelines.

698.24 ＋ 25.25 ＋ 157.08 ＋ 65.9 ＋ 54.2 ＋ 36.18 ＋ 26.2 ＋ 50 ＋ 38.68 ＝
1151.73 3320 ✕ 35 ％ 1162

✔ SELF-CHECK Complete the problems, then check your answers in the back of the book.

Would the Kwans be within the FHA recommended guidelines if they had a combined monthly take-home pay of:

1. $3100

2. $3600

PROBLEMS

	3.	4.	5.	6.	7.	8.
Monthly Net Pay	$1100	$3900	$880	$4284	$5439	$7942
Recommended FHA Maximum (round to $1)	?	?	?	?	?	?

9. Joshua and Peg Ryder.
   Monthly net pay is $1980.
      Housing expenses for June:
      Mortgage payment . . $328.65
      Insurance 18.50
      Real estate taxes 159.00
      Electricity 55.44
      Telephone service . . . 44.98
      Loan payment 82.35
   Find the total housing cost.
   Is it within the FHA
      recommendation?

10. Frank and Yvette Shelby.
    Monthly net pay is $2440.
       Housing expenses for March:
       Mortgage payment . . . $533.50
       Insurance 19.75
       Real estate taxes 132.40
       Electricity 75.80
       Telephone service 29.45
       Water 22.00
    Find the total housing cost.
    Is it within the FHA
       recommendation?

11. Manual and Irma Lopez.
    Monthly net pay is $4300.
    Housing expenses for July:
       Mortgage payment . . $932.80
       Insurance 28.25
       Real estate taxes 249.75
       Utilities 232.40
       Repairs 67.00
       Loan payment 74.60
    Find the total housing cost.
    Is it within the FHA
       recommendation?

12. Sally Pizzo.
    Monthly net pay is $22,900.
       Housing expenses for January:
       Mortgage payment . . $6174.00
       Insurance 146.33
       Real estate taxes 1100.00
       Utilities 439.40
       Repair driveway 240.00
    Find the total housing cost.
    Is it within the FHA
       recommendation?

13. Fara Pinkston recorded her housing expenses for the month of August: mortgage payment, $347.90; insurance, $17; taxes, $84; electricity, $64.40; phone service, $33.50; fuel, $98.25; water, $17.44; and repairs, $79.87. Her monthly take-home pay is $1990. What is her total monthly cost? Is it within the FHA recommendation?

14. Melvin Hayashi recorded his housing expenses for the month of December: mortgage payment of $548.36, $19.50 for insurance premium, $122.50 for real estate taxes, installment payment of $46.75 for refrigerator, $54.70 for electricity, $34.40 for telephone service, $86.70 for home heating oil, and $21.80 for water. His monthly take-home pay is $2500. What is his total monthly housing cost? Is it within the FHA recommendation?

15. Eric Solomon has a monthly net income of $4400. He keeps a record of his monthly housing expenses. A list of expenses for January is shown. Were his housing costs for January within the FHA guidelines?

Housing Expenses for January	
Mortgage payment	$972.45
Insurance	30.00
Real estate taxes	216.00
Electricity	95.30
Heating fuel	122.50
Telephone	49.48
Water	35.00
Loan payment	85.00
Total	?

Water/sewer charges	$ 292.00
Electricity	940.00
Telephone service	345.30
Water heater	490.32
Repair storm windows	580.10
New air conditioner	1458.68
Replace gutters	760.00
New lawn mower	579.20
Total	?

16. David and Helen Voss have a combined monthly net income of $4750. Their records show that for last year they paid $10,789.20 in mortgage payments, $281 for insurance premiums, and $2085 in real estate taxes. In addition, they had the expenses shown. What was their average monthly housing cost for last year? Was it within the FHA recommendation?

Use the table on page 645 to find the loan payment and the table on page 279 to find the insurance premium for the following.

17. A Brief Case Molly and Chris Spaulding recently purchased a brick house for $150,000. They made a 20% down payment and financed the remaining amount at 12% for 30 years. The tax rate in their area is 71.57 mills and the rate of assessment is 40%. They purchased a homeowner's insurance policy for the purchase price of the house. The fire protection in the neighborhood is rated class 9. For the month of August, they recorded the following housing expenses: $69.20 for electricity, $44.85 for telephone service, $18.80 for water, and $74.65 to repair a door. They have a combined monthly net income of $5400. What is their monthly mortgage payment? What is the monthly insurance premium? What are their monthly taxes? What was their total monthly housing cost for August? Is it within the FHA recommendation?

Which number is greater? (Skill 1)

18. 2109.8 or 2107.9   **19.** 7484.08 or 74,846.50   **20.** 534.76 or 544.71

21. 823.40 or 826.70   **22.** 63,879 or 63,396   **23.** 268.74 or 378.47

Add. (Skill 5)

24. 85.89 + 74.84 + 35.30 + 306.24

25. 456.26 + 24.98 + 24.5 + 9.39

26. 974.79 + 997.32 + 9.81 + 35.8

27. 527.86 + 368.2 + 41.88 + 9.2

Find the percentage. Round answers to the nearest hundredth. (Skill 30)

28. 35% of 8300   **29.** 40% of 4600   **30.** 42% of 7841

31. 35% of 5089.7   **32.** 30% of 986.44   **33.** 15% of 59,640

34. 0.3% of 88,000   **35.** 0.15% of 65,400   **36.** 0.04% of 23,200

Reviewing the Basics

Skills

(Skill 3)

Solve. Round answers to the nearest hundredth.

1. $744 + $821 + $50   **2.** $278 + $1435 + $594

3. $28,040 + $8152

(Skill 4)

4. $795 − $412   **5.** $56,000 − $42,400   **6.** $93,000 − $8794

(Skill 5)

7. $2941.93 + $2051.88 **8.** $819.54 + $583.32 **9.** $41.71 + $5.22

(Skill 6)

10. $723.12 − $90.07 **11.** $453.36 − $79.64 **12.** $531.12 − $379.84

(Skill 8)

13. 85.02 × 480   **14.** 3017.45 × 0.324   **15.** $4417 × 0.35

(Skill 30)

16. $44,000 × 25%   **17.** $87,500 × 3.25%   **18.** $4192.50 × 35%

Applications

(Application C)

Use the table to find the monthly payment.

19. The Jericho National Bank is willing to loan Max Glenn $80,000 at 12.50% for 20 years on the purchase of a home. What is the monthly payment?

MONTHLY PAYMENT FOR A $1000 LOAN			
Annual Interest Rate	Length of Loan (Years)		
	20	25	30
10.00%	$ 9.66	$ 9.09	$ 8.78
10.50%	9.99	9.45	9.15
11.00%	10.33	9.81	9.53
11.50%	10.67	10.17	9.91
12.00%	11.02	10.54	10.29
12.50%	11.37	10.91	10.68
13.00%	11.72	11.28	11.07
13.50%	12.08	11.66	11.46

20. The Reynolds are purchasing a $180,000 home. Everson Savings and Loan Association will finance the purchase at 13% for 25 years if the Reynolds can make a 20% down payment. What is the monthly payment?

Terms

Match each term with its definition on the right.

21. Mortgage loan

22. Closing costs

23. Real estate taxes

24. Premium

25. Principal

26. Utilities cost

a. an amount paid for an insurance policy

b. a loan whereby the lender has the right to sell the property if payments are not made

c. fees paid at the time documents are signed transferring ownership of a home

d. charges for public services

e. an amount owed upon which interest charged is calculated

f. fees collected on the ownership of property used to support the operation of government

Refer to your reference files in the back of the book if you need help.

Unit Test

Lesson 9-1

1. The Dixons are purchasing a $95,000 home. What is the amount of the mortgage loan needed to finance the purchase if they wish to make a 15% down payment?

Lesson 9-2

2. The Mejias have a $60,000 mortgage loan from Home Loan Bank at an interest rate of 11% for 25 years. Use the table on page 645 to find the monthly payment and the total interest charged.

Lesson 9-3

3. The McLaughlins were granted a $75,000 mortgage loan. At the time of the closing, they paid the closing costs shown. What is the total of the closing costs?

```
Credit Report:                      $50
Loan origination fee: 2.4% of loan
Abstract of title:               $215
Attorney fees:                   $200
Property tax:                 $122.50
```

Lesson 9-4

4. The Ramires obtained a 30-year, $85,000 mortgage loan at Swan Creek Bank at an interest rate of 10.5%. The monthly payment is $777.75. What is the new principal after the first monthly payment?

Lesson 9-5

5. The Hanlon County tax assessor stated that the market value of the Arjons' estate is $935,000. The rate of assessment in Dixon County is 30% of the market value. The tax rate is 67.23 mills. What is the annual real estate tax on the estate?

Lesson 9-6

6. The Rayans have insured their home for 90% of its replacement value of $85,000. The insurance company states that the amount of coverage on their personal property is 50% of the amount of coverage on their home. What is the amount of coverage on the Rayans' personal property?

Lesson 9-7

7. The replacement value of the Coreys' home is $60,000. They insured it for 100% of its replacement value. The townhouse is of masonry veneer and is rated in fire protection class 8. Use the table of annual premiums on page 279 to find the annual homeowner's policy premium.

Lesson 9-8

8. The Danelies have a combined monthly net income of $3155.00. Their housing expenses for June are listed below. Were their housing expenses for the month within the FHA recommendations?

Mortgage payment $600.85	Telephone $29.19
Insurance 34.35	Heating oil 84.21
Property taxes 176.67	Electricity 65.65
Water/sewer service 18.23	Furniture payment 49.95

A SPREADSHEET APPLICATION

Mortgage Payments

To complete this spreadsheet application, you will need the diskette Spreadsheet *Applications for Business Mathematics*, which accompanies this textbook.

Select option 9, Mortgage Payments, from the menu. Input the information in the following problems to find the monthly payment and the total amount paid for each mortgage loan.

1. Susan and Willard Weber.
   Home selling for $86,400.
   Terms: $9400 down payment.
   a. Loan at 13% for 25 years.
   b. Loan at 13.25% for 25 years.

2. Rudolf and Charlotte Hess.
   Home selling for $376,000.
   Terms: $74,000 down payment.
   a. Loan at 14% for 25 years.
   b. Loan at 13.50% for 30 years.

3. Mobile home selling for $39,000.
   Terms: $9900 down payment.
   a. Loan at 14.25% for 15 years.
   b. Loan at 15% for 12 years.

4. Mobile home selling for $33,500.
   Terms: $3350 down payment.
   a. Loan at 15% for 10 years.
   b. Loan at 14.50% for 12 years.

5. Estate for $440,000.
   Terms: 10% down payment.
   a. Loan at 12.75% for 30 years.
   b. Loan at 12.75% for 35 years.

6. Farmhouse for $175,000.
   Terms: 30% down payment.
   a. Loan at 9.25% for 25 years.
   b. Loan at 9% for 23 years.

7. Walt Tyson plans to purchase a home for $106,900. He has an $8900 down payment. First Federal will give him a loan at 14.25% for 25 years. Baltimore Trust Company will give him a loan at 14.75% for 20 years. Find the monthly payment and total amount paid for each loan.

8. Marge McCarthy plans to purchase a commercial property for $539,000. She has $39,000 down payment. Hamilton Mortgage Company will give her a loan at 12.51% for 20 years. Grogan Investors will give her a loan at 12.92% for 18 years. Find the monthly payment and total amount paid for each loan.

9. Agnes Golinto plans to purchase a duplex for $125,400. After a 15% down payment, she plans to get a 30-year loan. Find the monthly payment and total amount paid for a loan at 14% and a loan at 14.50%.

10. Ernie Thomas plans to purchase a condominium for $86,000. After a 30% down payment, he plans to finance the remainder for 20 years. Find the monthly payment and total amount paid for a 10% loan and a 10.25% loan.

▼

CAREER WISE

Mortgage Officer

Wu-Yi Wong is loan officer at a mortgage company, a company whose main goal is to make money available to people wishing to buy homes.

When someone has decided to purchase a particular house, he or she comes to Wu-Yi to apply for a long-term loan. Wu-Yi explains the terms of the loan, including the type of loan, rate, points involved, and closing costs. Wu-Yi also seeks information about the customer's credit history, employment status, and ability to make payments regularly.

In effect, Wu-Yi's company and the prospective home buyer will own the home jointly. As payments are made to the mortgage company, the company's ownership decreases and the home buyer's ownership increases.

OWNERSHIP OVER COURSE OF MORTGAGE

The graph at the right indicates how the share of ownership changes over the course of the mortgage. The graph is based on a 20-year mortgage with payments made monthly. Four different interest rates—8%, 10%, 12%, and 14%—are shown.

Interest rates change as the state of the economy changes. Wu-Yi has found that at some times of the year, many people seek loans, while at other times of the year, not many people can afford to apply for loans.

1. What is the percentage of equity a homeowner will have after 50% of term if the interest rate is 8%?

2. After what percent of term will the equity be 30% if the interest rate is 12%?

3. Copy the axes and the curves in the graph above. Sketch the curve that you think corresponds to a rate of 9%.

4. Use the graph to explain this statement: At any given percent of term, equity at 14% is always less than equity at 10%.

Insurance and Investments

Health and *life insurance* protect you and your dependents against financial losses in case of illness or death. *Whole life* and *universal life insurance* offer a savings feature. *Certificates of deposit, stocks,* and *bonds* are other types of *investments,* which often earn more interest than money in a bank account. A certificate of deposit is purchased for specific amounts that remain on deposit in a bank for a specified period of time. A stock is a *share* in a company. The company may pay you a return on your investment in the form of *dividends.* A bond is money you loan to a company for a specified period. When the loan is due, you are paid the *face value* of the bond, which includes the interest on your original investment.

O U T L I N E

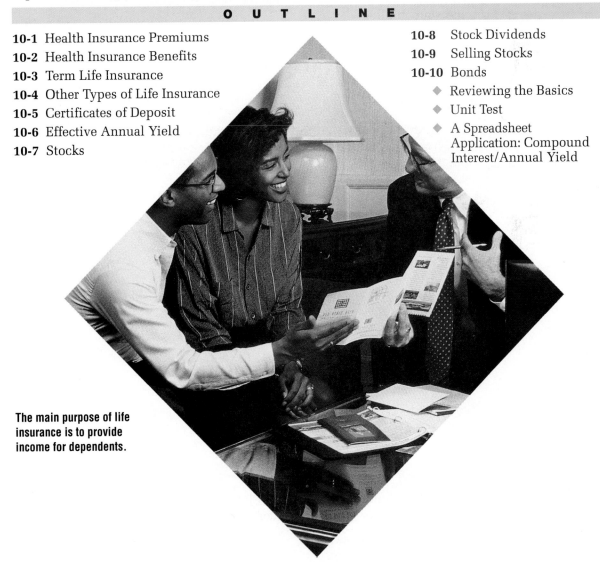

The main purpose of life insurance is to provide income for dependents.

10-1

Health Insurance Premiums

OBJECTIVE
Compute health insurance premiums.

An accident or illness could cut off your income, wipe out your savings, and leave you in debt. To protect against overwhelming medical expenses, many people have health insurance. A basic plan includes hospital and surgical-medical insurance. A comprehensive plan includes hospital, surgical-medical, and major medical insurance.

One way to get health insurance is by joining a group plan where you work. Your employer may pay part or all of the premium. Health insurance companies also offer nongroup plans for people not enrolled in group plans. You can choose an individual plan that covers only yourself or, for a larger premium, a family plan that covers yourself, your spouse, and your children.

Employee's Percent = 100% − Employer's Percent

Employee's Contribution = Total Premium × Employee's Percent

EXAMPLE   *Skills* 11, 30   *Application* K   *Term* Health insurance

Sean Watts is employed by Schwyn Company. He has a family membership in the group comprehensive medical insurance program. The annual premium includes $2523 for hospital insurance, $795 for surgical-medical insurance, and $138 for major medical insurance, for a total of $3456. Sean's employer pays 70% of the total cost. His contribution is deducted monthly from his paycheck. What is Sean's annual contribution? What is his monthly deduction?

SOLUTION

A. Find the **employee's percent.**
100% − 70% = 30% employee's percent

B. Find the **employee's contribution.**
Total Premium × Employee's Percent
$3456.00   ×   30%   = $1036.80 employee's contribution

C. Find the **employee's monthly deduction.**
Employee's Contribution ÷ 12
$1036.80   ÷ 12 = $86.40 monthly deduction

 100 $-$ 70 $=$ 30   3456 $\times$ 30 % 1036.8 $\div$ 12 $=$ 86.4

✓ SELF-CHECK   Complete the problems, then check your answers in the back of the book.

Annual premium includes $2408 for hospital, $804 for surgical-medical, and $168 for major medical. The company pays 80% of the cost.

1. Find the employee's total annual contribution.

2. Find the employee's monthly deduction.

	Plan	Annual Premium for			Total Annual = Premium	Employer's Percent	Employee's Annual Contribution
		Hospital +	Surgical-Medical +	Major Medical +			
3.	Single	$1980 +	$ 568 +	$112 =	$2660	60%	?
4.	Single	$1368 +	$ 365 +	$ 91 =	?	80%	?
5.	Family	$3066 +	$ 966 +	$168 =	?	75%	?
6.	Family	$2540 +	$ 730 +	$210 =	?	90%	?
7.	Family	$3150 +	$1125 +	$225 =	?	40%	?

8. Eric Ritter.
   Annual health insurance
      premium:
      Hospital $1984.
      Surgical-medical $568.
      Major medical $152.
   Employer pays 85% of cost.
   What is the total annual
      premium?
   How much does Eric pay
      annually?
   How much is deducted from his
      weekly paycheck?

9. Kelli Lenz.
   Annual health insurance
      premium:
      Hospital $1145.
      Surgical-medical $350.
      Major medical $75.
   Employer pays 65% of cost.
   What is the total annual
      premium?
   How much does Kelli pay
      annually?
   How much is deducted from her
      semimonthly paycheck?

10. Heather Farough has a family membership in the company's group
   medical insurance program. The total cost is $3817. Her employer pays
   75% of the total cost. Her contribution is deducted biweekly from her
   paycheck. How much is her annual contribution? How much is her
   biweekly deduction?

11. Rachel and Dustin Lutts are students at the state university. The com-
   prehensive health insurance rate is $37.50 per month for each student.
   The premiums are paid quarterly (every three months). What does
   Rachel's and Dustin's health insurance cost per quarter?

12. A Brief Case   Employees of the General Manufacturing Company may sign
   up for a group vision care health insurance at a cost of $4.00 per monthly
   pay period. The insurance actually costs the company $80 per employee
   per year. What percent of the cost of the insurance does the company pay?

MAINTAINING YOUR SKILLS   Look up the skills in parentheses if you need help or more practice.

Find the rate. **(Skill 31)**

13. What percent of $60 is $3?

14. What percent of $150 is $60?

15. What percent of $475 is $95?

16. What percent of $210 is $73.50?

Find the percentage. **(Skill 30)**

17. 367 × 55%

18. 940 × 14.2%

19. 4200 × 13.65%

20. 3500 × 4.81%

Health Insurance Benefits

OBJECTIVE

Compute the amount paid by the patient for medical care.

A comprehensive health insurance plan includes three kinds of insurance. Hospital insurance pays most of the cost of hospitalization, including a semiprivate room and most laboratory tests. Surgical-medical insurance pays your doctor's fee, up to a certain amount, for surgery. Major medical insurance is designed to protect you against the other costs of a serious and expensive illness. Most major medical policies have a deductible clause. A $250 deductible clause means that you must pay the first $250 of the amount not covered by your hospital and surgical-medical insurance. Many major medical policies also have a coinsurance clause. A typical one states that you must pay 20% of the amount remaining after you pay the deductible. Your insurance company pays the other 80%.

Amount Paid by Patient = Deductible + Coinsurance Amount

EXAMPLE   *Skills* 4, 30   *Application* A   *Term* Coinsurance clause

Brooke Kolodie's comprehensive medical insurance includes hospital, surgical-medical, and major medical insurance. The major medical policy has a $250 deductible and a 20% coinsurance clause. When she had surgery, her total bill for hospital care and physician's fee was $9350. The hospital and surgical-medical provisions of her insurance policy covered $6825. What amount did she pay?

SOLUTION

A. Find the **amount not paid by hospital and surgical-medical insurance.**
$9350 − $6825 = $2525 amount not paid

B. Find the **amount subject to coinsurance.**
$2525 − $250 deductible = $2275 subject to coinsurance

C. Find the **coinsurance amount paid by patient.**
$2275 × 20% = $455 coinsurance paid by patient

D. Find the **total amount paid by patient.**
Deductible + Coinsurance Amount
$250 + $455 = $705 total paid by patient

9350 − 6825 = 2525 − 250 = 2275 × 20 % 455 + 250 = 705

✔ SELF-CHECK   Complete the problems, then check your answers in the back of the book.

Krista Ridge had a total medical bill of $46,300. Her insurance covered $44,850. Her medical policy has a $200 deductible and a 20% coinsurance clause.

1. Find the amount subject to coinsurance.

2. Find the total amount Krista had to pay.

	Total Medical Bill	Amount Not Paid by Hospital and Surgical-Medical Insurance	Major Medical Deductible	Amount Subject to Coin-surance	Coin-surance Rate	Amount Paid by Patient	
						Coin-surance	Total
3.	$ 44,500	$ 1,900	$250	$1650	20%	$330	?
4.	$ 72,040	$ 2,970	$100	?	20%	?	?
5.	$148,215	$12,250	$250	?	20%	?	?
6.	$140,000	$ 1,050	$500	?	20%	?	?

F.Y.I.
In 1990, Americans spent $666 billion on health care.

7. Wanda Orsini.
Total bill is $48,240.
Amount paid by hospital and surgical-medical insurance is $44,790.
Major medical has a $250 deductible and 20% coin-surance clause.
How much is subject to coinsurance?
What is the total amount Wanda pays?

8. Kevin Bruce.
Total bill is $92,980.
Amount paid by hospital and surgical-medical insurance is $88,745.
Major medical has a $500 deductible and 20% coin-surance clause.
How much is subject to coinsurance?
What is the total amount Kevin pays?

9. Mike Rodriguez has comprehensive medical insurance that includes hospital, surgical-medical, and major medical insurance. The major medical policy has a $250 deductible and a 20% coinsurance clause. When he had surgery, his total bill for hospital care and physician's fee was $26,454. The hospital and surgical-medical provisions of his insurance policy covered $25,322. What amount did Mike pay?

10. Sofia Carbondale's comprehensive medical insurance includes hospital, surgical-medical, and major medical insurance. The major medical policy has a $400 deductible and a 20% coinsurance clause. When she had surgery, her total bill for hospital care and physician's fee was $15,372. The hospital and surgical-medical provisions of her insurance policy covered $13,509. What amount did she pay?

MAINTAINING YOUR SKILLS Look up the skills in parentheses if you need help or more practice.

Find the percentage. Round answers to the nearest hundredth. **(Skill 30)**

11. 20% of $2490   **12.** 31.4% of 952   **13.** $\frac{3}{4}$% of 120

Subtract. **(Skill 4)**

14. 978 − 865   **15.** 77,521 − 66,842   **16.** 997,341 − 942,876

17. 955 − 827   **18.** 82,321 − 32,966   **19.** 793,450 − 89,890

10-3

Term Life Insurance

OBJECTIVE

Use tables to compute the annual premium for term life insurance.

The main purpose of life insurance is to provide financial protection for your dependents in case of your death. Term life insurance is the least expensive form of life insurance that you can buy. You buy term insurance for a specified term, such as 5 years, or to a specified age. Unless you renew your policy at the end of each term, the insurance coverage ends.

If you should die during the term of the policy, your beneficiary, the person you name in the policy, will receive the face value of the policy. The face value is the amount of insurance coverage that you buy. The annual premium depends on your age at the time you buy the policy and the number of units. One unit of insurance has a face value of $1000. The annual premium for term life insurance usually increases with each new term.

ANNUAL PREMIUMS PER $1000 OF LIFE INSURANCE: 5-YR. TERM		
AGE	MALE	FEMALE
18	$ 2.64	$ 1.85
20	2.70	1.91
25	3.01	2.05
30	3.20	2.26
35	3.83	2.99
45	8.86	6.36
50	14.10	9.26
55	23.45	13.87

Annual Premium = Number of Units Purchased × Premium per $1000

EXAMPLE   *Skills* 11, 8   *Application* C   *Term* Term life insurance

Doug Mason is 30 years old. He wants to purchase a $40,000, 5-year term life insurance policy. What is his annual premium?

SOLUTION

A. Find the **number of units purchased.**
$40,000 ÷ $1000 = 40 units purchased

B. Find the **premium per $1000.**
Refer to the table above. (male, age 30) $3.20

C. Find the **annual premium.**
Number of Units Purchased × Premium per $1000
40   × $3.20 = $128.00 annual premium

40000 ÷ 1000 = 40 × 3.2 = 128

✓ SELF-CHECK   Complete the problems, then check your answers in the back of the book.

Use the table above to find the annual premium for a 5-year term policy.

1. $30,000, female, age 18.

2. $60,000, male, age 45.

Use the table on page 294 to answer the following.

	Insured	Age	Coverage	Number of Units	$\times$	Annual Premium per $1000	$=$	Annual Premium
3.	Gloria Carver	20	$ 10,000	10	$\times$	$1.91	$=$	?
4.	John O'Neill	45	$ 20,000	?	$\times$	?	$=$	?
5.	Kate Owens	35	$ 25,000	?	$\times$	?	$=$	?
6.	Dale Payne	30	$200,000	?	$\times$	?	$=$	?

7. Phil Davis's insurance policy. 5-year term life insurance. $75,000 coverage. He is 30 years old. What is his annual premium?

8. Marica Deerfoot's insurance policy. 5-year term life insurance. $120,000 coverage. She is 45 years old. What is her annual premium?

9. Joni Hauck wants to purchase a $15,000, 5-year term life insurance policy. She is 25 years old. What is her annual premium?

10. Peter and Edith Lichtner have 1 child. Peter is a career counselor and Edith is a child psychologist. Peter is 30 years old and Edith is 25 years old. Both want to purchase $40,000, 5-year term life insurance policies. What is Edith's annual premium? What is Peter's annual premium?

11. Sam and Kolleen Hastings have 2 children. Sam was 30 years old when he first purchased an $85,000, 5-year term life insurance policy. He will be 35 years old this year and his insurance agent has informed him that his policy comes up for renewal. What will be his annual premium if he renews his policy? What total amount did he pay during the previous 5-year term? What total amount will he pay for the next 5-year term?

Divide. Round answers to the nearest thousandth. (Skill 11)

12. 8.216 ÷ 6.12

13. 76.26 ÷ 0.14

14. 1.025 ÷ 0.05

15. 21,624 ÷ 1000

16. 93.40 ÷ 100

17. 18,400 ÷ 1000

18. 95,490 ÷ 10,000

19. 89.75 ÷ 10

20. 9.12 ÷ 100

Multiply. Round answers to the nearest hundredth. (Skill 8)

21. 34.362 × 100

22. 0.95 × 0.16

23. 3.49 × 0.035

24. 42.6 × 32.914

25. 0.052 × 1000

26. 3.481 × 1000

27. 73.2 × 1000

28. 460 × 3500

29. 5.62 × 0.101

Other Types of Life Insurance

Whole life insurance offers financial protection for your entire life. The premium usually remains the same each year. In addition to the face value that your beneficiary will receive, whole life insurance has a cash value and loan value. The cash value is the amount of money you will receive if you decide to cancel your policy. The loan value, usually the same amount as the cash value, is an amount that the insurance company will loan you if you request it. Limited-payment life insurance also offers lifetime protection. It has a cash value and a loan value. You pay premiums only for a specified number of years or until you reach a certain age. Universal life insurance is a combination of life insurance and a savings plan. The policy covers you for your entire life; however, you pay a minimum premium with anything over the minimum going into a savings account.

	ANNUAL PREMIUMS PER $1000 OF LIFE INSURANCE				MONTHLY PREMIUMS $50,000 UNIVERSAL LIFE
Age	Paid up at Age 65		Whole Life		Male or Female
	Male	Female	Male	Female	
20	$11.59	$ 9.61	$ 8.05	$ 6.29	$ 19.00
25	13.70	11.48	9.45	7.46	24.00
30	16.88	14.41	11.67	9.13	29.00
35	21.47	18.07	14.91	11.42	37.50
40	29.74	24.86	19.45	14.57	52.00
45	39.48	32.56	25.59	18.63	69.50
50	56.26	45.77	34.04	24.21	93.50
55	—	—	46.51	32.37	126.00

Annual Premium = Number of Units Purchased × Premium per $1000

EXAMPLE   *Skills* 11, 8   *Application* C   *Term* Whole life insurance

Phyllis Saul is 25 years old. She wants to purchase a whole life policy valued at $25,000. What is her annual premium?

SOLUTION

A. Find the **number of units purchased.**
$25,000 ÷ $1000 = 25 units purchased

B. Find the **premium per $1000.**
Refer to the table above. (female, age 25) $7.46

C. Find the **annual premium.**
Number of Units Purchased × Premium per $1000
        25   × $7.46 = $186.50 annual premium

$25000 \div 1000 = 25 \times 7.46 = 186.5$

✔ **SELF-CHECK**   Complete the problems, then check your answers in the back of the book.

Find the annual premium.

1. 30-year old male.
$70,000 whole life policy.

2. 40-year-old female.
Paid-up-at-age-65 policy of $90,000.

Use the table on page 296 to solve the following.

3. Ann Gosik's insurance policy. $50,000, whole life. She is 40 years old. What is her annual premium?

4. James Dolby's insurance policy. $35,000, paid up at age 65. He is 20 years old. What is his annual premium?

5. Terrance Gonzales is 30 years old and wants to purchase a $50,000 life insurance policy. He is considering a universal life insurance policy. What is his monthly premium?

6. Emerson Donohue is 40 years old. He wants to purchase a $50,000 universal life insurance policy. What is his monthly premium? How much will he pay in total premiums in a year?

7. Robert and Lucy Dubbs each purchase a $60,000 whole life insurance policy. Both are 25 years of age. What is their total annual premium? How much more is Robert's annual premium than his wife's?

8. Leona Sowinski purchased a $50,000 universal life insurance policy at the age of 20. What is her annual premium? If Leona pays $50 a month, how much is she saving annually?

9. A Brief Case Rather than make 1 annual payment, many insurance companies will allow you to make 2, 4, or 12 smaller payments. Due to the additional expense of collecting and handling the payments several times a year, a small fee is charged. Many companies use the guidelines shown below.

Mabel Brossis is 20 years old and has purchased a $40,000 whole life insurance policy. She wants to make quarterly payments. What are her quarterly payments? How much can she save in 1 year by paying the annual premium?

> **Optional Payment Plans**
> **Percent of Annual**
> **Premium**
>
> Semiannual Premiums = 50.5%
> Quarterly Premiums = 25.5%
> Monthly Premiums = 8.5%

Divide. Round answers to the nearest hundredth. **(Skill 11)**

10. $18.4 \div 0.032$

11. $47.614 \div 15.62$

12. $0.098 \div 1.9$

13. $344.62 \div 14.1$

14. $285 \div 25$

15. $30,344 \div 110$

16. $954 \div 9.25$

17. $9500 \div 6.50$

Multiply. Round answers to the nearest hundredth. **(Skill 8)**

18. 0.31×0.84 **19.** 7.81×8.1 **20.** 6.511×0.05 **21.** 0.371×1000

22. 814×12 **23.** 159×11 **24.** 2431×75 **25.** $95,175 \times 0.30$

Certificates of Deposit

OBJECTIVE

Use tables to compute interest on certificates of deposit.

One way to invest your money is to purchase a certificate of deposit (CD), which earns interest at a higher rate than a regular savings account. You buy CDs for specific amounts, such as $500 or $1000, and must leave the money on deposit for a specified time, ranging up to 30 years. You are penalized for early withdrawal. Most certificates of deposit earn interest compounded daily, monthly, or quarterly. Although banks use computers to calculate interest earned, you can use a table to compute the interest earned.

	AMOUNT PER $1.00 INVESTED, DAILY, MONTHLY, AND QUARTERLY COMPOUNDING					
Annual Rate	Interest Period — 1 year			Interest Period — 4 years		
	Daily	Monthly	Quarterly	Daily	Monthly	Quarterly
7.00%	1.072500	1.072290	1.071859	1.323094	1.322053	1.319929
7. 25%	1.075185	1.074958	1.074495	1.336389	1.335261	1.332961
7.50%	1.077875	1.077632	1.077135	1.349817	1.348599	1.346114
7.75%	1.080573	1.080312	1.079781	1.363380	1.362066	1.359388
8.00%	1.083277	1.082999	1.082432	1.377079	1.375666	1.372785
8.25%	1.085988	1.085692	1.085087	1.390916	1.389398	1.386306
8.50%	1.088706	1.088390	1.087747	1.404891	1.403264	1.399951
8.75%	1.091430	1.091095	1.090413	1.419008	1.417266	1.413723
9.00%	1.094162	1.093806	1.093083	1.433265	1.431405	1.427621
9.25%	1.096900	1.096524	1.095758	1.447666	1.445682	1.441647

Amount = Original Principal × Amount per $1.00
Interest Earned = Amount − Original Principal

EXAMPLE   *Skills* 2, 6, 8   *Application* C   *Term* Certificate of deposit

Paul Crates invests $4000 in a 1-year certificate of deposit that earns interest at an annual rate of 8.00% compounded monthly. How much interest will he earn at the end of 1 year?

SOLUTION

A. Find the **amount.**
Original Principal × Amount per $1.00
          $4000.00   ×   1.082999   = $4331.996 = $4332.00

B. Find the **interest earned.**
Amount   −   Original Principal
          $4332.00   − $4000.00   = $332.00 interest earned

4000 M+ × 1.082999 = 4331.996 − RM = 331.996

✔ SELF-CHECK   Complete the problems, then check your answers in the back of the book.

Find the interest earned.

1. $8000 CD for 1 year.
8.25% annual interest rate
compounded quarterly.

2. $50,000 CD for 4 years.
7.75% annual interest rate
compounded daily.

Use the table on page 298 to find the amount per $1.00 invested.

	Annual Rate	Interest Period	Original Principal ×	Amount per $1.00 =	Amount	Interest Earned
3.	7.00%	1 year quarterly	$ 4500 ×	1.071859 =	?	?
4.	9.25%	4 years daily	$ 18,000 ×	? =	?	?
5.	8.00%	4 years monthly	$ 9000 ×	? =	?	?
6.	9.25%	1 year quarterly	$140,000 ×	? =	?	?

F.Y.I.
A 3-month certificate of deposit carried an interest rate of 3.65% on February 4, 1992. The rate was 7.98% on February 6, 1991.

7. $3000 certificate of deposit. Interest period of 1 year. 8.75% annual interest rate. Compounded quarterly. What is the amount at maturity? What is the interest earned?

8. $3500 certificate of deposit. Interest period of 4 years. 7.50% annual interest rate. Compounded monthly. What is the amount at maturity? What is the interest earned?

9. Sophia Deles purchased a 1-year certificate of deposit for $500 that earns interest at a rate of 9.25% compounded daily. What is the amount of the CD at maturity? What is the interest earned?

10. Clifford and Hazel Ida purchased a 4-year certificate of deposit for $10,000. The CD earns interest at a rate of 8.75% compounded daily. What is the amount of the CD at maturity? What is the interest earned?

F.Y.I.
The average rate paid for a 1-year certificate of deposit was 3.92% on February 4, 1992.

11. The LaGelleys have $9500 that they want to invest in a certificate of deposit. Granite Trust offers a 4-year certificate that earns interest at a rate of 8.50% compounded quarterly. Hancock Cooperative Bank offers a 4-year certificate of deposit that earns interest at a rate of 8.25% compounded daily. What CD earns the most interest? By how much?

MAINTAINING YOUR SKILLS   Look up the skills in parentheses if you need help or more practice.

Round to the nearest ten. **(Skill 2)**

12. 78   **13.** 144   **14.** 3045   **15.** 337   **16.** 3559   **17.** 798   **18.** 5999

Subtract. **(Skill 6)**

19. 0.099 − 0.032   **20.** 921.7285 − 336.94   **21.** 5.435 − 0.794

Multiply. **(Skill 8)**

22. 0.073 × 8   **23.** 1.42 × 1543   **24.** 26.49 × 85.76   **25.** 7.61 × 1000

10-6

Effective Annual Yield

OBJECTIVE

Compute the effective annual yield.

Banks advertise not only the annual interest rates, but also the effective annual yields of their certificates of deposit. The annual yield is the rate at which your money earns simple interest in one year. The interest earned at the annual rate is affected by the frequency of compounding. The effective annual yield is not affected, so you can use it to compare investments.

$$\text{Effective Annual Yield} = \frac{\text{Interest for One Year}}{\text{Principal}}$$

EXAMPLE   *Skills* 11, 26   *Application* A   *Term* Annual yield

Randall Raye invested $5000 in a certificate of deposit for 3 years. The certificate earns interest at an annual rate of 9.25% compounded quarterly. What is the effective annual yield to the nearest hundredth of a percent?

S O L U T I O N   **A.** Find the **interest for one year.** (Refer to the table on page 298.)

Amount	–	Principal
($5000 × 1.095758)	–	$5000
$5478.79	–	$5000 = $478.79 interest

B. Find the **effective annual yield.**

Interest for One Year ÷ Principal
$478.79 ÷ $5000.00 = 0.095758 = 9.58%
effective annual yield

✔ SELF-CHECK   Complete the problems, then check your answers in the back of the book.

Find the effective annual yield. (Refer to the table on page 298.)

1. $10,000 certificate of deposit. 8.75% annual interest rate compounded monthly.

2. $2500 certificate of deposit. 9% annual interest rate compounded quarterly.

PROBLEMS

Use the table on page 298 to find the amount per $1.00 invested.

	Annual Rate	Interest Period	Original Principal	× Amount per $1.00	= Amount	Interest Earned	Effective Annual Yield
3.	7.25%	1 year quarterly	$6000	× 1.074495	= ?	?	?
4.	8.00%	1 year quarterly	$8400	× ?	= ?	?	?

	Annual Rate	Interest Period	Original Principal	×	Amount per $1.00	=	Amount	Interest Earned	Effective Annual Yield
5.	9.25%	1 year monthly	$5,000	×	?	=	?	?	?
6.	7.50%	1 year daily	$1,800	×	?	=	?	?	?

7. $9000 certificate of deposit.
9.25% annual interest rate.
Compounded quarterly.
4-year certificate.
What is the effective annual yield?

8. $4500 certificate of deposit.
8.75% annual interest rate.
Compounded daily.
$2\frac{1}{2}$-year certificate.
What is the effective annual yield?

9. Ollie Gibson invested $10,000 in a certificate of deposit of 4 years at an annual interest rate of 8.25% compounded daily. What is the interest earned for 1 year? What is the effective annual yield?

10. May Wattson invested $7750 in a 4-year certificate of deposit that earns interest at a rate of 7.75% compounded monthly. What is the interest earned for 1 year? What is the effective annual yield?

The interest in the Golden Dollar Accounts is compounded quarterly.

11. Paul Durant invested $4000 in a Golden Dollar Account that matures in 6 years. What is the interest rate that the bank will give? What is the interest earned for 1 year? What is the effective annual yield?

12. Ben Garison invested $15,000 in a Golden Dollar Account that matures in 1 year. What is the interest rate that the bank will give? What is the interest earned at maturity? What is the effective annual yield?

GOLDEN DOLLAR ACCOUNTS

Terms Available	Interest Rate	Minimum Deposit
1 yr.	7.75%	$500
2 – 3 yrs.	8.00%	$1000
4 – 5 yrs.	8.75%	$1500
6 – 7 yrs.	9.00%	$2000
8 – 10 yrs.	9.25%	$2500

13. A Brief Case Julie Whiteburn invested $2000 in a Golden Dollar Account for 4 years. What is the interest rate that the bank will give? What is the interest earned at maturity? What is the effective annual yield?

MAINTAINING YOUR SKILLS Look up the skills in parentheses if you need help or more practice.

Write as percents. Round answers to the nearest hundredth. **(Skill 26)**

14. 0.112741 **15.** 0.089146 **16.** 1.25642 **17.** 25.5 **18.** 2.0

Divide. **(Skill 11)**

19. 1500 ÷ 0.05 **20.** 28.42 ÷ 1.2 **21.** 679.2 ÷ 0.02 **22.** 3.60 ÷ 7.2

Stocks

You can invest your money in shares of stocks. Stock prices are published each day in the business sections of daily newspapers. Stock prices are usually quoted in eighths of a dollar. When you purchase a share of stock, you become a part owner of the corporation that issues the stock. You may receive a stock certificate as proof of ownership. The total amount you pay for the stock depends on the cost per share, the number of shares you purchase, and the stockbroker's commission.

Number of shares sold today, in hundreds

Highest and lowest prices paid during the day

Difference between last sale today and last sale yesterday

Abbreviation of company name

52-week High	Low	Stock	Div.	Yld.	PE	Sales 100s	High	Low	Last	Net Chg.
$39\frac{3}{4}$	$29\frac{1}{8}$	Bastek	2.24	6.9	6	68	$32\frac{3}{4}$	$32\frac{1}{4}$	$32\frac{1}{4}$	$-\frac{3}{4}$
$18\frac{3}{8}$	$14\frac{1}{2}$	BCA Co	1.24	8.5	6	187	15	$14\frac{5}{8}$	$14\frac{5}{8}$	$-\frac{3}{8}$
25	13	Beta	.28	2.1	12	432	$13\frac{3}{4}$	$13\frac{1}{4}$	$13\frac{1}{4}$	$-\frac{5}{8}$
$12\frac{1}{4}$	$9\frac{1}{8}$	BF Low	1	9.5	....	352	$10\frac{3}{4}$	$10\frac{1}{2}$	$10\frac{1}{2}$	$-\frac{1}{8}$
42	33	Bigley	1.82	5.5	6	845	$33\frac{1}{4}$	$32\frac{1}{2}$	$33\frac{1}{8}$	$+\frac{1}{8}$

Price paid for the last sale today

Cost of Stock = Number of Shares × Cost per Share

Total Paid = Cost of Stock + Commission

EXAMPLE *Skill* 14 *Application* A *Term* Stocks

Melanie Lambert purchased 100 shares of stock at $37\frac{3}{8}$ per share. Her stockbroker charged her a $25 commission. What is the total amount that she paid for the stock?

SOLUTION

A. Find the **cost of stock**.
Number of Shares × Cost per Share
100 × $37\frac{3}{8}$
100 × $37.375 = $3737.50 cost of stock

B. Find the **total paid**.
Cost of stock + Commission
$3737.50 + $25.00 = $3762.50 total paid

✔ **SELF-CHECK** Complete the problems, then check your answers in the back of the book.

Find the total paid.

1. 150 shares of stock at $58 per share. Commission of $75.

2. 400 shares of stock at $8\frac{1}{8}$ per share. Commission of $36.50.

Company	(Number of Shares	× Cost per Share	= Cost of Stock)	+ Commission	= Total Paid
3. AvGlo	(100	× $5\frac{1}{8}$	= $512.50)	+ $ 24.00	= ?
4. Peaseley	(1,000	× $40\frac{1}{2}$	= ?)	+ $250.00	= ?
5. ICN Br	(500	× $73\frac{1}{4}$	= ?)	+ $318.00	= ?
6. HM Co	(250	× $89\frac{5}{8}$	= ?)	+ $347.50	= ?

7. Trudy Fahringer.
500 shares of Air Waveson.
Cost per share is $24.30.
Commission is $122.
What is the cost of the stock?
What is the total paid?

8. David Daly.
215 shares of Atwood Tire.
Cost per share is $50.125.
Commission is $110.
What is the cost of the stock?
What is the total paid?

9. Enice Brudley purchased 300 shares of GKI Petroleum at $40.75 per share. A broker's commission of $183.38 was charged. What was the cost of the stock? What was the total paid?

10. Linda and Martin Sonoma purchased 300 shares of Hampton Publishing at $24\frac{3}{4}$ per share and 150 shares of A-1 Electronics at $28\frac{3}{4}$ per share. The broker charged a 1% commission. What was the total paid?

11. Dan Hostetler wants to purchase either 100 shares of Matell Scientific at $48\frac{1}{8}$ per share or 400 shares of Ballon Synergistics at $11\frac{7}{8}$ per share. Both companies show promising growth. The broker charges a 2% commission. Which purchase costs less? How much less?

F.Y.I.
On February 5, 1992, there were 262,440,000 shares of stock involved in active trading on the New York Stock Exchange.

12. Hamilton Investment Group charges a fee plus commission for stock transactions. Cindy Siegel purchases 500 shares of stock at $13\frac{3}{4}$ per share through Hamilton Investment Group. What is the cost of the stock? What is the total paid?

HAMILTON INVESTMENT GROUP	
Sales	Commission Rate
Up to $4999	1%
$5000–$9999	1.25%
Over $10,000	1.5%
LOWEST FEES IN TOWN $49.98 plus commission	

13. Albert Nash purchases 700 shares of stock at $33\frac{7}{8}$ per share through Hamilton Investment Group. What is the total paid?

MAINTAINING YOUR SKILLS   Look up the skills in parentheses if you need help or more practice.

Change the fractions to decimals. **(Skill 14)**

14. $15\frac{7}{8}$ **15.** $3\frac{1}{8}$ **16.** $9\frac{3}{4}$ **17.** $14\frac{3}{8}$ **18.** $22\frac{1}{2}$

Multiply. **(Skill 8)**

19. $45.50 × 400

20. $12.75 × 200

21. $57.375 × 400

22. $35.125 × 70

10-8

Stock Dividends

OBJECTIVE

Compute the annual yield and annual dividend of a stock investment.

When you buy stocks, you may receive dividends. Dividends are your return on your investment. You may receive an amount specified by the corporation for each share of stock that you own. The annual yield is the dividend expressed as a percent of the cost of the stock. The higher the annual yield, the greater the return on your investment.

Annual Dividend = Annual Dividend per Share × Number of Shares

$$\text{Annual Yield} = \frac{\text{Annual Dividend per Share}}{\text{Cost per Share}}$$

EXAMPLE   *Skills* 11, 26   *Application* A   *Term* Dividends, annual yield

Pam Schmidt bought 80 shares of stock at $42.50 per share. The company paid annual dividends of $2.75 per share. What is the annual dividend? What is the annual yield to the nearest hundredth of a percent?

SOLUTION   **A.** Find the **annual dividend.**
Annual Dividend per Share × Number of Shares
$2.75   ×   80   = $220.00 annual dividend

B. Find the **annual yield.**
Annual Dividend per Share ÷ Cost per Share
$2.75   ÷ $42.50 = 0.06470 = 6.47%
annual yield

2.75 ⊠ 80 ⊟ 220 2.75 ⊡ 42.5 ⊟ .064705

✔ SELF-CHECK   Complete the problem, then check your answer in the back of the book.

1. 180 shares at $4.70 per share. $0.25 annual dividend per share. Find the annual dividend and the annual yield.

PROBLEMS

	Annual Dividend per Share	÷	Cost per Share	=	Annual Yield
2.	$4.20	÷	$60.00	=	?
3.	$0.34	÷	$ 6.50	=	?
4.	$5.80	÷	$30.00	=	?
5.	$6.10	÷	$80.00	=	?

Round the annual yield to the nearest hundredth of a percent.

6. Purchase price: $44.35 per share. Dividends were $2.50 per share. What is the annual yield?

7. Purchase price: $81.375 per share. Dividends were $3.70 per share. What is the annual yield?

8. Christine Gony.
   Owns 400 shares of MPD Systems.
   Purchase price: $19\frac{3}{4}$ a share.
   Dividends were $1.10 per share.
   What were her annual dividends?
   What is the annual yield?

9. Woody Davenport.
   Owns 850 shares of Moran Funds.
   Purchase price: $17\frac{5}{8}$ a share.
   Dividends were $0.95 per share.
   What were his annual dividends?
   What is the annual yield?

10. Joyce Kronecki bought 350 shares of Apex Coporation stock at $9.75 per share. Last year the company paid annual dividends of $1.00 per share. What were her annual dividends? What is the annual yield?

11. Reed Hopkins purchased 400 shares of Computech stock at $30\frac{1}{8}$ per share. Recently the company paid annual dividends of $1.85 per share. What were his annual dividends? What is the annual yield?

12. Duane Hartley owns 2000 shares of Baywater Energy stock, which he purchased at $8\frac{7}{8}$. Recently he read that the average selling price of his stock was $18. The company paid annual dividends of $0.90 per share last year. What is the annual yield on his stock? For an investor who purchased the stock at $18 per share, what is the annual yield?

13. The Racitis own 300 shares of Balwin Aviation, which they purchased at $37.50 per share. What was the dividend per share paid by the company last year? What were the Racitis' annual dividends? What is the annual yield?

| 52-week | | | | | | | Net |
High	Low	Stock	Div.	High	Low	Close	Chg.
$12\frac{1}{4}$	$7\frac{7}{8}$	ASTEX	.50	$8\frac{3}{4}$	$8\frac{5}{8}$	$8\frac{3}{4}$	$+\frac{1}{4}$
$24\frac{1}{4}$	$5\frac{3}{8}$	BANLINE		$23\frac{1}{8}$	$22\frac{5}{8}$	$22\frac{7}{8}$	....
$42\frac{3}{4}$	23	BALW AVI	.32	$42\frac{3}{4}$	$42\frac{1}{4}$	$42\frac{3}{4}$	$+\frac{3}{4}$
$3\frac{1}{8}$	$1\frac{7}{8}$	BYTE TK		$2\frac{1}{4}$	$2\frac{1}{4}$	$2\frac{1}{4}$	....

Use the table on page 648 to find the amount per $1.00 of compound interest.

14. **A Brief Case**   Rick LaMar purchased a 1-year, $5000 certificate of deposit that earns interest at a rate of 8.75% compounded daily. He also owns 185 shares of River Corporation, which he purchased at $28\frac{1}{8}$. Last year the company paid annual dividends of $1.65 per share. What is the interest earned on his certificate of deposit? What is the effective annual yield on the CD? What were his dividends on his stock? What is the annual yield? Which investment produced a greater annual yield?

MAINTAINING YOUR SKILLS   Look up the skills in parentheses if you need help or more practice.

Divide. Round answers to the nearest hundredth. (Skill 11)

15. 24.0 ÷ 200   **16.** 32.61 ÷ 35   **17.** 0.587 ÷ 0.009   **18.** 16.7 ÷ 1.1

Write the decimals as percents. (Skill 26)

19. 0.2583   **20.** 0.017   **21.** 0.03126   **22.** 1.125   **23.** 0.13145

Selling Stocks

OBJECTIVE
Compute the profit or loss from a stock sale.

When you sell your stocks, the sale can result in either a profit or a loss. If the amount you receive for the sale minus the sales commission is greater than the total amount you paid for the stocks, you have made a profit. If the amount you receive minus the sales commission is less than the total paid, your sale has resulted in a loss.

Net Sale = Amount of Sale − Commission

Profit = Net Sale − Total Paid   Loss = Total Paid − Net Sale

EXAMPLE   *Skills* 1, 6, 8   *Application* A   *Term* Profit or loss

Bill Tennyson paid a total of $3738.43 for 75 shares of stock. He sold the stock for $51.50 a share and paid a sales commission of $39.45. What is the profit or loss from the sale?

SOLUTION

A. Find the **net sale.**
Amount of Sale − Commission
($51.50 × 75) − $39.45
$3862.50 − $39.45 = $3823.05 net sale

B. Is **net sale** greater than **total paid?**
Is $3823.05 greater than $3738.43? Yes

C. Find the **profit.**
Net Sale − Total Paid
$3823.05 − $3738.43 = $84.62 profit

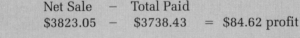

51.5 × 75 = 3862.5 − 39.45 = 3823.05 − 3738.43 = 84.62

✔ SELF-CHECK   Complete the problems, then check your answers in the back of the book.

Find the profit or loss.

1. Paid $1829 for 40 shares. Sold for $60 per share. Commission of $35.50.

2. Paid $24,000 for 1000 shares. Sold for $22 per share. Commission of $98.

PROBLEMS

	Total Paid	(Selling Price per Share × Number of Shares − Commission)			= Net Sale	Amount of Profit or Loss
3.	$4,000	($41 ×	100)	− $32.75 =	?	$67.25 profit
4.	$3,250	($18 ×	200)	− $44.90 =	?	?

	Total Paid	(Selling Price per Share	×	Number of Shares	−	Com-mission)	= Net Sale	Amount of Profit or Loss
5.	$5925	(52\frac{1}{4}$	×	100)	−	$59.60 =	?	?
6.	$ 380	(2\frac{1}{2}$	×	200)	−	$26.45 =	?	?

7. Wendy and Bob Serta.
Bought 47 shares of ITA stock.
Paid a total of $4512.85.
Sold at 93\frac{3}{4}$ per share.
Paid a $95 sales commission.
What was the profit or loss?

8. Sara and Eric Walton.
Bought 140 shares of Lap Top Stock.
Paid a total of $9788.50.
Sold at 81\frac{7}{8}$ per share.
Paid a 2% sales commission.
What was the profit or loss?

9. Brian Rowell owned 200 shares of Big Q stock for which he paid a total of $9187.50. He sold his stock at $58.25 per share and paid a sales commission of $150. What was the net amount of the sale? What was the profit or loss from the sale?

10. Ira and Eve Marcucci sold 120 shares of United Tool stock at 84\frac{3}{4}$ per share and paid a $110 sales commission. They had originally paid a total of $6984.15 for the stock. What was the net amount of the sale? What was the profit or loss from the sale?

11. Doris Hauler owned 300 shares of Watkins International stock for which she paid a total of $2781. She sold the stock at 3\frac{1}{8}$ per share and paid a 1.5% sales commission. What was the net amount of the sale? What was the profit or loss from the sale?

12. Bruce and Debi Maron sold 850 shares of Consumer Research stock at 14\frac{5}{8}$ per share. They paid a 2.5% sales commission. They had originally paid a total of $14,320.45 for the stock. What was the profit or loss?

13. Carl McCollun owned 80 shares of NewTech Computer stock for which he paid 24\frac{1}{2}$ per share plus a 2% commission. He sold at 48\frac{5}{8}$ and paid a 3% sales commission. What was the profit or loss from the sale?

14. The Montvilles purchased 350 shares of Software Design Inc. stock at 20\frac{3}{4}$ and sold 3 years later at 17\frac{7}{8}$. They paid a 1.25% sales commission when they purchased the stock and paid a 3.25% sales commission when they sold. What was the profit or loss from the sale?

MAINTAINING YOUR SKILLS   Look up the skills in parentheses if you need help or more practice.

Which number is greater? (Skill 1)

15. 381 or 381.6

16. $21.19 or $21.91

17. $219.84 or $218.94

Subtract. (Skill 6)

18. 88.35 − 81.15

19. 53.83 − 38.91

20. $8244.50 − $3398.49

Multiply. (Skill 8)

21. 471.2 × 0.035

22. 100 × 3.141

23. 15.058 × 7.2

10-10

Bonds

OBJECTIVE

Compute the annual interest and annual yield of a bond investment.

Many corporations and governments raise money by issuing **bonds.** When you invest in a bond, you do not become a part owner of the corporation. Instead, you are lending money to the corporation or government. In return for the loan, you are paid interest. On the date the bond matures, you will receive the **face value** of the bond. The face value is the amount printed on the bond.

Corporations usually issue bonds that mature in 10 to 30 years with face values that are multiples of $1000. Bonds may sell at a **discount** for less than the face value, or at a **premium** for more than the face value. The cost of a bond is usually a percent of the face value and is referred to as the **quoted price.** A price of 90 on a $1000 bond means that the bond sells for 90% of $1000, or $900. The interest that you receive from a bond is calculated on its face value.

Annual Interest = Face Value × Interest Rate

Bond Cost = Face Value × Percent

$$\text{Annual Yield} = \frac{\text{Annual Interest}}{\text{Bond Cost}}$$

EXAMPLE   *Skills* 30, 11   *Application* A   *Term* Bond

George Vanderhill purchased a $1000 bond at the quoted price of $89\frac{1}{2}$. The bond paid interest at a rate of $6\frac{1}{2}\%$. What is the annual yield, to the nearest hundredth of a percent?

SOLUTION   **A.** Find the **annual interest.**
Face Value × Interest Rate
$1000.00   × $6\frac{1}{2}\%$ = $65.00 annual interest

B. Find the **bond cost.**
Face Value × Percent
$1000.00   × $89\frac{1}{2}\%$ = $895.00 bond cost

C. Find the **annual yield.**
Annual Interest ÷ Bond Cost
$65.00   ÷ $895.00 = 0.07262 = 7.26% annual yield

1000 ⨯ 6.5 % 65   1000 ⨯ 89.5 % 895   65 ÷ 895 = .072625698

Find the interest, bond cost, and annual yield.

1. $1000 bond at $80\frac{1}{2}$. Pays 6% interest.

2. $10,000 bond at 92. Pays $7\frac{1}{4}$% interest.

PROBLEMS

	Face Value of Bond	Quoted Price	Cost of Bond	Interest Rate	Annual Interest	Annual Yield
3.	$10,000	95	$9,500	8%	$800	?
4.	$ 1,000	$85\frac{1}{2}$	?	$7\frac{1}{4}$%	?	?
5.	$10,000	$92\frac{3}{8}$	?	$8\frac{1}{2}$%	?	?
6.	$ 1,000	74.324	?	$9\frac{7}{8}$%	?	?

7. Celinda Vasquez.
Purchased a $1000 bond at $87\frac{1}{2}$.
Pays 4% annual interest.
What is the annual interest?
What was the cost of the bond?
What is the annual yield?

8. Giuseppe Caviness.
Purchased a $10,000 bond at $72\frac{5}{8}$.
Pays $6\frac{1}{4}$% annual interest.
What is the annual interest?
What was the cost of the bond?
What is the annual yield?

9. Sandy and Morry Doran purchased a $10,000 bond at a quoted price of $94\frac{3}{8}$. The bond pays annual interest at a rate of $3\frac{3}{4}$%. What is the annual interest earned? What was the cost of the bond? What is the annual yield?

10. The Sonoma Housing Authority is offering $5000 bonds that pay 7.4% annual interest. The quoted price of each bond is 92.128. Rollin Kowalski purchases 8 bonds through a broker who charges a 1.75% sales commission. What is the total cost of his purchase? What total interest will he earn yearly? What is the annual yield?

11. **A Brief Case**   Many newspapers give up-to-date bond information in the financial section. In the listing shown, the name of the corporation is given, then the annual interest rate the bond is paying, the year the bond matures, and so on. "08" means the bond matures in the year 2008.

Alex Wosick purchased six $1000 bonds of the Dyer Corporation at the closing price for the day. He had to pay a broker's commission of 2.25%. What was the total cost of his purchase? What will be his total yearly earnings? In what year will his bond mature?

Bonds			Cur. Yld.	Vol.	High	Low	Close
DICLO	$9\frac{3}{8}$	08	10.5	2	72	72	72
DYER	$10\frac{7}{8}$	09	8.7	10	$82\frac{1}{4}$	$82\frac{1}{4}$	$82\frac{1}{4}$
EANLO	$8\frac{3}{4}$	12	13.9	28	$71\frac{1}{2}$	70	$71\frac{1}{2}$
EGG CO	$12\frac{1}{4}$	10	9.0	3	$93\frac{5}{8}$	$93\frac{5}{8}$	$93\frac{5}{8}$

MAINTAINING YOUR SKILLS   Look up the skills in parentheses if you need help or more practice.

Find the percentage. **(Skill 30)**

12. $6\frac{1}{2}$% of $700   13. $93\frac{1}{4}$% of $15,000   14. $\frac{1}{2}$% of 40   15. 120% of 900

16. $15\frac{1}{4}$% of 350   17. 110% of 940   18. $2\frac{1}{2}$% of 82   19. 4.4% of 14,240

Divide. Round answers to the nearest hundredth. **(Skill 11)**

20. $2.22\overline{)56.80}$ 21. $19.4\overline{)608}$ 22. $0.004\overline{)96.28}$ 23. $14.10\overline{)48.215}$

Reviewing the Basics

Skills

(Skill 2)

Round to the nearest cent.

1. $74.135   **2.** $1.374   **3.** $318.334   **4.** $6.997

Solve. Round answers to the nearest cent.

(Skill 8)

5. $1411.25 × 30   **6.** $4.54 × 32   **7.** $838.06 × 205

8. $7.35 × 85   **9.** $72.50 × 41   **10.** $132.52 × 9

(Skill 11)

11. $908.14 ÷ 4   **12.** $43.20 ÷ 8   **13.** $983.62 ÷ 12

(Skill 30)

14. 42.3% of $4914.23   **15.** 6% of $180   **16.** 39.2% of $8974.17

Write as a decimal.

(Skill 14)

17. $\frac{5}{16}$ **18.** $\frac{7}{8}$ **19.** $1\frac{5}{8}$ **20.** $\frac{5}{32}$ **21.** $1\frac{1}{4}$ **22.** $3\frac{5}{10}$

Write as a percent.

(Skill 26)

23. 0.09   **24.** 0.16   **25.** 0.188   **26.** 0.0875   **27.** 1.79

Applications

(Application C)

Use the table to find the annual premium.

28. Tom Baker purchased a $25,000 whole life insurance policy at the age of 30. What is his annual premium?

29. Sandra Logan purchased a $40,000 paid-up-at-age-65 policy at the age of 30. What is her annual premium?

	ANNUAL PREMIUMS PER $1000 OF LIFE INSURANCE			
Age	Paid up at Age 65		Whole Life	
	Male	Female	Male	Female
20	$11.59	$9.61	$8.05	$6.28
25	13.70	11.48	9.45	7.46
30	16.88	14.41	11.67	9.13

Terms

Write your own definition for each term.

30. Term life insurance   **31.** Whole life insurance   **32.** Dividends

33. Certificate of deposit   **34.** Annual yield   **35.** Stocks

36. Profit or loss   **37.** Bonds   **38.** Health insurance

Refer to your reference files in the back of the book if you need help.

Unit Test

1. Russell Kirby is employed by Dental Associates. The annual premium for a comprehensive medical insurance program is $3160. His employer pays 70% of the cost. How much does he pay?

2. Nancy King has a medical bill of $26,000. Her hospital and surgical-medical insurance did not pay for $2800 of the total bill. Her major medical insurance has a $250 deductible and a 20% coinsurance clause. How much did she pay?

3. Monte Onstenk purchases a $60,000, 5-year term life insurance policy. The annual base premium is $14.10 per $1000. What is his annual premium?

4. Paul Kenyon purchased a $40,000 whole life insurance policy. The annual base premium is $19.45 per $1000. What is his annual premium?

5. Karolyn Cargill invested $8000 in a 1-year certificate of deposit. The CD earns interest at a rate of 10.00% compounded quarterly. How much interest will she earn on the date of maturity?

AMOUNT PER $1.00 INVESTED, COMPOUNDED QUARTERLY		
Annual Rate	Interest Period	
	1 year	4 Years
10.00%	1.103813	1.484506
10.25%	1.106508	1.499055
10.50%	1.109207	1.513738

6. Joe Ponderoza invested $4500 in a certificate of deposit that earns interest at a rate of 10.00% and matures after 4 years. Interest is compounded quarterly. What is the effective annual yield?

7. Leah O'Grady purchased 300 shares of Penn Brothers stock at $44\frac{3}{8}$. She was charged a $30 commission. What was the total amount that she paid for the purchase?

8. Linda Pitta owns 170 shares of Romez International for which she paid $28\frac{7}{8}$ per share. If the company pays dividends of $1.45 per share, what is the annual yield to the nearest hundredth of a percent?

9. Rick Reilly owns 70 shares of DuPere stock for which he paid a total of $4472. He sold the stock for $84 per share and paid a sales commission of $50. What was his profit or loss on the investment?

10. Denise Rutherford purchased a $1000 bond at the quoted price of $89\frac{3}{4}$. The bond pays interest at a rate of $4\frac{3}{4}$%. What is the annual yield to the nearest hundredth of a percent?

A SPREADSHEET APPLICATION

Compound Interest/ Annual Yield

To complete this spreadsheet application, you will need the template diskette for *Mathematics with Business Applications*. Follow the directions in the *User's Guide* to complete this activity.

Input the information in the following problems to find the amount, interest, and annual yield.

1. Investment: $5000.
Interest rate: 8%.
Compounded: quarterly.
Number of years: 10.
What is the amount?
What is the interest?
What is the annual yield?

2. Investment: $18,000.
Interest rate: 9%.
Compounded: monthly.
Number of years: 6.
What is the amount?
What is the interest?
What is the annual yield?

3. Investment: $1500.
Interest rate: 9.25%.
Compounded: quarterly.
Number of years: 4.
What is the amount?
What is the interest?
What is the annual yield?

4. Investment: $4000.
Interest rate: 8%.
Compounded: monthly.
Number of years: 12.
What is the amount?
What is the interest?
What is the annual yield?

5. Investment: $50,000.
Interest rate: 7.75%.
Compounded: daily.
Number of years: 4.
What is the amount?
What is the interest?
What is the annual yield?

6. Investment: $140,000.
Interest rate: 9.25%.
Compounded: quarterly.
Number of years: 1.
What is the amount?
What is the interest?
What is the annual yield?

7. Irving Pettit invested $5500 at 8.75% compounded daily for 6 years. What is the amount? What is the interest? What is the annual yield?

8. Kerri Keller invested $1000 at 6% compounded daily for 12 years. What is the amount? What is the interest? What is the annual yield?

9. Lyle Growden invested $5623.21 at 7.25% compounded monthly for 4 years. What is the amount? What is the interest? What is the annual yield?

10. Susie Bradford invested $6897 at 11.32% compounded monthly for 4 years. What is the amount? What is the interest? What is the annual yield?

11. At the end of the 4 years, Susie Bradford in problem 10 reinvested the amount at 10.72% compounded quarterly for 3 years. What is the amount? What is the interest? What is the annual yield?

12. Donald Huddleston has $10,000 to invest in a retirement account. He plans to leave the money on deposit for 30 years. Complete the chart below to help him decide which investment firm offers the best deal.

	Institution	Rate	Amount	Interest	Yield
a.	Farmers & Merchants	6.5%, annually	?	?	?
b.	First Federal S&L	6.5%, daily	?	?	?
c.	Metamora State Bank	7.5%, daily	?	?	?
d.	Century Insurance	8.5%, daily	?	?	?

Recordkeeping

Recordkeeping is a way for you to manage your money. By keeping track of your monthly *expenditures*, you can find out how you have spent your money and how much money you need to live on. Your next step is to prepare a *budget sheet* showing your monthly expenses. The purpose of a budget is to allow you to compare how much money you are spending with how much money you are earning.

The first step in planning a budget is to set realistic goals for yourself.

Average Monthly Expenditure

Your first step toward managing your money is to keep an accurate record of how you spend your money. Use a notepad to record any expenditures on the day you make them. At the end of the month, group them and total them. By keeping a record of your expenditures, you will be able to examine your spending habits.

$$\text{Average Monthly Expenditure} = \frac{\text{Sum of Monthly Expenditures}}{\text{Number of Months}}$$

EXAMPLE   *Skills* 5, 11   *Applications* A, Q   *Term* Expenditures

Sue and Bob Miller keep records of their expenditures. They want to know how much they spend each month, on the average. Here are their records for 3 months. What is their average monthly expenditure?

July		August		September	
mortgage payment	$675.00	mortgage payment	$675.00	mortgage payment	$675.00
grocery bill	$51.35	Beth's allowance	$32.00	electric bill	$51.42
Beth's allowance	$32.00	electric bill	$73.56	doctor bill	$35.00
electric bill	$71.47	restaurant	$27.80	cleaners	$17.65
dentist	$43.50	movies	$13.50	telephone bill	$32.75
telephone bill	$27.85	telephone bill	$26.45	gasoline	$16.75
gasoline	$15.60	donation	$25.00	grocery bill	$59.74
water/sewer bill	$31.45	grocery bill	$62.35	football game	$15.00
credit card payment	$41.74	personal expenses	$75.00	credit card payment	$71.46
baseball game	$19.50	credit card payment	$54.92	gasoline	$16.45
gift	$45.00	gasoline	$17.94	Beth's allowance	$32.00
clothing	$71.56	magazine subscription	$31.50	grocery bill	$56.74
car payment	$178.50	car payment	$178.50	car payment	$178.50
grocery bill	$63.70	grocery bill	$71.48	fuel oil	$78.75
TOTAL	$1368.22	TOTAL	$1365.00	TOTAL	$1337.21

SOLUTION   Find the **average monthly expenditure.**

Sum of Monthly Expenditures   ÷ Number of Months

$675.00 $1368.22 + $1365.00 + $1337.21 ÷   3 = $1356.81 average monthly expenditure

1368.22 + 1365 + 1337.21 = 4070.43 ÷ 3 = 1356.81

Find the average monthly expenditure.

1. January, $795; February, $776;   **2.** May, $1571.83; June, $1491.75;
   March, $751.   July, $1543.85; August, $1526.77.

PROBLEMS

Find the average monthly expenditure.

	3.	4.	5.	6.	7.
May	$640.00	$1178.50	$1789.75	$2311.75	$112.11
June	710.00	1091.80	1741.36	2210.91	97.13
July	700.00	1207.70	1707.85	2371.85	106.45
August	685.00	1197.80	1751.63	2353.67	121.85
September	705.00	1245.90	1811.75	2412.91	107.91
Total	?	?	?	?	?
Average	?	?	?	?	?

F.Y.I.
A typical family
of four with a net
income of $20,000
spends 48% of their
budget for food and
housing.

Use the Millers' records of monthly expenditures on page 316 to answer the following.

8. What is the Millers' average monthly expenditure for groceries?

9. Household costs include amounts for electric bills, telephone bills, water and sewer bills, home fuel oil bills, and so on. What is their average monthly expenditure for household expenses?

10. Entertainment expenses include amounts for restaurants, movies, and recreation. What is their average monthly expenditure for entertainment?

11. What do the Millers pay each month to repay their mortgage loan?

12. Transportation costs include car payments and amounts for gasoline, oil, repairs, and so on. What is their average monthly expenditure for transportation costs?

13. Can you determine how much the Millers save each month? Why or why not?

14. The Millers' records for October, November, and December show that their monthly expenditures totaled $1375.80, $1412.91, and $1512.18, respectively. What is their average monthly expenditure for the past 6 months?

MAINTAINING YOUR SKILLS   Look up the skills in parentheses if you need help or more practice.

Add. **(Skill 5)**

15. $716.45 + $820.97   **16.** $21.63 + $22.71 + $24.95

Find the average. **(Application Q)**

17. $1170, $1241, $1193, $1250   **18.** $17.91, $18.43, $16.25

Preparing a Budget Sheet

OBJECTIVE

Use records of past expenditures to prepare a monthly budget.

If you have records of your past expenditures, you can use them to prepare a budget sheet outlining your total monthly expenses. Living expenses vary from month to month and include amounts for food, utility bills, pocket money, and so on. Fixed expenses are expenses that do not vary from one month to the next. Annual expenses, such as insurance premiums and real estate taxes, occur once a year.

$$\text{Total Monthly Expenses} = \text{Monthly Living Expenses} + \text{Monthly Fixed Expenses} + \text{Monthly Share of Annual Expenses}$$

EXAMPLE *Skills* 5, 11 *Application* A *Term* Budget sheet

The Millers use records of their past expenditures to complete the budget sheet below. What is the total of their monthly expenses?

A MONEY MANAGER FOR _Sue and Bob Miller_   DATE _10/1/--_

MONTHLY LIVING EXPENSES		MONTHLY FIXED EXPENSES	
Food/Grocery Bill	$ _125.00_	Rent/Mortgage Payment	$ _675.00_
Household Expenses		Car Payment	$ _178.50_
Electricity	$ _70.00_	Other Installments	
Heating Fuel	$ _45.00_	Appliances	$ _____
Telephone	$ _30.00_	Furniture	$ _____
Water	$ _11.00_	Regular Savings	$ _75.00_
Garbage/Sewer Fee	$ _____	Emergency Fund	$ _50.00_
Other _____	$ _____	TOTAL	$ _978.50_
_____	$ _____	**ANNUAL EXPENSES**	
Transportation		Life Insurance	$ _575.00_
Gasoline/Oil	$ _25.00_	Home Insurance	$ _240.00_
Parking	$ _____	Car Insurance	$ _475.00_
Tolls	$ _____	Real Estate Taxes	$ _1215.00_
Commuting	$ _____	Car Registration	$ _26.50_
Other _____	$ _____	Pledges/Contributions	$ _100.00_
Personal Spending		Other_____	$ _____
Clothing	$ _30.00_	TOTAL	$ _2631.50_
Credit Payments	$ _60.00_	MONTHLY SHARE	
Newspapers, Gifts, Etc.	$ _25.00_	(Divide by 12)	$ _219.29_
Pocket Money	$ _57.00_	MONTHLY BALANCE SHEET	
Entertainment		Net Income	
Movies/Theater	$ _5.00_	(Total Budget)	$ _____
Sporting Events	$ _12.00_	Living Expenses:	$ _505.00_
Recreation	$ _____	Fixed Expenses:	$ _978.50_
Dining Out	$ _10.00_	Annual Expenses:	$ _219.29_
		TOTAL MONTHLY	
TOTAL	$ _505.00_	EXPENSES	$ _____
		BALANCE	$ _____

Find the **total monthly expenses.**

Monthly Living Expenses		Monthly Fixed Expenses		Monthly Share of Annual Expenses	
$505.00	+	$978.50	+	$219.29	= $1702.79
					total monthly expenses

505 $\boxed{+}$ 978.5 $\boxed{+}$ 219.29 $\boxed{=}$ 1702.79

✔ SELF-CHECK   Complete the problems, then check your answers in the back of the book.

Find the total monthly expenses.

1. Living, $670; fixed, $800; share of annual, $350.

2. Living, $475.75; fixed, $679.65; share of annual, $291.17.

PROBLEMS

Betty Kujawa is a landscaper. Walt Kujawa is a used car salesman. They complete the budget sheet shown using records of their past expenditures.

Use the budget sheet to answer the following.

3. What is the total of the Kujawas' monthly living expenses?

4. What is the total of their monthly fixed expenses?

5. What is the total of their annual expenses?

6. What must be set aside each month for annual expenses?

7. What is their total monthly expenditure?

8. Do the Kujawas live within their monthly net income?

9. What individual expenses would be difficult for the Kujawas to cut back on?

A MONEY MANAGER FOR _Walt and Betty Kujawa_   DATE _4/10/--_

MONTHLY LIVING EXPENSES		MONTHLY FIXED EXPENSES	
Food/Grocery Bill	$ _160.00_	Rent/Mortgage Payment .	$ _625.00_
Household Expenses		Car Payment	$ _____
Electricity	$ _45.00_	Other Installments	
Heating Fuel	$ _50.00_	Appliances	$ _____
Telephone	$ _35.00_	Furniture	$ _125.00_
Water	$ _24.50_	Regular Savings	$ _100.00_
Garbage/Sewer Fee . .	$ _____	Emergency Fund	$ _50.00_
Other _Cable TV_	$ _25.00_	TOTAL	$ _____
. . . .	$ _____	ANNUAL EXPENSES	
Transportation		Life Insurance	$ _840.00_
		Home Insurance	$ _____
Gasoline/Oil	$ _85.00_	Car Insurance	$ _750.00_
Parking	$ _5.00_	Real Estate Taxes	$ _____
Tolls	$ _10.00_	Car Registration	$ _52.00_
Commuting	$ _____	Pledges/Contributions . . .	$ _100.00_
Other _____	$ _____	Other_____	$ _____
Personal Spending		TOTAL	$ _____
Clothing	$ _40.00_	MONTHLY SHARE	
Credit Payments	$ _50.00_	(Divide by 12)	$ _____
Newspapers, Gifts, Etc.	$ _20.00_	MONTHLY BALANCE SHEET	
Pocket Money	$ _60.00_	Net Income	
Entertainment		(Total Budget)	$ _1800.00_
Movies/Theater	$ _10.00_	Living Expenses:	$ _____
Sporting Events	$ _20.00_	Fixed Expenses:	$ _____
Recreation	$ _12.00_	Annual Expenses:	$ _____
Dining Out	$ _100.00_	TOTAL MONTHLY	
		EXPENSES	$ _____
TOTAL	$ _____	BALANCE	$ _____

Nancy and Joe Thomas completed the budget sheet shown using records of their past expenditures.

Use the budget sheet to answer the following.

10. What is the total of their monthly living expenses?

11. What is the total of their monthly fixed expenses?

12. What is the total of their annual expenses?

13. What must be set aside each month for annual expenses?

14. What is their total monthly expenditure?

15. Do the Thomases live within their monthly net income?

16. What individual expenses would be difficult for the Thomases to cut back on?

A MONEY MANAGER FOR _Nancy and Joe Thomas_ DATE _7/20/--_	
MONTHLY LIVING EXPENSES	**MONTHLY FIXED EXPENSES**
Food/Grocery Bill .... $ _210.00_	Rent/Mortgage Payment . $ _715.20_
Household Expenses	Car Payment .... $ _____
Electricity .... $ _55.65_	Other Installments
Heating Fuel........ $ _63.75_	Appliances .... $ _57.75_
Telephone .... $ _21.47_	Furniture .... $ _110.80_
Water.............. $ _31.80_	Regular Savings .... $ _75.00_
Garbage/Sewer Fee .. $ _17.21_	Emergency Fund .... $ _50.00_
Other _Security_ .... $ _25.00_	TOTAL .... $ _____
_____ .... $ _____	**ANNUAL EXPENSES**
Transportation	Life Insurance......... $ _480.00_
Gasoline/Oil .... $ _60.00_	Home Insurance .... $ _180.00_
Parking .... $ _35.00_	Car Insurance......... $ _475.00_
Tolls .... $ _12.00_	Real Estate Taxes ..... $ _1200.00_
Commuting .... $ _20.00_	Car Registration .... $ _26.50_
Other _Misc._ .... $ _35.00_	Pledges/Contributions... $ _360.00_
Personal Spending	Other_____ .... $ _____
Clothing............ $ _100.00_	TOTAL .... $ _____
Credit Payments ..... $ _25.00_	MONTHLY SHARE
Newspapers, Gifts, Etc. $ _16.75_	(Divide by 12)........ $ _____
Pocket Money .... $ _32.00_	**MONTHLY BALANCE SHEET**
Entertainment	Net Income
Movies/Theater ...... $ _20.00_	(Total Budget) .... $ _1800.00_
Sporting Events ...... $ _20.00_	Living Expenses: $ _____
Recreation .... $ _15.00_	Fixed Expenses: $ _____
Dining Out .... $ _32.00_	Annual Expenses: $ _____
TOTAL .... $ _____	TOTAL MONTHLY EXPENSES .... $ _____
	BALANCE.......... $ _____

MAINTAINING YOUR SKILLS   Look up the skills in parentheses if you need help or more practice.

Add. **(Skill 5)**

17. $75 + $45 + $53 + $68

18. $475.80 + $519.20 + $647.80

19. $6.18 + $7.23 + $4.37 + $7.96

20. $71.14 + $86.23 + $64.91

21. $619.76 + $723.39 + $671.46

22. $1178.21 + $1371.89 + $1475.84

Divide. Round to the nearest cent. **(Skill 11)**

23. $241 ÷ 4

24. $1642.80 ÷ 3

25. $25.74 ÷ 4

26. $222.28 ÷ 3

Using a Budget

OBJECTIVE

Compare amount budgeted to actual expenditure and personal spending to "typical spending."

Once you have completed a budget sheet outlining your past expenditures, you can use it to plan for future spendings. You may want to prepare a monthly expense summary to compare the amounts that you spend to the amounts that you budgeted. When you draft a budget, you should include an emergency fund to provide for unpredictable expenses, such as medical bills and repair bills.

EXAMPLE   *Skills* 1, 6   *Term* Emergency fund

The Zornows kept accurate records of their expenditures. At the end of March, they prepared an expense summary. They had planned to spend $220 on groceries. They actually spent $231.85. How much more or less did they spend on groceries than they had budgeted for?

S O L U T I O N

A. Compare.
Is **amount spent** more or less than **amount budgeted?**
Is $231.85 more or less than $220.00?   More

B. Find the **difference.**
$231.85 − $220.00 = $11.85 more than amount budgeted

✔ SELF-CHECK   Complete the problems, then check your answers in the back of the book.

Find how much more or less the amount spent is than the amount budgeted.

1. Budgeted $167.80, spent $158.90. **2.** Budgeted $647.50, spent $671.92.

PROBLEMS

3. Mike Ogg's groceries for May.
Budgeted $176.80.
Actually spent $161.75.
Did he spend more or less
than budgeted?

4. May Church's telephone bill for June.
Budgeted $25.
Actually spent $33.78.
Did she spend more or less
than budgeted?

5. Mary Teal's water bill for June.
Budgeted $20.
Actually spent $24.79.
How much more or less did she
spend than budgeted?

6. Burke Long's credit payment for May.
Budgeted $45.
Actually spent $43.45.
How much more or less did he
spend than budgeted?

Here is the Kujawas' expense summary for the month of July. They want to compare what they had budgeted to what they actually spent.

Use the Kujawas' expense summary to answer the following.

7. Which household expenses for the month were more than the amount budgeted?

8. Did they spend more or less than the amount budgeted for household expenses for the month? By how much?

9. How much did they budget for transportation costs? Were the amounts spent for transportation during the month more or less than the amount budgeted? By how much?

EXPENDITURES FOR THE MONTH OF JULY		
Expenses	Amount Budgeted	Actual Amount Spent
Food	$160.00	$175.70
Household		
Electric bill	45.00	44.35
Telephone bill	35.00	41.20
Heating fuel	50.00	15.00
Water bill	24.50	31.70
Cable TV bill	25.00	25.00
Transportation		
Gasoline purchases	85.00	101.70
Parking/tolls	15.00	15.00
Personal		
Clothing	40.00	31.75
Credit payments	50.00	41.74
Newspapers, gifts	20.00	11.65
Pocket money	60.00	72.00
Entertainment		
Movies	10.00	5.00
Sporting events/recreation	32.00	32.00
Dining out	100.00	63.80
Fixed		
Rent	625.00	625.00
Furniture	125.00	125.00
Savings	100.00	100.00
Emergency fund	50.00	50.00
Life/car insurance premiums	132.50	132.50
Car registration	4.33	0
Pledges, contributions	8.33	0

10. Which personal expenses did they spend more on than they had budgeted?

11. Were their total personal expenditures for the month more or less than the amount budgeted? By how much?

12. Were their total entertainment expenditures for the month more or less than the amount budgeted? By how much?

13. Were there any monthly fixed expenses for which the Kujawas spent more than the amount budgeted?

14. What annual expenses occurred during the month? Was the amount set aside for annual expenses more or less than the amount actually spent for annual expenses?

15. What was the Kujawas' total expenditure for the month of July? Was this amount more or less than the amount they had originally budgeted? By how much?

16. Your total monthly expenditure will vary from month to month. During the winter months, your home heating bills may "push" your total monthly expenditure over the amount budgeted. In some months, spending will be less. Name some factors that might affect your spending for specific months.

Here is the Thomases' expense summary for December. They want to compare what they had budgeted to what they actually spent.

Use the Thomases' expense summary to answer the following.

17. Which household expenses for the month were more than the amount budgeted?

18. Did they spend more or less than the amount budgeted for household expenses for the month? By how much?

19. How much did they budget for transportation costs? Were the amounts spent for transportation more or less than the amount budgeted? By how much?

20. Which personal expenses were more than the amount budgeted?

EXPENDITURES FOR THE MONTH OF DECEMBER		
Expenses	Amount Budgeted	Actual Amount Spent
Food	$210.00	$235.80
Household		
Electric	55.65	59.90
Telephone	21.47	17.95
Heating fuel	63.75	74.85
Water bill	31.80	35.50
Garbage/sewer fee	17.21	17.21
Security	25.00	25.00
Transportation		
Gasoline/oil	60.00	54.75
Parking/misc.	102.00	105.70
Personal		
Clothing	100.00	125.00
Credit payments	25.00	25.00
Newspapers, gifts	16.75	9.75
Pocket money	32.00	45.00
Entertainment		
Movies	20.00	32.00
Sports events	35.00	38.50
Dining out	32.00	27.50
Fixed		
Mortgage payments	715.20	715.20
Loan payments	168.55	168.55
Savings	125.00	125.00
Life/home/car insurance	94.58	94.58
Property taxes	100.00	100.00
Car registration	2.21	0
Pledges/contributions	30.00	0

21. Were their total personal expenditures for the month more or less than the amount budgeted? By how much?

22. Were their total entertainment expenditures for the month more or less than the amount budgeted?

23. Were there any monthly fixed expenses for which the Thomases spent more than the amount budgeted?

24. What was the Thomases' total expenditure for the month of December? Was this amount more or less than the amount they had originally budgeted? By how much?

MAINTAINING YOUR SKILLS   Look up the skills in parentheses if you need help or more practice.

Which number is greater? **(Skill 1)**

25. $174.85 or $159.94   **26.** $35 or $37.19   **27.** $2215.73 or $2231.61

Subtract. **(Skill 6)**

28. $47.50 − $43.86   **29.** $171.84 − $165   **30.** $19.47 − $15.50

31. $1712.50 − $1697.43 **32.** $2179.84 − $2050 **33.** $3500 − $3147.81

Reviewing the Basics

Which number is greater?

(Skill 1)

1. $56.91 or $54.92   **2.** $124.61 or $122.93   **3.** $111.11 or $99.99

Solve.

(Skill 5)

4. $35.84 + $39.71 + $18.45 + $60.51 + $123.75

5. $85 + $157 + $31.71 + $90.08 + $141.72 + $74.87

6. $2.75 + $0.63 + $7 + $3.14 + $1.19 + $4.07

(Skill 6)

7. $371.84 − $296.79   **8.** $415.07 − $71.48

9. $1243.71 − $906.74

(Skill 11)

10. $634.29 ÷ 3   **11.** $419.84 ÷ 4   **12.** $894.84 ÷ 12

Write as a percent. Round to the nearest tenth of a percent.

(Skill 26)

13. 0.24718   **14.** 0.415612   **15.** 0.096782   **16.** 0.049951

Applications

(Application Q)

Find the average. Round answers to the nearest cent.

17. $147.85, $216.47, $312

18. $56.91, $62.54, $57.85, $60.19

19. $375, $419.81, $407.63, $384.91, $397, $411.71

20. $12.32, $14, $13.61, $10.71, $9.85, $10, $11.27, $12.50, $11.61

Terms

Use each term correctly in the paragraphs below. Then rewrite the paragraphs.

21. Budget sheet   **22.** Annual expenses   **23.** Monthly fixed expenses

24. Emergency fund   **25.** Living expenses   **26.** Expenditures

A _____ is used to summarize your spending habits and aid in planning for future spendings. Your budget is based on your take-home pay. To determine how you spend your money, keep records of your monthly _____ for three months to a year.

Real estate taxes, insurance payments, pledges, and contributions are examples of _____. Mortgage loan payments, rent, installment payments, and regular savings deposits are examples of _____. Amounts for food, clothing, pocket money, and utility bills are examples of _____. For your budget to succeed, an _____ is necessary to provide for unpredictable expenses, such as medical bills and repair bills.

Refer to your reference files in the back of the book if you need help.

Unit Test

Lesson 11-1

1. Thomas O'Keefe keeps records of his family's expenses. Total monthly expenses for September, October, and November were $1787.43, $1891.74, and $1811.12, respectively. What was the average monthly expenditure?

Lesson 11-2

2. Complete the budget sheet below. Find the total monthly living expenses, total monthly fixed expenses, total annual expenses, monthly share of annual expenses, and balance.

Use the budget sheet to answer the following.

3. What amount is set aside each month for annual expenses? What is the Bakers' total monthly expenditure?

Lesson 11-3

4. The Bakers had planned to spend $187.50 on transportation costs. They actually spent $212.45. How much more or less did they spend on transportation expenses than they had originally planned?

A MONEY MANAGER FOR _Pat and Charles Baker_   DATE _1/15/--_

MONTHLY LIVING EXPENSES		MONTHLY FIXED EXPENSES	
Food/Grocery Bill	$ _240.00_	Rent/Mortgage Payment	$ _575.00_
Household Expenses		Car Payment	$ _167.75_
Electricity	$ _61.50_	Other Installments	
Heating Fuel	$ _____	Appliances	$ _____
Telephone	$ _37.50_	Furniture	$ _45.00_
Water	$ _____	Regular Savings	$ _100.00_
Garbage/Sewer Fee	$ _____	Emergency Fund	$ _50.00_
Other _Child Care_	$ _60.00_	TOTAL	$ _____
_____	$ _____	ANNUAL EXPENSES	
Transportation		Life Insurance	$ _240.00_
Gasoline/Oil	$ _112.50_	Home Insurance	$ _____
Parking	$ _25.00_	Car Insurance	$ _567.50_
Tolls	$ _____	Real Estate Taxes	$ _____
Commuting	$ _____	Car Registration	$ _32.75_
Other _Repairs_	$ _50.00_	Pledges/Contributions	$ _120.00_
Personal Spending		Other _Renter's Ins._	$ _180.00_
Clothing	$ _90.00_	TOTAL	$ _____
Credit Payments	$ _50.00_	MONTHLY SHARE	
Newspapers, Gifts, Etc.	$ _15.00_	(Divide by 12)	$ _____
Pocket Money	$ _100.00_	MONTHLY BALANCE SHEET	
Entertainment		Net Income	
Movies/Theater	$ _20.00_	(Total Budget)	$ _2000.00_
Sporting Events	$ _15.00_	Living Expenses:	$ _____
Recreation	$ _20.00_	Fixed Expenses:	$ _____
Dining Out	$ _50.00_	Annual Expenses:	$ _____
TOTAL	$ _____	TOTAL MONTHLY EXPENSES	$ _____
		BALANCE	$ _____

A SPREADSHEET APPLICATION

Recordkeeping

To complete this spreadsheet application, you will need the diskette *Spreadsheet Applications for Business Mathematics,* which accompanies this textbook.

Input information in the following problem to find the total monthly budgeted amount, total monthly expenditures, and difference between amounts budgeted and spent for each month. Then answer the questions that follow.

MONTHLY BUDGET SHEET FOR JAMES AND SHIRLEY KUJKOWSKI

Budget Category	Amount Budgeted	April Actual Spent	Diff.	May Actual Spent	Diff.	June Actual Spent	Diff.	Three Months' Diff.
Groceries	$200.00	$195.00	?	$197.00	?	$206.00	?	?
Utilities	220.00	235.00	?	225.00	?	210.00	?	?
House payment	715.00	715.00	?	715.00	?	715.00	?	?
House insurance	30.00	30.00	?	30.00	?	35.00	?	?
Transportation	215.00	205.00	?	212.00	?	210.00	?	?
Clothing	40.00	25.50	?	45.00	?	48.00	?	?
Credit card payment	100.00	100.00	?	100.00	?	100.00	?	?
Entertainment	40.00	56.70	?	43.25	?	30.00	?	?
Dining out	50.00	42.85	?	61.95	?	51.85	?	?
Savings	200.00	200.00	?	180.00	?	200.00	?	?
Miscellaneous	100.00	89.80	?	107.90	?	93.45	?	?
TOTAL	?	?	?	?	?	?	?	?

1. What is the Kujkowskis' total amount budgeted?

2. What are the total expenditures for April?

3. What are the total expenditures for May?

4. What are the total expenditures for June?

5. Were they over or under their budget for April? By how much?

6. Were they over or under their budget for May? By how much?

7. Were they over or under their budget for June? By how much?

8. Were they over or under their budget for the three months combined? By how much?

9. For April, in which category did they differ most from their bud-geted amount? By how much?

10. For May, in which category did they differ most from their budgeted amount? By how much?

11. For June, in which category did they differ most from their budgeted amount? By how much?

12. Over the entire three months, in which category did they differ most from their budgeted amount? By how much?

13. Why did the actual amount of their house payment not change?

14. Explain why the actual amount of their house insurance payment went up.

15. Do you have any advice for the Kujowskis? Why or why not?

Financial Advisor

Olga Chebyshev works as a financial advisor at a medium-sized mail-order firm. She spends a great deal of her time compiling the book-keepers' financial records and making sense of the pages of data that result. Not only does she compare figures for the amount actually spent to figures for the projected expenditures, she also advises managers about ways that costs can be cut.

A computer network and an electronic spreadsheet program make much of the compilation work and analysis go more smoothly.

Of the actual and projected figures for production, advertising, marketing, facilities, and personnel, she decided to zero in on the costs of mailing the company's catalogs. The double bar graph below shows the actual cost figures (black bars) and projected cost figures (blue bars) for catalog mailings last year.

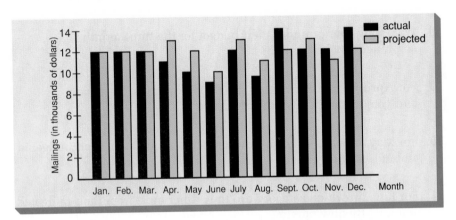

1. For which months did the actual and projected mailings agree?

2. For which months did actual mailings exceed projected mailings?

3. Estimate the projected cost of mailing the catalogs in July.

4. Estimate how much the actual cost of mailings exceeded the projected cost in November.

5. Estimate the total actual mailing costs for the year.

6. Estimate the total projected mailing costs for the year.

Cumulative Review
(Skills/Applications) Units 6–11

Skills

(Skill 1)

Find which number is greater.

1. 4.7 or 4.07   **2.** 27.18 or 27.174   **3.** 96.178 or 96.201

Round to the nearest cent.

(Skill 2)

4. $14.174   **5.** $219.676   **6.** $1.485   **7.** $74.997

Solve. Round to the nearest cent.

(Skill 3)

8. $74 + $15   **9.** $71 + $39   **10.** $78 + $50 + $95

11. $617 + $179 + $496   **12.** $956 + $78 + $491

(Skill 4)

13. $78 − $43   **14.** $219 − $74   **15.** $640 − $571

16. $7413 − $5527   **17.** $9142 − $7291   **18.** $8070 − $4793

(Skill 5)

19. $71.43 + $32.78   **20.** $49.56 + $17.72   **21.** $78.56 + $123.75

22. $84.32 + $66.75   **23.** $516.45 + $491.84   **24.** $718.45 + $709.91

(Skill 6)

25. $71.47 − $15.35   **26.** $65.86 − $41.93   **27.** $47.50 − $4.17

28. $91.86 − $35.46   **29.** $607.08 − $49.78   **30.** $2174.81 − $471.55

(Skill 8)

31. $365.85 × 92   **32.** $2.17 × 35   **33.** $51.76 × 330

34. $78.54 × 8.7   **35.** $8.75 × 6.8   **36.** $516.75 × 0.74

(Skill 11)

37. $112.80 ÷ 26   **38.** $15.81 ÷ 24   **39.** $417.93 ÷ 52

40. $956.74 ÷ 5.12   **41.** $75,750 ÷ 5.61   **42.** $2141.85 ÷ 74.6

Write as a decimal.

(Skill 14)

43. $\frac{7}{8}$ **44.** $\frac{9}{16}$ **45.** $\frac{21}{32}$ **46.** $\frac{145}{360}$

Write as a percent to the nearest tenth of a percent.

(Skill 26)

47. 0.748   **48.** 0.0971   **49.** 0.7137   **50.** 0.06449

Write as a decimal.

(Skill 28)

51. 45%   **52.** 14.7%   **53.** 212%   **54.** 6.7%

Solve. Round to the nearest cent.

(Skill 30)

55. 48% of $794   **56.** 53% of $145

57. 8.7% of $217   **58.** 17.8% of $93.47

59. 32% of $15.76   **60.** 6.5% of $119.79

61. 4.05% of $19.47   **62.** 0.4% of $71.89

(Application C)

Use the table to answer the following.

Monthly Payments $1000 Loan at 10% APR	
Term in Months	**Monthly Payment**
6	$171.56
12	87.91
18	60.05
24	46.14

63. What is the monthly payment on a $1000 loan for a term of 18 months?

64. What is the total amount repaid if the loan is repaid in 12 months?

65. For what term is the monthly payment lowest?

66. For what term is the total amount repaid lowest?

Use the table on page 645 to find the elapsed time in days.

(Application G)

67. From March 4 to March 30

68. From April 10 to July 20

69. From January 3 to February 20

70. April 15 to October 15

71. From October 16 to February 10

72. From July 20 to January 30

Write as a fraction of a year.

(Application J)

73. 4 months

74. 3 months

75. 90 days (ordinary year)

76. 180 days (ordinary year)

77. 182 days (exact year)

Find the mean. Round answers to the nearest cent.

(Application Q)

78. $64.74, $71.83, $95.74, $67.85

79. $7.15, $8, $6.45, $10, $7.15, $12.48

80. $4175.85, $7181.74, $5671.24

81. $4.15, $5, $7.20, $6.15, $4, $3.98, $6.20, $7, $6.45, $8.10

Terms

Write your own definition for each term.

82. Sales receipt

83. Finance charge

84. Previous balance

85. Unpaid balance

86. Rebate schedule

87. Discount loan

88. Installment loan

89. Interest rate

90. Sticker price

91. Rent

92. Deductible clause

93. Depreciation

94. Mortgage loan

95. Closing costs

96. Premium

97. Principal

98. Term life insurance

99. Whole life insurance

100. Bonds

101. Stocks

102. Budget sheet

103. Fixed expenses

104. Emergency fund

105. Living expenses

Cumulative Review Test
Units 6–11

Refer to your reference files in the back of the book if you need help.

Lesson 6-2

1. Ruth Warey has a charge account that charges 1.67% of the previous balance for a finance charge. Her statement shows purchases totaling $74.85, a previous balance of $294, and a payment of $100. What is the new balance?

Lesson 6-4

2. Adam North has a charge account that charges 1.75% of the unpaid balance. His statement shows purchases totaling $47.49, a previous balance of $374, and a payment of $70. Find the finance charge and the new balance.

Lesson 6-6

3. Find the average daily balance where new purchases are included when figuring the daily balances. Then find the finance charge and the new balance.

REFERENCE	POSTING DATE	TRANSACTION DATE	DESCRIPTION	PURCHASES & ADVANCES	PAYMENTS & CREDITS
13100	9/14	8/23	Wesford Clothes	104.50	
20046	9/14	8/27	Record Mart	35.60	
54545	9/17		PAYMENT		30.00

BILLING PERIOD	PREVIOUS BALANCE	PERIODIC RATE	AVERAGE DAILY BALANCE	FINANCE CHARGE
9/1-9/30	$245.20	1.5%	?	?

PAYMENTS & CREDITS	PURCHASES & ADVANCES	NEW BALANCE	MINIMUM PAYMENT	PAYMENT DUE
$30.00	$140.10	?	$20.00	10/21

Lesson 7-1

4. Alice Russel's bank granted her a single-payment loan of $6500 for 90 days at an annual interest rate of 10%. Her bank charges ordinary interest. What is the maturity value of Alice's loan?

Lesson 7-3

5. Tony Bonfiglio obtained an installment loan of $500 to purchase a stereo set. The annual percentage rate is 18%. He plans to repay the loan in 12 months. Use the table to find the finance charge.

APR	Term in Months	If You Finance . . .			
		$200	$500	$1000	$1500
		Your Monthly Payments Are			
	6	35.10	87.76	175.52	263.28
18%	12	18.33	45.84	91.68	137.52
	18	12.76	31.90	63.80	95.70
	24	9.98	24.96	49.92	74.88

Lesson 8-2

6. Amy Green is interested in a luxury car that has a base price of $27,895, options totaling $6174.95, and a destination charge of $475. She has read that the dealer's cost is about 80% of the base price and 75% of the price of the options. What is the estimated dealer's cost?

7. David Scott uses his car to drive to and from work. His insurance coverage includes 50/100 bodily injury and $100,000 property damage. His driver-rating factor is 4.10. He has $50-deductible comprehensive and $100-deductible collision insurance coverage on his car. His car is in age group A and rating group 8. What is his annual premium?

Property Damage Limits	Bodily Injury Limits		
	25/50	25/100	50/100
$ 25,000	$103.20	$114.00	$119.20
50,000	107.20	118.00	123.20
100,000	110.00	121.20	126.00
Age Group	Insurance-Rating Group		
	6	7	8
	Comprehensive $50-ded.		
A	$32.80	$35.60	$39.60
B	30.80	33.20	36.80
	Collision $100-ded.		
A	$84.00	$83.20	$102.40
B	72.80	90.80	108.40

8. The Sturgeons obtained a 30-year, $70,000 mortgage loan at First Federal Bank at an interest rate of 10.5%. Find the monthly payment. What is the new principal after the first monthly payment? Use the monthly payment table on page 645.

9. The Hocking County tax assessor stated that the market value of the King estate is $975,800. The rate of assessment in Hocking County is 35% of market value. The tax rate is 74.85 mills. What is the annual real estate tax on the King estate?

10. Debbie Young is employed by A.I.M. Ind., Inc. The annual premium for comprehensive medical insurance is $2870. Her employer pays 50% of the cost. How much does Debbie pay?

11. Vicky Osinski purchaseed a whole life insurance policy with a face value of $75,000. The annual base premium is $22.68 per $1000. What is her annual premium?

12. Tom Green invested $8000 in a certificate of deposit that earns interest at a rate of 10.00% and matures after 4 years. Interest is compounded quarterly. What is the effective annual yield?

AMOUNT PER $1.00 INVESTED, COMPOUNDED QUARTERLY		
Annual Rate	Interest Period	
	1 Year	4 Years
10.00%	1.103813	1.258577
10.25%	1.106508	1.271224
10.50%	1.109207	1.283998

13. Manny Diaz purchased 150 shares of Petrie stock at $5\frac{7}{8}$. He was charged at $40 commission. What was the total amount that he paid for the purchase?

14. Cynthia Pinziotti keeps records of family expenses. Total monthly expenses for September, October, and November were $2147.40, $2271.85, and $2141.47, respectively. What is the average monthly expenditure?

▼ A SIMULATION

Buying a Condominium

You have decided to purchase a condominium. You are chief systems programmer for Datacrunch, Inc., and your gross pay is $36,000 a year. You have $14,000 in a savings account.

Before you look at condominiums, you decide to figure out how much you can afford to spend.

Your bank suggests that you spend not more than 28% of your gross pay on principal, interest, and taxes.

Most condominium management companies in your area want a down payment of 20% of the selling price. You decide that you can use $12,000 of your savings account as a down payment.

Write your name on the Datacrunch badge.

Use this work sheet to figure out how much you can afford to spend for your condominium.

1.	Amount you can afford to spend per year on principal, interest, and taxes	?
2.	Amount you can afford to spend per month on principal, interest, and taxes	?
3.	Maximum selling price on which you can afford to make a 20% down payment	?

2 Employer's name, address, and ZIP code		6 Statutory employee ☐	Deceased ☐	Pension plan ☐	Legal rep. ☐	942 emp. ☐	Subtotal ☐	Deferred compensation ☐	Void ☐
DATACRUNCH, INC.		7 Allocated tips				10 Advance EIC payment			
		9 Federal income tax withheld 7124.00				10 Wages, tips, other compensation 36,000.00			
3 Employer's identification number	4 Employer's state I.D. number	11 Social security tax withheld 2754.00				12 Social security wages 36,000.00			
5 Employee's social security number 123-45-6789		13 Social security tips				14 Nonqualified plans			
19 Employee's name, address, and ZIP code		15 Dependant care benefits				16 Fringe benefits incl. in Box 10			
		17				18 Other			
20	21	22				23			

24 State income tax 1035.00	25 State wages, tips, etc. 36,000.00	26 Name of state	27 Local income tax 900.00	28 Local wages, tips, etc. 36,000.00	29 Name of locality

Which Condominium?

After looking at several condominiums, you narrow your selection to two, one in the city and one in a suburb. The advertising booklet for each condominium gives the base price and the additional charges for extra features.

One-Bedroom Unit
$56,000 850 sq ft

Plush carpeting: 80 sq yd @ $6.75 per sq yd extra cost

Country Squire Estates
Fireplace: stone $3300
Refrigerator with ice maker   790
Microwave oven: built in   280

Townview
One-Bedroom Unit $52,000 790 sq ft
Fireplace: brick
Refrigerator with ice maker $3500
Microwave oven: built in   860
Plush carpeting: 72 sq yd @   250
$8.50 per sq yd extra cost

Choose one of these condominiums. You will keep the same condominium for the rest of this simulation. You decide to purchase all the extra features listed in the advertising booklet.

Use the prices in the booklet to find the total cost of your condominium.

4. Condominium you have chosen	?
5. Base price	?
6. Fireplace	?
7. Refrigerator	?
8. Microwave oven	?
9. Carpeting: ___ sq yd × $ ___	?
10. Total cost of condominium	?

A SIMULATION
(CONTINUED)

Your Net Worth

After you choose the condominium you want, you fill out a mortgage application. The lender will check to make sure that you have as much money as you say you do, and that you have paid your debts on time.

The application also asks for a statement of assets and liabilities. You have $14,000 in a savings account and $950 in a checking account. You own a car worth $9000, on which you still owe $3000. Your life insurance has a net cash value of $3000. You have personal property, such as furniture and jewelry, with a total value of $6350. You owe $380 on your credit card.

Use this information to fill out the loan application and find your net worth.

RESIDENTIAL LOAN APPLICATION

Borrower		
Name		
Address		
Employer		
Position or Title		
Gross Annual Income		

	Assets		Liabilities	
Description		**Cash or Market Value**	**Description**	**Unpaid Balance**
Cash			Installment debts	?
Checking account		?	Real estate loans	
Savings account		?	Automobile loans	?
Automobile		?	Other debt	
Life insurance net cash value		?		
Furniture and personal property		?		
Total assets		?	Total liabilities	?
Net Worth (Total Assets Minus Total Liabilities)				?

11. What are your total assets?

12. What are your total liabilities?

13. What is your net worth?

A SIMULATION
(CONINTUED)

Bank Charges and Taxes

Now that you have selected the condominium you would like to buy, you will have to calculate what it will cost you each month.

14. Look back at question 10 to find the total selling price of your condominium. If you make a down payment of 20% of the selling price, what is the mortgage loan amount?

Both development companies arrange financing. The mortgage for Townview is 11% for 25 years. The mortgage for Country Squires is 10.50% for 30 years.

15. Use the mortgage table (below left) to find the monthly payment per $1000 financed. Then calculate your monthly payment for principal and interest.

MONTHLY MORTGAGE PAYMENTS
PRINCIPAL AND INTEREST PER $1000 FINANCED

Interest Rate	15 Years	20 Years	25 Years	30 Years
10.50%	11.06	9.99	9.45	9.15
11.00%	11.37	10.33	9.81	9.53
11.50%	11.69	10.67	10.17	9.91
12.00%	12.01	11.02	10.54	10.29
12.50%	12.33	11.37	10.91	10.68

REAL ESTATE TAX RATES

Town	Rate of Assessment	Tax Rate per $1000
Sparta	63%	$109.40
Springfield	80%	96.20
Springton	75%	101.15
Stafford Valley	50%	136.50
Strawberry	100%	88.60

As a condominium owner, you will have to pay real estate taxes. Townview is in Springfield. Country Squires is in Stafford Valley.

Use the real estate tax table (above right) to find your tax.

16.	Value of your condominium	?
17.	Rate of assessment	?
18.	Assessed value: ____% × $____	?
19.	Tax rate per $1000	?
20.	Real estate taxes per year	?
21.	Real estate taxes per month	?

22. What is your total monthly cost for principal, interest, and taxes? Is this amount less than 28% of your gross income?

A SIMULATION
(CONTINUED)

Fees, Insurance, and Utilities.

Your condominium management company charges you an annual fee. This fee covers the cost of trash pickup, snow removal, landscaping, maintenance, management costs, and insurance on the buildings.

The Townview fee is $0.60 per square foot per year. The Country Squires fee is $0.75 per square foot per year.

23. Look at the advertising booklets on page 334 to find the number of square feet in your condominium. What is the condominium fee per year? What is the fee per month?

The condominium fee includes insurance on the buildings. You have to get separate insurance on your furniture and other possessions. Your insurance company has a basic annual rate of $95. There are extra charges for insurance on certain valuable items.

Use these work sheets to calculate the cost of your insurance.

	Item	Value	Annual Rate per $100	Annual Rate
24.	Camera	$400	$ 1.65	?
25.	Bicycle	250	10.00	?
26.	Jewelry	350	1.60	?

27.	Basic rate per year	?
28.	Extra charges per year	?
29.	Total rate per year	?
30.	Insurance cost per month	?

Townview uses electricity for heating, cooking, and lighting. It is estimated that the unit will use 9750 kW·h per year. The cost of electricity is $0.11 per kW·h.

Country Squires uses gas for heating and cooking. The estimated annual consumption of gas is 745 CCF at $0.65 per CCF. The unit would also use 2600 kW·h of electricity per year at $0.10 per kW·h.

Each unit will use about 8000 cubic feet of water per year. In Springfield, water costs $9 per 1000 cubic feet. Sewer charges, which are based on 70% of water usage or 5600 cubic feet, are $14 per 1000 cubic feet. In Stafford Valley, water and sewer charges are 50% more than in Springfield.

Find your utility costs.

31.	Annual electricity cost	?
32.	Annual gas cost	?
33.	Annual water and sewer cost	?
34.	Total annual utilities cost	?
35.	Monthly utilities cost	?

36. What is your total monthly cost for fees, insurance, and utilities?

A SIMULATION
(CONTINUED)

Summary

Summarize your results so far.

37. Your condominium	?
38. Total cost	?
39. Your net worth	?
40. Monthly cost for principal, interest, and taxes	?
41. Monthly cost for fees, insurance, and utilities	?
42. Total monthly cost	?

A common rule of thumb is that your monthly housing cost should be less than 35% of your monthly net pay. Your net pay is your gross pay minus deductions for taxes and FICA. Shown below is the stub attached to your monthly paycheck.

Use the stub to calculate your monthly net pay.

DATACRUNCH, INC.					
EARNINGS	AMOUNT	DEDUCTIONS		YEAR TO DATE	
REGULAR	3000 00	FICA TAX FED. TAX STATE TAX LOCAL TAX	229 00 596 50 86 25 75 00	GROSS PAY FICA TAX FED. TAX STATE TAX LOCAL TAX	3000.00 229.50 596.00 86.25 75.00
				FOR PERIOD ENDING	
				JANUARY 31, 19--	
				NET PAY	

43. Monthly net pay	?
44. 35% of monthly net pay	?

45. Is your total monthly housing cost less than 35% of your monthly net pay?

46. Your monthly housing expenses are your mortgage, taxes, condominium fee, insurance, and utilities. Compare expenses with someone who chose the other condominium. For each expense, which condominium costs more?

47. If you had another chance to decide about buying a condominium, what might you do differently? Why?

PART

3

Business Applications

In Part 3, you will learn how to apply mathematics to actual business practices. Different departments of one business are covered in Part 3. You will learn firsthand how mathematics is used in each department and how each department contributes to the entire operation of the business.

The following definitions provide an overview of the activities involved in the operations of the departments within a business. The lessons within each unit explain in detail how business mathematics is used to carry out these activities.

Personnel   The administrative activities of hiring staff, developing wage and salary scales, and providing benefits for employees within your business.

Production   The activities involved in manufacturing and packaging the items that your business sells.

Purchasing   The activities involved in buying goods at discount prices from your suppliers.

Sales   The activities involved in determining the selling price of your product in the marketplace.

Marketing   1. The activities concerned with researching buyer preference.   2. The activities concerned with communicating information about your product to prospective buyers.

Warehousing and Distribution   The activities of storing, shipping, and maintaining accurate inventory records of your products.

Services   1. The activities related to using buildings, equipment, or utilities from other businesses for a fee.   2. The activities related to hiring employees of another business to do work for your business.

Accounting   The activities related to monitoring normal expenses of your business, such as payroll and depreciation.

Accounting Records   The activities of maintaining and analyzing documents that show the income, expenses, and value of your business.

Financial   The activities concerned with managing your business's money, including paying taxes, borrowing, and investing.

Personnel

The personnel department of your business fills openings by *recruiting* new employees through advertising and interviews. The department also handles *wage and salary scales* for each position and *employee benefits,* such as disability, health, and life insurance and paid vacations and holidays.

Part of the cost of running a business goes to employee benefits. Employee benefits add real dollars to your paycheck.

Hiring New Employees

To fill openings in your business, you may **recruit** new employees. The cost of recruiting includes advertising fees, interviewing expenses, such as travel expenses, and hiring expenses, such as moving expenses. In addition, you may pay an employment agency or your own employees to locate job candidates.

$$\text{Total Recruiting Cost} = \text{Advertising Expense} + \text{Interviewing Expense} + \text{Hiring Expense}$$

EXAMPLE Skill 5 Application A Term Recruiting costs

The Talbot Manufacturing Company is searching for a person to head the production department. The personnel department placed advertisements for a total cost of $2495 and employed the Empire Executive Search Company to locate candidates. Empire recommended Alice Welch, Thomas Snow, and Victor Grabowski. Talbot paid the candidates' travel expenses for interviews.

Alice Welch   Total travel expenses: $317
Thomas Snow   Total travel expenses: $435
Victor Grabowski   Total travel expenses: $474

After the interviews, Talbot hired Victor Grabowski at an annual salary of $54,900. Talbot paid these expenses to hire him:

Moving expenses $1200
Sale of home (real estate broker's fee) $9470
Empire's Fee (25% of Victor's first-year salary) $13,725

What was the total expense of recruiting Victor Grabowski?

SOLUTION Find the **total recruiting cost.**

$$\text{Advertising Expense} + \text{Interviewing Expense} + \text{Hiring Expense}$$
$$\$2495 + (\$317 + \$435 + \$474) + (\$1200 + \$9470 + \$13,725)$$
$$\$2495 + \$1226 + \$24,395 = \$28,116 \text{ total recruiting cost}$$

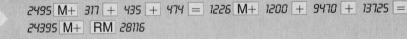
2495 M+ 317 + 435 + 474 = 1226 M+ 1200 + 9470 + 13725 = 24395 M+ RM 28116

Complete the problems, then check your answers in the back of the book.

Find the total recruiting cost.

1. Advertising expenses, $2150; interviewing expenses, $245 and $315; hiring expenses, $1240 and $2170.

2. Advertising expenses, $1975; interviewing expenses, $470 and $860; hiring expenses $475, $8600, and $11,415.

	Advertising Cost	+	Interviewing Cost	+	Hiring Cost	=	Total Recruiting Cost
3.	$ 420.70	+	$ 41.20	+	$ 517.20	=	?
4.	$ 315.85	+	$ 79.80	+	$ 847.72	=	?
5.	$ 789.16	+	$415.25	+	$1213.49	=	?
6.	$1412.71	+	$614.91	+	$1971.44	=	?
7.	$8761.43	+	$971.84	+	$3147.43	=	?

8. Wearite Tires recruits district sales manager.
Advertising cost: $917.45.
Interviewing cost: $694.74.
Hiring cost: $2191.47.
What is the total recruiting cost?

9. Primo Company recruits data processing manager.
Advertising cost: $1475.00.
Interviewing cost: $861.79.
Hiring cost: $3791.86.
What is the total recruiting cost?

10. The Mobile Communications Company hired Marilyn Curtiss as its new national distribution manager at an annual salary of $54,950.

Advertising costs: $2247.50
Interviewing expenses: Marilyn Curtiss, $147.43; Tom Hart, $216.94
Finder Agency fee: 20% of first year's salary

What was the total cost of hiring the national distribution manager?

11. Novi Discount Brokers hired the Wall Street Search Service to locate candidates for the position of manager, investment bonds. The agency's fee is 25% of the first year's salary if one of its candidates is hired. Novi also ran several advertisements at a total cost of $816.40. Novi interviewed three people:

David Gold	Nancy Cooper	Henry Little
Applied through agency	Answered advertisement	Applied through agency
Travel costs: $148.75	Travel costs: $216.40	Travel costs: $171.80

Novi hired Henry Little at an annual salary of $74,760. Novi paid his moving expenses of $419.20 and his real estate broker's fee of 7% to sell his $149,00 home. What was the recruiting cost?

MAINTAINING YOUR SKILLS Look up the skill in parentheses if you need help or more practice.

Add. **(Skill 5)**

12. $148 + $74 + $865

13. $615 + $419 + $1291

14. $1484 + $815 + $11,650

15. $2241 + $915 + $14,542

16. $715.80 + $523.40 + $3120.50

17. $746.50 + $319.20 + $4314.70

18. $619.24 + $141.17 + $2512.64

19. $1171.89 + $471.75 + $12,147.94

Administering Wages and Salaries

OBJECTIVE

Compute the new salary after merit increase and cost-of-living adjustment.

Your business may have wage and salary scales for the positions in the company. You can use the information to compare various jobs or to estimate the cost of giving employees cost-of-living adjustments or merit increases. A cost-of-living adjustment is a raise in your salary to help you keep up with inflation. A merit increase is a raise in your salary to reward you for the quality of your work.

$$\begin{array}{ccccc} \text{New} \\ \text{Salary} \end{array} = \begin{array}{c} \text{Present} \\ \text{Salary} \end{array} + \begin{array}{c} \text{Cost-of-Living} \\ \text{Adjustment} \end{array} + \begin{array}{c} \text{Merit} \\ \text{Increase} \end{array}$$

EXAMPLE *Skill* 30 *Application* C *Term* Salary scale

Elaine Taylor is a heavy equipment operator, level 2, for Construction Inc. The executive board of Construction Inc. voted to give all employees a cost-of-living adjustment of 4.1%. In addition, Elaine was awarded a merit increase of 2.8% for excellent work during the year. What will Elaine's salary be for the coming year?

Job Level	Equipment Maintenance	Lt. Equipment Operator	Hvy. Equipment Operator	Crew Supervisor
1	$18,000	$19,845	$22,975	$27,900
2	18,900	20,837	24,125	29,295
3	19,845	21,880	25,330	30,760
4	20,837	22,975	26,600	32,300

S O L U T I O N

A. Find the **cost-of-living adjustment.**
$24,125.00 × 4.1% = $989.125 = $989.13 cost-of-living adjustment

B. Find the **merit increase.**
$24,125.00 × 2.8% = $675.50 merit increase

C. Find the **new salary.**

$$\begin{array}{ccccc} \text{Present} \\ \text{Salary} \end{array} + \begin{array}{c} \text{Cost-of-Living} \\ \text{Adjustment} \end{array} + \begin{array}{c} \text{Merit} \\ \text{Increase} \end{array}$$
$24,125.00 + $989.13 + $675.50 = $25,789.63 new salary

24125 ⊠ 4.1 % = 989.125 24125 ⊠ 2.8 % = 675.5 24125 + 989.13 + 675.5
= 25789.63

✔ SELF-CHECK Complete the problem, then check your answer in the back of the book.

1. A light equipment operator, level 1, receives a 4% cost-of-living adjustment and a 2% merit increase. Find the new salary.

	Present Salary +	Cost-of-Living Adjustment	+	Merit Increase	=	New Salary
2.	$13,400 +	$ 520	+	$ 780	=	?
3.	$15,650 +	$ 620	+	$ 750	=	?
4.	$19,860 +	$ 924	+	$1110	=	?
5.	$27,847 +	$1365	+	$1087	=	?
6.	$39,517 +	$1615	+	$1244	=	?

Level	Systems Analyst	Senior Systems Analyst
1	$19,600	$27,600
2	22,100	32,100
3	25,100	37,600

7. Norm Young.
Systems analyst, level 2.
Cost-of-living adjustment: 4%.
What is his new salary?

8. May Song.
Senior systems analyst, level 1.
Cost-of-living adjustment: 5%.
Merit increase: 3.6%.
What is her new salary?

9. Beatrice Apptou.
Senior systems analyst, level 3.
Merit increase: 7.6%.
What is her new salary?

10. Robert Moore, systems analyst, level 1, received a 4.7% cost-of-living adjustment and a 4.5% merit increase. What is his new salary?

Refer to the table on page 344 for problems 11 and 12.

11. Ruth Tomasi is a crew supervisor, level 4. She receives a 4.8% cost-of-living adjustment and a 7.4% merit increase. What is her new salary?

12. Mike Rossi is in equipment maintenance, level 3. He receives a 3.7% cost-of-living adjustment and a 0.7% merit increase. What is his new salary?

13. Jim O'Reilly works 40 hours a week for the Metro Delivery Company. He earns $7.15 an hour.
  a. What is Jim's annual gross pay?
  b. If Jim receives a 4.9% merit increase, what will be his new annual gross pay?
  c. What would be his new hourly rate?

MAINTAINING YOUR SKILLS Look up the skills in parentheses if you need help or more practice.

Find the present salary by using the table on page 344. **(Application C)**

14. Heavy equipment operator, level 3

15. Light equipment operator, level 4

16. Crew supervisor, level 2

Find the percentage. **(Skill 30)**

17. $17,740 × 4.5%   **18.** $21,510 × 8.6%   **19.** $15,100 × 3.1%

20. $34,191 × 2.4%   **21.** $38,964 × 5.7%   **22.** $64,315 × 1.8%

Employee Benefits

OBJECTIVE

Compute the rate of employee benefits based on annual gross pay.

Your business may offer several **employee benefits.** Employee benefits include health and dental insurance, life insurance, pensions, paid vacations and holidays, unemployment insurance, and sick leave. The total of the benefits may be figured as a percent of annual gross pay.

$$\text{Rate of Benefits} = \frac{\text{Total Benefits}}{\text{Annual Gross Pay}}$$

EXAMPLE   *Skills* 30, 31   *Application* A   *Term* Employee benefits

The personnel department of the Commercial Credit Company is preparing annual reports on employee benefits. Tamara Rey's total annual benefits are what percent of her annual salary?

Tamara Rey Annual Salary: $18,720.00	Weekly Salary: $360.00
Benefits	
Vacation: 2 weeks × $360.00	$ 720.00
Holidays: 8 days × ($360.00 ÷ 5)	576.00
Health insurance: 12 months × $147.80	1773.60
Dental insurance: 12 months × $18.74	224.88
Sick leave: 30 days × ($360.00 ÷ 5)	2160.00
Unemployment insurance: 4.6% of $18,720.00	861.12
Social security: 7.65% of $18,720.00	1432.08
Total	$7747.68

SOLUTION

Find the **rate of benefits.**

Total Benefits ÷ Annual Gross Pay

$7747.68   ÷ $18,720.00   = 0.413 = 41% rate of benefits

7747.68 ÷ 18720 = 0.4138

✔ SELF-CHECK   Complete the problem, then check your answer in the back of the book.

1. Annual salary is $32,000. Total benefits are $9600. Find the rate of benefits.

PROBLEMS

	Total Benefits	÷	Annual Gross Pay	=	Rate of Benefits
2.	$ 5937	÷	$23,748	=	?
3.	$15,610	÷	$44,600	=	?

Total Benefits	÷	Annual Gross Pay	=	Rate of Benefits
4. $ 2940	÷	$14,700	=	?
5. $11,392	÷	$35,600	=	?
6. $ 9860	÷	$24,650	=	?

7. Stock clerk's annual salary
is $11,860.
Benefits:
  Vacation $228.
  Holidays $365.
  Health insurance $1275.
  Unemployment insurance $545.56.
  Social security $907.29.
  Compensation insurance $486.26.
What are the total benefits?
What is the rate of benefits?
(Round to nearest tenth percent.)

8. Teacher's annual salary
is $19,740.
Benefits:
  Retirement $1727.25
  Holidays $789.60.
  Group life insurance $347.50.
  Sick leave $2961.
  Health insurance $271.80
What are the total benefits?
What is the rate of benefits?
(Round to nearest tenth percent.)

F.Y.I.
In 1992, a new college
graduate in business
administration was
paid, on the average,
$25,653 in salary
and $31,350 in total
compensation.
(annual gross pay
and total benefits)

9. a. Complete the benefits chart for the French Coffee Shoppe
employees. (Round to the nearest cent.)
  b. What is the rate of benefits for each employee? (Round to the
nearest whole percent.)

Position	Annual Wage	Two-Week Vacation	3.6% Compensation Insurance	10-Day Sick Leave	7.65% Social Security	Total Benefits
Manager	$30,680	?	?	?	?	?
Service	$11,752	?	?	?	?	?
Cook	$14,872	?	?	?	?	?
Dishwasher	$ 7488	?	?	?	?	?

10. a. Complete the benefits chart for the Pathology Lab staff.
  b. What is the rate of benefits for each employee? (Round to the
nearest tenth percent.)
  c. How much does the lab pay in benefits for the 3 employees?
  d. How much more would it cost the lab to give each employee
a 3-week vacation?

Position	Annual Wage	Two-Week Vacation	8 Holidays	4.6% Unemployment Insurance	7.65% Social Security	Total Benefits
Lab technician	$20,488	?	?	?	?	?
Lab analyst	$23,400	?	?	?	?	?
Receptionist	$ 9360	?	?	?	?	?

MAINTAINING YOUR SKILLS Look up the skills in parentheses if you need help or more practice.

Find the percentage. **(Skill 30)**

11. $24,700 × 7.51%   **12.** $18,960 × 4.6%   **13.** $26,418 × 3.6%

Find the rate. Round answers to the nearest tenth percent. **(Skill 31)**

14. $6500 ÷ $32,500   **15.** $6720 ÷ $16,800

12-4

Disability Insurance

OBJECTIVE

Compute disability benefits under independent retirement systems and under social security.

Disability insurance pays benefits to individuals who must miss work due to an illness or injury. Short-term disability is covered by an employer or is obtained from private insurance companies. Long-term or permanent disability coverage is provided by social security or by an independent retirement system. Most independent retirement systems compute disability benefits based on a percent of the final average salary. To determine the annual disability benefit, multiply the sum of the years of service and the expected retirement age minus the present age times the rate of benefits times the final average salary. The monthly disability benefit is the annual benefit divided by 12.

$$\begin{array}{c}\text{Annual}\\\text{Disability}\\\text{Benefit}\end{array} = \left(\begin{array}{c}\text{Years}\\\text{Worked}\end{array} + \begin{array}{c}\text{Expected}\\\text{Retirement}\\\text{Age}\end{array} - \begin{array}{c}\text{Present}\\\text{Age}\end{array}\right) \times \begin{array}{c}\text{Rate}\\\text{of}\\\text{Benefits}\end{array} \times \begin{array}{c}\text{Final}\\\text{Average}\\\text{Salary}\end{array}$$

EXAMPLE   *Skills* 5, 6, 30   *Application* A   *Term* Disability insurance

Alicia Wormsley had worked at Northern State University for 21 years when she became permanently disabled and could not continue to work. Alicia was 52 years of age and had planned to retire in 13 years at Northern State's normal retirement age of 65. Her final average salary was $38,740. Northern State's rate of benefits is 2%. What is Alicia's monthly disability benefit?

SOLUTION   **A.** Find the **annual disability benefit.**

$$\left(\begin{array}{c}\text{Years}\\\text{Worked}\end{array} + \begin{array}{c}\text{Expected}\\\text{Retirement}\\\text{Age}\end{array} - \begin{array}{c}\text{Present}\\\text{Age}\end{array}\right) \times \begin{array}{c}\text{Rate}\\\text{of}\\\text{Benefits}\end{array} \times \begin{array}{c}\text{Final}\\\text{Average}\\\text{Salary}\end{array}$$

$$(21 + 65 - 52) \times 2\% \times \$38{,}740$$
$$34 \qquad\qquad \times 0.02 \times \$38{,}740$$
$$= \$26{,}343.20 \text{ annual disability benefit}$$

B. Find the **monthly disability benefit.**

Annual disability ÷ 12
$$\$26{,}343.20 \quad ÷ \; 12 \; = \$2195.266 = \$2195.27 \text{ monthly}$$
$$\text{disability benefit}$$

21 + 65 − 52 = 34 × .02 × 38740 = 26343.2 ÷ 12 = 2195.266667

✓ SELF-CHECK   Complete the problem, then check your answer in the back of the book.

1. Final average salary is $47,800, years worked are 15, retirement age is 60, your age is 50, and rate of benefits is 1.8%. Find the annual and monthly disability benefits.

If an employee paid social security taxes for 5 of the 10 years before becoming disabled, the person would be eligible to receive disability benefits. Benefits are based on the table shown. To use the table, find the age and earnings closest to the age and earnings of the person.

APPROXIMATE MONTHLY DISABILITY BENEFITS
(For Worker Becoming Disabled in 1992 With Steady Earnings)

Worker's Age	Worker's Family	Worker's Earnings In 1991					
		$10,000	$20,000	$30,000	$40,000	$50,000	$54,600 or More¹
25	Worker only	$492	$ 761	$ 996	$1122	$1244	$1250
	Worker, spouse, and child²	714	1142	1494	1684	1867	1875
35	Worker only	487	751	988	1112	1221	1224
	Worker, spouse, and child²	701	1126	1483	1686	1832	1837
45	Worker only	486	749	985	1082	1147	1149
	Worker, spouse, and child²	699	1124	1478	1622	1721	1723
55	Worker only	488	753	971	1037	1082	1083
	Worker, spouse, and child²	703	1129	1458	1556	1623	1624
64	Worker only	495	765	980	1037	1076	1077
	Worker, spouse, and child²	718	1148	1471	1556	1615	1616

[1] Use this column if you earn more than the maximum social security earnings base.
[2] Equals the maximum family benefit.

Note: The accuracy of these estimates depends on the pattern of the worker's actual past earnings.

EXAMPLE *Skill* 1 *Application* C *Term* Disability insurance

An employee, age 46, paid social security taxes for 18 years. The worker's last annual earnings were $22,000. What will the worker and the worker's spouse and child receive for a monthly disability benefit?

S O L U T I O N Referring to the table, the worker's age is closest to 45 and the last annual earnings are closest to $20,000. The approximate monthly disability benefit for the disabled worker, spouse, and child is $1124.

✔ SELF-CHECK Complete the problem, then check your answer in the back of the book.

2. Worker's age is 57. Last annual earnings were $29,900. Find the worker's approximate monthly disability benefit.

In problems 3–8, find the annual and monthly disability benefits.

	(Years Worked	+	Expected Retirement Age	−	Present Age)	×	Rate of Benefits	×	Final Average Salary	=	Annual Disability Benefit	Monthly Disability Benefit
3.	(20	+	65	−	60)	×	2.0%	×	$40,000	=	?	?
4.	(26	+	60	−	54)	×	2.1%	×	$56,000	=	?	?
5.	(14	+	62	−	44)	×	1.8%	×	$35,700	=	?	?
6.	(16	+	65	−	37)	×	2.0%	×	$38,450	=	?	?
7.	(3	+	70	−	31)	×	1.75%	×	$29,840	=	?	?
8.	(30	+	65	−	54)	×	2.1%	×	$71,910	=	?	?

In problems 9–14, use the social security table to find the approximate monthly disability benefit.

	Worker's Age	Worker's Family	Earnings	Monthly Benefit
9.	64	Worker only	$48,950	?
10.	35	Worker, spouse, and child	$18,880	?
11.	54	Worker, spouse, and child	$40,000	?
12.	63	Worker only	$65,760	?
13.	36	Worker, spouse, and child	$26,970	?
14.	24	Worker only	$12,400	?

15. Marci Unger, age 58.
Worked 28 years.
Expected retirement age: 67.
Rate of benefits: 2.0%.
Final average salary: $64,845.
What is the annual disability benefit?
What is the monthly disability benefit?

16. Mark LaVine, age 58.
Worked 23 years.
Expected retirement age: 60.
Rate of benefits: 3.0%.
Final average salary: $47,147.
What is the annual disability benefit?
What is the monthly disability benefit?

17. Joe Hogan, age 31.
Annual earnings of $33,960.
What will Joe, his wife, and child receive in social security disability benefits?

18. Lisa Minatel, age 60.
Annual earnings of $37,000.
What will she receive in social security disability benefits?

19. Paul Thornton had worked for the state for 18 years. He suffered a stroke and became disabled at age 49. His final average salary was $36,947.80. Normal retirement age is 65. The rate of benefits is 2.1%.
a. What is Paul's annual disability benefit?
b. What is Paul's monthly disability benefit?

20. Theresa Gasiorowski had worked for Central State University for 13 years when she suffered a heart attack and became disabled. Theresa was 54 years of age and had planned to retire at the normal retirement age of 60. Her final average salary was $41,247.86. Central State's rate of benefits is 2%. What is Theresa's monthly disability benefit?

21. Richard Wexler, age 58, had been an employee of Chambers, Inc., for 16 years when he became permanently disabled. He was covered under social security for all 16 years. His annual earnings last year were $41,870. Find the approximate monthly disability benefit for Richard, his wife, and child.

22. Yolanda Winters, age 27, had been an employee of Etna-North Manufacturing for 7 years when she became permanently disabled. She was covered under social security for those 7 years. Her annual earnings last year were $14,560. Find the approximate monthly disability benefit she would receive.

23. George Hercule, age 23, had been an employee of Blendo Company for 5 years when he became permanently disabled. He was covered under social security for those 5 years. His annual earnings last year were $21,870. Find the approximate monthly disability benefit he would receive.

24. Henrietta Jordon, age 60, had been an employee of Giant Department Stores for 38 years when she became permanently disabled. She was covered under social security for all 38 years. Her annual earnings last year were $54,640. Find the approximate monthly disability benefit for Henrietta, her husband, and child.

F.Y.I.
In 1989, the average monthly benefit a disabled worker received from social security was $556.

MAINTAINING YOUR SKILLS   Look up the skills in parentheses if you need help or more practice.

Add. **(Skill 5)**

25. 18 + 17

26. 21 + 12

27. 7 + 32

28. 21 + 9

Solve. **(Skills 5, 6)**

29. 17 + (65 − 56)

30. 9 + (60 − 31)

31. 23 + (62 − 54)

32. 16 + (70 − 41)

Find the percentage. Round to the nearest cent. **(Skill 30)**

33. 2% of $47,965

34. 2.1% of $36,741

35. 1.8% of $29,176

36. $51,416 × (20 × 2%)

37. $48,615 × (18 × 2.1%)

12-5

Travel Expenses

OBJECTIVE
Compute the total business travel expense.

If you travel for your business, you will probably be reimbursed, or paid back, for all authorized expenses during your trip. Travel expenses usually include transportation, lodging, and meals.

$$\begin{array}{c} \text{Total Travel} \\ \text{Expense} \end{array} = \begin{array}{c} \text{Cost of} \\ \text{Transportation} \end{array} + \begin{array}{c} \text{Cost of} \\ \text{Lodging} \end{array} + \begin{array}{c} \text{Cost of} \\ \text{Meals} \end{array} + \begin{array}{c} \text{Additional} \\ \text{Costs} \end{array}$$

EXAMPLE   *Skills* 8, 5   *Application* A   *Term* Travel expenses

The accounting department of the Diversified Sales Company will reimburse Roger Martin for attending a 3-day marketing conference. Roger drove to the conference, so Diversified will pay him $0.22 per mile. Roger's expenses included:

Tolls: $2.20
Hotel: $94.50 per night for 2 nights
Conference registration: $45
Meals (including tips):

Thursday		Friday		Saturday	
lunch	$ 9.47	breakfast	$ 8.50	breakfast	$ 7.89
dinner	$19.65	lunch	$15.00	lunch	$11.48
		dinner	$ 24.70		

Mileage: 240 miles round-trip

What is the total cost of sending Roger to the conference?

SOLUTION

A. Find the **cost of transportation.**
(240 × $0.22) + $2.20
$52.80   + $2.20 ....$ 55.00

B. Find the **cost of lodging.**
2 × $94.50 ....$189.00

C. Find the **cost of meals.**
$9.47 + $19.65 + $8.50 + $15.00 + $24.70 +
$7.89 + $11.48 ....$ 96.69

D. Find the **additional costs.**
Conference registration ....$ 45.00
Total Travel Expense $385.69

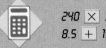

240 × .22 = 52.8 + 2.2 = 55 M+ 2 × 94.5 = 189 M+ 9.47 + 19.65 +
8.5 + 15 + 24.7 + 7.89 + 11.48 = 96.69 M+ 45 + RM = 385.69

1. Cost of transportation, $476; cost of lodging, $219; cost of meals, $147.80; additional costs, $89. Find the total travel expense.

PROBLEMS

Complete the table for these business trips.

	2.	**3.**	**4.**	**5.**	**6.**	**7.**
Name	T. Wills	B. Henry	M. Goss	V. Tarski	T. Lanza	B. Pappas
Miles Traveled	48	80	240	180	417	623
Cost at $0.21/mile	$10.08	?	?	?	?	?
Meals	$27.80	$46.90	$ 71.95	$ 70.40	$238.51	$291.94
Hotel Room	$65.48	0	$189.40	$210.00	$314.90	$516.85
Total Expenses	?	?	?	?	?	?

8. Tomas Rogers.
Airfare: $488.
Hotel: 3 nights at $74.50 each
Meals: $117.45
Registration: $75
What is the total travel expense?

9. Carol Cipriani.
Train: $147.85.
Hotel: 2 nights at $110 each.
Meals: $71.85.
Conference registration: $85
What is the total travel expense?

10. Tim Kandis is a troubleshooter for dpa, Inc. This month his travel expenses included airplane fares of $217.60 and $147.80, 156 miles of driving at $0.23 per mile, taxicab fares of $7.25 and $4.50, and meals totaling $76.85. What was the total travel expense for the month?

11. Meg Larson, a sales representative for Curry Corporation, flew to New York to make a sales presentation. Airfare was $215. Meg rented a car for 3 days for $21.40 a day plus $0.32 a mile. She drove a total of 70 miles. Meg's hotel bill was $124.50 a night for 2 nights. Her meals cost $11.90, $24.85, $9.76, $14.91, $27.80, $9.80, and $14.90. What was Meg's total travel expense?

MAINTAINING YOUR SKILLS   Look up the skills in parentheses if you need help or more practice.

Multiply. **(Skill 8)**

12. 350 × $0.21 **13.** 900 × $0.23 **14.** 620 × $0.22 **15.** 145 × $0.21

16. 540 × $0.20 **17.** 450 × $0.22 **18.** 422 × $0.23 **19.** 2 × $74.90

20. 3 × $110.80 **21.** 5 × $85.71 **22.** 3 × $93.47 **23.** 4 × $86.90

Add. **(Skill 5)**

24. $55 + $190 + $100

25. $74 + $87 + $219

26. $74.85 + $217.47 + $117.95

27. $184.73 + $67.52 + $347.85

Employee Training

Your business may pay the expenses involved in training employees. Your company may send you to special job-related programs or may offer special training programs within the company. Expenses for training during regular work hours include the cost of release time. When you are granted release time, you are paid your regular wages or salary while you are away from your job.

$$\text{Total Training Costs} = \text{Cost of Release Time} + \text{Cost of Instruction} + \text{Additional Costs}$$

EXAMPLE   *Skills* 8, 5   *Application* A   *Term* Release time

The Acme Manufacturing Company chose 8 employees to attend a training program in the company. The employees were paid their regular wages while attending the two-day program. Their combined wages amounted to $512 per day. The production control manager, who earns $168 per day, was the course instructor. Refreshments were brought in twice a day at a cost of $45.70 per day. Supplies and equipment for the program amounted to $35 per person. What was the total cost for the training program?

SOLUTION

A. Find the **cost of release time.**
2 days × $512.00 = $1024.00 release time

B. Find the **cost of instruction.**
2 days × $168.00 = $336.00 instruction cost

C. Find the **additional costs.**
(2 × $45.70) + (8 × $35.00)
$91.40 + $280.00 = $371.40 additional costs

D. Find the **total training cost.**

| Cost of Release Time | + | Cost of Instruction | + | Additional Costs | |
| $1024.00 | + | $336.00 | + | $371.40 = $1731.40 total training cost |

2 ✕ 512 = 1024 M+ 2 ✕ 168 = 336 M+ 2 ✕ 45.7 = 91.4 M+
8 ✕ 35 = 280 M+ RM 1731.40

✔ SELF-CHECK   Complete the problem, then check your answer in the back of the book.

1. Cost of release time, $1247; cost of instruction, $250; additional costs, $140. Find the total training cost.

	2.	3.	4.	5.	6.	7.
Number of Days	1	1	4	3	2	5
Daily Cost of Release Time	$417	$545	$345	$716	$ 96	$176
Daily Cost of Instruction	$ 75	$250	$175	$200	$150	$100
Daily Cost of Supplies	$ 50	$120	$100	$ 30	$ 25	$ 15
Total Training Cost	?	?	?	?	?	?

8. 3 physicians.
Attend 2-day seminar.
$1027 per day total for
    release time.
$175 per day total for instruction.
$25 per day total for supplies.
What is the total training cost?

9. 5 marketing representatives.
Attend 3-day workshop.
$930 per day total for
    release time.
$85 per person for instruction.
$285 per person total travel expense.
What is the total training cost?

10. The payroll department is sending 8 payroll clerks to a 2-day seminar
on the new tax law. Three of the clerks earn $52 per day, while the
other 5 earn $60 per day. Registration costs $150 per person. Materials
cost $25 per person. What will the 2-day seminar cost the payroll
department?

11. The Laser Research Company is sending their 5-person research team
to a 2-day conference to learn a new automated lab procedure. One
researcher earns $157.80 daily, two earn $136.80 daily, and two earn
$119.20 daily. What will the 2-day conference cost the Laser Research
Company in lost productivity for the 5 researchers?

12. Six employees of the Solar Energy Company underwent a 5-day sales
training program. Their wage rate averaged $9.65 per hour for an 8-
hour day. The sales manager, at a daily salary of $146.53, conducted
the 5-day session. Refreshments cost $37.50 per day. Sales kits were
provided for each of the 6 employees at a cost of $36.45 each. What
was the total cost of the sales training program?

MAINTAINING YOUR SKILLS Look up the skills in parentheses if you need help or more practice.

Multiply. **(Skill 8)**

13. 4 × $145

14. 5 × $112

15. 8 × $74

16. 3 × $716

17. 2 × $217

Divide. Round answers to the nearest cent. **(Skill 10)**

18. $47,650 ÷ 52

19. $34,840 ÷ 52

20. $28,645 ÷ 260

Add. **(Skill 5)**

21. $1076 + $345 + $130

22. $916 + $718 + $215

23. $1240.76 + $337.50 + $27.85

24. $3167.25 + $175.86 + $47.95

Reviewing the Basics

Skills

(Skill 5)

(Skill 8)

(Skill 10)

(Skill 30)

Solve. Round to the nearest cent.

1. $14,716 + $588.64

2. $35,640 + $1603.80 + $712.80

3. $643.75 × 2

4. 315 × $0.21

5. $42,650 ÷ 52

6. $26,790 ÷ 260

7. 7.65% of $31,745

8. 4.6% of $17,895.75

Round answers to the nearest tenth of a percent.

(Skill 31)

9. $14,716 is what percent of $45,617?

10. $4174.76 is what percent of $17,817.91?

Applications

(Application C)

Use the table to answer.

11. What is the base salary for a research assistant with 2 years of experience and a master's degree?

12. What is the base salary for a research assistant with 1 year of experience and a bachelor's degree?

Years Experience	RESEARCH ASSISTANT—BASE SALARY Degree	
	Bachelor's	Master's
0	$25,125	$27,500
1	26,380	28,875
2	27,700	30,300
3	29,100	31,800

Terms

Use each term in one of the sentences.

13. Employee benefits

14. Release time

15. Disability insurance

16. Recruiting costs

17. Salary scale

18. Travel expenses

a. The personnel department may use a ___?___ to compare the salaries of various jobs.

b. ___?___ may include the cost of advertising in the newspaper.

c. Paid vacations and holidays are ___?___ that most businesses give their employees.

d. When training sessions are held during work hours, the cost of ___?___ is one of the expenses.

e. If a job requires travel, the employee may be reimbursed for ___?___, which include transportation, lodging, and meals.

f. A ___?___ is a raise in an employee's salary.

g. ___?___ pays benefits to an individual who must miss work due to an illness or injury.

Refer to your reference files in the back of the book if you need help.

Unit Test

Lesson 12-1

1. The Hi-Tech Corporation hired Albert Jent for a new clerk-typist position at an annual salary of $14,500. The hiring costs included:

   Advertising: $315
   Interviews: Janet Gray, $75; Albert Jent, $85
   Office Agency fee: 20% of first year's salary

   What was the total cost of hiring Albert?

Lesson 12-2

2. Tonia Irwin is a statistician, level 3, for the Heritage Insurance Corporation. Tonia receives a 4.1% cost-of-living increase and a 3.4% merit increase. What is her new salary?

Level	Statistician
1	$24,450
2	25,917
3	27,472

Lesson 12-3

3. Martha Hiller works for the General Gravel Corporation. She earns $14,820 per year. The corporation provides these benefits for Martha. To the nearest percent, what is the rate of benefits?

   Vacation: 2 weeks
   Holidays: 8 days
   Compensation insurance: 3.7%
   Unemployment insurance: 4.7%
   Social security: 7.65%

Lesson 12-4

4. Vic Iagulli had worked at Eastern State University for 16 years when he suffered a stroke and became permanently disabled. Vic was 49 years of age and had planned to retire at the normal retirement age at Eastern of 60. His final average salary was $32,417.80. Eastern's rate of benefits is 2.1%. What is Vic's monthly disability benefit?

Lesson 12-5

5. Ursula Swede attended a professional conference last month. The payroll department reimbursed Ursula for these expenses:

   Airfare: $315.60
   Hotel: 2 nights at $87.95 each
   Meals: $117.80
   Registration: $55

   What was the total travel expense?

Lesson 12-6

6. The personnel department of the Krio Company is sending 3 of its employees to a local 2-day training program. The 3 employees' combined wages total $207.80 per day. The company paid the registration fee of $90 for each person. The company also paid for the employees' lunches for the 2 days at a total cost of $72.75. What was the total cost for the Krio Company to send their employees to the training program?

A SPREADSHEET APPLICATION

Personnel

To complete this spreadsheet application, you will need the diskette *Spreadsheet Applications for Business Mathematics,* which accompanies this textbook.

Select option 12, Personnel, from the menu. Input the information in the following problems to find the individual benefits, total benefits, and rate of benefits.

Title	Annual Wage	Vacation	3.6% Comp. Ins.	8–day Sick Lv.	7.65% Soc. Sec.	Total Benefits	Rate of Benefits
Dir.	$72,600	(3 wks.) ?	?	?	?	?	? %
Asst. Dir.	48,900	(3 wks.) ?	?	?	?	?	? %
Adm. Asst.	42,300	(2 wks.) ?	?	?	?	?	? %
1st Clk.	27,650	(2 wks.) ?	?	?	?	?	? %
2nd Clk.	18,540	(2 wks.) ?	?	?	?	?	? %
Total	?	$?	?	?	?	?	? %

a. What are the total benefits for the director?

b. What is the rate of benefits for the assistant director?

c. What is the total compensation insurance collected for these 5 employees?

d. How much social security is withheld from the 1st clerk's pay? How much is contributed by the company?

e. How much more would it cost to give the administrative assistant a 3-week vacation?

Title	Annual Wage	Vacation	6 Paid Holidays	4.6% Unemp. Ins.	7.65% Soc. Sec.	Total Benefits	Rate of Benefits
Mgr.	$45,000	(4 wks.) ?	?	?	?	?	? %
Asst. Mgr.	37,500	(4 wks.) ?	?	?	?	?	? %
Ser. Tech.	28,500	(3 wks.) ?	?	?	?	?	? %
Ser. Tech.	24,000	(3 wks.) ?	?	?	?	?	? %
Janitor	18,000	(2 wks.) ?	?	?	?	?	? %
Total	?	$?	?	?	?	?	? %

a. What is the rate of benefits for the manager?

b. What are the total benefits for the janitor?

c. What does vacation time cost for these 5 employees?

d. What is the total unemployment insurance collected for these 5 employees?

e. How much social security was withheld from the service technician earning $28,500 per year? How much was contributed by the employer?

f. How much could be saved by reducing the paid holidays to 5 days rather than 6?

g. How much could be saved by reducing the assistant manager's vacation to 3 weeks rather than 4 weeks?

▼ A SIMULATION

Applying for a Job

Suppose you have decided to get a summer job to earn extra money. To decide what kind of work you might like to do, you talk to friends, family, teachers, and others. There are many factors to consider. Here are some of them.

Compensation	Transportation	Interest
hourly pay	location	interesting work
benefits	public transportation	work related to
overtime pay	need for car	career goals
tips	cost	
commission		

Hours	Conditions
regular hours	duties
night work	indoor work
weekend work	outdoor work
holiday work	clean, quiet environment
overtime work	

In comparing the wages of different jobs, consider costs too. For example, if you have to spend a lot of money for automobile expenses to drive to and from a job, you may end up with less money to spend than if you had a lower-paying job closer to home.

1. Look at the list of factors above. You may consider other factors important. List the ten factors that are most important to you, with the most important one first. (If you have never had a job, you may want to discuss the factors with someone who has.)

Calculate the spendable income for each job.

	Hourly Rate	Hours per Week	Gross Pay	Deductions	Net Pay	Transportation Cost	Spendable Income
2.	$4.85	20	$97	$24.24	$72.76	$ 5	?
3.	$5.00	15	?	$18.24	?	0	?
4.	$6.25	40	?	$69.13	?	$15	?
5.	$8.50	36	?	$83.77	?	$36	?

Finding a Job

There are many ways to find a job. Relatives and friends may know of job openings. Schools, churches, and other organizations may be able to help. Some cities have programs that find jobs for young people. State employment services list job openings. Newspaper help-wanted ads may be a good source. Some young people do odd jobs, such as gardening and painting, for their neighbors. If you want to do a particular kind of work, start early when writing or calling possible employers. You may also want to place a "Position Wanted" ad in a neighborhood newspaper.

6. Find the gross weekly pay for each of the jobs listed in the ads.

7. If you could have any of these jobs, which one would you choose? Why? Keep in mind the factors on page 359. If you have trouble choosing between two jobs, write down the reasons you like each one and compare them. This may help you narrow your choice to one job.

HELP WANTED

CAR WASHER $4.90/hr. Tues–Sat, 9–6, 1 hr. lunch. Apply in person to Happy Time Auto Wash, 2167 Elm St.

CASHIER and sandwhich maker. 11–2 and 4–8, Mon. –Fri. $4.75/hr. Call 555-3597 for interview appointment.

CLERK TYPIST Must type at least 50 wpm and be willing to be trained as data input operator. 8–4:30, Mon.–Fri. $240/week. Assured Products, 621 East St., 555-3418 ex. 43.

COOK'S ASSISTANT for summer camp. No experience necessary, will train. 7hr./day, 7 days/wk. Meals, lodging, $200/wk. Green Mt. Camp, Woodstock, Vermont.

RECREATION AIDE for city parks. Supervise games, teach crafts. Tues–Sat, 10–6. $5/hr. Experience w/children required. Park Dept, 555-5000 ex. 104.

KENNEL HELP No experience necessary, love of animals needed. Clean, feed, exercise, bathe animals. 7–3:30 with 1/2 hr. lunch, Mon.–Fri. $4.90/hr. Wagon Wheel Kennels, 555-4225 before noon.

SWIMMING POOL SERVICE Cleaning & servicing pools, will train. $7.50/hr., 8–5 with 1 hr. lunch, Mon.–Fri. E-Z Pool Service, 555-8216

A SIMULATION
(CONTINUED)

Job Application

When you apply for a job, you will be asked to fill out a job application form. You should be prepared to complete all items on the form. Be sure to receive permission to use the names of your personal character references.

Here is a sample of a job application form. Write down your answers as if you were filling out the form.

EMPLOYMENT APPLICATION

Name _____

Address _____

Home Phone _____

Social Security Number _____

Date of Birth _____

In case of emergency, notify _____

Address _____

EDUCATIONAL BACKGROUND

Name and Address of School	Dates Attended		Date Graduated	Major Area of Study
	From	To		
Elementary School				
High School				
College				
Other				

EMPLOYMENT

Name and Address of Employer	Dates Employed		Position	Reason for Leaving
	From	To		

REFERENCES

Name _____ Address _____ Phone _____

Name _____ Address _____ Phone _____

Name _____ Address _____ Phone _____

MISCELLANEOUS

Hobbies _____

Hours Available for Interview _____

Number of Days per Week You Wish to Work: Minimum _____ Maximum _____

Are You Willing to Work Evenings? _____ Weekends? _____

A SIMULATION
(CONTINUED)

Job Interviews

Usually an employer interviews many applicants for a job opening. It is important that you dress neatly, speak clearly, and are able to answer the interviewer's questions. Many people are nervous before and during an interview. It is helpful to know what type of questions an interviewer might ask. Here are some of the most common questions.

Write your answers to these questions.

8. What kind of work are you most interested in? Why?

9. What are your future vocational or educational plans?

10. What do you know about our company? Why do you think you might like to work for us?

11. What jobs have you held? What did you learn from them?

12. What courses in school did you like best? least? Why?

13. What is your average grade in school? Do you think you have done the best academic work of which you are capable? If not, why not?

14. In what school activities have you participated? Which did you like most? Why? Have you been an officer in any activity?

15. How do you spend your spare time? What are your hobbies?

16. Do you prefer working with others or by yourself?

17. What is your major strength? What is your major weakness?

18. What have you done that shows initiative and willingness to work?

19. What qualifications do you have that make you feel you will be successful in your field?

An interview is an opportunity for you to ask questions about the job and the company. You may want to ask about pay, benefits, duties, tuition credit, opportunities for advancement, or other factors.

20. Write five specific questions you might ask if you were being interviewed for the job you chose on page 360.

If there is time, you and someone in your class might take turns interviewing each other.

Production

The production department of your business makes, or *manufactures,* and packages the products you sell. The costs of manufacturing an item include both materials and labor. For you to break even, the income from your sales must cover your manufacturing costs. Sales beyond the *break-even point* result in profits. You may use *quality controls* to check for defective items during manufacturing. *Time studies* can tell you how much time is needed to make one item.

O U T L I N E

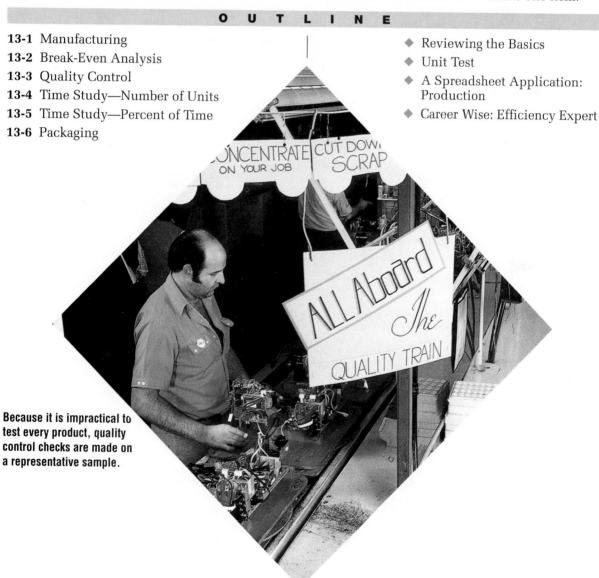

Because it is impractical to test every product, quality control checks are made on a representative sample.

Manufacturing

OBJECTIVE

Compute the prime cost of manufacturing an item.

The cost of manufacturing an item depends, in part, on the direct material cost and the direct labor cost. The direct material cost is the cost of the goods that you use to produce the item. The direct labor cost includes the wages paid to the employees who make the item. The prime cost is the total of the direct material cost and the direct labor cost. The prime cost is frequently expressed on a per-unit basis.

$$\begin{array}{ccc} \text{Prime Cost} & = & \text{Direct Material Cost} & + & \text{Direct Labor Cost} \\ \text{per Item} & & \text{per Item} & & \text{per Item} \end{array}$$

EXAMPLE   *Skills* 11, 12   *Application* A   *Term* Prime cost

Electric Supply Inc. produces aluminum circuit housings. The machine operator stamps 20 housings from each strip of aluminum. Each strip costs $0.90. The operator can stamp 720 housings per hour. The direct labor charge is $12.50 per hour. To the nearest tenth of a cent, what is the prime cost of manufacturing a housing?

SOLUTION

A. Find the **direct material cost per item.**
   $0.90 ÷ 20 = $0.045 direct material cost

B. Find the **direct labor cost per item.**
   $12.50 ÷ 720 = $0.0173 = $0.017 direct labor cost

C. Find the **prime cost per item.**
   Direct Material Cost + Direct Labor Cost
      $0.045   + $0.017 = $0.062 or 6.2¢ cost per housing

✔ SELF-CHECK   Complete the problems, then check your answers in the back of the book.

1. 40 brackets are made from a strip of metal costing $0.80. What is the direct material cost per item?

2. The direct material cost per item is $0.04. The direct labor cost per item is $0.18. Find the prime cost per item.

PROBLEMS

Round answers to the nearest tenth of a cent.

	Cost per Strip	Pieces per Strip	Direct Material Cost per Piece	Labor Cost per Hour	Pieces per Hour	Direct Labor Cost per Piece	Prime Cost
3.	$0.80	10	$0.08	$11.70	1000	$0.012	?
4.	$0.75	25	?	$13.50	1200	?	?
5.	$0.70	50	?	$17.45	800	?	?

6. 36 boxes per sheet of cardboard.
Cost: $3.60 per sheet.
Direct labor charge: $9.60
  per hour.
200 boxes cut per hour.
What is the prime cost per box?

7. 4200 washers per steel strip.
Cost: $8.75 per strip.
Direct labor charge: $12.65
  per hour.
13,500 washers punched per hour.
What is the prime cost per washer?

8. The Donnely Manufacturing Company manufactures plastic table-cloths. Each roll of printed plastic yields 120 tablecloths. Each roll costs $11.70. The direct labor charge is $10.20 per hour. The machine operator can cut and fold 90 tablecloths per hour. What is the prime cost of manufacturing each tablecloth?

9. The Northern Aluminum Company manufacturers aluminum products. One of their machines stamps housings for smoke alarms from strips of aluminum. Each strip of aluminum costs $0.90. The machine stamps 40 housings from each strip. The machine operator produces 10 housings per minute. The direct labor cost is $14.76 per hour. What is the prime cost of manufacturing each housing?

10. Sue Clark is a machine operator at Modern Plastics, Inc. She molds 65 buckets from one container of molding plastic. Each container costs $2.98. Sue's machine molds one bucket every 4 seconds. The direct labor cost is $13.65 per hour. What is the prime cost of manufacturing one bucket?

11. A strip of heavy gauge tin makes 115 switch plates. The cost per strip is $0.98. The direct labor cost is $11.67 per hour. 1060 plates are stamped per hour. What is the prime cost of manufacturing each switch plate?

12. A strip of aluminum makes 20 fan blades. The aluminum costs $1.12 per strip. The direct labor cost is $15.65 per hour. One blade is stamped every 5 seconds. What is the prime cost of manufacturing each fan blade?

F.Y.I.
In fiscal 1991, the American Greeting Card Company earned $82.5 million on revenues of $1.43 billion.

MAINTAINING YOUR SKILLS Look up the skills in parentheses if you need help or more practice.

Round answers to the nearest tenth of a cent. **(Skill 2)**

13. $0.21486

14. $0.05512

15. $0.00707

16. 34.19¢

17. 7.241¢

18. 1.987¢

Divide. Round answers to the nearest tenth of a cent. **(Skill 11)**

19. $0.90 ÷ 18

20. $0.95 ÷ 5

21. $0.92 ÷ 12

22. $0.85 ÷ 12

23. $1.47 ÷ 10

24. $1.76 ÷ 5

25. $17.85 ÷ 470

26. $9.65 ÷ 330

27. $17.90 ÷ 435

28. $7.15 ÷ 120

29. $11.72 ÷ 510

30. $21.93 ÷ 1475

13-2

Break-Even Analysis

OBJECTIVE
Compute the break-even point in the number of manufactured units.

Your business can prepare a break-even analysis to determine how many units of a product must be made and sold to cover production expenses. The break-even point is the point at which income from sales equals the cost of production. Units sold after that point will result in a profit for your business.

To calculate the break-even point, you must know the total fixed costs, the variable costs per unit, and the selling price per unit. Fixed costs include rent, salaries, and other costs that are not changed by the number of units produced. Variable costs include the cost of raw materials, the cost of packaging, and any other costs that vary directly with the number of units produced.

$$\text{Break-Even Point in Units} \ = \ \frac{\text{Total Fixed Costs}}{\text{Selling Price} \ - \ \text{Variable Costs}}$$
$$\text{per Unit} \qquad \text{per Unit}$$

EXAMPLE   *Skills* 6, 11   *Application* A   *Term* Break-even point

Token Metal Products, Inc., manufactures can openers. They plan to manufacture 750,000 hand-held can openers to be sold at $0.44 each. The fixed costs are estimated to be $142,570. Variable costs are $0.19 per unit. How many can openers must be sold for Token Metal Products, Inc., to break even?

S O L U T I O N   Find the **break-even point in units.**

$$\begin{array}{c}\text{Total Fixed} \\ \text{Costs}\end{array} \div \left(\begin{array}{cc}\text{Selling Price} & \text{Variable Costs} \\ \text{per Unit} & \text{per Unit}\end{array}\right)$$

$142,570.00 ÷   ($0.44   – $0.19)
$142,570.00 ÷ $0.25   =   570,280 break-even point in units

.44 ⊟ .19 ⊟ 0.25 M+ 142570 ÷ RM ⊟ 570280

✔ SELF-CHECK   Complete the problems, then check your answers in the back of the book.

Find the break-even point in units.

1. Selling price is $0.79.
Variable cost per unit is $0.29.
Total fixed costs are $87,500.

2. Selling price is $1.49.
Variable cost per unit is $0.74.
Total fixed costs are $124,500.

	Total Fixed Costs	÷	(Selling Price per Unit	−	Variable Costs per Unit)	=	Break-Even Point in Units
3.	$ 95,000	÷	($ 1.79	−	$ 0.79)	=	?
4.	$ 148,000	÷	($ 0.59	−	$ 0.34)	=	?
5.	$ 225,000	÷	($ 2.29	−	$ 1.79)	=	?
6.	$ 478,400	÷	($ 3.49	−	$ 2.69)	=	?
7.	$ 1,470,000	÷	($174.99	−	$ 98.75)	=	?
8.	$12,476,500	÷	($417.79	−	$279.81)	=	?

9. TenCo tennis rackets.
   Fixed costs: $740,000.
   Selling price per racket: $74.29.
   Variable cost per racket: $48.76.
   What is the break-even point
      in units?

10. Farm Supply rural mailboxes.
   Fixed costs: $192,800.
   Selling price per mailbox: $39.99.
   Variable cost per mailbox: $21.43.
   What is the break-even point
      in units?

11. Today's Music produces traditional CDs. The fixed costs total $2,417,950. The selling price per disc is $14.95. The variable cost per disc is $7.48. What is the break-even point in number of discs?

12. Superior Custom Homes builds a limited number of luxury homes. The fixed costs of the operation total $2,157,750. The base selling price per custom home is $479,500. The base variable cost per custom home is $431,550. What is the break-even point in number of custom homes?

13. True Bounce basketballs are manufactured by General Sports, Inc. They have total fixed costs of $3,110,400. The variable cost per basketball is $18.47. The selling price per basketball is $24.95. What is the break-even point in number of basketballs?

14. The Alumni Association plans to produce and market monogrammed sweatshirts. They anticipate selling each sweatshirt for $24.95. Their variable costs per sweatshirt include $13.19 for supplies and $7.47 per shirt for direct labor. The fixed costs of the operation total $7450. How many monogrammed sweatshirts does the Alumni Association need to sell to break even?

MAINTAINING YOUR SKILLS   Look up the skills in parentheses if you need help or more practice.

Subtract. **(Skill 6)**

15. $14.79 − $9.43   **16.** $1.89 − $0.97   **17.** $14.49 − $7.74

18. $7.19 − $5.54   **19.** $147.47 − $109.88   **20.** $8129.00 − $6417.48

Divide. Round answers to the nearest whole number. **(Skill 11)**

21. $47,500 ÷ $0.25   **22.** $163,540 ÷ $0.20   **23.** $216,491 ÷ $0.79

Quality Control

OBJECTIVE

Compute the percent of defective goods and determine if the process is in or out of control.

If your business deals with mass production, you need a quality control inspector to check the items that are manufactured. The inspector may examine a specified number of items. If the size of the item is incorrect or if the item is broken or damaged, it is classified as defective. The inspector computes the percent of the sample that is defective and plots the percent on a quality control chart. The quality control chart shows the percent of defective products that is allowable. If the actual percent of defective items is greater than the percent allowable, the process is said to be "out of control." Production must then be stopped and corrected.

$$\text{Percent Defective} = \frac{\text{Number Defective}}{\text{Total Number Checked}}$$

EXAMPLE *Skills* 1, 31 *Application* N *Term* Quality control

Bob Atkinson is a quality control inspector for Material Plating, Inc. At 9 a.m., he checked 25 stamped brackets produced during the previous hour and found 2 defective brackets. The process is in control if 5% or less of the sample is defective. Is the process in control or out of control? Plot the percent defective on a quality control chart.

SOLUTION **A.** Find the **percent defective.**

Number Defective ÷ Total Number Checked

2 ÷ 25 = 0.08 = 8% defective

Process is out of control.

B. Plot the **percent defective** on a quality control chart.

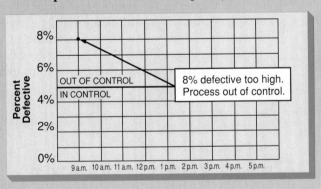

---

1. Number defective is 5.
Number checked is 200.
What is the percent defective?

2. Number defective is 3.
Number checked is 50.
What is the percent defective?

If more than 5% of the sample is defective, the process is out of control.

	Number Defective	÷	Number Checked	=	Percent Defective	In or Out of Control?
3.	2	÷	50	=	?	?
4.	1	÷	50	=	?	?
5.	4	÷	50	=	?	?
6.	0	÷	50	=	?	?
7.	3	÷	50	=	?	?
8.	5	÷	50	=	?	?

9. Sweaters.
100 in sample.
5 defective.
In control if 6% or less defective.
Is the process in or out of control?

10. Plastic flashlights.
25 in sample.
1 defective.
In control if 5% or less defective.
Is the process in or out of control?

11. Alice McHenry is a quality control inspector for Current Cassettes, Inc. The process is in control if 6% or less of each sample is defective. Alice checked a sample of 50 cassettes and found 4 defective cassettes. What percent of the sample is defective? Is the process in or out of control?

12. Tucker Jensen, a quality control inspector for Office Products Co., checked a sample of 250 ballpoint pens. Fourteen of the pens were defective. If more than 6% of the sample is defective, the process is out of control. What percent of the sample is defective? Is the process in or out of control?

For Exercises 13–17, draw a quality control chart, compute the percent defective in each sample, and plot the percents on the chart.

13. Every 2 hours, Bev Wallace checks samples from a punch press. Each sample contains 50 pieces. The operation is in control if 5% or less of each sample is defective. Bev found the following:

Time	9 a.m.	11 a.m.	1 p.m.	3 p.m.	5 p.m.
Number Defective	1	0	3	2	2

14. Every hour, 10 compact disc players are taken off the production line at Modern Technologies and checked to see if they are defective. If more than 10% are defective, production is out of control. The number of defective players for the 9 a.m. to 5 p.m. checks are:

Time	9 a.m.	10 a.m.	11 a.m.	12 noon	1 p.m.	2 p.m.	3 p.m.	4 p.m.	5 p.m.
Number Defective	0	1	0	2	1	0	1	0	2

15. Every 3 hours, Tom Smith pulls 25 bicycle rims off the assembly line and checks their diameter. If more than 6% of the rims are found to be defective, the assembly line is shut down. Determine whether the process is in or out of control at each time check. Draw a quality control chart.

Time	9:30 a.m.	12:30 p.m.	3:30 p.m.	Day
Number Defective	0	1	1	Monday
	1	0	0	Tuesday
	1	2	2	Wednesday
	0	0	1	Thursday
	1	1	3	Friday

16. Nora Stamos randomly checks 75 transistors 10 times a day. If more than 4% are defective, the production is stopped and corrections are made. Complete the table and draw a quality control chart.

Checkpoint No.	1	2	3	4	5	6	7	8	9	10
Number Defective	1	2	1	0	2	1	3	4	1	0
Percent Defective	?	?	?	?	?	?	?	?	?	?

17. Albert Burns checks circuit board production at different times during the day. He will check a different number each time. If more than 5% are defective, he will close down production and have corrections made. The circuit boards are checked electronically. Complete the table and draw a quality control chart.

Time	9:40 a.m.	10:15 a.m.	11:50 a.m.	1:30 p.m.	3:30 p.m.	4:30 p.m.
Number Checked	60	50	80	100	40	50
Number Defective	3	2	3	4	2	2
Percent Defective	?	?	?	?	?	?

MAINTAINING YOUR SKILLS   Look up the skills in parentheses if you need help or more practice.

Find the rate. Round answers to the nearest tenth of a percent. **(Skill 31)**

18. 2 is what percent of 50?

19. 6 is what percent of 100?

20. 4 is what percent of 25?

21. 2 is what percent of 60?

22. 3 is what percent of 40?

23. 4 is what percent of 150?

24. 4 is what percent of 130?

25. 2 is what percent of 200?

Which number is greater? **(Skill 1)**

26. 5.3% or 5%

27. 6% or 5.2%

28. 3% or 4%

29. 6% or 6.7%

30. 4.3% or 4%

31. 5% or 0%

32. 2% or 2.5%

33. 5% or 5%

Time Study—
Number of Units

OBJECTIVE

Use time-study results to compute how many units can be produced.

Your business may conduct a time study to determine how long a particular job should take. A time study involves watching an employee complete a job, recording the time required for each task, and calculating the average time for each task. You can use the averages to determine how many units a worker can produce in a fixed period of time.

$$\text{Number of Units} = \frac{\text{Actual Time Worked}}{\text{Average Time Required per Unit}}$$

EXAMPLE   *Skills* 5, 11   *Application* Q   *Term* Time study

General Lamps, Inc., did a time study of Beth Peters's job as a carton packer. Beth's times were recorded. The averages were calculated.

$$\frac{\text{Sum of times}}{\text{Number of observations}}$$

Task	#1	#2	#3	#4	#5	Average Time
Pick up carton	3.9	5.4	3.9	4.4	3.9	4.3 sec.
Fill carton	12.0	13.0	14.5	12.5	11.0	12.6 sec.
Apply glue	14.5	12.5	13.0	13.5	13.5	13.4 sec.
Close carton	5.1	4.4	4.6	4.3	5.1	4.7 sec.
Remove filled carton	5.5	5.5	5.5	5.5	5.5	5.5 sec.

Observations in Seconds

If Beth gets a 10-minute break each hour, how many cartons can she fill per hour?

SOLUTION

A. Find the **average time required per unit.**
   4.3 + 12.6 + 13.4 + 4.7 + 5.5 = 40.5 seconds

B. Find the **actual time worked per hour.**
   (60 − 10) minutes × 60 seconds = 3000 seconds

C. Find the **number of units per hour.**
   Actual Time Worked ÷ Average Time Required per Unit
   3000 ÷ 40.5 = 74.07 = 74 units per hour

4.3 + 12.6 + 13.4 + 4.7 + 5.5 = 40.5 M+ 60 − 10 = 50 × 60 = 3000 ÷ RM = 74.074

✔ SELF-CHECK   Complete the problems, then check your answers in the back of the book.

Find the number of units per hour.

1. Average time is 6 minutes per unit. 48 minutes worked per hour.

2. Average time is $\frac{1}{4}$ hour per unit. $7\frac{1}{2}$ hours worked per day.

	Actual Time Worked	÷ Average Time per Unit	=	Number of Units
3.	48 minutes (× 60) ÷	43.5 seconds	=	?
4.	50 minutes (× 60) ÷	39.7 seconds	=	?
5.	55 minutes (× 60) ÷	65.6 seconds	=	?
6.	45 minutes (× 60) ÷	5.5 seconds	=	?

7. Bagging and carrying out groceries. Average time is 11.42 minutes per customer. 50 minutes worked per hour. How many customers can be served per hour?

8. Selling theater tickets. Average time is 21.4 seconds per customer. 50 minutes worked per hour. How many customers can be served per hour?

9. Ben Krieger prepared a time study to determine the average time required to make a cash withdrawal at the 24-hour automatic teller machine.

	Time in Seconds				
Task	**#1**	**#2**	**#3**	**#4**	**#5**
Insert card in machine	1.0	2.0	1.5	2.5	1.5
Type in code	17.0	11.5	16.0	15.5	14.5
Specified amount wanted	7.4	6.9	7.1	7.0	7.1
Remove cash	1.5	1.0	2.0	1.5	1.5
Remove card	1.2	1.4	1.0	1.1	1.3

a. What is the average time required for each task?
b. What is the average time required to make a cash withdrawal?
c. How many cash withdrawals can be made at a machine in 1 hour?

10. The American Pump Company prepared a time study of Shirley Monroe's job as an assembler. How many units can Shirley complete in 1 hour if she takes one 10-minute break during the hour?

	Time in Seconds				
Task	**#1**	**#2**	**#3**	**#4**	**#5**
Move water pump to filter	3.7	4.2	3.9	3.8	3.7
Insert filter coil	5.1	5.2	4.9	5.2	5.1
Affix gasket	4.8	4.4	4.7	4.6	4.8
Hand thread bolts	6.1	6.3	6.3	6.2	6.4
Machine tighten bolts	6.1	6.0	6.2	6.3	6.2

MAINTAINING YOUR SKILLS Look up the skills in parentheses if you need help or more practice.

Add. (Skill 5)

11. 3.6 + 3.8 + 3.7 + 3.6 + 3.7

12. 14.2 + 14.1 + 13.8 + 14.0 + 13.9

13. 7.2 + 7.4 + 7.1 + 6.8 + 7.2

14. 1.1 + 0.9 + 0.8 + 1.2 + 1.0

Divide. Round answers to the nearest whole number. (Skill 11)

15. 3600 ÷ 4.5

16. 3000 ÷ 2.4

17. 3300 ÷ 15.2

18. 8 ÷ 0.12

13-5

Time Study—
Percent of Time

OBJECTIVE

Use time-study results to compute the percent of time spent on each task.

You can use a time study to determine what percent of an employee's time is spent on various activities during a workday.

$$\text{Percent of Time Spent on Activity} = \frac{\text{Time Spent on Activity}}{\text{Total Time}}$$

EXAMPLE   *Skills* 5, 31   *Application* A   *Term* Time study

Mike Reese works in the mail room of a large office. A time study showed that he spent his time on these activities. What percent of his time does Mike spend picking up and delivering mail?

Activity	Hours
Sorting mail	2.5
Picking up and delivering mail	3.0
Talking with employees	0.5
Taking coffee breaks	0.5
Making special deliveries	1.5
Total	8.0

SOLUTION   Find the **percent of time spent on activity.**

Time Spent on Activity ÷ Total Time
3.0 ÷ 8.0 = 0.375 = 37.5% of time spent on activity

✔ SELF-CHECK   Complete the problems, then check your answers in the back of the book.

1. In the example above, what percent of his time does Mike spend sorting mail?

2. In the example above, what percent of his time does Mike spend making special deliveries?

PROBLEMS

Round answers to the nearest hundredth of a percent.

	Activity	Hours	÷	Total Time	=	Percent of Time
3.	Loading furniture	3.00	÷	8	=	?
4.	Delivering furniture	4.50	÷	8	=	?
5.	Lunch	1.00	÷	8	=	?
6.	Returning truck	0.50	÷	8	=	?

7. Barb Lane, carton packer.
Fills cartons 2.5 hours per day.
Works 8 hours per day.
What percent of her day is spent filling cartons?

8. Amy Narcussi, realtor.
Answers phone 1 hour per day.
Works 8 hours per day.
What percent of her day is spent answering the phone?

9. Wilma Cole, receptionist. Answers telephones 2.8 hours per day. Works 7.5 hours per day. What percent of her day is spent answering telephones?

10. Paul Hong, college professor. Is in the classroom 3 hours per day. Works 8.5 hours per day. What percent of his day is spent in the classroom?

11. Ben Lucas is a maintenance programmer for the Ace Corporation. Ben works 7.5 hours per day. He spends about 2.25 hours each day testing programs on which he has made changes. What percent of Ben's day is spent testing these programs?

12. Ruth Loeb is a clerk at Violet's Dress Shoppe. She works 8 hours each day. A time study recorded her activities. What percent of the 8 hours does Ruth spend on each activity?

Activity	Hours
Assisting customers	4.2
Processing sales	1.1
Restocking racks	0.9
Pricing merchandise	0.8
Break and miscellaneous	1.0
Total	8.0

13. Don Kurezlenski delivers bread for the City Bakery Company. He works 9 hours each day. A time study showed these figures. What percent of Don's workday is spent on each of the activities?

Activity	Hours
Loading truck	1.3
Driving truck	3.1
Delivering bread	2.2
Recording sale	0.5
Checking out	0.3
Break and miscellaneous	0.6
Lunch	1.0
Total	9.0

MAINTAINING YOUR SKILLS Look up the skills in parentheses if you need help or more practice.

Add. **(Skill 5)**

14. 1.7 + 2.4 + 3.1 + 0.8

15. 2.1 + 1.8 + 3.3 + 0.9 + 1.9

16. 3.4 + 1.8 + 1.6 + 0.7

17. 3.2 + 2.1 + 1.6 + 1.2 + 0.9

Find the rate. Round answers to the nearest hundredth of a percent. **(Skill 31)**

18. 1.6 is what percent of 7.5?

19. 2.1 is what percent of 9?

20. 0.9 is what percent of 10?

21. 1.7 is what percent of 8?

Packaging

OBJECTIVE
Compute the dimensions of packaging cartons.

The production of your company's merchandise ends with packaging, placing the product in a container for shipment. The container eases handling and prevents breaking. The package may also identify the product and show it attractively. The size of the package depends on the size of the finished product.

Dimensions: Length, Width, Height

EXAMPLE *Skills* 16, 20 *Term* Packaging

The Cleaning Genie Company manufactures a cleaning solution that is sold in bottles with a diameter of 7 cm and a height of 20 cm. The company plans to package 12 bottles to a carton for shipping. The carton is made of 0.5-cm thick corrugated cardboard with 0.3-cm cardboard partitions. What are the dimensions of the package that the Cleaning Genie Company needs?

SOLUTION

A. Find the **length.**
4 bottles × 7 cm in diameter	28.0 cm
3 partitions × 0.3 cm wide	0.9 cm
2 ends × 0.5 cm wide	1.0 cm
Length	29.9 cm

B. Find the **width.**
3 bottles × 7 cm in diameter	21.0 cm
2 partitions × 0.3 cm	0.6 cm
2 ends × 0.5 cm	1.0 cm
Width	22.6 cm

C. Find the **height.**
1 bottle × 20 cm	20.0 cm
2 top flaps × 0.5 cm	1.0 cm
2 bottom flaps × 0.5 cm	1.0 cm
Height	22.0 cm

D. Find the **dimensions.**
Length: 29.9 cm; Width: 22.6 cm; Height: 22.0 cm

4 × 7 = 28 M+ 3 × .3 = 0.9 M+ 2 × .5 = 1 M+ RM 29.9 CM 3 × 7 = 21 M+ 2 × .3 = 0.6 M+ 2 × .5 = 1 M+ RM 22.6 CM 1 × 20 = 20 M+ 2 × .5 = 1 M+ 2 × .5 = 1 M+ RM 22

1. 4 packages, each 5 inches wide, next to each other. Each end is $\frac{1}{4}$-inch cardboard. How wide is the carton?

2. 6 packages, each 12 cm long, next to each other. Each end is 0.5-cm cardboard. How long is the carton?

PROBLEMS

Find the dimensions of the carton.

Height:

3. Top and bottom of 0.4 cm = ?

4. Height = ?

5. Total height = ?

Width:

6. 3 packages, each 8 cm wide = ?

7. 2 inserts, each 0.2 cm = ?

8. 2 ends, each 0.4 cm = ?

9. Total width = ?

Length:

10. 4 packages, each 12 cm long = ?

11. 3 inserts, each 0.2 cm = ?

12. 2 ends, each 0.4 cm = ?

13. Total length = ?

14. Dimensions?

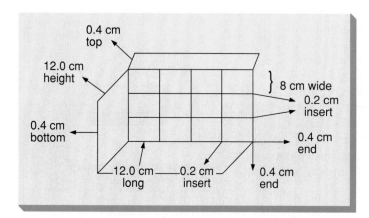

Draw a sketch for each carton in Exercises 15–19.

15. The Clean Air Corporation packages 3 of these humidifiers per carton. The humidifiers are placed next to each other with a $\frac{3}{4}$-inch thick piece of cardboard between them. The carton is made of $\frac{5}{16}$-inch thick corrugated cardboard. What are the dimensions of the carton in which the humidifiers are packed?

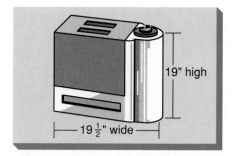

16. The Canned Food Company packages 12 cans of mixed vegetables to a package. Each can is 9 cm in diameter and 12 cm high. The carton is 0.5 cm thick and has 0.2-cm-thick partitions positioned between the cans.

   a. What are the dimensions of the carton if the cans are packaged in 1 row with 12 cans in the row?

   b. What are the dimensions of the carton if the cans are arranged in 3 rows with 4 cans in each row?

17. The Major Radio Corporation packages 6 of their weather-alert radios in 1 carton. The carton is made of $\frac{3}{8}$-inch-thick corrugated cardboard. Partitions between the radios are $\frac{1}{4}$-inch-thick.

a. What are the dimensions of the carton if the weather-alert radios are arranged in 2 rows of 3, with each radio on its side?

b. What are the dimensions of the carton if the weather-alert radios are arranged in 1 row of 6, with each radio on its side?

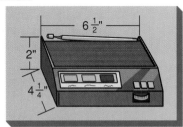

18. These 100-capacity tackle boxes are packaged 12 to a carton. The box is $10\frac{3}{4}$ in long, 6 in wide, and $6\frac{1}{2}$ in high. The carton is constructed of $\frac{1}{4}$-inch-thick cardboard. Cardboard spacers $\frac{1}{8}$-inch-thick are positioned between the boxes. What are the dimensions of the carton if the boxes are arranged in 2 layers, with each layer consisting of 2 rows of 3 boxes?

19. These boxed hand warmers are packaged 24 to a carton. The carton is made of 0.3 cm cardboard. There are no partitions.

a. What are the dimensions of the carton if the boxes are arranged standing up as shown, in 4 rows of 6 hand warmers?

b. What other arrangements could be used to package the hand warmers? Draw sketches and give the dimensions of the cartons.

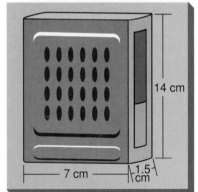

MAINTAINING YOUR SKILLS Look up the skills in parentheses if you need help or more practice.

Add. (Skill 16)

20. $\frac{1}{2} + \frac{1}{8}$ **21.** $\frac{1}{8} + \frac{3}{16}$ **22.** $\frac{1}{4} + \frac{3}{16}$ **23.** $\frac{1}{8} + \frac{7}{16}$

24. $\frac{1}{2} + \frac{1}{4}$ **25.** $\frac{3}{4} + \frac{9}{16}$ **26.** $\frac{3}{4} + \frac{3}{8} + \frac{1}{8}$

27. $\frac{1}{8} + \frac{1}{16} + \frac{3}{16}$ **28.** $\frac{3}{4} + \frac{7}{8} + \frac{3}{16}$ **29.** $\frac{1}{2} + \frac{3}{8} + \frac{7}{16}$

Multiply. (Skill 20)

30. $6 \times \frac{1}{2}$ **31.** $4 \times \frac{1}{8}$ **32.** $5 \times \frac{1}{16}$ **33.** $4 \times \frac{3}{8}$ **34.** $6 \times \frac{3}{4}$

35. $6 \times \frac{5}{8}$ **36.** $4 \times \frac{1}{4}$ **37.** $4 \times \frac{1}{2}$ **38.** $6 \times \frac{5}{16}$ **39.** $8 \times \frac{1}{4}$

40. $6 \times \frac{3}{8}$ **41.** $4 \times \frac{9}{16}$ **42.** $2 \times \frac{5}{32}$ **43.** $10 \times \frac{7}{8}$ **44.** $4 \times \frac{7}{32}$

Reviewing the Basics

Skills

(Skill 2)   Round answers to the nearest cent.

1. $0.1751   **2.** $4.0447   **3.** $147.7158

Solve.

(Skill 5)   **4.** 7.1 + 8.4 + 5.9 + 3.5   **5.** 78.4 + 11.7 + 9.6 + 7.9 + 84.7

(Skill 6)   **6.** $1.71 − $1.43   **7.** $74.47 − $56.92   **8.** $216.71 − $149.84

(Skill 11)   **9.** $146.70 ÷ 1600   **10.** $74,750 ÷ $0.56   **11.** $1,714,700 ÷ $2.10

(Skill 16)   **12.** $\frac{3}{4} + \frac{7}{16} + 5$ **13.** $7\frac{3}{4} + \frac{9}{16} + \frac{1}{8}$ **14.** $2\frac{3}{8} + \frac{1}{4} + 3$

(Skill 20)   **15.** $6 \times \frac{3}{8}$ **16.** $4 \times \frac{7}{16}$ **17.** $6 \times 5\frac{3}{4}$

(Skill 31)   **18.** 7 is what percent of 25?   **19.** 4.5 is what percent of 14?

Applications

Use the graph to answer.

(Application N)   **20.** What was the income for March?

21. What was the income for June?

Find the mean.

(Application Q)   **22.** 7.2, 6.7, 7.1, 6.8

23. 17.5, 15.8, 16.5, 16.8, 16.1

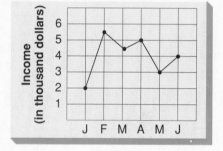

Terms

Match each term with its definition on the right.

24. Prime cost

25. Break-even point

26. Quality control

27. Time study

28. Packaging

a. a periodic inspection of items manufactured

b. the total of the direct labor cost and the direct material cost

c. the cost of wages paid to employees

d. the point at which income from sales equals the cost of production

e. determines how long a particular job should take

f. the final step in the production of a company's product

Refer to your reference files in the back of the book if you need help.

Unit Test

Lesson 13-1

1. Cosmo Steel produces cable from scrap steel that it buys for $248 per ton. The company produces 4 rolls of cable from each ton. The direct labor charge is $18.75 an hour. The employees produce 3 rolls of cable per hour. To the nearest tenth of a cent, what is the prime cost of manufacturing 1 roll of cable?

Lesson 13-2

2. Central Publishing Company plans to produce do-it-yourself books to be sold at $14.95 each. The fixed costs are estimated at $180,800. The variable costs are $10.43 per book. What is the estimated break-even point in units?

Lesson 13-3

3. A quality control inspector for the Hu-Day Company found 3 out of 65 glass containers with cracks during the 5 p.m. inspection. The process is in control if the percent defective is 5% or less. Is the process in control or out of control?

Lesson 13-4

4. Family Food Stores made a time study of its stock operation. The store recorded the times that Kelly Booth spent shelving the contents of cartons.

Task	Observations in Seconds				
	#1	#2	#3	#4	#5
Pick up carton	5.0	5.5	4.0	5.5	4.5
Open carton	14.0	14.5	15.5	14.0	15.0
Place contents on shelf	243.0	254.0	252.5	243.5	254.5
Discard carton	5.0	5.5	5.0	5.0	5.5

If Kelly takes a 5-minute break each hour, how many cartons can she unpack in 1 hour?

Lesson 13-5

5. Thomas Kuback is a medical office assistant for Family Practice, Inc. Thomas prepared a time study of his activities during the day. What percent of his time does Thomas spend retrieving files?

Activity	Hours
Interviewing patients	1.5
Processing records	2.5
Retrieving files	1.5
Lunch	1.0
Patient consultation	1.5

Lesson 13-6

6. Pro-Weight, Inc., ships its industrial weight containers in wooden boxes. The containers are arranged in 1 layer. Each box is made of $\frac{1}{4}$-inch-thick wood with $\frac{1}{8}$-inch-thick wood spacers. Each container is 5 inches high and $2\frac{1}{2}$ inches in diameter. What are the dimensions of each wooden box?

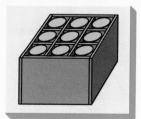

A SPREADSHEET APPLICATION

Production

To complete this spreadsheet application, you will need the diskette *Spreadsheet Applications for Business Mathematics,* which accompanies this textbook.

Input the information in the following problems to find the percent defective and whether the process is in control or out of control.

	Number Tested	Number Defective	Percent Defective	Defective Parts Allowable (%)	Process In or Out of Control
1.	200	6	?	4%	?
2.	300	15	?	5%	?
3.	60	3	?	4%	?
4.	50	2	?	5%	?
5.	200	7	?	3%	?
6.	425	9	?	2%	?
7.	150	9	?	5%	?
8.	250	19	?	6%	?
9.	500	3	?	1%	?
10.	125	4	?	3%	?

11.

Day	Number Tested	Number Defective	Percent Defective	Defective Parts Allowable (%)	Process In or Out of Control
Mon.	180	9	?	4%	?
Tue.	180	7	?	4%	?
Wed.	190	7	?	4%	?
Thur.	180	6	?	4%	?
Fri.	180	8	?	4%	?

12. In #1, what percent defective parts were found?

13. In #1, is the process in or out of control?

14. In #2, what percent defective parts were found?

15. In #2, is the process in or out of control?

16. In #3, what percent defective parts were found?

17. In #3, is the process in or out of control?

18. What would happen in a situation such as that described in #3?

19. In #4, what percent defective parts were found?

20. In #4, is the process in or out of control?

21. What would happen in a situation such as that described in #4?

22. In #5, what percent defective parts were found?

23. In #5, is the process in or out of control?

24. In #6, is the process in or out of control?

25. In #7, is the process in or out of control?

26. In #8, is the process in or out of control?

27. In #9, is the process in or out of control?

28. In #10, is the process in or out of control?

29. Describe what happened with respect to being in or out of control in #11.

30. With respect to being in or out of control, why is what happened in #11 not uncommon?

Efficiency Expert

Joshua Gomperts works as an efficiency expert for a mail-order sales company. Part of his job is to find ways to carry out the company's goals in an efficient manner. He also conducts time-management training sessions and discussion groups to get employee ideas and promote employee satisfaction.

To develop efficient procedures, Joshua gathers data regarding tasks and the time it takes to complete tasks. To evaluate the data, he uses statistical measures, such as averages. He also suggests how to arrange, label, and locate goods so that shippers can find them more easily and process orders more quickly.

Joshua prepared a survey for the 8 employees in the shipping department. Each employee was timed over a 5-week period and times needed to complete each task were averaged. The results are shown in the table at the left and the pie chart at the right.

TIME SPENT PROCESSING AN ORDER	
Gather goods ordered	0.75 hours
Package the goods	0.25 hours
Prepare the invoice	0.33 hours
Ship the order	0.25 hours
Total	1.58 hours

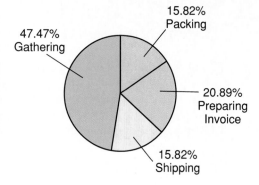

1. Which tasks take the same amount of time to complete? What percent of 100%?

2. True or false: The time needed to gather the goods is about half of all the time it takes to process an order.

3. Some employees recommended using small carts to get around the inventory area. If managers agree to this, do you think the 47.47% slice of the pie will get smaller?

4. A new computer system will reduce the time needed to prepare an invoice to 0.1 hours. What percent of the new total time is this? Will the invoice prep slice of the pie be smaller?

Purchasing

The purchasing department orders merchandise for your business from suppliers, such as distributors, wholesalers, or manufacturers. Most suppliers sell goods to businesses at a *discount,* or a reduction in price. The lower your *net price* (the price you actually pay for an item), the higher your profit will be when you sell the item. Suppliers offer three kinds of discounts: *trade, chain,* and *cash.*

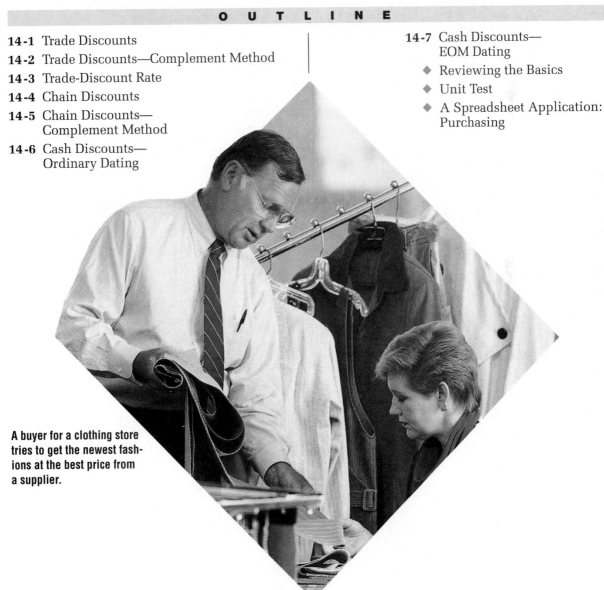

A buyer for a clothing store tries to get the newest fashions at the best price from a supplier.

14-1

Trade Discounts

OBJECTIVE

Compute the trade discount and the net price.

Most suppliers provide a catalog from which your business can choose items to purchase. The catalog includes a description and the list price, or catalog price, of each item that the supplier sells. The list price is generally the price at which you may sell the item. You usually purchase the item at a trade discount, which is a discount from the list price. The trade discount is often expressed as a percent. The net price is the price you actually pay for the item.

Trade Discount = List Price × Trade-Discount Rate

Net Price = List Price − Trade Discount

EXAMPLE Skills 30, 6   *Application* A   *Term* Trade discount

Johnson Bicycle Shop is purchasing 10-speed, 27-inch Trailblazer bicycles from a wholesaler. The list price in the wholesaler's catalog is $845. Johnson receives a 40% trade discount. What is the net price of each bicycle?

SOLUTION

A. Find the **trade discount.**

List Price × Trade-Discount Rate

$845.00 × 40% = $338.00 trade discount

B. Find the **net price.**

List Price − Trade Discount

$845.00 − $338.00 = $507.00 net price

845 × 40 % 338 845 − 338 = 507

✔ SELF-CHECK   Complete the problems, then check your answers in the back of the book.

1. List price: $380.
   Trade-discount rate: 30%.
   Find the trade discount.

2. List price: $380.
   Trade discount: $114.
   Find the net price.

PROBLEMS

Round answers to the nearest cent.

	3.	4.	5.	6.	7.	8.	9.
List Price	$140.00	$240.00	$400.00	$360.00	$174.00	$216.80	$94.75
Trade-Discount Rate	25%	10%	35%	15%	32%	27%	9%
Trade Discount	$35.00	?	?	?	?	?	?
Net Price	?	?	?	?	?	?	?

10. $47 list price for wok. 25% trade discount. What is the net price?

11. $219 list price for compact disc player. 42% trade discount. What is the net price?

12. An Easy-Wake clock radio carries a catalog price of $39.75. The Easy-Wake Company offers a 15% trade discount. What is the net price?

13. Giant discount stores purchases hand-held electronic calculators from Ace Wholesalers. Find the trade discount and net price for each calculator.

	Item	List	Trade-Discount Rate
a.	Basic	$19.95	20%
b.	Scientific	$24.49	23%
c.	Paper-tape	$34.95	28%

14. Hero, Inc., manufactures automobile parts and accessories. A portion of its "Parts and Accessories Price List" for auto dealers is shown. Hero offers the trade discounts listed to its dealers. Find the dealer price for each item.

Class	Part Number	Suggested List Price	Trade Discount
A	D2UZ 8200-B	$89.60	42%
HX	D27Z 82-A	$57.20	38%
CQ	COYT 8A178	$ 2.90	45%
D	C7ZZ 8B90	$ 7.80	30%

15. Commercial Paper Company offers a 55% discount to commercial users of their products. Pro Printers, Inc., placed this order. What is the net price?

Quantity	Units	Description	Price per Unit
25	SHEETS	H1163 32 × 40 14 PLY #33 MAT	$217.60
		Cut to 150 PCS. 11 × 14	
		125 PCS. 11 × 12	
		TOTAL	?

16. Alvin Sales Agency receives a 32% trade discount on farm equipment from Land Equipment, Inc. What is the net price for each order?

Quantity	Stock No.	Description	Unit Price
6	1-1	Heavy nuts	$0.45
4	1-2	Starlock washer	$0.20
2	61 HOB	Blades	$7.45

MAINTAINING YOUR SKILLS   Look up the skills in parentheses if you need help or more practice.

Find the percentage. Round answers to the nearest cent. **(Skill 30)**

17. $47.80 × 10%

18. $91.40 × 18%

19. $416.60 × 20%

20. $27.79 × 35%

21. $279.49 × 50%

22. $2178.99 × 42%

Subtract. **(Skill 6)**

23. $47.80 − $4.78

24. $416.60 − $83.32

25. $279.49 − $139.75

Trade Discounts— Complement Method

OBJECTIVE

Compute the net price using the complement method.

The complement method is another method of finding the net price. When you receive a trade discount, you can subtract the discount rate from 100%. This gives you the complement of the trade-discount rate, or the percent you actually pay, sometimes called the net price rate. Multiply the complement by the list price to find the net price.

Net Price = List Price × Complement of Trade-Discount Rate

EXAMPLE | *Skills* 30, 6 *Application* L *Term* Complement method

Johnson Bicycle Shop is also purchasing 3-speed, 26-inch Tour bicycles, which have a list price of $578. Johnson receives a trade discount of 40%. What is the net price of each bicycle?

SOLUTION

A. Find the **complement of trade-discount rate.**
100% − 40% = 60% complement of trade-discount rate

B. Find the **net price.**
List Price × Complement of Trade-Discount Rate
$578.00 × 60% = $346.80 net price

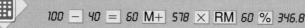
100 − 40 = 60 M+ 578 × RM 60 % 346.8

✔ SELF-CHECK

Complete the problems, then check your answers in the back of the book.

1. What is the complement of 30%?

2. List price: $380.
Trade-discount rate: 30%.
Find the net price.

PROBLEMS

Use the complement method to solve each problem. Round to nearest cent.

	3.	4.	5.	6.	7.	8.	9.
List Price	$120.00	$85.00	$27.00	$74.65	$317.48	$4.73	$417.89
Trade-Discount Rate	25%	20%	35%	15%	24%	18%	8%
Complement	75%	?	?	?	?	?	?
Net Price	?	?	?	?	?	?	?

10. $48.77 list price for sprinkler. 32% trade discount. What is the net price?

11. $2178.90 list price for personal computer. 27% trade discount. What is the net price?

12. Baxter Sales Agency receives a 27% trade discount on these items. What will Baxter pay for each order?

Quantity	Stock No.	Description	Unit Price	Total List
2	84-D	DAXCO cutter-complete	$316.45	$632.90
5	84-06	Blades	$ 7.35	$ 36.75

13. Exact-Fit Automotive, Inc., receives a 41% trade discount on these items. What is the net price for each?

Number	Description	Quantity	List	Total
ED-1	Water pump	2	$37.43	$ 74.86
400	Headlight	4	$ 6.95	$ 27.80
10S	Shock absorber	8	$17.75	$142.00
PF-91	Sound module	3	$54.70	$164.10

14. dba Charles receives a 54% trade discount from Commercial Clothes Company. Find the net price for each total listed.

Order No.	Descript.	Style No.	No. of Units	Unit Price	Total
11260467	Blazer	87024	2	$79.45	$158.90
11260467	Jacket	87026	5	$64.75	$323.75
11260467	Shirt	87233	9	$29.79	$268.11
11260467	Shirt	87237	5	$24.49	$122.45
11260467	Shirt	87533	6	$19.99	$119.94

15. The Star Sports Shop purchases fish finders from Icthos, Inc., for a list price of $179.79. Star Sports Shop receives a 28% trade discount.
a. What net price does the Star Sports Shop pay for the fish finders?
b. What is the dollar value of the discount?

16. Do-It-All Lumber Co. receives a 19% discount from Wholesale Lumber Supply. Find the net price for each amount on the invoice.

Fixture Location	Description	Dept.	Qty.	Price	Amount
NA3	Fastway	4	1	$8.25	$ 8.25
EQ25	ITE	4	3	7.54	22.62
BR15	Rite Corp.	4	6	1.94	11.64
762OC	EFCO	4	2	2.42	4.84

MAINTAINING YOUR SKILLS   Look up the skills in parentheses if you need help or more practice.

Find the complement. **(Application L)**

17. 50%   **18.** 40%   **19.** 22%   **20.** 19%

21. 8%   **22.** 26%   **23.** 42%   **24.** 37%

Find the percentage. Round to nearest cent. **(Skill 30)**

25. $43.80 × 60%   **26.** $149.45 × 78%   **27.** $91.79 × 81%

28. $9.89 × 74%   **29.** $4171.84 × 63%   **30.** $191.99 × 50%

Trade-Discount Rate

OBJECTIVE
Compute the trade-discount rate.

Some wholesalers and manufacturers publish only the list prices and net prices in their catalogs. If you know both of these prices, you can calculate the trade-discount rate.

Trade Discount = List Price − Net Price

$$\text{Trade-Discount Rate} = \frac{\text{Trade Discount}}{\text{List Price}}$$

EXAMPLE   *Skills* 6, 31   *Application* A   *Term* Trade-discount rate

The Auto Parts Manufacturing Company shows a $56 list price in its catalog for a water pump, #GF147. The net price to the retailer is $42. What is the trade-discount rate?

SOLUTION

A. Find the **trade discount.**
List Price − Net Price
  $56.00 − $42.00 = $14.00 trade discount

B. Find the **trade-dicount rate.**
Trade Discount ÷ List Price
  $14.00   ÷ $56.00 = 25% trade-discount rate

56 M+ − 42 = 14 ÷ RM 56 % 25

✔ SELF-CHECK   Complete the problems, then check your answers in the back of the book.

1. List price: $140.
Net price: $84.
Find the trade discount.

2. Trade discount: $56.
List price: $140.
Find the trade-discount rate.

PROBLEMS

Round the trade-discount rate to the nearest tenth of a percent.

	3.	4.	5.	6.	7.	8.	9.
List Price	$5.00	$75.00	$60.00	$50.00	$112.60	$97.49	$5147.58
Net Price	$4.00	$50.00	$42.00	$27.50	$ 74.79	$68.78	$4258.69
Trade Discount	$1.00	?	?	?	?	?	?
Trade-Discount Rate	?	?	?	?	?	?	?

10. List price is $198.49.
Net price is $149.27.
Estimate the trade discount
and rate.
Compute the trade-
discount rate.

11. List price is $102.79.
Net price is $78.56.
Estimate the trade discount
and rate.
Compute the trade-
discount rate.

12. Beauty Shop Wholesalers, Inc., offers discounts on most items they sell. What is the trade-discount rate for each of the items listed?

Item	List	Net
Complexion brushes	$15.79	$11.68
Lighted make-up mirror	$59.99	$41.45
Cleansing brushes	$ 9.79	$ 6.11
4-way make-up mirror	$47.49	$24.73

13. Auto Paint Supply, Inc., varies the rate of discount with the item purchased. This invoice was received by Nu-Car Service. What is the trade-discount rate for each gross amount?

USE INVOICE NUMBER ON ALL CORRESPONDENCE			ALL ITEMS NET UNLESS OTHERWISE SPECIFIED			
Quantity Shipped	Ordered	Description	Unit Price	Gross Amount	Disc.	Net Amount
4	4	0001-WH-TOUCH-UP	$3.98	$15.92	$5.57	$10.35
2	2	0001-BL-TOUCH-UP	$3.98	$ 7.96	$1.99	$ 5.97
2	2	0001-AVO-TOUCH-UP	$4.98	$ 9.96	0	$ 9.96
4	4	0001-GLD-TOUCH-UP	$4.98	$19.92	$3.98	$15.94

14. Find the trade-discount rate for each gross amount listed on this invoice.

Quantity	Product Code	Price per Unit	Gross Amount	Net Amount
140	00010	$ 7.99	$1118.60	$1006.74
160	00015	$19.67	$3147.20	$2045.68
110	00018	$24.84	$2732.40	$1366.20
100	00022	$36.71	$3671.00	$2845.03
200	00024	$14.35	$2870.00	$1937.25
24	00500	$ 8.43	$ 202.32	$ 157.18
48	00502	$ 6.20	$ 297.60	$ 201.40

MAINTAINING YOUR SKILLS Look up the skills in parentheses if you need help or more practice.

Subtract. **(Skill 6)**

15. $78.00 − $54.60

16. $16.40 − $9.84

17. $162.70 − $122.03

18. $45.84 − $34.38

19. $1075.79 − $788.32

20. $2471.83 − $1971.92

Find the rate. Round answers to the nearest tenth of a percent. **(Skill 31)**

21. $23.40 ÷ $78.00

22. $6.56 ÷ $16.40

23. $40.67 ÷ $162.70

24. $214.91 ÷ $1073.78

25. $1171.81 ÷ $11,721.49

26. $1.78 ÷ $3.54

Chain Discounts

OBJECTIVE

Compute the final net price after a chain of discounts.

A supplier may offer chain discounts to sell out a discontinued item or to encourage you to place a larger order. A chain discount is a series of trade discounts. For example:

35% less 20% less 15%

This is often written as 35/20/15.

This does not mean 35% + 20% + 15% off.

35% discount is deducted from list price to yield first net price.

20% of first net price is deducted from first net price to yield second net price.

15% of second net price is deducted from second net price to yield final net price.

EXAMPLE   *Skills* 30, 2, 6   *Application* A   *Term* Chain discount

The Green Lawn Corporation produces a $2\frac{1}{2}$- horsepower lawn mower, which is priced at $695. A new model is ready for production, so the $2\frac{1}{2}$-horsepower model is to be discontinued. Green Lawn is offering a chain discount of 20% less 10%. What is the final net price of the lawn mower?

SOLUTION

A. Find the **first discount.** ➡
List Price × Discount Rate
$695.00 × $20% = $139.00

B. Find the **first net price.**
List Price − Discount
$695.00 − $139.00 = $556.00

C. Find the **second discount.** ➡
$556.00 × 10% = $55.60

D. Find the **final net price.**
$556.00 − $55.60 = $500.40
final net price

695 ✕ 20 % 139 695 − 139 = 556 ✕ 10 % 55.6 556 − 55.6 = 500.4

✔ SELF-CHECK   Complete the problems, then check your answers in the back of the book.

List price is $400. Chain discount is 30% less 25%.

1. Find the first net price.

2. Find the final net price.

PROBLEMS

	3.	4.	5.	6.	7.
List Price	$500	$780	$1240	$475	$2170
Chain Discount	30% less 20%	25% less 15%	20% less 15%	40% less 30%	40% less 25%
First Discount	$150	$195	?	?	?
First Net Price	$350	$585	?	?	?
Second Discount	?	?	?	?	?
Final Net Price	?	?	?	?	?

8. $74.60 list price for clock radio. 30/10 chain discount. What is the net price?

9. $125.75 list price for cordless phone. 40/20 chain discount. What is the net price?

10. Storm-Tite Inc. offers Mark Stores chain discounts of 30/15. What is the net price for each item?

Item	List Price
Storm door	$147.80
Storm window	$ 79.85
Foil-backed insulation	$ 14.50/roll

11. The General Saw Corporation offers trade discounts and additional discounts to encourage large orders. What is the net price per item if each invoice total is high enough to obtain the additional discount?

Item	List Price	Trade Discount	Additional Discount
Masonry drill	$98.80	50%	5% (invoice total over $250)
Raw-tip drill	$76.75	40%	10% (invoice total over $500)
Hole saw	$62.53	30%	20% (invoice total over $1000)

12. The Wholesale Supply Co. offers chain discounts of 35% less 20% less 15% to sell out a discontinued item. Find the final net price for an order of $1460.

13. What is the final net price for each of these overstocked items from the Everything Automotive Parts list?

Part Number	Suggested List Price	Chain Discount
B7S	$914.80	40% less 25% less 15%
C37X	$247.50	42% less 20% less 10%
173A	$ 76.67	35% less 15% less 5%
B62Y	$ 8.54	25% less 10% less 7%

MAINTAINING YOUR SKILLS Look up the skills in parentheses if you need help or more practice.

Round to the place value indicated. **(Skill 2)**

14. $17.71155 (nearest cent)

15. $113.7051 (nearest dollar)

16. 17.98% (nearest tenth of a percent)

17. $8178.1449 (nearest cent)

18. 71.47% (nearest percent)

19. $51.476 (nearest whole number)

Subtract. **(Skill 6)**

20. $17.60 − $5.28

21. $94.35 − $37.74

22. $214.80 − $21.48

23. $567.88 − $141.97

24. $1774.86 − $591.62

25. $24,599.99 − $11,069.55

Find the percentage. Round to nearest cent. **(Skill 30)**

26. $217.80 × 35%

27. $670.45 × 25%

28. $814.86 × 15%

29. $3417.56 × 30%

30. $7147.63 × 20%

31. $56,147.71 × 10%

14-5

Chain Discounts— Complement Method

OBJECTIVE

Compute the final net price using the complement method.

The complement method applied to trade discounts can also be applied to chain discounts. First subtract each discount from 100% to find the complements. Then multiply the complements to find the percent that you actually pay. This percent is called the net-price rate. To find the net price, multiply the net-price rate by the list price.

You can also use the complement method to find the discount. Find the complement of the net-price rate by subtracting it from 100%. That percent is called the single equivalent discount (SED). Multiply the SED by the list price to find the discount. You can check by subtracting the net price from the list price.

Net-Price Rate = Product of Complements of Chain-Discount Rates

Net Price = List Price × Net-Price Rate

SED = Complement of Net-Price Rate

Discount = List Price × SED

EXAMPLE   *Skills* 30, 2, 6   *Application* A   *Term* Chain discount

The Green Lawn Corporation manufactures an 8-horsepower riding mower, which is priced at $1450. A new model is ready for production, so the 8-horsepower model is to be discontinued. Green Lawn is offering a chain discount of 30% less 20% less 10%. What is the net price of the mower? What is the SED and the discount?

SOLUTION

A. Find the **net-price rate.**
Product of Complements of Chain-Discount Rates
70% × 80% × 90% = 50.4% net-price rate

B. Find the **net price.**
List Price × Net-Price Rate
$1450.00 × 50.4% = $730.80 net price

C. Find the **SED.**
Complement of the Net-Price Rate
100% − 50.4% = 49.6% SED

D. Find the **discount.**
List Price × SED
$1450.00 × 49.6% = $719.20 discount
Check: $1450.00 − $730.80 = $719.20

✔ SELF-CHECK   Complete the problems, then check your answers in the back of the book.

List price is $560. Chain discount is 30% less 10%.

1. Find the net-price rate and the net price.   **2.** Find the SED and the discount.

	3.	**4.**	**5.**	**6.**	**7.**	**8.**
List Price	$620.00	$140.00	$436.00	$1237.00	$147.80	$96.46
Chain Discounts	30% less 20%	20% less 15%	40% less 20%	40% less 30%	30% less 20% less 10%	20% less 10% less 5%
Net Price	$347.20	$95.20	?	?	?	?
SED	?	?	?	?	?	?
Discount	?	?	?	?	?	?

9. $2150 list price for personal computer.
15/10 chain discount.
What is the net price?
What is the discount?

10. $74.85 list price for calculator.
20/5 chain discount.
What is the net price?
What is the discount?

11. Outfitters, Inc., offers Clark's Clothes Co. chain discounts of 25/10. What is the net price for each item? What is the SED?

Item	List Price
Storm boots	$149.79
Storm coat	$219.49
Storm gloves	$ 29.99

12. The Globe Corporation offers trade discounts and additional discounts to encourage large orders. What is the net price per item and net discount if each invoice total is high enough to obtain the additional discount?

Item Number	List Price	Trade Discount	Additional Discount
14ZB	$119.25	40%	10% (invoice total over $500)
11TC	$ 84.76	20%	20% (invoice total over $1000)
93MR	$ 51.18	30%	5% (invoice total over $250)

13. What is the net price and net discount for each of these overstocked items from the Car Catalog price list?

Item	Suggested List Price	Chain Discount
Shock 147	$ 78.45	40% less 25% less 10%
Bumper 96	$341.82	30% less 20% less 5%
Rim 412	$ 58.37	25% less 15% less 5%
Quarter 753	$147.86	20% less 10% less 5%

MAINTAINING YOUR SKILLS   Look up the skills in parentheses if you need help or more practice.

Round to the nearest cent. **(Skill 2)**

14. $19.7171   **15.** $114.555   **16.** $912.4142

17. $2481.5764   **18.** $14.7210

Find the complement. **(Application L)**

19. 30% **20.** 20% **21.** 25% **22.** 5% **23.** 10% **24.** 15% **25.** 7%

Find the percentage. **(Skill 31)**

26. $347.80 × 50.4% **27.** $81.70 × 43.2% **28.** $591.45 × 53.375%

Cash Discounts— Ordinary Dating

OBJECTIVE

Compute the cash price when discount is based on ordinary dating.

You receive an invoice for each purchase you make from a supplier. The invoice lists the quantities and costs of the items purchased. To encourage prompt payment, the supplier may offer a cash discount if the bill is paid within a certain number of days. The exact terms of the discount are stated on the invoice. Many suppliers use ordinary dating. For example:

2/10, net 30   or   **2/10, n/30**

2% cash discount if paid within 10 days of date of invoice.

Net price must be paid within 30 days of date of invoice.

Cash Discount = Net Price × Cash-Discount Rate

Cash Price = Net Price − Cash Discount

EXAMPLE   *Skills* 30, 6   *Application* A   *Term* Ordinary dating

This is part of the invoice that The Lighting Store received for a shipment of lamps. What is the cash price of the lamps if the bill is paid within 10 days?

Oak Hill Lighting, Inc.			SHIP TO: THE LIGHTING STORE INVOICE NO: 4-3467		
DATE June 5, XX	ORDER NO. LH 3019		TERMS 2/10, NET 30		ACCT. NO. 712E 4
STYLE	COLOR	QUANTITY	PRICE		AMOUNT
A9407	BGE	8	$37.25		$298.00
A9841	GRN	6	42.15		252.90
J2113	NVY	10	26.40		264.00
J2114	GRN	8	26.40		211.20
				TOTAL	$1026.10

SOLUTION

A. Find the **last day a discount can be taken.**
Date of invoice: June 5
Terms: 2/10, net 30
Discount can be taken until June 15, 10 days from date of invoice.

B. Find the **cash discount.**
Net Price × Cash-Discount Rate
$1026.10 × 2% = $20.522 = $20.52 cash discount

C. Find the **cash price.**
Net Price − Cash Discount
$1026.10 − $20.52 = $1005.58 cash price

1026.1 × 2 % 20.522 1026.1 − 20.52 = 1005.58

✔ SELF-CHECK   Complete the problems, then check your answers in the back of the book.

Terms are 3/10, n/30. Net price is $640.

1. Find the cash discount.   **2.** Find the cash price.

	Invoice Date	Terms	Date Paid	Net Price	Cash Discount	Cash Price
3.	3/19	2/10, n/30	3/28	$ 740.00	$14.80	?
4.	5/6	3/10, n/30	5/13	$ 516.40	?	?
5.	9/3	5/15, n/30	9/23	$ 348.64	?	?
6.	11/22	7/10, n/30	12/1	$4178.45	?	?

F.Y.I.

The phrase "cash discount terms of 2/10, net 30, ROG" means that the discount period starts with the Receipt Of the Goods.

7. $6715.80 net price for hardware. Terms are 4/10, n/30. Date of invoice is October 7. Date paid is October 14. What is the cash price?

8. $614.85 net price for auto parts. Terms are 5/15, net 30. Date of invoice is December 11. Invoice is paid on December 23. What is the cash price?

9. The net price of goods from Fashion Footwear, Inc., to the Mall Shoe Store was $916.78. The invoice was paid on February 28. According to the terms on the invoice, what was the cash price?

DATE SHIPPED	VIA	TERMS: 2%-20 DAYS	SALESPERSON	DATE OF INVOICE
2/24/-	OUR DELIVERY		65	2/23/-

10. The net price of goods from Women's Wear, Inc., to Violet's is $4217.86. If the invoice is paid on December 11, what is the cash price?

CUSTOMER NUMBER 1238	SHIPPING DATE 11-22-XX		INVOICE DATE 12-05-XX	
CUST. ORDER NO.	CUST. ORDER DATE	TERMS	TYPE	SHPG. ORDER NO.
211	11-15-XX	8/10, n/30	02	847710

11. What is the cash price for these materials from Ace Tool & Die, Inc., to Summit Machine Shop? The invoice is paid on November 14.

CUSTOMER'S COPY INVOICE NO. NO. 4559	DATE NOV. 7, 19XX			TERMS 1%-10 days Net 30 days		
YOUR ORDER NO.	PIECES	PATTERN NO.	WEIGHT	PRICE	AMOUNT	
7552	240	641-7		$0.61	$146.40	
1319	157	320-5		0.53	83.21	
					$229.61	

MAINTAINING YOUR SKILLS   Look up the skills in parentheses if you need help or more practice.

Subtract. **(Skill 6)**

12. $516 − $10.32

13. $916.70 − $27.50

14. $3178.42 − $127.14

15. $119.20 − $11.92

16. $9784.52 − $489.21

17. $96.20 − $0.48

Find the percentage. Round to nearest cent. **(Skill 30)**

18. $415.00 × 2%

19. $147.84 × 3%

20. $916.74 × 8%

21. $714.50 × 1%

14-7

Cash Discounts— EOM Dating

OBJECTIVE

Compute the cash price when discount is based on end-of-month dating.

Many wholesalers use **end-of-month dating** when granting cash discounts. With this method of dating, your business receives a cash discount if you pay for your merchandise within a certain number of days after the end of the month. For example:

6/15 EOM ◄—— A 6% cash discount if bill is paid within 15 days after the end of the month in which the invoice is issued.

Cash Discount = Net Price × Cash-Discount Rate

Cash Price = Net Price − Cash Discount

EXAMPLE   *Skills* 30, 6   *Application* A   *Term* End-of-month dating

Bob's Drugstore received this invoice from General Wholesalers. What is the last day that Bob's can take advantage of the 4% cash discount? What is the cash price?

DATE	INVOICE NO.	ACCT. NO.	STORE NO.	TERMS	VENDOR NO.
JUL 12	213788-6	208 712	34	4/10 EOM	12/7

SALESPERSON	VIA		ORDER REG. NO.
38L	OV DIRECT TRUCK		0547-6

CUSTOMER ORDER NO.	STYLE	QUANTITY	PRICE	AMOUNT
L3075	02673	10	$17.80	$178.00
L3076	02677	15	24.17	362.55
L3077	02681	25	7.46	186.50
			TOTAL	$727.05

SOLUTION

A. Find the **last day a discount can be taken**.
Date of invoice: July 12
Terms: 4/10 EOM
Discount can be taken until August 10, 10 days after the end of July.

B. Find the **cash discount**.
Net Price ×   Cash-Discount Rate
$727.05   ×   4%   = $29.082 = $29.08 cash discount

C. Find the **cash price**.
Net Price −   Cash Discount
$727.05 − $29.08 = $697.97 cash price

---

396 ◆ Unit 14 Purchasing

An invoice dated November 7 has terms of 2/10 EOM. Net price is $7100. The invoice is paid December 9.

1. Find the last day a discount can be taken.

2. Find the cash price.

PROBLEMS

Invoice Date	Terms	Date Paid	Net Price	Cash Discount	Cash Price
3. Jan. 21	2/10 EOM	Feb. 6	$ 615.00	$12.30	?
4. April 14	3/15 EOM	May 12	$ 1417.80	?	?
5. Sep. 2	4/10 EOM	Oct. 9	$ 7184.73	?	?
6. Dec. 19	5/10 EOM	Jan. 13	$16,517.84	?	?

7. Net price is $916.40.
   Terms are 1/10 EOM.
   Invoice date is September 20.
   Invoice is paid on October 9.
   What is the cash price?

8. Net price is $7641.60.
   Terms are 6/10 EOM, n/30 EOM.
   Invoice date is May 20.
   Invoice is paid on June 2.
   What is the cash price?

9. An invoice from Gentlemen's Clothiers, Inc., to Ducky's Tux Rentals shows a net price of $761.87. The terms are 5/15 EOM, n/30 EOM. The invoice is dated August 14. The invoice is paid September 15. What is the cash price?

10. An invoice from Krio Frozen Foods Corporation to Barr Wholesalers carries a net price of $26,715.81. The terms are 7/10 EOM. The invoice is dated Feb. 20. The invoice is paid March 7. What is the cash price?

11. An invoice from Classic Clothes to the Clothes Palace has a list price of $10,719.75. After a trade discount of 20%, the net price is $8575.80. A portion of the invoice is shown. What is the cash price if the invoice is paid on January 10?

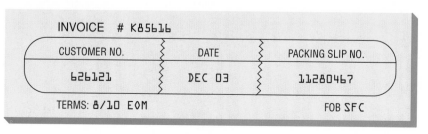

INVOICE # K85616

CUSTOMER NO.	DATE	PACKING SLIP NO.
626121	DEC 03	11280467

TERMS: 8/10 EOM   FOB SFC

Subtract. **(Skill 6)**

12. $821.00 − $16.42   **13.** $1417.80 − $14.18   **14.** $8614.50 − $258.44

Find the percentage. **(Skill 30)**

15. $821.00 × 2%   **16.** $8614.50 × 3%   **17.** $12,643.17 × 8%

18. $821.00 × 98%   **19.** $8614.50 × 97%   **20.** $1417.80 × 99%

Reviewing the Basics

Skills

(Skill 1)

Round to the nearest cent.

1. $9187.7649   **2.** $514.4551   **3.** $71.8686   **4.** $217.6054

Solve. Round to the nearest cent.

(Skill 6)

5. $54.73 − $41.95   **6.** $4716.23 − $917.85   **7.** $347.80 − $69.56

8. $7168.42 − $143.37   **9.** $43,517.93 − $1305.54   **10.** $571.84 − $22.87

(Skill 30)

11. $179.47 × 3%   **12.** $781.46 × 20%   **13.** $297.97 × 2%

14. $8654.50 × 8%   **15.** $6247.75 × 90%   **16.** $787.14 × 50.4%

Round to the nearest tenth of a percent.

(Skill 31)

17. $4.17 is what percent of $20.85?

18. $72.73 is what percent of $289.99?

19. $216.43 is what percent of $649.29?

20. $1516.74 is what percent of $3794.49?

Applications

(Application L)

Find the complement.

21. 30%   **22.** 20%   **23.** 68.4%   **24.** 3%   **25.** 50.4%

Terms

Use each term in one of the sentences.

26. Trade discount

27. Chain discount

28. Complement method

29. Ordinary dating

30. End-of-month dating

31. Cash price

a. *3/5, net 30* is an example of ___?___.

b. The ___?___ is a way to find the net price when a trade-discount rate is given.

c. A series of discounts on an item is called a ___?___.

d. *9/5 EOM* is an example of ___?___.

e. Chain discounts are applied to the list price to get the net price. Cash discounts are applied to the net price to get the ___?___.

f. A discount from the catalog price is called a ___?___.

g. The catalog price is the ___?___.

Refer to your reference files in the back of the book if you need help.

Unit Test

1. Pet Pride purchases its pet supplies from Wholesale Suppliers, Inc. The list price on one invoice was $714.86. Pet Pride receives a 20% trade discount. What is the net price for that invoice?

Lesson 14-2

2. Penn Skate purchases Ambassader roller skates that have a list price of $129.79 a pair. Penn Skate receives a trade discount of 25%. What is the net price that Penn Skate pays for each pair of skates? Use the complement method to solve.

Lesson 14-3

3. Heat Mfg., a manufacturer of small heaters, lists the price of its top-of-the-line heater as $179.99. The net price to the retailer is $116.99. What is the trade-discount rate?

Lesson 14-4

4. The Fitness Company manufactures an all-purpose exercise set that is priced at $947.50. When it introduced its new aerobic model, The Fitness Company offered a chain discount of 30% less 20% on the old model. What is the final net price of the all-purpose exercise set?

Lesson 14-5

5. Air-Hi produces a sports shoe that is priced at $114.49. When the new model was introduced, the $114.49 shoe was discounted at 15% less 5%. What is the net price of the old model using the complement method?

Lesson 14-6

6. The Capital Theatre received an invoice dated September 13 from Theatre Supply. The invoice carried terms of 2/10, net 30. The net price of the invoice was $561.87. What was the cash price of the invoice if it was paid on September 20?

Lesson 14-7

7. Custom Computer received this invoice from Hi-Tech, Inc. What is the last day that Custom Computer can make use of the 6% cash discount? What is the cash price?

DATE	INVOICE NO.	ACCT. NO.	STORE NO.	TERMS	VENDOR NO.
July 10	181818	75239	07	6/10 EOM	16

CUSTOMER ORDER NO.	TYPE	QUANTITY	PRICE	AMOUNT
F78906	08765	20	$10.56	$211.20
F90875	98765	35	6.98	244.30
L56329	09453	15	23.74	356.10
			TOTAL	$811.60

A SPREADSHEET APPLICATION

Purchasing

To complete this spreadsheet application, you will need the template diskette for *Mathematics with Business Applications*. Follow the directions in the *User's Guide* to complete this activity.

Input the information in the problems to find the net price and the cash price. Then answer the questions that follow.

	List Price	Chain Discounts	Net Price	Terms	Date of Invoice	Date Paid	Cash Price
1.	$ 500	30% less 20% less 10%	?	2/10, n/30	Jan. 23	Jan. 30	?
2.	$ 460	25% less 20%	?	net 30	Apr. 14	May 11	?
3.	$3820	15%	?	3/15, n/30	Nov. 11	Nov. 25	?
4.	$ 987	20% less 10% less 5%	?	2/10 EOM	Feb. 3	Mar. 9	?
5.	$ 987	25% less 10%	?	2/10, n/30	Feb. 3	Mar. 1	?
6.	$4750	30% less 10%	?	8/20, n/50	June 13	July 1	?
7.	$4750	38%	?	3/10, n/30	Mar. 23	Apr. 7	?
8.	$ 500	10% less 20% less 30%	?	3/15 EOM	May 21	June 3	?
9.	$8259	12.5% less 20% less 50%	?	2/30, n/60	Aug. 23	Sep. 23	?
10.	$8259	65%	?	2/30, n/60	Sep. 23	Oct. 23	?

11. What is the net price in #1?

12. What is the cash price in #1?

13. What is the cash price in #2?

14. What is the net price in #3?

15. What is the cash price in #3?

16. What is the net price in #4?

17. What is the net price in #5?

18. In #4 and #5, the list price is the same and the sum of the chain discounts is the same. Why isn't the net price the same? Can you make a general statement about which net price will be greater if the sum of the chain discounts is the same?

19. In #4, you get the cash discount by paying on March 9; in #5, you don't get the cash discount and you paid on March 1. Why?

20. By looking at the net prices for #6 and #7, compare chain discounts of 30% less 10% and a single trade discount of 38%.

21. When is the net price due in #7 if the cash discount is not taken?

22. What is the net price in #8?

23. Both #1 and #8 result in the same net price. How can this be?

24. What is the net price in #9?

25. What is the net price in #10?

26. How can the net prices in #9 and #10 be the same?

27. What is the cash price in #9?

28. What is the cash price in #10?

29. Both #9 and #10 have the same cash discount terms. Why do you get the cash discount in #10 but not in #9?

15

Sales

After your business has purchased goods from suppliers, wholesalers, distributors, or manufacturers, you must determine the *retail price,* or selling price, of the product. You want to sell the product at a higher price than its *cost,* or the amount you paid for it. The difference between the cost and the selling price is your *markup,* or *gross profit.* When your markup is larger than your overhead or operating expenses, you make a *net profit.*

O U T L I N E

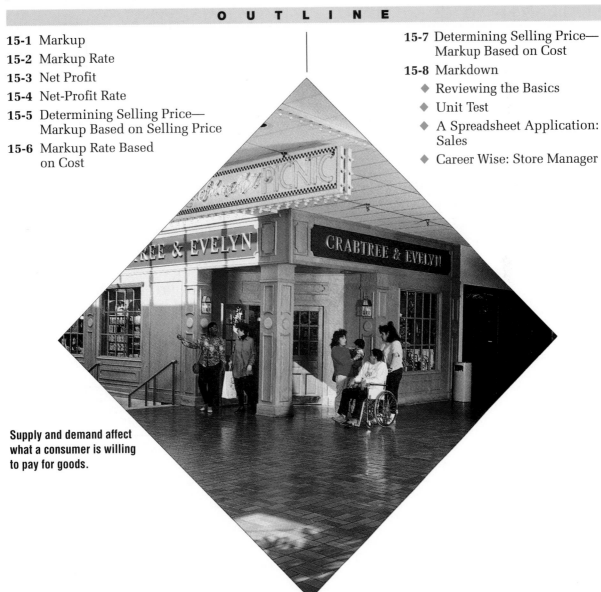

**Supply and demand affect
what a consumer is willing
to pay for goods.**

15-1

Markup

The **cost** of a product is the amount that your business pays for it. The cost includes expenses such as freight charges. Your business sells the product at a **selling price,** or **retail price,** which is higher than the cost. The **markup** is the difference between the selling price and the cost. The markup may also be called the **gross profit** or **profit margin.**

Markup = Selling Price − Cost

EXAMPLE   *Skill 6   Application A   Term Markup*

Discount Electronics, Inc., purchased cable-ready TV sets for $117.83 each. Discount Electronics sells the TVs for $199.99 each. What is the markup on each TV?

SOLUTION   Find the **markup.**

Selling Price − Cost
$199.99 − $117.83 = $82.16 markup

199.99 − 117.83 = 82.16

✔ SELF-CHECK   Complete the problems, then check your answers in the back of the book.

Find the markup.

1. Selling price is $19.49. Cost is $7.63

2. Selling price is $545.45. Cost is $272.72.

PROBLEMS

	Item	Selling Price	−	Cost	=	Markup
3.	Automobile	$16,497.00	−	$13,197.60	=	?
4.	Telephone	$ 49.79	−	$ 27.84	=	?
5.	Watch	$ 97.87	−	$ 32.98	=	?
6.	Book	$ 24.97	−	$ 18.43	=	?
7.	Snowblower	$ 417.79	−	$ 235.84	=	?
8.	Rosebush	$ 9.97	−	$ 6.68	=	?
9.	Bicycle	$ 219.79	−	$ 148.93	=	?
10.	Eraser	$ 0.35	−	$ 0.27	=	?

11. Shoes.
Cost is $31.48.
Selling price is $79.59.
What is the markup?

12. Grape jam.
Cost is $0.56.
Selling price is $0.89.
What is the markup?

13. Basketball.
Cost is $12.78.
Selling price is $17.49.
What is the markup?

14. Cassette deck.
Cost is $17.97.
Selling price is $49.79.
What is the markup?

15. Mostly Kitchens, Inc., purchases paper towel holders at a cost of $7.14. Mostly Kitchens sells the paper towel holders for $10.99 each. What is the markup?

16. General Department Stores purchased bedroom curtains from The Linen Company for $6.19. General Department Stores sold the curtains for $9.79. What was the markup on the bedroom curtains?

17. Infinity Computers, Inc., advertised this monitor in yesterday's
paper. The cost of the monitor
to Infinity Computers is
$297.88. What is the markup on
the monitor?

ENHANCED
GRAPHICS
MONITOR
$429.49
Limited Supply

18. Theresa Oakley operates a mini-engine repair shop. Recently she bought an older lawn mower for $20 at a local auction. She repaired the lawn mower for a cost of $9.17 in parts. She painted and polished the lawn mower for a cost of $6.43. When the lawn mower was finished, Theresa sold it for $75. What was the markup on the lawn mower?

19. Tasty Bakery Company sells chocolate eclairs for $15.48 a dozen. It costs Tasty Bakery $0.74 to produce each eclair.
a. What is the markup on a dozen eclairs?
b. What is the markup on each eclair?

20. Country Produce buys lettuce by the crate. Each crate contains 24 heads of lettuce. The cost of each crate to Country Produce is $11.76. Country Produce sells the lettuce for $0.64 per head.
a. What is the markup per head of lettuce?
b. What is the markup per crate of lettuce?

MAINTAINING YOUR SKILLS   Look up the skill in parentheses if you need help or more practice.

Subtract. **(Skill 6)**

21. $54.45 − $43.56

22. $1.74 − $1.57

23. $216.96 − $130.18

24. $491.79 − $418.02

25. $24.49 − $23.27

26. $643.89 − $321.94

27. $14.79 − $11.54

28. $4217.83 − $2319.81

29. $9179.84 − $4773.52

15-2

Markup Rate

OBJECTIVE

Compute the markup as a percent of the selling price.

The markup rate is the markup expressed as a percent. Businesses usually express the markup as a percent of the selling price.

$$\text{Markup Rate} = \frac{\text{Markup}}{\text{Selling Price}}$$

EXAMPLE *Skills* 6, 31 *Application* A *Term* Markup rate

Roy's Florist buys roses for $10.99 a dozen. It sells them for $18.95 a dozen. What is the markup rate based on the selling price?

SOLUTION **A.** Find the **markup.**

Selling Price – Cost

$18.95 – $10.99 = $7.96 markup

B. Find the **markup rate** based on the selling price.

Markup ÷ Selling Price

$7.96 ÷ $18.95 = 0.420 = 42% markup rate

18.95 M+   − 10.99 = 7.96 ÷ RM 18.95 = 0.420

✔ SELF-CHECK Complete the problems, then check your answers in the back of the book.

Find the markup and the markup rate.

1. Selling price is $49.79.
Cost is $34.85.

2. Selling price is $249.19.
Cost is $161.97.

PROBLEMS

Round answers to the nearest tenth of a percent.

3. Video cassette recorder.
Cost is $124.78.
Selling price is $249.99
What is the markup rate based on the selling price?

4. PC desk.
Cost is $74.63.
Selling price is $119.79.
What is the markup rate based on the selling price?

5. Cookware.
Cost is $37.87.
Selling price is $59.79.
What is the markup rate based on the selling price?

6. Tricycle.
Cost is $65.91
Selling price is $98.25.
What is the markup rate based on the selling price?

7. Pete Kraemer is a buyer for The Sleep Store. He purchases Majestic twin-size box springs for $86.74 each. The Sleep Store sells them for $167.49 each. What is the markup for each box spring? What is the markup rate based on the selling price?

8. Rodolfo's purchases silk shirts for $18.43 each. The selling price of each shirt is $42.50. What is the markup rate based on the selling price of each shirt?

9. Automobile batteries cost Main Auto Supply $86.94. Main sells the batteries for $129.47. What is the markup rate based on the selling price?

10. Convenient Carryout purchases donuts for 90¢ a dozen. The selling price of the donuts is $1.99 per dozen. What is the markup rate based on the selling price?

11. Town Auto Sales pays $13,596, including freight, for the new model Tracker. Town Auto Sales sells each new Tracker for $16,995. What is the markup rate based on the selling price?

12. Video Rentals sells used cassettes for $24.95. The cost to Video Rentals is $17.86 per cassette. What is the markup rate based on the selling price?

13. A series L gas water heater sells for $219.29 at Carl's Appliances. Each water heater costs Carl's $152.90. What is the markup rate based on the selling price?

14. Dock 2 Imports held a special sale on wicker furniture. The selling price of each item was $99.95. What was the markup rate based on the selling price for each item?

ITEM NUMBER	DESCRIPTION	COST
WT145	Table	$65.49
WCR74	Chair	58.43
WCE04	Chaise	74.97

MAINTAINING YOUR SKILLS   Look up the skills in parentheses if you need help or more practice.

Subtract. **(Skill 6)**

15. $25.45 − $19.32

16. $79.49 − $63.59

17. $119.29 − $83.50

18. $272.21 − $163.33

19. $2618.45 − $1309.23

20. $24,749 − $18,066.71

Find the rate. Round answers to the nearest tenth of a percent. **(Skill 31)**

21. $6.13 ÷ $25.45

22. $35.79 ÷ $119.29

23. $1309.22 ÷ $1309.23

24. $216.45 ÷ $541.13

25. $4176.48 ÷ $8352.96

26. $10,874 ÷ $16,311

15-3

Net Profit

OBJECTIVE
Compute the net profit in dollars.

The markup on the products that you sell must cover your overhead or operating expenses. These expenses include wages and salaries of employees, rent, utility charges, and taxes.

The exact overhead expense on each item sold is difficult to determine accurately. You may approximate the overhead expense of each item. For example, if your total overhead expenses are 40% of total sales, you may estimate the overhead expense of each item sold to be 40% of its selling price. When the markup of an item is greater than its overhead expense, you make a net profit on the item.

Overhead = Selling Price × Overhead Percent
Net Profit = Markup − Overhead

EXAMPLE   *Skills* 30, 2   *Application* A   *Term* Net profit

Jaworski Sport Shop purchases 15-foot rowboats for $44.98 each. Jaworski sells the boats for $89.99 each. Lou Jaworski estimates his overhead expenses to be 40% of the selling price of his merchandise. What is the net profit on each 15-foot rowboat?

SOLUTION

A. Find the **markup.**
   Selling Price − Cost
   $89.99 − $44.98 = $45.01 markup

B. Find the **overhead.**
   Selling Price × Overhead Percent
   $89.99 × 40% = $35.996 = $36.00 overhead

C. Find the **net profit.**
   Markup − Overhead
   $45.01 − $36.00 = $9.01 net profit

✔ SELF-CHECK   Complete the problems, then check your answers in the back of the book.

Find the markup, overhead, and net profit.

1. Selling price is $140.
   Cost is $56.
   Overhead is 50% of selling price.

2. Selling price is $964.
   Cost is $578.40.
   Overhead is 35% of selling price.

PROBLEMS

	Markup	−	(Selling Price	×	Overhead Percent)	=	Net Profit
3.	$ 20	−	($ 40	×	30%)	=	?
4.	$108	−	($180	×	40%)	=	?
5.	$ 81	−	($270	×	20%)	=	?

Estimate the net profit based on the estimated overhead.

6. Power rake costs $49.74.
Selling price is $98.79.
Overhead is $19.64.
What is the net profit?

7. Light fixture costs $31.78.
Selling price is $89.79.
Overhead is $24.56.
What is the net profit?

8. Sweater costs $47.84.
Selling price is $119.49.
Overhead is 30% of selling price.
What is the net profit?

9. Pen/pencil set costs $41.89.
Selling price is $74.74.
Overhead is 35% of selling price.
What is the net profit?

10. Scroll Hardware Company purchases Allegré door hardware sets for $29.86 each. Scroll Hardware charges customers $49.99 to have the hardware installed. The labor and other overhead total $11.75 for each set installed. What is the net profit per set?

11. Storm King, Inc., purchases storm doors at a cost of $41.74 each. The selling price of an installed storm door is $119.50. The overhead for the storm door is estimated to be 40% of the selling price. What is the net profit?

12. Mary Callas is a buyer for Classic Shoes. She purchased some shoes for a cost of $13.89 a pair. Her store sells them for $29.99 a pair. Operating expenses are estimated to be 30% of the selling price. What is the net profit per pair of shoes?

13. Custom Comp, assemblers of specialized central processing units, delivered a system to Universal Delivery, Inc., for a selling price of $1,078,450. The cost of the unit to Custom Comp was $416,785. The system design, programming, and other overhead expenses were estimated to be 38.5% of the selling price. What net profit did Custom Comp make on the sale?

14. Thomas Gibson manages Craft Cave. He buys E-Z Trim knives at $11.28 a dozen. He sells them at $3.38 a pair. The overhead is estimated to be 25% of the selling price. What is the net profit per pair?

Round answers to the nearest cent. **(Skill 2)**

15. $1.111

16. $4.545

17. $0.4581

18. $9.8949

19. $7.6661

20. $96.4125

Subtract. **(Skill 6)**

21. $71.86 − $45.47

22. $9.78 − $9.43

23. $47.81 − $21.93

Find the percentage. Round answers to the nearest cent. **(Skill 30)**

24. $160 × 30%

25. $47.79 × 35%

26. $178.88 × 25%

27. $219.79 × 37.5%

28. $1789.49 × 40%

29. $18,749.45 × 31.4%

15-4

Net-Profit Rate

OBJECTIVE
Compute the net profit as a percent of the selling price.

You may want to know the **net-profit rate** of an item your business sells. The net-profit rate is net profit expressed as a percent of the selling price of the item.

$$\text{Net-Profit Rate} = \frac{\text{Net Profit}}{\text{Selling Price}}$$

EXAMPLE   *Skills* 6, 30, 31   *Application* A   *Term* Net-profit rate

East Imports sells a wok for $49.95. The cost of the wok to East Imports is $23.74. East Imports estimates the overhead expenses on the wok to be 30% of its selling price. What is the net-profit rate based on the selling price of the wok?

SOLUTION

A. Find the **net profit**.

Markup	–	Overhead		
($49.95 – $23.74)	–	($49.95 × 30%)		
$26.21	–	$14.99	=	$11.22 net profit

B. Find the **net-profit rate**.

Net Profit	÷	Selling Price		
$11.22	÷	$49.95	= 0.2246 =	22.5% net-profit rate

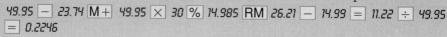

49.95 − 23.74 M+ 49.95 × 30 % 14.985 RM 26.21 − 14.99 = 11.22 ÷ 49.95 = 0.2246

✔ SELF-CHECK   Complete the problems, then check your answers in the back of the book.

Find the markup, overhead, net profit, and net-profit rate.

1. Selling price is $119.50.
Cost is $44.78.
Overhead expense is 40% of selling price.

2. Selling price is $224.50.
Cost is $102.60.
Overhead expense is 35% of selling price.

PROBLEMS

Round answers to the nearest percent.

	Markup –	(Selling Price ×	Overhead Percent) =	Net Profit	Net-Profit Rate
3.	$ 40.00 –	($ 50.00 ×	30%) =	?	?
4.	$ 35.00 –	($ 70.00 ×	25%) =	?	?
5.	$ 81.47 –	($ 149.70 ×	35%) =	?	?
6.	$ 191.75 –	($ 449.49 ×	20%) =	?	?
7.	$7366.92 –	($8475.75 ×	40%) =	?	?

Round answers to the nearest tenth of a percent.

8. Selling price is $22.49.
Markup is $11.45.
Overhead is $5.47.
What is the net-profit rate?

9. Selling price is $9.98.
Markup is $5.31.
Overhead is $2.47.
What is the net-profit rate?

10. Snow sled costs $21.74.
Selling price is $47.50.
Overhead is 40% of selling price.
What is the net-profit rate?

11. End table costs $55.67.
Selling price is $129.99.
Overhead is 15% of selling price.
What is the net-profit rate?

12. Court-Time, Inc., sells these
tennis rackets. The cost to
Court-Time is $37.48 per racket.
Overhead is estimated to be
20% of the selling price. What
is the net-profit rate?

Clean up with savings!
Tennis rackets $64.99
at Court-Time, Inc.

13. Value Discount Shops sell G.S. sweaters for $27.79 each. The markup
on each sweater is $12.85. Overhead is estimated to be 30% of the sell-
ing price. What is the net-profit rate?

14. Dixie Deli purchases apple butter for $1.03 a jar. It sells the apple
butter for $1.79 a jar. Dixie Deli estimates the overhead at 25% of
the selling price. What is the net-profit rate on each jar?

15. Elaine Luttrell manufactures hunting decoys. She pays $4.79 for the
parts for each decoy. Elaine assembles the decoys and sells them to
sporting goods stores for $14.85. She estimates the overhead to be
37.5% of the selling price of the decoys. What is the net-profit rate
per decoy?

16. The Sevas Die Shop purchases steel bars for $9.85 per bar. An automat-
ic machine cuts and forms a bar into 30 steel pins. The 30 pins are
packaged individually and marked to sell at 79¢ each. Overhead
expenses are estimated to be 12.5% of the selling price per pin. What is
the net-profit rate per pin?

17. Video Show, Inc., buys video cassettes for $52.56 a carton. A carton
contains 24 cassettes. Video Show sells the cassettes for $4.95 each or
$109.99 per carton. Overhead is estimated to be 32% of selling price.
a. What is the net-profit rate on the sale of 1 cassette?
b. What is the net-profit rate when a carton of cassettes is sold?

MAINTAINING YOUR SKILLS   Look up the skills in parentheses if you need help or more practice.

Find the percentage. Round answers to the nearest cent. **(Skill 30)**

18. $49.49 × 30%   **19.** $9.78 × 20%   **20.** $134.49 × 25%

21. $749.79 × 15%   **22.** $1278.58 × 12.5%   **23.** $8645.15 × 37.5%

Find the rate. Round answers to the nearest tenth of a percent. **(Skill 31)**

24. $5.86 ÷ $19.79   **25.** $41.16 ÷ $99.49   **26.** $67.50 ÷ $449.99

Determining Selling Price—
Markup Based on Selling Price

OBJECTIVE

Compute the selling price of an item based on cost and markup rate.

You can use records of past sales and expenses to plan the markup rate needed to cover overhead expenses and yield a profit. You can use the cost of an item and the desired markup rate based on selling price to figure the best selling price.

$$\text{Selling Price} = \frac{\text{Cost}}{\text{Complement of Markup Rate}}$$

EXAMPLE *Skills* 28, 11   *Application* L   *Term* Markup rate

VJ's Sport Store knows from past expense records that it must aim for a markup that is 40% of the selling price of its merchandise. The store received a shipment of running shoes at a cost of $38.99 per pair. What is the minimum selling price that the store should charge?

		% of Total Sales
Sales for Month:	$42,000	Sales
Cost of goods sold:	25,200	60%
Overhead expenses:	8,400	20%
Profit:	8,400	20%
	$42,000	100%

SOLUTION

A. Find the **complement of markup rate.**
   100% − 40% = 60% complement of markup rate

B. Find the **selling price.**
   Cost ÷ Complement of Markup Rate
   $38.99 ÷   60%   = $64.983 = $64.98 selling price

✔ SELF-CHECK   Complete the problems, then check your answers in the back of the book.

Find the selling price.

1. Cost is $345.
   Markup is 25% of selling price.

2. Cost is $1.40.
   Markup is 30% of selling price.

PROBLEMS

	3.	4.	5.	6.	7.	8.
Cost	$6.50	$120.00	$86.74	$420.00	$47.84	$212.60
Markup Rate (Selling Price Base)	35%	20%	50%	45%	37.5%	12.5%
Complement of Markup Rate	65%	?	?	?	?	?
Selling Price (Nearest Cent)	?	?	?	?	?	?

9. Gold chain.
   Cost is $7.48.
   Markup is 60% of selling price.
   What is the selling price?

10. Stemware.
    Cost is $5.16.
    Markup is 40% of selling price.
    What is the selling price?

11. Paperback Paupers marks up paperbacks 35% of their selling price. A paperback costs Paupers $2.57. What is the selling price of the book?

12. April's Shower purchases curtains from the manufacturer and marks them up 55% of the selling price. April's pays $8.10 per pair for a lined pair. What is the selling price for the curtain?

F.Y.I.
A markup of 50% of cost is the same as a markup of $33\frac{1}{3}$% of selling price.

13. Wade Charles prices items at BarBells, Inc. BarBells has a storewide policy to mark up each item 53% of the selling price. What selling price does he place on these items?

ITEM	COST
Adjustable bench	$23.50
Weight rod	9.40
Weight belt	11.75

14. Hillary Lee manages the craft department of Super Craft, Inc. What selling price will she place on each of these items?

LINE	ITEM CODE	DESCRIPTION	UNIT	STORE COST	MARKUP RATE SELLING PRICE BASE
14	063614	TAPE MASKING 300 IN ROLL	RL	0.545	55 %
15	071358	PROTRACTOR W/RULER PKG	EA	0.105	50%
16	031615	TWINE COTTON 4-PLY 430 FT	EA	0.302	57.5%
17	062824	CHALK WHITE 12-STICK PKG	PK	0.149	60%
18	120057	MODEL CEMENT PLASTIC 1 OZ	EA	0.218	48.2%

15. **A Brief Case** Suppose Hillary Lee in problem 14 decided to mark up all the items 55% of the selling price. If you purchase 1 of each item, how much more or less do you pay?

MAINTAINING YOUR SKILLS Look up the skills in parentheses if you need help or more practice.

Find the complement. **(Application L)**

16. 35%   17. 20%   18. 50%   19. 40%   20. 37.5%   21. 47.6%

Divide. Round answers to the nearest cent. **(Skill 11)**

22. $23.00 ÷ 65%   23. $48.00 ÷ 80%   24. $117.40 ÷ 50%

25. $471.83 ÷ 60%   26. $978.56 ÷ 62.5%   27. $1748.72 ÷ 52.4%

28. $7.53 ÷ 64.7%   29. $71.46 ÷ 37.8%   30. $12,317.87 ÷ 41.3%

15-6

Markup Rate Based on Cost

OBJECTIVE
Compute the markup rate based on cost.

Your business may use the cost of a product as the base for the markup rate. The markup rate in supermarkets is relatively low. For example, milk may be marked up 5% of the cost and other dairy products 20% to 30% of the cost. The markup of clothing may be from 80% to 140% of the cost. The markup of jewelry may be 100% or more of the cost.

$$\text{Markup Rate} = \frac{\text{Markup}}{\text{Cost}}$$

EXAMPLE *Skills* 6, 31 *Application* A *Term* Markup

Dinettes Inc. purchases a 5-piece dinette set for $180 from the manufacturer. It sells the 5-piece set for $288. What is the markup rate based on cost?

SOLUTION

A. Find the **markup.**
Selling Price − Cost
   $288.00 − $180.00 = 108.00 markup

B. Find the **markup rate** based on cost.
Markup ÷ Cost
   $108.00 ÷ $180.00 = 0.6 = 60% markup rate based on cost

$$288 - 180 = 108 \div 180 = 0.6$$

✔ SELF-CHECK Complete the problems, then check your answers in the back of the book.

Find the markup rate based on cost.

1. Selling price is $97.50.
Cost is $58.50.

2. Selling price is $148.
Cost is $74.

PROBLEMS

Round answers to the nearest tenth of a percent.

	(Selling Price	− Cost	= Markup)	÷ Cost	= Markup Rate
3.	($ 1.75	− $ 1.25 =	?)	÷ $ 1.25	= ?
4.	($ 50.00	− $ 20.00 =	?)	÷ $ 20.00	= ?
5.	($ 85.50	− $ 47.50 =	?)	÷ $ 47.50	= ?
6.	($791.00	− $316.40 =	?)	÷ $316.40	= ?
7.	($ 1.04	− $ 0.13 =	?)	÷ $ 0.13	= ?

8. Selling price is $226.64. Cost is $141.65. What is the markup rate based on cost?

9. Selling price is $18.65 each. Cost is $7.46 each. What is the markup rate based on cost?

10. The Teen Jeans Co. buys 805 jeans at $11.99 per pair. They mark up each pair $7.99. What is the markup rate based on cost?

11. Quentin Clark, a salesperson at Doolittle's Department Store, sold a pair of golf shoes for $84.95. The shoes cost Doolittle's $27.84 a pair. What is the markup rate based on cost?

12. Trudi Alvarez is a buyer for the Small Tots Shoppe. She purchased an assortment of sunsuits for $2.74 each. They were marked to sell for $5.48 each. What is the markup rate based on cost?

13. Craig Nielson works at his parents' outdoor market. He sells tomatoes at $1.19 a pound. The tomatoes cost $0.39 a pound. What is the markup rate based on cost?

14. Martha Muntz is in charge of newspapers and magazines at Cramer's News. What is the markup rate based on cost for each item?

ITEM	COST	SELLING PRICE
Local Sunday paper	$1.50	$2.50
Map of city	0.50	1.25
Out-of-state Sunday paper	2.50	6.50
Antique magazine	1.25	2.75

15. Discount Electronics, sells quality stereo components. The following list shows the cost and selling price of various components. Find the markup rate based on cost for each item.

ITEM	COST	SELLING PRICE
P636 Receiver	$167.96	$419.89
P552 CD player	224.69	374.49
P9191 Cassette deck	199.99	399.99
P88A Speakers (pair)	286.72	819.19

MAINTAINING YOUR SKILLS   Look up the skills in parentheses if you need help or more practice.

Subtract. **(Skill 6)**

16. $74.80 − $37.40

17. $149.49 − $59.80

18. $19.19 − $12.38

19. $274.49 − $164.69

20. $4738.50 − $3174.74

21. $25,789.49 − $20,631.59

Find the rate. Round answers to the nearest tenth of a percent. **(Skill 31)**

22. $89.69 ÷ $59.80

23. $109.80 ÷ $164.69

24. $5157.90 ÷ $20,631.59

Determining Selling Price—Markup Based on Cost

OBJECTIVE

Compute the selling price based on cost and markup rate.

You can use the cost of an item and the desired markup rate based on cost to figure the selling price of an item.

Markup = Cost × Markup Rate

Selling Price = Cost + Markup

EXAMPLE   *Skills* 30, 5   *Application* A   *Term* Markup rate

Wholesale Jewelers, Inc., sells electronic digital watches to jewelry stores for $18.45 each. Wholesale Jewelers calculates the suggested retail price and attaches it to each watch. The retail price is computed by marking up the cost to the jewelry store by 160%. What is the suggested retail selling price?

SOLUTION

A. Find the **markup.**
   Cost   ×   Markup Rate
   $18.45 ×   160%   = $29.52 markup

B. Find the **selling price.**
   Cost   +   Markup
   $18.45 + $29.52 = $47.97 selling price

18.45 ☒ 160 % 25.92 18.45 ☐ 25.92 ☐ 47.97

✔ SELF-CHECK   Complete the problems, then check your answers in the back of the book.

Find the markup and the selling price.

1. Cost is $50.
Markup is 70% of cost.

2. Cost is $140.
Markup is 50% of cost.

PROBLEMS

	3.	**4.**	**5.**	**6.**	**7.**	**8.**
Cost	$45.00	$96.49	$86.40	$16.40	$751.80	$14.24
Markup Rate	80%	100%	150%	225%	87.5%	400%
Markup	$36.00	$96.49	?	?	?	?
Selling Price	?	?	?	?	?	?

9. Handsaw.
Cost is $11.40.
Markup is 65% of cost.
What is the selling price?

10. Camera.
Cost is $37.48.
Markup is 180% of cost.
What is the selling price?

11. The Sports Wholesale Company purchases weight racks directly from the manufacturer for $10.80 each. The weight racks are marked up 210% of cost and sold to retail sporting good stores. What is the selling price of each weight rack?

12. The Haas Door Co. produces truck dock door seals. Their 9' × 10' seal costs $280 to produce. Markup is 75% of cost. What is the selling price?

13. XYZ Appliances buys Circle Clean washers from a distributor for $147.85 each. The markup is 100% of cost. What is the selling price?

14. Surface Combustion, Inc., calculates the cost of manufacturing a particular furnace to be $1214.78. Surface marks up each furnace 120% based on cost. What is the selling price?

15. The Pet-Agree Co. buys dog collars from a manufacturer for $0.18 each. Pet-Agree marks up each collar 356% of cost. What is the selling price?

16. The Libbey Glass Company manufacturers stemware. Cost per gross (144 items) is $66.24. The stemware is sold at a markup of 115% based on cost.
a. What is the selling price for a dozen stemware?
b. What is the selling price for a single stemware?

17. Cook-n-Serve carries fancy oven mits that cost $0.86 a pair. Cook-n-Serve operates on a markup of 166% of cost.
a. What is the selling price?
b. What is the markup as a percent of the selling price?

18. Alvin Exporters buys engine gaskets from a U.S. manufacturer for $13.32 a dozen. Alvin marks up the gaskets 124% based on cost. The overhead is estimated to be 30% of the selling price.
a. What is the selling price per gasket?
b. What is the markup as a percent of the selling price?
c. What is the net profit?
d. What is the net profit as a percent of the selling price?

MAINTAINING YOUR SKILLS   Look up the skills in parentheses if you need help or more practice.

Add. (Skill 5)

19. $85 + $44   **20.** $144.47 + $185.52   **21.** $1474.87 + $444.62

Find the percentage. Round answers to the nearest cent. (Skill 30)

22. $60.00 × 30%   **23.** $240.00 × 75%   **24.** $179.49 × 100%

25. $18.60 × 120%   **26.** $2417.60 × 150%   **27.** $4.32 × 187%

28. $48.76 × 237.5%   **29.** $112.47 × 112.5%   **30.** $0.72 × 86.4%

15-8

Markdown

OBJECTIVE

Compute the markdown in dollars and as a percent of the regular selling price.

Your business may sell some merchandise at sale prices to attract customers or to make room for new merchandise. The markdown, or discount, is the difference between the regular selling price of an item and its sale price. The markdown rate is the markdown expressed as a percent of the regular selling price of the item.

Markdown = Regular Selling Price − Sale Price

$$\text{Markdown Rate} = \frac{\text{Markdown}}{\text{Regular Selling Price}}$$

EXAMPLE Skills 31, 6 Application A Term Markdown rate

Ski's Sport Shop sells cross-country skis at a regular selling price of $98.49. For one week only, Ski's has marked down the price to $68.94. What is the markdown rate?

SOLUTION

A. Find the **markdown.**

Regular Selling Price − Sale Price
$98.49 − $68.94 = $29.55 markdown

B. Find the **markdown rate.**

Markdown ÷ Regular Selling Price
$29.55 ÷ $98.49 = 0.3000 = 30% markdown rate

98.49 − 68.94 = 29.55 ÷ 98.49 = 0.30

✔ SELF-CHECK Complete the problems, then check your answers in the back of the book.

Find the markdown and markdown rate.

1. Regular selling price is $80. Sale price is $60.

2. Regular selling price is $174.79. Sale price is $104.87.

PROBLEMS

Round answers to the nearest tenth of a percent.

	(Regular Price −	Sale Price =	Markdown) ÷	Regular Price	=	Markdown Rate
3.	($ 25.00	− $ 20.00 =	?	)÷ $ 25.00	=	?
4.	($ 99.99	− $ 64.99 =	?	)÷ $ 99.99	=	?
5.	($ 247.87	− $ 149.99 =	?	)÷ $ 247.87	=	?
6.	($8674.50	− $4337.25 =	?	)÷ $8674.50	=	?

7. Regular price is $4.49.
   Sale price is $1.29.
   What is the markdown rate?

8. Regular price is $564.79.
   Sale price is $499.99.
   What is the markdown rate?

9. Regular price is $750.
   Sale price is $500.
   What is the markdown rate?

10. Regular price is $79.29.
    Sale price is $71.29.
    What is the markdown rate?

11. A dozen eggs can be purchased for 59¢ with a coupon, 79¢ without. What is the markdown rate if the coupon is used?

12. Country Furniture has marked down this group of furniture.
    a. What is the markdown rate on each item?
    b. Based on markdown rate, which is the best buy?
    c. Based on dollar markdown, which is the best buy?

	Reg.	Sale
Sofa	$979	$489
Love seat	$749	$359
Chair	$499	$269
Tables, ea.	$219	$ 99

F.Y.I.
A markdown of 25% means you pay 75%.

13. Free-form patio blocks are marked down.
    a. Find the markdown rate for each.
    b. Find the best buy, based on dollar markdown.
    c. Find the best buy, based on percent markdown.

FREE-FORM PATIO BLOCKS

	Reg.	Sale
Outside radius	$5.45	$3.49
Border block	$4.45	$3.15
Inside radius	$4.95	$3.37
24" x 24" block	$6.45	$4.15
16" x 16" block	$5.25	$3.44

14. **A Brief Case**   The Lion Store is participating in a national promotion of Port-a-Vac vacuum cleaners. The lowest-priced vacuum cleaner is marked down $5.25 to a sale price of $64.74. The middle-priced vacuum cleaner is marked down from $129.99 to $115.99. The top-priced vacuum cleaner is marked down $30.00 from the regular price of $149.99. What is the markdown rate for each?

MAINTAINING YOUR SKILLS   Look up the skills in parentheses if you need help or more practice.

Subtract. **(Skill 6)**

15. $74.99 − $14.99   16. $7.25 − $2.15   17. $42.49 − $8.99

Find the rate. (Round to nearest tenth percent.) **(Skill 31)**

18. $15.00 ÷ $50.00   19. $4.99 ÷ $24.99   20. $5.25 ÷ $25.99

21. $71.79 ÷ $179.79   22. $15.52 ÷ $44.49   23. $9.00 ÷ $19.99

24. $124.50 ÷ $829.29   25. $269.50 ÷ $489.49

26. $5.78 ÷ $17.35   27. $16.50 ÷ $329.39

28. $186.88 ÷ $747.50   29. $8818.33 ÷ $26,455.00

Reviewing the Basics

(Skill 2)

Round answers to the nearest tenth of a percent.

1. 10.15%   **2.** 24.99%   **3.** 147.864%   **4.** 7.4996%

Solve.

(Skill 5)

5. $117.89 + $47.93   **6.** $523.47 + $67.84

7. $916.47 + $56.38   **8.** $2.16 + $4.19 + $219.29

(Skill 6)

9. $78.47 − $24.73 **10.** $217.25 − $31.45 **11.** $1417.45 − $221.84

Round answers to the nearest cent.

(Skill 11)

12. $24.50 ÷ 60%   **13.** $419.71 ÷ 55%   **14.** $947.84 ÷ 67.5%

Write as a decimal.

(Skill 28)

15. 50%   **16.** 65%   **17.** $12\frac{1}{2}$%   **18.** 47.4%   **19.** $5\frac{1}{4}$%

Round answers to the nearest cent.

(Skill 30)

20. 64% of $245.00   **21.** 120% of $667.50   **22.** 15.6% of $14.78

Round answers to the nearest tenth of a percent.

(Skill 31)

23. $7.96 is what percent of $39.79?

24. $211.89 is what percent of $847.48

Applications

(Application L)

Find the complement.

25. 40%   **26.** 65%   **27.** 2%   **28.** 23.6%

Terms

Match each term with its definition on the right.

29. Markup

30. Markdown

31. Net profit

32. Net-profit rate

33. Markdown rate

a. the difference between the regular selling price and the sale price of a product

b. the net profit expressed as a percent of the selling price of a product

c. the difference between the markup and the overhead when the markup is greater than the overhead

d. the amount that your business pays for a product

e. the discount expressed as a percent of the regular selling price of a product

f. the difference between the selling price and the cost of a product

Refer to the reference files in the back of the book if you need help.

Unit Test

1. The Deckmasters Company buys an outdoor light fixture for $27.87. The company sells the fixture for $67.89. What is the markup?

2. The Fruit Barn buys oranges for $0.33 a dozen. It sells them for $1.29 a dozen. What is the markup rate based on the selling price?

3. Pipe Outfitters, Inc., an oil equipment wholesaler, purchases piping for $217.85 a length. It sells the lengths for $447.50. Pipe Outfitters estimates the overhead expenses for each length to be 30% of the selling price. What is the estimated net profit on each length?

4. Secure-It, Inc., sells a dead bolt lock for $79.49. It costs Secure-It $41.67 to purchase the lock. The overhead is estimated to be 20% of the selling price of the lock. What is the estimated net-profit rate based on the selling price?

5. Office Supplies, Inc., operates on a markup of 60% of the selling price of its merchandise. During December, the shop bought the supply of notebooks listed on the invoice. What selling price will Office Supply charge per notebook?

SHIP TO: OFFICE SUPPLIES, INC.	
ITEM DESCRIPTION	COST
007-31 Notebooks	$1.48 ea.

6. Tia's, a shoe boutique, sells Italian-made shoes for $164.49. Tia's purchases the shoes for $97.88. To the nearest tenth of a percent, what is the markup rate based on cost?

7. During the winter, Sun Grown buys oranges from overseas growers. Sun Grown determines the retail selling price of the oranges by marking up the cost by 125%. The fruit company bought its first shipment of winter oranges for $0.48 per dozen. What is the retail selling price per dozen of this shipment?

8. Northside Auto Supply, an automobile supply shop, buys its products from Federal Auto Parts, Inc. Because of a downturn in sales, Federal has decided to reduce the price of its rebuilt generator. The regular selling price is $47.38. The generator's new selling price is $37.90. To the nearest tenth of a percent, what is the markdown rate?

A SPREADSHEET APPLICATION

Sales

To complete this spreadsheet application, you will need the diskette *Spreadsheet Applications for Business Mathematics,* which accompanies this textbook.

Select option 15, Sales, from the menu. Input the information in the following problems to find the cost (C), markup (MU), selling price (SP), percent markup based on cost (%M/C), percent markup based on selling price (%M/SP), markdown (MD), percent markdown (%MD), or sale price (SALE). Then answer the questions that follow.

	Cost	Mark-Up	Selling Price	%M/C	%M/S	%MD	MD	Sale
1.	$ 45.00	$?	$?	?	70%	20%	$?	$?
2.	124.50	?	?	?	50%	30%	?	?
3.	35.75	?	?	210%	?	25%	?	?
4.	124.50	?	?	100%	?	50%	?	?
5.	?	?	79.49	300%	?	25%	?	?
6.	?	?	145.78	85%	?	?	72.89	?
7.	?	?	12.96	?	25%	?	1.30	?
8.	?	?	4985.00	?	12%	?	997.00	?
9.	?	35.80	?	200%	?	?	?	42.96
10.	?	437.76	?	?	40%	?	?	711.36

11. What is the selling price in #1?

12. What is the sale price in #1?

13. What is the selling price in #2?

14. What is the selling price in #4?

15. The cost is the same in #2 and #4, but the markup in #2 is 50% of the selling price and in #4 it is 100% of the cost; yet the selling price is the same in #2 and #4. Why?

16. How can the markup possibly be more than 100% in #3?

17. What is the markdown in #3?

18. What is the sale price in #3?

19. What is the cost in #5?

20. What is the percent markup based on selling price in #5?

21. In #5, the percent markup based on selling price and the percent markdown are the same (25%). The sale price equals the cost ($19.87). Will this always happen? Why or why not?

22. What is the markup in #6?

23. What is the percent markdown in #6?

24. What is the sale price in #6?

25. What is the cost in #7?

26. What is the sale price in #7?

27. Why are there percents over 100% for the percent markup based on cost, but none over 100% for the percent markup based on selling price?

28. What is the percent markup based on cost in #8?

29. What is the selling price in #9?

30. What is the cost in #10?

31. What is the percent markdown in #10?

▼ CAREER WISE

$ $

Store Manager

Jamal Mizrahi is the store manager at a fast-food franchise. As owner and manager, he purchases food supplies from the parent company. According to company policy and local conditions, he marks up the foods he sells. Beverages, sandwiches, side orders, and desserts each get different markups.

At the store, Mr. Mizrahi has given many local high school teenagers their first job. They learned how to "ring up a sale," as well as about paychecks and deductions.

The table and pie chart below show how the 50 largest franchises in the United States are divided. For example, there are 3 hotel-motel chains in the top 50 franchises. This is $\frac{3}{50} = 6\%$ of the top 50 franchises.

50 LARGEST FRANCHISES	
Fast foods	15
Auto rental, maintenance, supplies	6
Cleaning services	5
Printing/mailing	5
Personal fitness	4
Real estate	4
Hotels-motels	3
Convenience stores	2
Haircutters	1
Drug stores	1
Electronics	1
Hardware	1
Home decorating	1
Tax preparation	1

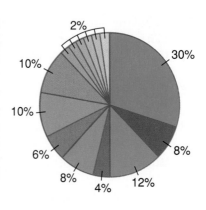

1. What percent of the pie is represented by the fast-food industry?

2. The 10% slices correspond to what two industries?

3. The 2% slices of the pie together equal the slice of which one industry?

4. True or false: From the data, you can tell how many tax preparation franchises there are. Discuss your answer.

16

Marketing

The marketing department of your business estimates the number, the price, and the amount of sales you can expect to make on your product. This *market share* can be based on *opinion* *surveys* of people who buy your product and *sales projections,* or estimates of future sales. The marketing department also handles the advertising costs for your product.

Sales projections have an impact on a company's plans for growth.

Opinion Surveys

OBJECTIVE

Compute the rate of a particular response in an opinion survey.

When you develop a new product, you will want to know how well it is likely to sell. You may conduct a product test by asking a group of people to try the product. You may hire an opinion research firm that specializes in product testing to conduct the test. Panels of volunteers try the product and respond to questions in an opinion survey. The opinion research firm tabulates the answers to the questions and submits tables of results to you.

$$\text{Percent of Particular Response} = \frac{\text{Number of Times Particular Response Occurs}}{\text{Total Number of Responses}}$$

EXAMPLE   *Skill* 31   *Application* C   *Term* Opinion survey

Countryside Cereal Company conducted an opinion survey of 2600 people for its new Good Morning Cereal. This table shows the responses of the 2600 people in the survey.

Opinion	Under 18	Age Group 18 to 40	40 or Over	Total
Excellent	200	430	1450	2080
Good	25	108	215	348
Fair	10	12	99	121
Dislike	5	10	36	51
Total	240	560	1800	2600

What percent of the total responses were for "good"?

SOLUTION Find the **percent of particular response.**

Number of Times Particular Response Occurs	÷	Total Number of Responses	
348	÷	2600	= 0.133 = 13%

percent of "good" responses

348 ÷ 2600 = 0.1338

✔ SELF-CHECK   Complete the problems, then check your answers in the back of the book.

1. 1500 out of 2000 people surveyed own computers. What percent of the total surveyed own computers?

2. 460 out of 500 people surveyed like Mintie toothpaste. What percent of the total surveyed like it?

	Product in Survey	Number of "Good" Responses	÷	Total Number of Responses	=	Percent of "Good" Responses
3.	Ice cream	320	÷	400	=	?
4.	Soup	75	÷	200	=	?
5.	Cleanser	600	÷	1000	=	?
6.	Movie	350	÷	1250	=	?
7.	Calculator	2047	÷	2300	=	?

8. Survey for orange juice.
375 responses.
180 "fair" responses.
What is the percent of "fair" responses?

9. Survey for silicone caulking.
525 responses.
50 negative responses.
What is the percent of negative responses?

10. Fisher Motors conducted a survey on its dealer service. The people surveyed were asked to choose one answer for this question: "If you DO NOT usually go to the dealer from whom you bought your car, why?"

The choices and number of responses received for each were:

　37 Moved away from vicinity　　　110 Crowded service department

　14 Disliked quality of service　　　14 Location not convenient

　95 Service charges too high　　　30 Some other reason

What is the percent of each response?

11. Before beginning production of its new vacuum cleaner, the Devaney Company conducted an opinion survey of 1740 people. When asked if they would purchase the product, the people gave these responses:

	Age Group				
Response	Under 25	25–34	35–49	50 or over	Total
Definitely	53	151	126	75	?
Probably	75	184	203	140	?
Possibly	87	135	130	99	?
No	91	57	54	80	?
Total	306	527	513	394	1740

a. Find the total for each response.
b. Based on 1740 responses, what is the percent of each response?
c. What is the percent of "definitely" responses in each age group?
d. What age group is most likely to purchase the product?

Find the percentage. Round to the nearest tenth. **(Skill 30)**

12. 8% of 70　　**13.** 70% of 80　　**14.** 30% of 500　　**15.** 25% of 1200

Sales Potential

OBJECTIVE

Compute the annual sales potential of a new product.

Before your business mass produces a new product, you may try to determine the product's **sales potential.** The sales potential is an estimate of the sales volume of a product during a specified period of time. You may manufacture a small number of the product for a selected group of people to try. This group is called a **sample.** The sales potential of your new product is based on: (1) the percent of the people in the sample who would purchase your product, (2) an estimate of the size of the **market,** and (3) the average number of times that an individual might purchase this type of product during a specified period of time. The market is the total number of people who might purchase the type of product that you make.

$$\frac{\text{Annual Sales}}{\text{Potential}} = \frac{\text{Estimated}}{\text{Market Size}} \times \frac{\text{Individual Rate}}{\text{of Purchase}} \times \frac{\text{Percent of}}{\text{Potential Purchasers}}$$

EXAMPLE   *Skills* 31, 30   *Application* A   *Term* Sales potential

The Ruston Corporation has developed a sun tan cream called Lite Stuff. Ruston chose a sample of teenagers to try the new product. Of the 3000 people in the sample, 1200 said they would purchase Lite Stuff. The Ruston Corporation estimates that there are 2,000,000 teenagers that buy sun tan creams. Ruston's surveys indicate that each teenager purchases about 3 tubes of sun tan cream per year. What is the sales potential for Lite Stuff for 1 year?

SOLUTION

A. Find the **percent of potential purchasers.**
   1200 ÷ 3000 = 0.40 = 40% potential purchasers

B. Find the **annual sales potential.**

Estimated Market Size		Individual Rate of Purchase		Percent of Potential Purchasers	
2,000,000	×	3	×	40%	= 2,400,000 annual potential sales

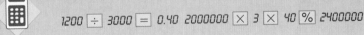

$1200 \boxed{\div} 3000 \boxed{=} 0.40 \ 2000000 \boxed{\times} 3 \boxed{\times} 40 \boxed{\%} 2400000$

✔ SELF-CHECK   Complete the problem, then check your answer in the back of the book.

1. Out of 2400 secretaries in a sample, 720 preferred a new computer printer ribbon. The estimated market size is 100,000. Each secretary uses about 30 ribbons a year. What is the annual sales potential?

	2.	3.	4.	5.	6.
Number in Sample	400	500	1500	2000	2000
Number of Potential Purchasers	100	10	300	240	50
Percent of Potential Purchasers	25%	?	?	?	?
Estimated Market Size	800,000	1,800,000	5,700,000	25,000,000	32,600,000
Individual Rate of Purchase	2 cans	once	once	4 boxes	10 issues
Annual Sales Potential	?	?	?	?	?

7. Child's car seat.
1200 people in sample.
360 would purchase seat.
Market size about 3,000,000.
Each family buys once.
What is the annual sales
potential?

8. Word-processing package.
8700 people in sample.
870 would purchase product.
Market size about 5,000,000.
Each person buys about 1 per year.
What is the annual sales
potential?

9. Glo dishwashing detergent is tested by 1000 people. Eighty-seven said they would buy it. The estimated market size is 4,500,000. The company estimates that each person would buy the detergent 12 times a year. What is the annual sales potential?

10. The National Optometrics Company is marketing a new, softer, more pliable disposable contact lens. Out of a sample of 1200 users, 150 preferred the new lenses. There is an estimated total market of 1,500,000 contact lens users. The average contact lens wearer would purchase 12 lenses per year. What is the sales potential for the new lenses for 1 year?

11. The Automotive Institute has developed a new spark plug for automobiles that will increase gas mileage. The marketing department had a sample of 2000 automobile owners use the new spark plug. Sixty owners responded favorably and stated that they would buy the spark plug. There are approximately 15,000,000 automobiles that could use this special spark plug. An average number of 4 spark plugs would be purchased per automobile per year. What is the sales potential for the new spark plug for 1 year?

MAINTAINING YOUR SKILLS Look up the skill in parentheses if you need help or more practice.

Find the percentage. Round answers to the nearest hundredth. **(Skill 30)**

12. $4\frac{1}{4}\%$ of 96

13. $\frac{3}{8}\%$ of 120

14. 125% of $4140

15. 15.5% of 74.3

16-3

Market Share

OBJECTIVE

Compute the market share of a new product.

To find out how well your product is selling in the marketplace, you may want to find your product's market share. Market share is that percent of the total market that purchases your product instead of a competitor's. You can calculate your percent of the total market sales by using either the number of units sold or the dollar value of sales.

$$\text{Market Share} = \frac{\text{Total Product Sales}}{\text{Total Market Sales}}$$

EXAMPLE Skill 31 Application A Term Market share

Amdex, an air conditioner manufacturer, sold 1,200,000 air conditioners during the year. During the same period, a total of 8,000,000 air conditioners were purchased in the entire U.S. market. What was Amdex's market share for the year?

SOLUTION Find the **market share.**

Total Product Sales ÷ Total Market Sales

1,200,000 ÷ 8,000,000 = 0.15 = 15% market share

✔ SELF-CHECK Complete the problems, then check your answers in the back of the book.

Find the market share.

1. Product sales total 2,000,000.
   Market sales total 20,000,000.

2. Product sales total $4,000,000.
   Market sales total $180,000,000.

PROBLEMS

	Company	Total Product Sales	÷	Total Market Sales	=	Market Share
3.	Diwan	$8,000,000	÷	$20,000,000	=	?
4.	Morrill	$ 700,000	÷	$42,000,000	=	?
5.	Hersh	$1,400,000	÷	$ 5,000,000	=	?
6.	TBC	$9,000,000	÷	$45,000,000	=	?

7. Typewriters.
   Product sales total
   1,200,000 units.
   Market sales total
   20,000,000 units.
   What is the market share?

8. Lawn mowers.
   Product sales total
   750,000 units.
   Market sales total
   5,000,000 units.
   What is the market share?

9. Light bulbs.
Product sales total $8,000,000.
Market sales total $80,000,000.
What is the market share?

10. Carpets.
Product sales total $30,000,000.
Market sales total $900,000,000.
What is the market share?

11. Radial Tires, Inc., sells approximately 4,000,000 automobile tires per year. There are approximately 44,000,000 automobile tires sold per year in the entire market. What is Radial Tires' market share?

12. Tiger Paperbacks Inc. sells approximately 9,000,000 paperback books per year. Sales for the entire paperback book market total approximately 30,000,000 books per year. What is Tiger Paperbacks' market share of paperbacks?

13. The Mesquite Grill Company sells about $3,100,000 in outdoor grills annually. The total annual sales of outdoor grills in the entire market are about $8,500,000. What is Mesquite's market share for outdoor grills?

14. The Quality Linen Sheet Company has sales totaling approximately $1,400,000 in fitted sheets. Total sales of fitted sheets by all companies are approximately $5,350,000. What is Quality Linen's market share of fitted sheets?

15. Redbo Shoes had sales totaling approximately $975,000 in basketball shoes last year. Last year sales of all basketball shoes totaled approximately $12,450,000. What market share of basketball shoes did Redbo Shoes have?

16. Last year there were total sales of approximately $8,500,000 in dining room furniture. St. Helena Chair Company's sales of dining room furniture totaled approximately $900,000 last year. What was St. Helena Chair Company's market share of dining room furniture?

17. Chateau Brothers, a company that manufactures games, sells an electronic video game called Match. Last year sales for Match totaled approximately $732,000. Total market sales for electronic video games were approximately $15,350,000. What market share did Chateau have?

18. Mountainside Nursery provides a landscaping service in Fresno County. Mountainside's landscaping business totaled approximately $789,400 for the year. The landscaping business for Fresno County totaled about $1,340,000 for the year. What market share did Mountainside have?

MAINTAINING YOUR SKILLS   Look up the skills in parentheses if you need help or more practice.

Find the rate. Round answers to the nearest tenth of a percent. **(Skill 31)**

19. $n\%$ of $84 = 21$

20. $n\%$ of $50 = 35$

21. $n\%$ of $45 = 50$

22. $n\%$ of $36 = 90$

Divide. Round answers to the nearest tenth. **(Skill 10)**

23. $920 \div 40$

24. $1145 \div 85$

25. $868 \div 1224$

26. $821 \div 15$

27. $546 \div 42$

Sales Projections

OBJECTIVE

Use a graph to compute projected sales.

A **sales projection** is an estimate of the dollar volume or unit sales that might occur during a future time period. A sales projection is usually based on past sales. You may use a sales projection to plan for production or purchasing. New products or changing economic conditions may result in sales figures that differ from the figures you projected.

You may use a graph to project a rough estimate of future sales:

1. Construct a graph of past sales.
2. Draw a straight line from the first year of data, approximately through the "middle" of the data, to the year for which the projection is being made.
3. Read the number or dollar value, the sales projection.

EXAMPLE   *Application* N   *Term* Sales projection

The marketing department of Stanley Stores, Inc., wishes to project sales for the year 2000. The sales history is:

Year	1986	1987	1988	1989	1990	1991	1992
Sales (in millions)	$2.5	$2.0	$3.0	$4.3	$3.3	$5.0	$4.5

Using a graph, what sales projection might the marketing department make for 2000?

SOLUTION

A. Graph the sales from 1986 through 1992.

B. Starting from sales in 1986, draw a straight line through the "middle" of the data to 2000.

C. Read the sales projection for 2000.

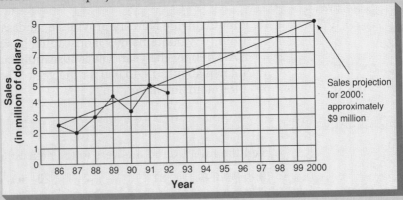

Sales projection for 2000: approximately $9 million

✔ **SELF-CHECK**   Complete the problem, then check your answer in the back of the book.

1. Refer to the graph above. Read the approximate sales projections for 1993, 1995, and 1997.

Use the graph to estimate the sales projections for problems 2–7.

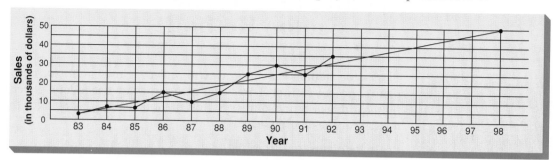

	2.	**3.**	**4.**	**5.**	**6.**	**7.**
Year	1993	1994	1995	1996	1997	1998
Sales Projection	?	?	?	?	?	?

Construct a graph for each problem and draw a straight line through the "middle" of the data to project sales.

8. Sales records for Target Sporting Goods show this data:

Year	1983	1984	1985	1986	1987	1988	1989	1990	1991	1992
Sales (in millions)	$4.4	$4.9	$5.0	$4.5	$5.1	$5.4	$5.0	$5.5	$5.7	$5.0

What sales might be projected for 1999?

9. The sales history of Bi-Low Markets shows:

Year	1965	1970	1975	1980	1985	1990
Sales (in millions)	$12	$14	$11	$12	$15	$14

Project sales for 1995, 2000, and 2005.

10. Digico manufactures computers. Production records for the first half of the year show this information for the Model XT360 computer:

Month	Jan.	Feb.	Mar.	Apr.	May	June
Production (in thousands of units)	14	13	12	13	14	13

Project production quantities for July, August, and September.

Multiply. Round answers to the nearest thousandth. **(Skill 8)**

11. $442.86
$\times$ 100

12. $3240.03
$\times$ 1000

13. 0.044
$\times$ 400

14. 0.00362
$\times$ 16.02

15. 0.725
$\times$ 143

Write each number in words. **(Skill 1)**

16. 407   **17.** 52   **18.** 0.6   **19.** 3104   **20.** 16.246

Sales Projections— Factor Method

OBJECTIVE

Use the factor method to compute projected sales.

The factor method is another way to project sales. The factor is your company's present market share. Your company may use federal government publications or other sources to find the total sales projected for the entire market for the coming year.

Projected Sales = Projected Market Sales × Market-Share Factor

---

EXAMPLE   *Skill* 30   *Application* A   *Term* Factor method

SafeAway Food Stores have a 4% share of the market for food sales in the Chicago Heights area. Food sales in the Chicago Heights area for next year are estimated to be $28,400,000. What is the projected sales figure for SafeAway Food Stores in the Chicago Heights area for next year?

SOLUTION

Find the **projected sales.**
Projected Market Sales × Market-Share Factor
    $28,400,000   ×   4%   = $1,136,000 projected sales

28400000 ⊠ × 4 % 1136000

✔ SELF-CHECK   Complete the problems, then check your answers in the back of the book.

Find the projected sales.

1. Market-share factor: 7%.
   Projected market sales: $6,300,000.

2. Market-share factor: 14%.
   Projected market sales: $10,250,000.

PROBLEMS

	Company	Projected Market Sales	×	Market Share	=	Projected Sales
3.	ABS	$10,000,000	×	4%	=	?
4.	Webb	$32,400,000	×	8%	=	?
5.	Mairs	$54,000,000	×	7.5%	=	?
6.	Beck	$70,000,000	×	3.2%	=	?

7. Long-Life Lumber Distributors.
   18% share of market.
   $1,800,000 estimated market total for next year.
   What is Long-Life's sales projection?

8. Quick Copy Machines.
   15% share of market.
   $9,100,000 estimated market total for next year.
   What is Quick Copy's sales projection?

9. Bonaventura Bakery Inc. now has 13% of the market for Watson County. The total estimated sales of the county for next year are $3,000,000. What sales volume should Bonaventura Bakery project for next year?

10. Silver Oaks Bowling Lanes has traditionally had 5% of the bowling business in the East Delta area. The total estimated bowling business in this area for next year is $14,900,000. What bowling business can Silver Oaks project for next year?

11. Elliot's Auto Repair does 24% of the auto repair business in the town of Northport. The estimated auto repair business for next year is $542,600. What sales volume can Elliot's project for next year?

12. This year enrollment at Silverado Central University (SCU) accounted for 55% of all university students in the greater metropolitan area. Next year the total university enrollment in the greater metropolitan area is expected to be about 28,000 students. What enrollment figures can SCU project for next year?

13. Based on past records, First National Savings Bank determined that it has a market share of 9.5% of all the savings accounts in Butler County. Estimated savings accounts in Butler County for next year are 470,000. Project the number of savings accounts that First National Savings Bank will have next year.

14. A Brief Case   Many retail businesses increase their market share by opening additional stores.

a. With their 4 outlets, Valley Dairy Outlets now do 12% of the dairy business in Leland County. Estimated dairy products sales in Leland County for next year are $1,826,000. Valley is adding 1 more outlet next year and hopes to increase its market share to 18%. What is Valley's projected increase in sales for all 5 outlets?

b. Eagle's 3 stores now do 15% of the bicycle sales business in North County. Estimated bicycle sales in the North County market area for next year are $2,920,000. Eagle is adding 2 stores and estimates the 2 stores will have total sales of $292,000 next year. What will be Eagle's projected market share next year?

Find the percentage. Round answers to the nearest hundredth. **(Skill 30)**

15. 5.8% of 410	**16.** 9.81% of $32,500	**17.** 4.1% of $420,000
18. 145% of 700	**19.** $\frac{5}{8}$ % of $800	**20.** 0.8% of $60,000
21. 2000% of $120	**22.** 1% of $85,000	**23.** 15% of $120

Newspaper Advertising Costs

OBJECTIVE

Compute the cost of advertising in a newspaper.

You may advertise your products or services in a newspaper to increase your sales. The cost of the advertisement depends on whether it is in color or black and white and on the amount of space it uses. The amount of space is figured in column inches. Generally, newspapers charge a certain rate for each column inch. Some newspapers offer a reduced rate if you contract weekly for a specified number of column inches.

Advertisement Cost = Number of Column Inches × Rate per Column Inch

EXAMPLE   *Skills 8, 2   Application A*

Northshore Real Estate, Inc., contracted with *Hamilton News* for 126 inches of advertising each week. Northshore plans to advertise a new subdivision in the daily newspaper. The advertisement is the equivalent of 21 column inches. What is the cost of the advertisement each time it appears?

HAMILTON NEWS ADVERTISING RATES

	Per Column Inch	
	Daily	Sunday
Noncontract rates	$45.54	$55.28
Weekly Contract rates		
1 inch	35.57	43.52
2 inches	35.42	43.36
4 inches	35.28	43.36
8 inches	35.15	43.07
16 inches	34.90	42.80
31 inches	34.65	42.53
63 inches	34.40	42.26
94 inches	34.15	41.99
126 inches	33.90	41.72
252 inches	33.65	41.46

SOLUTION   Find the **advertisement cost.**

Number of Column Inches × Rate per Column Inch

21   × $33.90 = $711.90 advertisement cost

✓ SELF-CHECK   Complete the problems, then check your answers in the back of the book.

Use the table above to find the advertisement costs for Carpet Store. Carpet Store has contracted for 94 inches per week.

1. An advertisement in the Sunday paper is 12 column inches.

2. An advertisement in the daily paper is 10 column inches.

Use the table on page 436 to find the advertisement cost per run.

	Weekly Contract	Edition	Number of Column Inches ×	Rate per Column Inch =	Advertisement Cost
3.	16 inches	Daily	15	× $34.90	= ?
4.	31 inches	Daily	9	× ?	= ?
5.	252 inches	Sunday	40	× ?	= ?
6.	94 inches	Sunday	120	× ?	= ?
7.	No Contract	Daily	5	× ?	= ?

8. MJ Sporting Goods.
Weekly contract for 63 inches.
26-inch ad on Sunday.
What is the cost of the
advertisement?

9. Maxson Auto Sales.
Weekly contract for 31 inches.
8-inch daily ad.
What is the cost of the
advertisement?

10. SuperValue Supermarket has a weekly contract for 126 inches of advertising. In the Wednesday paper, SuperValue has an advertisement
equivalent to 105 column inches. What is the cost of the advertisement?

11. The Aspen Ski Lodge does not have a contract for advertising. Sunday's
paper will carry an advertisement for the lodge of 12 column inches.
How much does the advertisement cost?

12. A Brief Case Jo-Del
Corporation placed this
advertisement in the
Monday *Hamilton News.*
The advertisement is
2 columns wide and
2 inches deep.
a. For how many
column inches
will Jo-Del be
charged?
Columns
Wide × Inches
Deep = ?

BATHTUB REFINISHING

Bathtub worn out?
Hard to clean?
Rough and pitted?
Let JO-DEL restore
it to its original
beauty right on the
premises.

Any Color
All Work Guaranteed

JO-DEL Corporation
1430 Broadway
Easton

VISIT our display room
Hours: 9:00–5:00

b. Jo-Del does not have a contract. What will the advertisement cost?

Multiply. Round answers to the nearest thousandth. **(Skill 8)**

13. 4.9 × 1.8

14. 0.76 × 0.12

15. 0.003 × 10.6

16. 63.05 × 0.007

17. 0.982 × 0.15

18. 0.14 × 300

19. 66.00 × 0.25

20. 0.3498 × 25

16-7

Television Advertising Costs

OBJECTIVE

Compute the cost of advertising on television.

You may advertise your products or services on television. The cost of a television advertisement depends on the time of day, program ratings, and the length of the advertisement. Television commercials are generally 10, 30, or 60 seconds long. The cost of a 10-second advertisement is usually one-half the cost of a 30-second advertisement. The cost of a 60-second advertisement is usually twice the cost of a 30-second advertisement.

$$\text{Cost of 10-Second Ad} = \frac{1}{2} \times \text{Cost of 30-Second Ad}$$

$$\text{Cost of 60-Second Ad} = 2 \times \text{Cost of 30-Second Ad}$$

EXAMPLE Skill 20 Application A

The marketing department of the Mcphee Company plans to advertise several new cosmetics on network television. The advertising campaign calls for these commercials:

Number	Length	Time
3	30-second	Daytime
2	10-second	Daytime
2	30-second	Prime time
4	60-second	Prime time

The rates are $8400 for a 30-second daytime commercial and $38,000 for a 30-second prime-time commercial. What is the total cost for this advertising campaign?

SOLUTION

A. Find the **cost of 30-second ads.**
Daytime: $8400 × 3 .... $ 25,200
Prime time: $38,000 × 2 .... $ 76,000

B. Find the **cost of 10-second ads.**
$\frac{1}{2}$ × Cost of 30-Second Ad
Daytime: ($\frac{1}{2}$ × $8400) × 2 .... $ 8400

C. Find the **cost of 60-second ads.**
2 × Cost of 30-Second Ad
Prime time: (2 × $38,000) × 4 .... $304,000

D. Find the **sum of the costs.** .... $ 413,600

8400 ×3 = 25200 M+ 38000 × 2 = 76000 M+ .5 × 8400 × 2 = 8400
M+ 2 × 38000 × 4 = 304000 M+ RM 413600

1. A 30-second ad costs $1200. Find the total cost for four 60-second ads.

2. A 30-second ad costs $35,000. Find the total cost for four 30-second ads and five 10-second ads.

PROBLEMS

	Rate per 30-Second Ad	Number of 10-Second Ads	Number of 30-Second Ads	Number of 60-Second Ads	Total Cost
3.	$ 1,000	4	3	3	?
4.	$ 500	3	4	0	?
5.	$30,000	5	5	2	?
6.	$ 6,500	0	4	3	?

F.Y.I.
Television network affiliated stations' average earnings amounted to $14.6 million in 1991.

7. Daytime local television. 60-second ad for dog food. 30-second ad costs $400. What is the cost of the 60-second ad?

8. Prime-time national television. 10-second ad for radio station. 30-second ad costs $21,400. What is the cost of the 10-second ad?

9. The manufacturers of Brighten mouthwash have planned an advertising campaign that includes these commercials:

Number	Length	Time
4	30-second	Daytime
3	10-second	Daytime
5	30-second	Prime time
6	60-second	Prime time

F.Y.I.
A 60-second ad for a recent Super Bowl game cost approximately $1 million.

The rates are $2800 for a 30-second daytime commercial and $13,500 for a 30-second prime-time commercial. What is the cost of the advertising campaign?

10. Pacafic New Car Sales is interested in sponsoring the sports coverage on the local television station. The rate per 30-second commercial is $475. Pacafic New Car Sales contracts for ten 30-second advertisements, five 10-second advertisements, and five 60-second advertisements. What is the total cost for these advertisements?

11. Alamo Amusement Park has planned a special television campaign. It will use a daytime show for twenty 10-second ads and five 30-second ads. It will use a prime-time show for ten 30-seconds ads and five 60-second ads. The rates are $250 per 30-second daytime ad and $1500 per 30-second prime-time ad. What is the cost of the television campaign?

MAINTAINING YOUR SKILLS   Look up the skill in parentheses if you need help or more practice.

Multiply. Express the answers in lowest terms. **(Skill 20)**

12. $\frac{1}{3} \times \frac{3}{5}$

13. $\frac{3}{12} \times \frac{4}{8}$

14. $6 \times 8\frac{1}{7}$

15. $2\frac{3}{7} \times 1\frac{3}{8}$

16. $4 \times 16\frac{1}{4}$

Pricing

You must set the selling prices of your products high enough to cover expenses and to make a profit. You may determine the selling price of an item by estimating the net profit for each of several possible selling prices and choosing the selling price that will result in the highest profit. In manufacturing, the cost per item varies with the number manufactured and ultimately sold. Often, the higher the selling price of an item, the fewer the items that will be sold.

$$\begin{matrix}\text{Possible} \\ \text{Net Profit}\end{matrix} = \left(\begin{matrix}\text{Selling Price} \\ \text{per Unit}\end{matrix} - \begin{matrix}\text{Total Cost} \\ \text{per unit}\end{matrix}\right) \times \begin{matrix}\text{Estimated} \\ \text{Unit Sales}\end{matrix}$$

EXAMPLE   *Skills* 6, 8   *Application* A   *Term* Net profit

Gamma Electronics manufactures preprinted circuit boards. Gamma has a fixed overhead of $120,000. The variable costs to produce the circuit boards are $1.50 per unit. To determine the best selling price, Gamma estimates the number of units that could be sold at various selling prices. What selling price will maximize Gamma's profits?

| | | | Fixed Costs / Estimated Unit Sales | Fixed Costs per Unit + Variable Costs per Unit | |
Selling Price per Unit	Estimated Unit Sales	Fixed Costs	Fixed Costs per Unit	Variable Costs per Unit	Total Costs per Unit
$ 8.00	20,000	$120,000	$ 6.00	$1.50	$ 7.50
11.00	15,000	120,000	8.00	1.50	9.50
15.00	10,000	120,000	12.00	1.50	13.50
22.00	6,000	120,000	20.00	1.50	21.50

SOLUTION   **A.** Find the **possible net profit** for each selling price.

	$\left(\begin{matrix}\text{Selling Price} \\ \text{per Unit}\end{matrix}\right.$		$\left.\begin{matrix}\text{Total Cost} \\ \text{per Unit}\end{matrix}\right)$	$\times$	$\begin{matrix}\text{Estimated} \\ \text{Unit Sales}\end{matrix}$		
(1)	($ 8.00	−	$ 7.50)	×	20,000	=	$10,000
(2)	($11.00	−	$ 9.50)	×	15,000	=	$22,500
(3)	($15.00	−	$13.50)	×	10,000	=	$15,000
(4)	($22.00	−	$21.50)	×	6000	=	$ 3000

B. Find the **greatest possible profit.** $22,500

C. Find the **best selling price.** $11.00

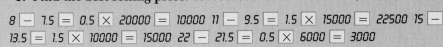

8 − 7.5 = 0.5 × 20000 = 10000   11 − 9.5 = 1.5 × 15000 = 22500   15 −
13.5 = 1.5 × 10000 = 15000   22 − 21.5 = 0.5 × 6000 = 3000

Complete the problems, then check your answers in the back of the book.

Selling Price per Unit	Estimated Unit Sales	Fixed Costs	Fixed Costs per Unit	Variable Costs per Unit	Total Cost per Unit
$3.00	200,000	$200,000	$1.00	$0.05	$1.05
$4.50	150,000	$200,000	$1.33	$0.05	$1.38

1. Find the possible net profit for each selling price.

2. Which selling price yields the greatest possible profit?

PROBLEMS

Round answers to the nearest cent.

	Selling Price per Unit	Estimated Unit Sales	Fixed Costs	Fixed Costs per Unit	Variable Costs per Unit	Total Cost per Unit	Possible Net Profit
3.	$45.00	10,000	$120,000	$12.00	$21.00	$33.00	?
4.	$65.50	8000	$120,000	$15.00	$21.00	$36.00	?
5.	$75.00	7000	$120,000	?	$21.00	?	?
6.	$85.00	6000	$120,000	?	$21.00	?	?

7. Complete the table for the Wood Specialties Company. Which selling price yields the greatest possible profit? How many units should be produced?

Selling Price per Unit	Estimated Unit Sales	Fixed Costs	Fixed Costs per Unit	Variable Costs per Unit	Total Cost per Unit	Possible Net Profit
$ 3.75	950,000	$500,000	?	$1.05	?	?
$ 4.50	700,000	$500,000	?	$1.05	?	?
$ 8.50	300,000	$500,000	?	$1.05	?	?
$10.25	250,000	$500,000	?	$1.05	?	?

8. Software, Inc., developed these cost and price figures for a new video game. Which combination should the company choose to handle the smallest number of items and to exceed $500,000 in possible profit?

Selling Price per Unit	Estimated Unit Sales	Fixed Costs	Fixed Costs per Unit	Variable Costs per Unit	Total Cost per Unit	Possible Net Profit
$19.95	70,000	$500,000	?	$11.75	?	?
$29.95	60,000	$500,000	?	$11.75	?	?
$39.95	50,000	$500,000	?	$11.75	?	?
$49.95	40,000	$500,000	?	$11.75	?	?
$59.95	30,000	$500,000	?	$11.75	?	?

MAINTAINING YOUR SKILLS

Look up the skill in parentheses if you need help or more practice.

Subtract. **(Skill 6)**

9. 649.91 − 73.89 **10.** 12.9 − 1.023 **11.** 78 − 32.8 **12.** 34.12 − 31

Reviewing the Basics

Skills

(Skill 6)

Solve.

1. $9.00 − $4.25

2. $99.40 − $34.17

3. $419.70 − $332

(Skill 8)

4. 442 × $0.895

5. $8.74 × 16,000

6. 2.1 × 96.8

(Skill 6)

7. 275.75 − 45.4

8. 36.2 − 0.138

9. 2.5 − 1.95

(Skill 20)

10. $\frac{1}{2}$ × $471

11. $\frac{1}{2}$ × $18,483

12. $\frac{1}{4}$ × $26,360

(Skill 30)

13. 35% of 19,840

14. 21% of $16,820,000

Applications

(Skill 31)

Round answers to the nearest tenth of a percent.

15. 1122 is what percent of 3670? **16.** $27,320 is what percent of $83,580?

(Application C)

17. What is the total of the responses? "Very interested" is what percent of the total responses?

Opinion	City	Suburbs	Rural	Total
Very interested	210	1521	2040	?
Interested	515	1235	1326	?
Slightly interested	960	752	1158	?
Not interested	340	430	281	?
Total	?	?	?	?

(Application N)

18. What production level is projected for Lime Rock Mineral Company for 1996?

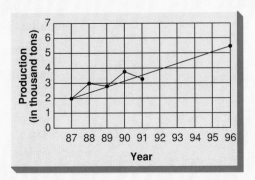

Terms

Write your own definition for each term.

19. Opinion survey

20. Sales potential

21. Market share

22. Sales projection

23. Factor method

24. Net profit

Refer to your reference files in the back of the book if you need help.

Unit Test

Lesson 16-1

1. Willshire High School conducted an opinion survey of the student council's performance. Of the 500 students surveyed, 90 gave an "average" response. What percent of the total gave "average" responses?

Lesson 16-2

2. Of a sample of 800 swimming pool owners, 60 said they would buy Leisure Living's new chemical for swimming pools. There is a market of about 8,000,000 potential users. Each user would purchase the chemical twice per year. What is the sales potential for the chemical for 1 year?

Lesson 16-3

3. Last year Madsen Wheel Works sold 736,000 wagons. Approximately 1,330,000 wagons were sold in the entire market last year. What was Madsen's market share?

Lesson 16-4

4. Use a graph to project Suntyme's sales for 1995 based on this sales history.

Year	1989	1990	1991	1992
Sales (in millions)	$3.0	$4.0	$3.1	$4.9

Lesson 16-5

5. Willshire Clothing has 8.7% of the Willow Run metropolitan clothing market. Next year's clothing sales in the area are estimated to be $18,500,000. Use the factor method to project Willshire's sales for next year.

Lesson 16-6

6. Samson's Music Studio has a contract for 64 column inches of advertising in the *Herald* this week. What is the cost of a 40-column-inch advertisement placed on Sunday?

Per Column Inch

Contract Rates	Daily	Sunday
16 inches	$35.80	$45.80
64 inches	35.00	43.20
126 inches	34.00	42.70

Lesson 16-7

7. Countryside Cereal plans to advertise on television. Costs for 30-second commercials are $1000 for daytime and $8000 for prime time. The cost of a 10-second commercial is $\frac{1}{2}$ the cost of a 30-second commercial. The cost of a 60-second commercial is 2 times the cost of a 30-second commercial. What is the total advertising cost?

Number	Length	Time
5	10-second	Daytime
4	30-second	Prime time
2	60-second	Prime time

Lesson 16-8

8. Complete the chart. Which selling price will maximize profits?

Selling Price per Unit	Estimated Unit Sales	Fixed Costs	Fixed Costs per Unit	Variable Costs per Unit	Total Costs per Unit	Possible Net Profit
$20,000	16	$200,000	$12,500	$2500	?	?
$25,000	14	$200,000	?	$2500	?	?
$27,500	12	$200,000	?	$2500	?	?

A SPREADSHEET APPLICATION

Sales Analysis

To complete this spreadsheet application, you will need the diskette *Spreadsheet Applications for Business Mathematics,* which accompanies this textbook.

Input the information in the following problems to find the break-even point, gross sales, and gross profit.

1. Use the FIXED COSTS and VARIABLE COSTS given in the program. Change the SELLING PRICE and EXPECTED SALES to the amounts indicated.

	10—SELLING PRICE	11—EXPECTED SALES	BREAK-EVEN POINT	GROSS SALES	GROSS PROFIT
a.	$10.95	17,000	?	?	?
b.	11.45	15,500	?	?	?
c.	11.95	15,000	?	?	?
d.	12.45	14,000	?	?	?
e.	12.95	13,583	?	?	?
f.	13.45	11,500	?	?	?

2. Sobo Electronic Calculator Company developed these cost and price figures for their new Model 101 calculator.

```
FIXED COSTS:
1-RENT          2-TAXES        3-MARKETING       4-MANUFACTURING
10,200          62,400          75,000            81,500


VARIABLE COSTS:
5-SUPPLIES      6-LABOR  7-UTILITIES  8-PACKAGING  9-SHIPPING
1.50             3.75      0.85         0.25         0.25
```

Complete the table below to determine the best selling price and greatest profit.

	10—SELLING PRICE	11—EXPECTED SALES	BREAK-EVEN POINT	GROSS SALES	GROSS PROFIT
a.	$6.95	900,000	?	?	?
b.	7.25	800,000	?	?	?
c.	7.45	750,000	?	?	?
d.	8.45	400,000	?	?	?
e.	8.95	200,000	?	?	?
f.	9.45	50,000	?	?	?

Cumulative Review
(Skills/Applications) Units 12–16

Skills

(Skill 2)

Round answers to the nearest cent.

1. $7.848   **2.** $31.4545   **3.** 79.9¢   **4.** 12.3¢

Round answers to the nearest tenth of a percent.

5. 7.47%   **6.** 34.66%   **7.** 125.24%   **8.** 212.625%

Solve. Round to nearest cent.

(Skill 5)

9. $74.85 + $125.14   **10.** $517.87 + $493.34

11. $19.47 + $34.58 + $7.86   **12.** $4.98 + $7.49 + $2.19

13. $1478.85 + $2715.66   **14.** $916.43 + $791.56 + $288.72

(Skill 6)

15. $74.49 − $41.33   **16.** $21.19 − $7.79   **17.** $46.22 − $18.79

18. $741.13 − $564.45   **19.** $9164.55 − $7217.79

20. $16,415.50 − $3283.00   **21.** $31,415.74 − $23,567.79

(Skill 8)

22. $2.19 × 6   **23.** $74.49 × 4   **24.** $61.84 × 12

25. $19.17 × 0.06   **26.** $141.49 × 1.5   **27.** $71.89 × 0.85

(Skill 11)

28. $47.85 ÷ 12   **29.** $617.81 ÷ 144   **30.** $416.43 ÷ 0.6

31. $19.48 ÷ 0.75   **32.** $4217.55 ÷ 1.2   **33.** $27.45 ÷ 0.45

(Skill 16)

34. $\frac{1}{2} + \frac{1}{4}$ **35.** $\frac{3}{8} + \frac{1}{4}$ **36.** $\frac{3}{4} + \frac{7}{8}$

37. $1\frac{1}{2} + 2\frac{3}{8}$ **38.** $3\frac{1}{4} + 1\frac{7}{8}$ **39.** $\frac{1}{2} + 1\frac{5}{8} + 3\frac{1}{4}$

(Skill 20)

40. $\frac{1}{3} \times \frac{5}{8}$ **41.** $\frac{3}{4} \times \frac{8}{9}$ **42.** $\frac{1}{4} \times \frac{6}{7}$

43. $\frac{1}{3} \times$ $42   **44.** $\frac{1}{4} \times$ $36   **45.** $\frac{1}{2} \times$ $56

Write as a decimal.

(Skill 28)

46. 7%   **47.** 30%   **48.** 25%   **49.** 10%

50. $1\frac{1}{2}$%   **51.** 5.5%   **52.** $4\frac{1}{4}$%   **53.** 0.3%

Solve. Round answers to the nearest cent or tenth of a percent.

(Skill 30)

54. 5% of $174.50   **55.** 30% of $417.80   **56.** $1\frac{1}{4}$% of $241.70

57. 40% of $617.84   **58.** 25% of $1471.46   **59.** 6% of $249.99

(Skill 31)

60. ?% of $5000 = $200   **61.** ?% of 12 = 3

62. ?% of $19.79 = $0.40   **63.** ?% of $416.78 = $162.73

Use the table to answer the following.

(Application C)

64. What is the regular price of a package of 100 mL beakers?

65. What is the sale price of a package of 250 mL beakers?

PACKAGE OF 12 BEAKERS		
Size (mL)	Regular Price	Sale Price
50	$ 6.75	$ 6.24
100	9.75	8.75
250	15.75	14.24

66. For 50 mL beakers, how much less is the sale price than the regular price?

67. For 100 mL beakers, the sale price is what percent of the regular price?

Find the complement.

(Application L)

68. 40%   **69.** 25%   **70.** 12.5%   **71.** 5%

Use the graph to answer the following.

(Application N)

72. How many units were sold in 1988?

73. In what year were the least units sold?

74. How many more units were sold in 1989 than in 1987?

75. What is the sales projection for 1994?

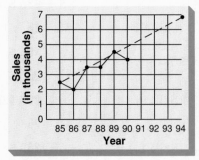

Find the mean.

(Application Q)

76. 10, 12, 17   **77.** 18, 12, 11, 11

78. 174, 155, 163, 148   **79.** 4.2, 1.6, 2.1, 4.0, 3.5

80. $14.51, $13.48, $14.76   **81.** $28,748, $30,615, $33,517

Write your own definition for each term.

82. Employee benefits   **83.** Release time   **84.** Salary scale

85. Travel expenses   **86.** Disability insurance   **87.** Prime cost

88. Break-even point   **89.** Quality control   **90.** Time study

91. Trade discount   **92.** Chain discount   **93.** Trade-discount rate

94. End-of-month dating   **95.** Markup   **96.** Markup rate

97. Net profit   **98.** Markdown rate   **99.** Opinion survey

100. Sales potential   **101.** Market share   **102.** Sales projection

Refer to your reference files in the back of the book if you need help.

Cumulative Review Test
Units 12–16

Lesson 12-2

1. Sue Clark is a systems analyst for the Second National Bank. She presently earns $32,480 annually. Sue will receive a 4.8% cost-of-living increase and a 4.5% merit increase. What will Sue's new salary be?

Lesson 12-3

2. Tony Brown works for the Stark Corporation. He earns $21,840 per year. Stark provides these benefits for Tony. To the nearest percent, what is the rate of benefits?

Vacation: 2 weeks
Holidays: 8 days
Compensation insurance: 4.4%
Unemployment insurance: 3.6%
Social security: 7.65%

Lesson 12-5

3. Meg Hart is a travel consultant. Her travel expenses this month include: airfare $1474, taxi fares $71.65, meals $416.50, hotels $821.37, and miscellaneous $61.40. What is Meg's total travel expense?

Lesson 12-6

4. The personnel department of the Tru-Door Company is sending 5 of its employees to a local 2-day training program. The 5 employees' combined wages total $386 per day. The company paid the registration fee of $50 for each person. The company also paid for the employees' lunches for the 2 days at a total cost of $82.40. What was the total cost for the Tru-Door Company to send their employees to the training program?

Lesson 13-1

5. Kasko Steel produces cable from scrap steel that it buys for $226 per ton. The company produces 4 rolls of cable from each ton. The direct labor charge is $38.78 an hour. The employees produce 2 rolls of cable per hour. To the nearest tenth of a cent, what is the prime cost of manufacturing 1 roll of cable?

Lesson 13-4

6. Giant Food Mart made a time study of its stock department. The store recorded the times that Joe Martin spent unpacking and shelving cartons of canned goods.

Task	Observations in Seconds				
	#1	**#2**	**#3**	**#4**	**#5**
Open carton	4.0	4.5	5.5	4.0	5.0
Remove cans	13.0	14.0	12.5	13.5	14.5
Place cans on shelf	12.0	11.6	12.8	12.2	11.4
Discard carton	5.0	5.5	5.0	5.0	5.5

If Joe takes a 10-minute break each hour, how many cartons can he unpack and shelve in 1 hour?

7. Colonial Candle Company ships its top-of-the-line fancy candles in wooden boxes. The candles are arranged in 1 layer. Each box is made of $\frac{1}{8}$-inch thick wood with $\frac{1}{16}$-inch thick wood spacers. Each candle is 8 inches high and $3\frac{1}{4}$ inches in diameter. What are the dimensions of each wooden box?

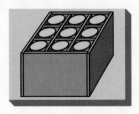

8. Sun-Like, Inc. lists the price of its Sure-Tan Lamp as $47.49. The net price to the retailer is $37.99. What is the trade-discount rate?

9. Gendron Wheel Co. produces a bicycle that is priced at $217.86. When a new model was introduced, the $217.86 bike was discounted at 30% less 15%. What is the net price of the old model using the complement method?

10. DP Associates received an invoice from Commercial Computer Supply dated September 13. The invoice carried terms of 3/10, net 30. The net price of the invoice was $7419.80. What was the cash price of the invoice if it was paid on September 22?

11. Eagle-Hawk, Inc., a fireplace equipment wholesaler, purchases fireplace inserts for $147.85 each and sells them for $199.79. The company estimates the overhead expenses for each insert to be 20% of the selling price. What is the estimated net profit on each insert?

12. Tree Grown determines the retail selling price of the oranges by marking up the cost by 140%. The fruit company bought its first shipment of winter oranges for $0.66 per dozen. What is the retail selling price per dozen of this shipment?

13. Of a sample of 1200 swimming pool owners, 360 said they would buy Litehouse Living's new chemical for swimming pools. There is a market of about 4,000,000 potential users. Each user would purchase the chemical once per year. What is the sales potential for the chemical for 1 year?

14. Main Street Meat Market has a contract for 24 column inches of advertising at $35.15 per column inch in the *Herald* this year. What is the cost of an 8-column-inch advertisement?

15. Complete the chart for Sailaway Corporation. Which selling price will maximize Sailaway's profits?

Selling Price per Unit	Estimated Unit Sales	Fixed Costs	Fixed Costs per Unit	Variable Costs per Unit	Total Costs per Unit	Possible Net Profit
$38,000	15	$200,000	$13,333.33	$4200	?	?
45,000	13	200,000	?	4200	?	?
53,000	9	200,000	?	4200	?	?

UNIT

17

Warehousing and Distribution

The warehousing and distribution department keeps a record of all merchandise, or *inventory,* stored in your warehouse. *Inventory cards* show the number of each item in stock, the number of incoming items (*receipts*), and the number of outgoing items (*issues*). Your inventory must be large enough to meet your sales needs. The size and value of your inventory will affect your insurance and taxes and the worth of your business. The warehousing and distribution department is also responsible for shipping your product.

Retail stores periodically conduct an inventory of their stock on hand.

17-1

Storage Space

OBJECTIVE
Compute the total storage space.

Your business needs warehouse or storage space in which to keep raw materials or products until you are ready to use them. You may need space in a large warehouse or only a small stockroom, depending on the size and quantity of the items you are storing.

Storage Space = Volume per Item × Number of Items

EXAMPLE Skill 8 Application Z Term Storage space

The PC-View Corporation manufactures personal computer monitors. Each monitor is packaged in a carton measuring 2 feet long, 1.5 feet wide, and 2 feet high. How many cubic feet of space does PC-View need to store 800 monitors?

SOLUTION **A.** Find the **volume per item.**
2 ft × 1.5 ft × 2 ft = 6 ft^3 volume per item

B. Find the **storage space.**
Volume per Item × Number of Items
6 ft^3 ×   800 items = 4800 ft^3 storage space

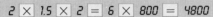

$2 \boxed{\times} 1.5 \boxed{\times} 2 \boxed{=} 6 \boxed{\times} 800 \boxed{=} 4800$

✔ SELF-CHECK Complete the problems, then check your answers in the back of the book.

Find the storage space needed.

1. Dimensions of item:
2 ft × 3 ft × 3.5 ft.
700 items.

2. Dimensions of item:
7 cm × 8 cm × 1 cm.
1000 items.

PROBLEMS

	Carton Dimensions				Number	Storage
	(Length	× Width	× Height	= Volume)	× of Items	= Space
3.	(6.0 in	× 8.5 in	× 10.0 in =	?) ×	500	= ?
4.	(2.5 ft	× 1.5 ft	× 1.0 ft =	?) ×	200	= ?
5.	(1.2 m	× 0.6 m	× 0.4 m =	?) ×	500	= ?
6.	(43 cm	× 24 cm	× 12 cm =	?) ×	340	= ?
7.	(2.1 yd	× 1.8 yd	× 1.4 yd =	?) ×	120	= ?
8.	(2.2 ft	× 1.5 ft	× 1.1 ft =	?) ×	1200	= ?

---

450 ◆ Unit 17 Warehousing and Distribution

9. Washstand cupboard.
Carton is 3 ft by 1.8 ft by 2.5 ft.
100 cartons to be stored.
How much storage space is
   required?

10. Table lamp.
Carton is 0.75 m by 0.6 m by 0.8 m.
700 cartons to be stored.
How much storage space is
   required?

11. Green Shade study lamps are packed 2 in a carton. The dimensions of the carton are 2.4 feet by 1.2 feet by 1.8 feet. The School Supply Store wants to order 200 pairs of lamps. How much storage space will the School Supply Store need to store the lamps?

12. Family Drugs has ordered 350 cartons of tape holders. Each carton of holders measures 1.6 feet by 0.8 feet by 0.5 feet. How much storage space is required?

13. Craft Electronics produces console television sets. The sets are packaged in cartons measuring 4 feet long, 2.5 feet wide, and 3 feet high. How much storage space is required for 1750 television sets?

14. The London Furniture Company packs its Queen Anne tables in cartons measuring 1.75 meters long, 0.8 meters wide, and 1.2 meters high. Empire Importing has ordered 500 tables. How much storage space is needed?

15. The Great Western power lawn mower is housed in a carton measuring 2.5 feet by 1.3 feet by 0.8 feet. General Lawn Company ordered 875 Great Western lawn mowers. How much storage space is needed?

16. Bananas are delivered to grocery stores in crates measuring 1 yard by 1.5 feet by 12 inches. How many cubic feet of storage space are required for 25 crates of bananas?

17. Fireplace hardware sets are delivered in boxes that are 0.8 meters by 25 centimeters by 15 centimeters. How many cubic meters of storage space are required to store 750 sets?

18. A Brief Case Twenty-four clock radios are packed in a carton measuring 3 feet long, 2.4 feet wide, and 1.6 feet high. How much storage space is required for 200 cartons of 24 clock radios each? How much space is required for 3600 clock radios packaged in cartons of 24 clock radios each?

MAINTAINING YOUR SKILLS   Look up the skill in parentheses if you need help or more practice.

Multiply. **(Skill 8)**

19. 2 ft $\times$ 3 ft $\times$ 5 ft

20. 1.4 m $\times$ 1.0 m $\times$ 0.8 m

21. 8 in $\times$ 6 in $\times$ 4 in

22. 18 mm $\times$ 15 mm $\times$ 9 mm

23. 9.7 cm $\times$ 7.5 cm $\times$ 6.8 cm

24. 2.1 yd $\times$ 0.75 yd $\times$ 1.5 yd

25. 45 ft^3 $\times$ 400

26. 6.5 m^3 $\times$ 620

27. 112 in^3 $\times$ 250

Taking an Inventory

OBJECTIVE

Compute total inventory.

An inventory shows the number of each item that your business has in stock. You may use an inventory card to keep a record of the items on hand. The number of incoming items, called receipts, are added to the previous inventory. The number of outgoing items, called issues, are subtracted from the previous inventory. This is frequently done on a computer.

Inventory = Previous Inventory + Receipts − Issues

EXAMPLE

Skills 3, 4 *Application* A *Term* Inventory

The inventory card for Wholesale Paint Supply shows that 80 gallons of white latex paint, stock number WL-9, were in inventory on February 1. Each week, the inventory record is updated. How many gallons were on hand March 1?

SOLUTION

ITEM: *Latex paint, white*
STOCK NUMBER: *WL-9*

Inventory at beginning of week

Week of	Opening Balance	Receipts (Quantity IN)	Issues (Quantity OUT)	Inventory at End of Week
Feb 1	80	40	62	58
Feb 8	58	0	25	33
Feb 15	33	80	28	85
Feb 22	85	40	37	88
Mar 1	88			

Number on hand March 1

80 + 40 − 62 = 58 − 25 = 33 + 80 − 28 = 85 + 40 − 37 = 88

✔ SELF-CHECK

Complete the problems, then check your answers in the back of the book.

Find the inventory.

1. Previous inventory, 400; receipts, 150; issues, 200.

2. Previous inventory, 220, receipts, 70; issues 50.

PROBLEMS

Complete the inventory record.

	Month	Opening Balance	+	Receipts (Quantity IN)	−	Issues (Quantity OUT)	=	Number on Hand
3.	January	300	+	120	−	80	=	?
4.	February	340	+	80	−	100	=	?
5.	March	?	+	120	−	100	=	?
6.	April	?	+	0	−	80	=	?

7. Inventory of garden shovels.
80 on hand at beginning of week.
Received 40 during the week.
Issued 25 on Friday.
How many on hand at end
of week?

8. Inventory of 12-oz cans of
baked beans.
Started the week with 148 cases.
Received 40 cases during the week.
Shipped out 65 cases during
the week.
How many on hand at end
of week?

9. The Key Auto Supply Company takes an inventory on a monthly basis.
What is the balance on hand for each item?

Item Number	Description	Beginning Balance	Quantity IN	Quantity OUT	Balance on Hand
X 23	10 W-30 Oil	147 qt	50 qt	78 qt	?
W 15	Washer solvent	91 gal	36 gal	25 gal	?
C 5	5-gallon can 30 W oil	17 cans	12 cans	3 cans	?

10. How many of each item does the central supply department have on
hand at the end of this month?

STOCK NO.	DESCRIPTION	OPENING BALANCE	+ RECEIPTS	– ISSUES	= ON HAND
11398	STPLR-RS153	46	0	10	?
11402	BOX STPLS-RT11	117	0	50	?
11610	BOX PA CLPS-L3	18	144	21	?
11682	BOX PA CLPS-L7	7	36	9	?

11. How many dictionaries does the bookstore have in stock on each date?

LAST ACTION DATE	RECEIPTS	ISSUES	ON HAND
9/3	225		225
9/7		37	?
9/10		24	?
9/15		35	?
9/20	144		?
9/25		43	?

12. The Stock Status Summary for all of the stores in the Cushing
Department Stores chain is printed monthly. How many of each item
are in stock?

DATE 8-28		STOCK STATUS SUMMARY					
MATER'L CLASS	STOCK NUMBER	DESCRIPTION	OPENING BALANCE	TRANSACTIONS			
				RECEIVED	ISSUED	ON HAND	
174	146301	MARKER-FP	117		26	?	
174	146334	MARKER-W/2	34	12		?	
175	121	NOTEBOOK	987	144	347	?	
175	6841	NOTEBOOK	1413	72	465	?	

13. Whitewater Supply Company carries this inventory card on inflatable vests. Find the balance for each date.

Part No. IV–17	Date	Receipts	Issues	Balance
Description:	6/1	76		76
Inflatable	6/12		36	?
vest	6/18	144		?
Location: 24	6/26		58	?

14. Complete this hardware supply inventory record for a deadbolt lock set.

Date	Receipts	Shipments	Balance
9/1			110
9/9	100		?
10/12		75	?
10/30		100	?
11/15	200		?
12/11		95	?

15. An inventory record for Sunrise Wholesale Pool Supply shows a receipt of 100 folding canvas lounge chairs on 4/1. Another shipment of 125 arrived on 4/15. One hundred ten chairs were shipped out on 5/10. Fifty arrived on 6/2 and 75 were shipped out on 6/17. Find the balance as of 6/17.

16. On January 1, Sport Wholesalers started with an inventory of 175 basketballs. Shipments of 45, 76, and 25 were sent out on 1/13, 2/18, and 3/11, respectively. Receipts of 36, 144, and 72 were received on 1/20, 2/25, and 3/15, respectively. Find the balance as of 3/15.

17. The Car Care Co. started on August 1 with a balance of 47 cases of car wax. On August 6, 24 cases were shipped out. On August 16, 40 cases were received. On August 27, 36 cases were shipped out, and on August 31, 50 cases were received. Prepare an inventory control card showing the balance after each transaction.

18. A Brief Case The Wholesale Hardware Supply Company began July 1 with a balance of 74 gas chain saws. It received shipments of 30 and 24 chain saws on July 20 and September 20, respectively. It made shipments of 24, 40, and 36 chain saws on July 17, August 22, and September 30, respectively. Prepare an inventory control card showing the balance after each transaction.

MAINTAINING YOUR SKILLS Look up the skills in parentheses if you need help or more practice.

Add. **(Skill 3)**

19. 125 + 45 + 25 **20.** 144 + 72 + 36 **21.** 50 + 35 + 48

Subtract. **(Skill 4)**

22. 144 − 50 **23.** 72 − 48 **24.** 286 − 144 **25.** 7146 − 5268

26. 9178 − 493 **27.** 14,147 − 9749 **28.** 26,147 − 19,464

Valuing an Inventory

OBJECTIVE

Use the average-cost method to compute the inventory value.

You must calculate the value of your inventory when you purchase insurance, pay taxes, or compute the worth of your business. The **average-cost method** is one way of calculating the value of an inventory. Because the cost of incoming items may change, you calculate the value of the inventory based on the average cost of the goods you received.

Inventory Value = Average Cost per Unit × Number on Hand

EXAMPLE   *Skills* 8, 11   *Application* Q   *Term* Average-cost method

Wholesale Paint Supply is valuing its inventory of white latex paint. On March 1, Wholesale Paint Supply had 88 gallons of white latex paint on hand. What is the value of the inventory on March 1?

ITEM: Latex paint, white
STOCK NUMBER: WL-9

Number on hand at beginning of week →

Average cost in January →

Week of		Receipts	Unit Cost	Total Cost
Feb 1	Opening Balance:	80	$2.15	$172.00
		40	2.25	90.00
Feb 8		0	- -	- -
Feb 15		80	2.30	184.00
Feb 22		40	2.40	96.00
	Total	240		$542.00

SOLUTION

A. Find the **average cost per unit.**

Total Cost of Units ÷ Number Received
$542.00 ÷ 240 = $2.2583 = $2.26 average cost per unit

B. Find the **inventory value.**

Average Cost per Unit × Number on Hand
$2.26 × 88 = $198.88 inventory value

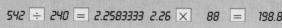

542 ÷ 240 = 2.2583333 2.26 × 88 = 198.88

---

100 items purchased at $4.00 each. 60 items purchased at $3.80 each.

1. Find the average cost per unit.

2. Find the inventory value if 50 items were on hand.

3. Top-Mart is valuing its inventory of popcorn poppers. On June 1, Top-Mart had 23 poppers on hand.
   a. What is the average cost per popper?
   b. What is the value of the inventory of poppers on June 1?

Week of	Receipts	Unit Cost	Total Cost
May 1	40	$2.10	$84.00
May 8	36	2.15	77.40
May 15	24	2.20	52.80
May 22	36	2.15	77.40
May 29	12	2.25	27.00
Total	148		$318.60

4. Sure Sound, Inc., is valuing its inventory of AM/FM cassette portable players. On April 1, Sure Sound had 53 players on hand.
   a. What is the average cost per player?
   b. What is the value of the inventory of players on April 1?

Date	Receipts	Unit Cost	Total Cost
January 12	50	$6.30	$315.00
February 11	65	6.10	396.50
March 21	80	5.85	468.00
Total	195		$1179.50

For problems 5 and 6, complete the tables and answer the questions.

5. a.

Date	Receipts	Unit Cost	Total Cost
3/10	1120	$0.25	?
4/22	940	0.28	?
Total	?		?

   b. Average cost per unit: ___?___.
      Number on hand on 4/30: 960.
   c. Value of inventory: ___?___.

6. a.

Date	Receipts	Unit Cost	Total Cost
12/18	650	$12.10	?
1/10	500	13.31	?
2/6	450	13.98	?
Total	?		?

   b. Average cost per unit: ___?___.
      Number on hand on 2/8: 674.
   c. Value of inventory: ___?___.

7. Records for Kazrich Market show this opening balance and theses receipts for Hearthside soup in November. At the end of the month, 31 cans were on hand. What is the value of the inventory?

Date	Receipts	Unit Cost
11/1 Opening bal.	94	$0.32
11/10	72	0.30
11/17	36	0.32
11/24	48	0.31

8. TBA Inc's inventory records for packages of windshield wipers show this information. At the end of the month, 79 packages were on hand. Find the value of the inventory.

Date	Receipts	Unit Cost
5/1	93	$3.43
5/19	120	3.56
5/31	115	3.63

9. The Leisure Company, a swimming pool supply company, uses the FIFO (first in, first out) method of valuing inventory. The FIFO method assumes that the first items received are the first items sold.

Date	Receipts	Unit Cost	Total Cost
5/1	100	$4.65	?
5/8	65	4.75	?
5/19	145	4.90	?
Total	?		?

a. Complete The Leisure Company's records for receipts of tubs of chlorine.

As of May 20, The Leisure Company had sold 160 of the tubs of chlorine.

b. Use the FIFO method to calculate the cost of the 160 tubs sold.

c. Use the FIFO method to calculate the value of the 150 tubs in stock.

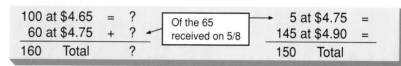

```
100 at $4.65  =  ?                        →   5 at $4.75   =
 60 at $4.75  +  ?  ←  Of the 65            145 at $4.90   =
                        received on 5/8
160   Total      ?                         150   Total
```

10. A Brief Case Another method used to value an inventory is the LIFO (last in, first out) method. The LIFO method assumes that the last items received are the first items shipped.
a. Use the LIFO method to calculate the value of The Leisure Company's 150 tubs in stock in problem 9 above.
b. Use the average-cost method to calculate the value of The Leisure Company's 150 tubs in stock in problem 9 above.

11. In a rising economy (costs continue to rise), which of the three methods of valuing an inventory discussed in this lesson will result in the:
a. Highest valued inventory? Why?
b. Lowest valued inventory? Why?

12. In a declining economy (costs continue to decline), which of the three methods of valuing an inventory will result in the:
a. Highest valued inventory? Why?
b. Lowest valued inventory? Why?

13. Does the average-cost method of valuing an inventory always result in a value between the value determined using FIFO and LIFO?

MAINTAINING YOUR SKILLS Look up the skills in parentheses if you need help or more practice.

Multiply. **(Skill 8)**

14. 50 × $8.40

15. 100 × $17.85

16. 72 × $36.90

17. 144 × $141.73

18. 36 × $0.87

19. 119 × $47.87

20. 216 × $247.81

21. 483 × $79.86

Divide. Round answers to the nearest cent. **(Skill 11)**

22. $1046.25 ÷ 125

23. $438.70 ÷ 28

24. $314.91 ÷ 92

25. $30,613.74 ÷ 216

26. $357.81 ÷ 410

27. $9012.31 ÷ 187

28. $13,111.54 ÷ 53

29. $56,918.43 ÷ 719

17-4

Carrying an Inventory

OBJECTIVE

Compute the annual cost of carrying an inventory.

Your business must keep a sufficient inventory of goods to meet production or sales needs. Based on past records, you may estimate the annual cost of maintaining, or carrying, the inventory at a certain level for the coming year. The annual cost of carrying the inventory includes taxes, insurance, storage fees, handling charges, and so on. The annual cost is often expressed as a percent of the value of the inventory.

Annual Cost of Carrying Inventory = Inventory Value × Percent

EXAMPLE Skill 30 Application A

Steel Facilitators, Inc., produces washers, nuts, bolts, and similar items. The company maintains an inventory of metal sheets and bars to be used in production. The company also maintains an inventory of finished products to fill customers' orders. Steel Facilitators estimates the annual cost of maintaining the inventory to be 25% of the value of the inventory. This year the company plans to maintain an inventory totaling $500,000 in value. What amount should be set as the estimated cost of carrying the inventory for the year?

SOLUTION Find the **annual cost of carrying inventory.**
Inventory Value × Percent
$500,000 × 25% = $125,000 annual cost of carrying inventory

 500000 × 25 % 125000

✔ SELF-CHECK Complete the problems, then check your answers in the back of the book.

Find the annual cost of carrying an inventory.

1. Inventory value is $200,000. Estimated annual cost is 30% value.

2. Inventory value is $85,000. Estimated annual cost is 20% of value.

PROBLEMS

	3.	4.	5.	6.	7.	8.
Value of Inventory	$960.00	$14,000.00	$7800.00	$418.40	$1347.86	$5147.95
Percent	25%	35%	15%	22%	28.5%	32.5%
Annual Cost of Carrying Inventory	?	?	?	?	?	?

9. Coral Fabric Company.
Maintains $16,800 inventory.
Costs about 25% of value
   to maintain.
What is the approximate annual
   cost of carrying the inventory?

10. The Steel Assembly Company.
Maintains $82,000 inventory.
Costs about 32% of value
   to maintain.
What is the approximate annual
   cost of carrying the inventory?

11. Central Drug Stores estimates the cost of carrying its inventory of goods
to be 18% of the value of the merchandise. About how much does it
cost Central Drug Stores annually to carry a $21,560 inventory?

12. Ace Iron and Steel, Inc., carries a $64,800 inventory. The company
estimates the annual cost of carrying the inventory at 27% of the value
of the inventory. What is the approximate annual cost of carrying the
inventory?

13. Baker Wholesale Grocers, Inc., estimates the annual cost of carrying its
inventory at 26% of the value of the inventory. The 26% is broken
down as follows:

Type of Expense	Percent
Spoilage and physical deterioration	7.0%
Interest	11.0%
Handling	4.0%
Storage facilities	2.0%
Transportation	1.0%
Taxes	0.5%
Insurance	0.5%
Total	26.0%

a. Baker generally carries an inventory valued at $37,470. What is the
approximate annual cost for each expense of carrying the inventory?

b. Baker was able to reduce its inventory to one half of its value and
still meet customer orders. About how much did Baker save on the
total cost of carrying its inventory?

14. In problem 13, if Baker carried an inventory valued at $100,000, what
would be the approximate annual cost for each expense of carrying the
inventory?

MAINTAINING YOUR SKILLS   Look up the skill in parentheses if you need help or more practice.

Multiply. **(Skill 30)**

15. $120,000 × 20%

16. $96,400 × 25%

17. $146,500 × 30%

18. $196,750 × 15%

19. $71,630 × 35%

20. $347,840 × 28%

21. $516,780 × 37.5%

22. $7893 × 23.4%

23. $11,746 × 12.5%

24. $2,147,894 × 16%

25. $714,816 × 31.7%

26. $486.92 × 41.3%

Transportation by Air

OBJECTIVE
Compute the total shipping cost by air freight.

Your business may ship goods to its customers by air freight. Either you or the customer pays the shipping charges. The total cost of shipping goods by air freight includes pickup and delivery to the airport, air freight charges to the city of destination, local delivery from the airport to the final destination, and federal tax, where it applies.

$$\text{Total Shipping Cost} = \text{Pickup Charge} + \text{Air Freight Charge} + \text{Delivery Charge} + \text{Federal Tax}$$

EXAMPLE   *Skills* 5, 8, 10, 30   *Application* A   *Term* Air freight

The Innovative Auto Company is shipping automotive equipment by air freight. The packaged equipment weighs 650 pounds. Express Pickup Service charges $6.85 for each 100 pounds or fractional part to transport the equipment to the airport. The air freight rate is $67.50 per 100 pounds, charged on the actual number of pounds shipped. Speedy Delivery Service charges $9.10 per 100 pounds, in increments of 50 pounds, to deliver the equipment. In addition, there is a 5% federal tax on the air freight charge because it involves interstate transportation. What is the total cost of shipping the equipment?

SOLUTION

A. Find the **pickup charge.**
$6.85 × (700 ÷ 100) = $ 47.95 pickup charge

B. Find the **air freight charge.**
$67.50 × (650 ÷ 100) = $438.75 air freight charge

C. Find the **delivery charge.**
$9.10 × (650 ÷ 100) = $ 59.15 delivery charge

D. Find the **federal tax** of 5%.
$438.75 × 5% = $21.9375 = $ 21.94 federal tax
  $567.79 total shipping cost

700 ÷ 100 × 6.85 = 47.95 M+ 650 ÷ 100 × 67.5 = 438.75 M+ 650 ÷ 100 × 9.1 = 59.15 M+ 438.75 × 5 % 21.9375 M+ RM 567.7875

✔ SELF-CHECK   Complete the problems, then check your answers in the back of the book.

1. Federal tax is 6% of the air freight charge.
Air freight charge is $246.50.
Find the federal tax.

2. Pickup charge is $50.40.
Air freight charge is $378.
Delivery charge is $67.50.
Federal tax is $20.79.
Find the total shipping cost.

Goods	Cost of Pickup	+	Cost of Air Freight	+	Cost of Delivery	+	Federal Tax	=	Total Cost
3. Auto parts	$102.00	+	$474.00	+	$63.00	+	$23.70	=	?
4. Arc welders	$138.60	+	$828.00	+	$73.20	+	$45.54	=	?
5. Cameras	$ 25.00	+	$285.60	+	$37.50	+	$17.14	=	?
6. Books	$ 76.50	+	$378.00	+	$67.50	+	$24.57	=	?

For problems 7 and 8, charges are per 100 pounds or fractional part.

7. 400 pounds of battery chargers. $8.50 per 100 pounds for pickup. $65.50 per 100 pounds for air freight. $7.50 per 100 pounds for delivery. $5\frac{1}{2}$% federal tax on air freight charge. What is the total shipping cost?

8. 540 pounds of 10-speed bicycles. $5.60 per 100 pounds for pickup. $62 per 100 pounds for air freight. $6.40 per 100 pounds for delivery. $6\frac{1}{2}$% federal tax on air freight charge. What is the total shipping cost?

F.Y.I.
About 11.5% of airline revenues are derived from air freight and express package services.

9. A crate of musical instruments that weighs 470 pounds is shipped by air freight. The costs per 100 pounds or fractional part are: $5.00 for pickup, $78 for air freight, and $7.50 for delivery. A 5% federal tax on the air freight charge is added. What is the total shipping cost?

10. New Century is shipping 1460 pounds of prototype CAD-CAM equipment by air freight. Pickup is $6.50 per 100 pounds or fractional part. Delivery is $5.65 per 100 pounds or fractional part. The air freight is $64.50 per 100 pounds, charged on the actual number of pounds shipped. There is a 4% tax on the air freight and delivery charges. What is the total shipping cost?

11. AB Consultants, Inc., is shipping 48 pounds of computer tapes by air freight from Detroit to Tampa. Reliable Service charges $6.85 per 100 pounds, in increments of 50 pounds, to transport the tapes to the airport. ATP Air Freight charges $0.395 per pound, actual weight, with a minimum charge of $30. Central Carrier charges $9.10 per 100 pounds or fractional part to deliver the tapes. There is a 6% federal tax on the air freight. What is the total shipping cost?

Find the percentage. Round answers to the nearest cent. (Skill 30)

12. $417.80 × 5%

13. $1216.90 × 6%

14. $641.85 × $5\frac{1}{2}$%

15. $347.84 × $6\frac{1}{2}$%

Add. (Skill 5)

16. $54.50 + $317.80 + $41.70 + $15.89

17. $31.75 + $743.86 + $71.80 + $44.63

Transportation by Truck

OBJECTIVE

Compute the total shipping cost by truck.

Your business may ship its goods by truck. The basic rate for shipping by truck depends on the classification of the goods that are shipped. The classification is determined by the type and value of the goods. The basic rates differ according to the weight of the goods and the distance that they are being shipped, and they may also vary from carrier to carrier. The basic rates are often given per 100 pounds. Most companies charge only for the actual weight of the goods being shipped.

Total Shipping Cost = Weight × Basic Rate

EXAMPLE   *Skills* 8, 10   *Application* C   *Term* Basic rate

The King Spark Plug Co. is shipping 4470 pounds of spark plugs (class 100) via Motor Express. Motor Express will pick up the spark plugs at the manufacturing plant and deliver them to an automobile parts distributor. The distance is 325 miles. What is the total shipping cost?

MOTOR EXPRESS...BASIC RATES PER 100 POUNDS... CLASS 100 ITEMS

Weight Group (in pounds)	Distance (in miles)				
	1–100	101–200	201–300	301–400	401–500
0–500	$24.35	$28.17	$33.77	$36.67	$40.10
501–1000	19.81	22.54	26.51	29.28	32.08
1001–2000	14.37	16.63	19.85	21.61	23.66
2001–5000	11.94	13.81	16.55	17.95	19.65
5001–10,000	8.54	9.87	11.77	12.82	14.04

SOLUTION

A. Find the **basic rate.** (Refer to the table above.)
Weight group: 2001–5000 pounds
Distance: 301–400 miles.........................$17.95 basic rate

B. Find the **weight** in hundreds of pounds.
4470 ÷ 100 = 44.7 weight

C. Find the **total shipping cost.**
Weight × Basic Rate
44.7 × $17.95 = $802.365 = $802.37 total shipping cost

4470 ÷ 100 = 44.7 × 17.95 = 802.365

✔ SELF-CHECK   Complete the problems, then check your answers in the back of the book.

1. Using the table above, find the basic rate to ship 420 pounds 425 miles.

2. Find the total shipping cost.

Use the table on page 462 to find the basic rates.

	3.	**4.**	**5.**	**6.**	**7.**	**8.**
Weight (pounds)	9470	620	1476	3719	381	2147
Distance (miles)	125	251	18	376	391	474
Basic Rate	$9.87	$26.51	?	?	?	?
Total Shipping Cost	?	?	?	?	?	?

9. 2470 pounds of ski boots. Transported 315 miles. What is the total shipping cost?

10. 176 pounds of garden hose. Transported 437 miles. What is the total shipping cost?

F.Y.I.
The vast network of highways, roads, and streets in the United States handles 135 million automobiles, 44 million trucks, and 1 million buses per day (*Transport Topics*, June 24, 1991).

11. A shipment of ski jackets weighing 384 pounds is transported by truck from the Out Country Clothing plant to its Waterside warehouse. The distance is 278 miles. What is the shipping cost?

12. A shipment of brooms and brushes is sent by truck from Green Bay, Wisconsin, to St. Louis, Missouri. The distance is 445 miles. The brooms and brushes weigh 2447 pounds. What is the shipping cost?

13. A 641-pound shipment of seat covers is transported from The Cover Factory, Inc., to the national distribution center in Chicago. The distance is 124 miles. What is the shipping charge?

14. A 500-pound shipment of darkroom equipment and chemicals is taken from Rochester, New York, to Cincinnati, Ohio. The distance is 491 miles. What is the shipping charge?

15. A shipment of drive shafts is being sent from Replacement Parts' central warehouse to a local distribution center in Vancouver, British Columbia. The distance is 454 miles. The drive shafts weigh a total of 6418 pounds. What is the shipping cost?

MAINTAINING YOUR SKILLS   Look up the skill in parentheses if you need help or more practice.

Multiply. **(Skill 8)**

16. 4.70 × $32.13

17. 11.60 × $20.40

18. 41.73 × $15.30

19. 14.84 × $31.02

20. 35.72 × $12.84

21. 17.13 × $12.66

22. 7.56 × $27.81

23. 20.72 × $17.64

24. 81.48 × $10.25

25. 9.12 × $33.00

26. 7.73 × $25.41

27. 1.43 × $19.35

28. 19.11 × $22.62

29. 97.87 × $12.51

30. 61.47 × $8.43

31. 41.67 × $10.38

32. 0.71 × $29.58

33. 3.97 × $37.11

Reviewing the Basics

Skills

(Skill 2)

Round answers to the nearest hundred.

1. 462   **2.** 3475   **3.** 1147   **4.** 8169

Solve.

(Skill 3)

5. 47 + 64   **6.** 473 + 349   **7.** 719 + 582

(Skill 4)

8. 76 − 35   **9.** 417 − 249   **10.** 219 − 135

(Skill 8)

11. $47.52 × 3.4   **12.** $71.83 × 12.44   **13.** $89.91 × 431.51

Round answers to the nearest cent.

(Skill 10)

14. $217.84 ÷ 19   **15.** $4175 ÷ 48   **16.** $1241.84 ÷ 36

(Skill 30)

17. 5.5% of $189.97   **18.** 35% of $256.73   **19.** 1.75% of $47.79

Applications

(Application C)

Find the shipping charge.

20. Weight: 3478 pounds.
Distance: 1243 miles.

21. Weight: 8217 pounds.
Distance: 715 miles.

BASIC RATES PER 100 POUNDS

Weight Group (in pounds)	Distance (in miles)	
	500–1000	1001–1500
2001–5000	$31.12	$39.84
5001–10,000	26.50	35.25

Find the average cost to the nearest cent.

(Application Q)

22. 80 units at $3.50 each and
75 units at $3.65 each.

23. 215 units at $0.72 each and
100 units at $0.75 each.

Find the volume.

(Application Z)

24. Box is 6 ft by 3 ft by 8.5 ft.   **25.** Box is 15 m by 12 m by 7 m.

Terms

Match each term with its definition on the right.

26. Inventory

27. Average-cost method

28. Air freight

29. Basic rate

30. Storage space

a. a way of calculating an inventory's value

b. the number of each item you have in stock

c. the charge per 100 pounds to ship goods by truck

d. a way of shipping goods to customers

e. a place in which to keep products until they are needed

f. includes taxes, storage fees, and handling costs

Refer to your reference files in the back of the book if you need help.

Unit Test

Lesson 17-1

1. The Data-Base Equipment Company manufactures laser printers. Each printer is stored in a box measuring 2 feet high, 1.5 feet wide, and 2.5 feet deep. How many cubic feet of space does Data-Base need to store 744 laser printers?

Lesson 17-2

2. How many StraightAway models does Luxury Motor Coach have on its lot on October 1?

ITEM: StraightAway
STOCK NUMBER: JR 2201F

Month of	Opening Balance	Receipts	Issues	Inventory at End of Month
August	42	5	12	?
September	?	6	18	?
October	?			

Lesson 17-3

3. The Farm Store took inventory of its heavy-duty tarps. On March 1, the Farm Store had 40 heavy-duty tarps in stock. Using the average-cost method, what is the value of the inventory on March 1?

ITEM: Heavy Duty Tarps
STOCK NUMBER: 3251

Date		Receipts	Unit Cost	Total Cost
Feb 1	Opening Balance	50	$ 9.87	?
Feb 18		25	10.12	?
Feb 27		40	10.18	?
	Total	?		?

Lesson 17-4

4. City Cleaners maintains a soap inventory valued at $14,700. The cost of maintaining the inventory is approximately 20% of the value of the inventory. What is the approximate annual cost of carrying this inventory?

Lesson 17-5

5. Cloud Aeronautics is shipping 186 pounds of flight instruments by air freight. The costs per 100 pounds or fractional part are: $6.76 for pick-up, $21.45 for air freight, and $5.75 for delivery. A 5% federal tax is added to the air freight charge. What is the total cost to deliver the instruments?

Lesson 17-6

6. The Barstow Company is shipping 846 pounds of automotive equipment. Commercial Trucking will handle the shipment. The distance is 180 miles. What is the total shipping cost?

COMMERCIAL TRUCKING CLASS 100 ITEMS BASIC RATES PER100 POUNDS			
Weight Group (in pounds)	Distance (in miles)		
	1–100	101–200	201–300
0–500	$22.65	$25.60	$28.55
501–1000	$20.95	$23.30	$25.30

A SPREADSHEET APPLICATION

Warehousing and Distribution

To complete this spreadsheet application, you will need the template diskette for *Mathematics with Business Applications.* Follow the directions in the User's Guide to complete this activity.

Input the information in the following problems to find the balance on hand for each date, the average cost per unit, and the value of the inventory using the average-cost method. Then answer the questions that follow.

Date	Unit Cost	Receipts (Quantity IN)	Issues (Quantity OUT)	Balance on Hand
10/01	$17.45	100		?
10/05			75	?
10/09	17.20	80		?
10/12	17.00	250		?
10/19			98	?
10/26			123	?
10/29	16.50	100		?
10/30			54	?

What is the average cost per unit?

What is the inventory value using the average-cost method?

1. What is the balance on hand after 10/01?

2. What is the balance on hand after 10/05?

3. What is the value of the 80 items received on 10/09?

4. What is the balance on hand after 10/12?

5. What is the balance on hand after 10/26?

6. What is the ending balance on 10/30?

7. What is the average cost per item?

8. How did the computer calculate the average cost per unit?

9. What is the inventory value using the average-cost method?

10. How did the computer calculate the inventory value using the average cost per unit?

11. Suppose the unit cost for each receipt (10/01, 10/09, 10/12, and 10/29) equaled $17.00. What would the average cost per unit equal?

12. In the situation described in #11, what would be the inventory value using the average-cost method?

13. In the given situation, a declining economy, where unit costs are decreasing, would FIFO or LIFO result in the higher valued inventory? Why?

14. In the given situation, a declining economy, where unit costs are decreasing, would FIFO or LIFO result in the lower valued inventory? Why?

15. What is the value of this inventory using FIFO?

16. What is the value of this inventory using LIFO?

Shipping Clerk

Mail-order shopping has become big business. Rosa DiPace knows this well, since she works for a mail-order house that boasts it can ship within 24 hours of the time the order is received. Hundreds of credit card orders are placed each day.

Rosa, along with others, is responsible for making sure that the package is complete, the invoice is correct, and that shipping schedules are met. Her company is liable for the goods until they are secured aboard the trucks that deliver them to customers.

The step graph shows the shipping cost as a function of the amount of the purchase. The graph shows, for example, that it costs $5.50 to ship an order valued at between $20.00 and $29.99.

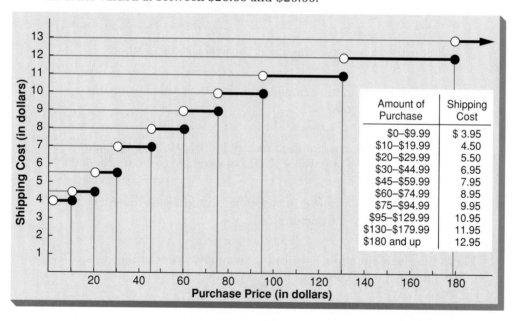

Amount of Purchase	Shipping Cost
$0–$9.99	$ 3.95
$10–$19.99	4.50
$20–$29.99	5.50
$30–$44.99	6.95
$45–$59.99	7.95
$60–$74.99	8.95
$75–$94.99	9.95
$95–$129.99	10.95
$130–$179.99	11.95
$180 and up	12.95

1. True or false: It costs the same to ship an order valued at $110 as it does an order valued at $100.

2. Find the difference between shipping an order valued at $48 and an order valued at $26.

3. True or false: The cost of shipping an order valued at $320 is twice that of an order valued at $160.

4. How much less does it cost to ship one order valued at $210 than an order valued at $100 and an order valued at $110?

18

Services

Rent for your office space, maintenance costs of the building, and costs of utilities are all part of business *services.* Many businesses hire professional services for building maintenance and telephone installation and repair. In addition, your business may hire professional *consultants* to advise you on particular problems.

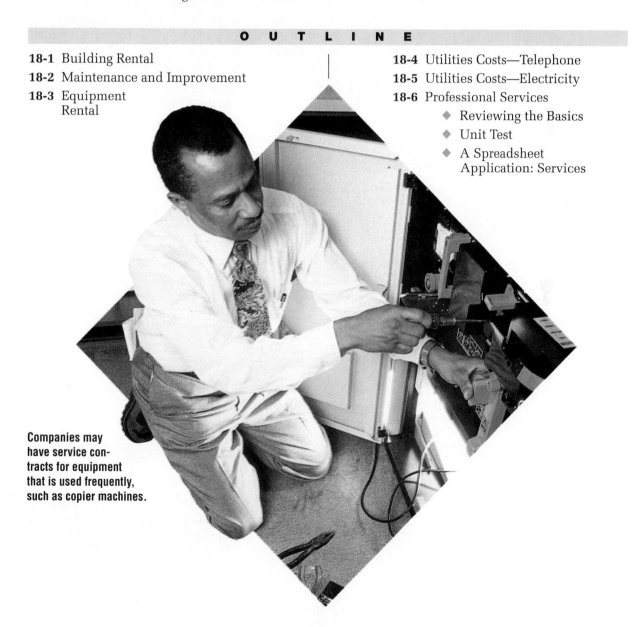

Companies may have service contracts for equipment that is used frequently, such as copier machines.

Building Rental

OBJECTIVE

Compute the monthly rental charge.

Your business may rent or lease a building or a portion of a building, usually on an annual basis. The building owner may charge a certain rate per square foot per year. Your total monthly rental charge depends on the number of square feet that your business occupies.

$$\text{Monthly Rental Charge} = \frac{\text{Annual Rate} \times \text{Number of Square Feet}}{12}$$

EXAMPLE Skills 10, 2 Application X Term Rent

Ajax Assemblers, Inc., rents a portion of a building owned by The Gray Company. The floor space of Ajax's portion of the building measures 80 feet by 60 feet. The Gray Company charges an annual rate of $5.00 per square foot. To the nearest dollar, what is Ajax's monthly rental charge?

SOLUTION

A. Find the **number of square feet.**
   Length × Width
   80 ft × 60 ft = 4800 ft^2

B. Find the **monthly rental charge.**

$$\left(\begin{array}{c} \text{Annual} \\ \text{Rate} \end{array} \times \begin{array}{c} \text{Number of} \\ \text{Square Feet} \end{array} \right) \div 12$$

   ($5.00 × 4800) ÷ 12

   $24,000.00 ÷ 12 = $2000.00 monthly rental charge

   80 × 60 = 4800 × 5 = 24000 ÷ 12 = 2000

✓ SELF-CHECK Complete the problems, then check your answers in the back of the book.

Find the number of square feet and the monthly rental charge.

1. 100 ft × 50 ft at $4.00 per square foot annually.

2. 125 ft × 45 ft at $3.50 per square foot annually.

PROBLEMS

Find the monthly rental charge to the nearest dollar.

	Dimensions =	(Number of Square Feet	×	Annual Rate per Square Foot)	÷ 12 =	Monthly Rental Charge
3.	15 ft by 30 ft =	(?	×	$ 8.00)	÷ 12 =	?
4.	40 ft by 20 ft =	(?	×	$ 7.50)	÷ 12 =	?
5.	25 ft by 60 ft =	(?	×	$10.00)	÷ 12 =	?

6. *Westside Herald* warehouse. Dimensions are 45 feet by 50 feet. Annual rental charge is $3.40 per square foot. What is the monthly rental charge?

7. Earring Tree mall store. Dimensions are 15 feet by 30 feet. Annual rental charge is $10.75 per square foot. What is the monthly rental charge?

8. The Miller Manufacturing Company is considering the rental of additional manufacturing space at $3.80 per square foot per year. The space Miller wants to rent measures 80 feet by 120 feet. What monthly rent will Miller pay for the additional space?

9. The Cave has rented additional mall space to expand its arcade operation. The space measures 30 feet by 40 feet and rents for $15.60 per square foot per year. What monthly rent does the Cave pay for the additional space?

10. The Flower Shoppe is opening a store in the warehouse district. The rent is $8.75 per square foot per year plus 5% of the store's gross sales. The area of the store is 2000 square feet. If The Flower Shoppe had $180,000 in gross sales the first year, what monthly rent will it pay?

11. The Luncheonette rents a 30-foot by 50-foot area at the Nottingham Mall. The Luncheonette pays $9.20 per square foot per year plus $2\frac{1}{2}$% of gross sales. Last year The Luncheonette had $225,000 in gross sales. What was its monthly rent?

12. The local office of Wall Medical Clinic is planning to open a branch office in the Heatherdowns area. The office in the advertisement suits its needs. What is the cost per square foot per year for this office?

> **HEATHERDOWNS AREA**
> OFFICE SPACE
> 1400 sq ft, carpeted and paneled, gas heat and air-conditioned. $750 per mo.
> 555-6619

13. Some businesses rent office space for 2 or 3 years at a time. Often, the lease for the office space includes a **rent escalation clause.** The rent escalation clause states that the rent may increase by a certain percent after a certain period of time.

Robon Legal Clinic rents a suburban office space for $8.75 per square foot per year. The dimensions of the office are 35 feet by 40 feet. Robon signed a 2-year lease with a 6% rent escalation clause. The rent will increase by 6% for the second year of the lease. What is the monthly rental charge for each year of the lease?

MAINTAINING YOUR SKILLS   Look up the skill in parentheses if you need help or more practice.

Divide. Round answers to the nearest dollar. **(Skill 10)**

14. $18,900 ÷ 12

15. $25,800 ÷ 12

16. $14,949 ÷ 12

17. $9070 ÷ 12

18. $31,645 ÷ 12

19. $216,847 ÷ 12

18-2

Maintenance and Improvement

OBJECTIVE

Compute the total building maintenance charge.

If your business owns a building, you will need people to clean and maintain the building. You may have your own maintenance department, or you may hire a service firm. The total cost of a particular maintenance job generally includes a labor charge and a materials charge. The labor charge is the cost of paying the people who do the job and is calculated on an hourly basis for each service person.

Total Charge = Labor Charge + Materials Charge

EXAMPLE   *Skill* 8   *Application* A   *Term* Labor charge

Central Law Offices, Inc., hired Commercial Painting Service to paint its offices. Four painters worked 23 hours each to complete the job. The regular hourly rate for each painter is $19. All work was done on weekends, so the painters were paid time and a half. The painters used 42 gallons of paint, for which they charged $15.75 per gallon. What was the total charge for painting the offices?

S O L U T I O N

A. Find the **labor charge.**
($19.00 × 1.5) × 23 hrs. × 4 = $2622.00 labor charge

B. Find the **materials charge.**
42 × $15.75 = $661.50 materials charge

C. Find the **total charge.**
Labor Charge + Materials Charge
$2622.00 + $661.50 = $3283.50 total charge

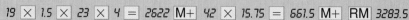

19 $\times$ 1.5 $\times$ 23 $\times$ 4 $=$ 2622 M+ 42 $\times$ 15.75 $=$ 661.5 M+ RM 3283.5

✔ SELF-CHECK   Complete the problems, then check your answers in the back of the book.

Find the total charge.

1. 4 people worked 10 hours each at $6.50 per hour. Materials cost $420.

2. 3 people worked 6 hours each at $8.00 per hour. 9 gallons of paint at $14.85/gal.

PROBLEMS

$\Big($ Time Required	$\times$	Number of Employees	$\times$	Hourly Rate	$=$	Labor Charge $\Big)$	$+$	Materials Charge	$=$	Total
3. (8 hours	×	2	×	$10.50 =		?)	+	$425.00	=	?
4. (3 hours	×	12	×	$ 5.75 =		?)	+	0	=	?
5. ($4\frac{1}{2}$ hours	×	5	×	$ 9.80 =		?)	+	$947.56	=	?

6. 2 people replace fluorescent lights.
Hourly rate of $4.75 each.
Job takes 4 hours.
What is the labor charge?

7. 4 people clean office area.
Hourly rate of $5.37 each.
Job takes 12 hours.
What is the labor charge?

8. Dr. Alice Desmond is moving from her present office to another office in the same building. It takes 4 people $3\frac{3}{4}$ hours to complete the move. The hourly rate per person is $12.80. What is the total charge?

9. City Cleaners hired 2 carpenters to remodel its store. The carpenters earned $14.95 each per hour. Each carpenter worked $21\frac{1}{2}$ hours. The materials charge was $1617.48. Find the total charge.

10. The National Freight Company used 6 loads of crushed stone on a terminal lot at $92.75 a load. Ten people were each paid $5.15 an hour to spread the stone. It took 4 hours. What was the total charge?

11. Scott's Supermarket hired 5 people to refinish its floor. The regular hourly rate for each person was $6.15. Because the floor was refinished on Sunday, each worker was paid double time. The job took $5\frac{1}{2}$ hours to complete. The materials charge was $371.85. What was the total charge?

12. Miller's Department Store had a minor fire in its store. Miller's received these estimates for repair work.

Contractor	Hourly Rate	Time Required	Materials Charge
Carpenter	$17.15	$12\frac{1}{2}$ hours	$949.73
Electrician	$18.65	13 hours	$221.87
Painter	$ 9.85	$12\frac{1}{4}$ hours	$214.46

Miller's must decide if it should hire these 3 people or have its own employees repair the damage. Materials would cost Miller's a total of $1517.45. It would take 5 Miller's employees 13 hours each to do the job. Three of the employees each earn $9.40 an hour. The other 2 employees each earn $6.97 an hour. If the price is the only consideration, what should Miller's do?

13. General Janitorial Service cleans the Second National Bank daily from Monday through Saturday. One General employee works at the bank for 3 hours each day. The employee is paid $4.60 per hour on weekdays and time and a half on Saturdays. When calculating Second National's total charge, General adds 75% of the cost of labor to cover its overhead. How much does General charge the Second National Bank per week?

MAINTAINING YOUR SKILLS Look up the skill in parentheses if you need help or more practice.

Multiply. **(Skill 8)**

14. $3.85 × 5 × 4

15. $9.75 × 5 × 3

16. $7.43 × 2.5 × 4

17. $8.45 × $2\frac{1}{4}$ × 5

18. $6.53 × $3\frac{3}{4}$ × 6

19. $5.17 × $4\frac{1}{2}$ × 2

Equipment Rental

Your business may find that it is more economical to rent than to buy certain equipment or furniture. Generally, the total cost of renting items is determined by the length of time for which you rent them. Some states charge a use tax on items that are rented.

Total Rental Cost = (Rental Charge × Time) + Tax

EXAMPLE   *Skill* 30   *Application* A   *Term* Rent

Tax-Aide, Inc., is renting new furniture for a small, temporary office. Office Rental Company charges 10% of the list price of new furniture per month. The list prices of the pieces that Tax-Aide is renting are:

Item	List Price
Desk, 60 in x 30 in	$259.95
File cabinet, 4-drawer	119.95
Swivel armchair	184.45
2 guest chairs ($97.45 ea)	194.90
Total	$759.25

In addition to the rental charge, there is a 5% tax. What is the total cost of renting the furniture for 4 months?

SOLUTION

A. Find the **rental charge per month.**
10% of list price total
$759.25 × 10% = $75.925 = $75.93 rental charge

B. Find the **tax.**
($75.93 × 4) × 5%
$303.72 × 5% = $15.186 = $15.19 tax

C. Find the **total rental cost.**
(Rental Charge × Time) + Tax
($75.93 × 4) + $15.19
$303.72 + $15.19 = $318.91 total rental cost

759.25 ⊠ 10 % 75.925 75.93 ⊠ 4 ⊟ 303.72 M+ ⊠ 5 % 15.186 ⊞ RM
318.906

✔ SELF-CHECK   Complete the problems, then check your answers in the back of the book.

Find the total rental cost.

1. Rental charge: 10% of $640.
Tax rate: 6%.
Time: 3 months.

2. Rental charge: 8% of $1470.
Tax rate: 5.5%.
Time: 5 months.

Item Price	List Charge	Monthly Charge	Monthly Time	Tax	Cost	Rental
3. Word processor	$ 947.80	10% of list	4 months	4%	?	?
4. File cabinet	$ 97.45	11% of list	12 months	6%	?	?
5. Fax machine	$1141.80	9% of list	8 months	$6\frac{1}{4}\%$	?	?

6. Backhoe rented for 3 weeks.
Rental charge is $1475 per week.
Tax is 5%.
What is the total rental cost?

7. Jackhammer rented for 7 days.
Rental charge is $175.95 per day.
No tax.
What is the total rental cost?

8. The list price of a personal computer is $2248. The monthly rental charge is $10\frac{1}{2}\%$ of the list price. There is a $5\frac{1}{4}\%$ tax per month. What is the total rental charge for 1 year?

9. The Strong Cement Company plans to rent a bulldozer for 2 days. The rental charge is $212.45 per day plus a flat fee of $45 for delivery and pickup. There is no tax. What is the total rental cost?

10. The Legal Clinic is renting additional furniture for 6 months. The monthly rental charge is 9.5% of the list price. There is an 8% use tax. What is the total rental cost of the following:

Item	List Price
3 desks with chairs	$329.95 each
10 guest chairs	69.25 each
1 typewriter	712.45 each

11. Harris and Sons, plumbers, rent a ditcher as they need it. The daily rental charge is $197.47. There is a $4\frac{1}{2}\%$ tax. What is the total cost of renting the ditcher for 10 days?

12. Airline Travel needs an additional 4-drawer file cabinet for 6 months. Airline received cost information from 2 rental agencies and 1 used-equipment store. If price is the only consideration, which company should Airline contact?

CA Furniture Rental
Per month:
• $16.45 rental charge
• 5% insurance charge
• 5% use tax

Direct Office Rental
Per month:
• $17.49 rental charge
• No insurance charge
• 5% use tax

Economy Office Furniture
• $119.49 purchase price
• 5% sales tax
• $12.50 delivery charge

MAINTAINING YOUR SKILLS   Look up the skill in parentheses if you need help or more practice.

Find the percentage. Round answers to the nearest cent. **(Skill 30)**

13. 5% of $516 **14.** 6% of $48.40 **15.** $5\frac{1}{2}\%$ of $146.30 **16.** 10% of $271.84

18-4

Utilities Costs—Telephone

OBJECTIVE

Compute the monthly telephone cost.

To operate your business, you will need several utilities. Utilities are public services, such as telephone, electricity, water, and gas. Each utility uses a different cost structure for charging its customers. For example, the basic monthly charge for your telephone service depends on the number of incoming lines, the type of equipment, and the type of service that you have. If you make more calls than the number included in your basic monthly charge, you must pay an additional amount. A federal excise tax is also added to your telephone charge each month.

$$\begin{array}{c} \text{Total Cost} \\ \text{for Month} \end{array} = \begin{array}{c} \text{Basic Monthly} \\ \text{Charge} \end{array} + \begin{array}{c} \text{Cost of} \\ \text{Additional Calls} \end{array} + \begin{array}{c} \text{Federal} \\ \text{Excise Tax} \end{array}$$

EXAMPLE   *Skills* 5, 30   *Application* A   *Term* Basic monthly charge

Andy's Laundry has 1 incoming telephone line. The basic monthly charge for this line is $26.15. The basic monthly charge includes 73 outgoing local calls per month. The cost of additional outgoing local calls is $0.08 each. Andy's made 92 local calls last month. A 3% federal excise tax is added to the bill. What is the total cost of Andy's telephone service for the month?

SOLUTION

A. Find the **cost of additional calls.**

(92 − 73)   × $0.08

     19   × $0.08   = $1.52 cost for additional calls

B. Find the **federal excise tax.**

($26.15 + $1.52)   ×   3%

     $27.67   ×   3% = $0.8301 = $0.83 excise tax

C. Find the **total cost for month.**

$$\begin{array}{c} \text{Basic Monthly} \\ \text{Charge} \end{array} + \begin{array}{c} \text{Cost of} \\ \text{Additional Calls} \end{array} + \begin{array}{c} \text{Federal} \\ \text{Excise Tax} \end{array}$$

     $26.15   + $1.52   + $0.83 = $28.50 total cost

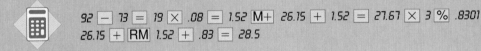

92 − 73 = 19 × .08 = 1.52 M+ 26.15 + 1.52 = 27.67 × 3 % .8301
26.15 + RM 1.52 + .83 = 28.5

✓ SELF-CHECK   Complete the problems, then check your answers in the back of the book.

Find the total cost for the month.

1. Base charge of $26.15.
    10 additional calls at $0.08 each.
    3% federal tax.

2. Base charge of $28.45.
    30 additional calls at $0.09 each.
    3% federal tax.

	Basic Monthly Charge	Additional Local Calls	Cost per Additional Call	Cost of All Additional Calls	3% Federal Excise Tax	Total Cost for Month
3.	$26.15	15	$0.08	$1.20	$0.82	?
4.	$28.75	8	$0.09	?	?	?
5.	$42.35	21	$0.10	?	?	?
6.	$15.25	0	$0.08	?	?	?

Add a 3% federal excise tax to find the total cost for the month.

7. Spotlight Discount has 1 telephone. Basic monthly charge is $26.15. Includes 73 outgoing local calls. Each additional call costs $0.08. Made 115 outgoing local calls. What is the total cost for the month?

8. Public Pharmacy has 1 telephone. Basic monthly charge is $27.50. Includes 80 outgoing local calls. Each additional call costs $0.09. Made 42 outgoing local calls. What is the total cost for the month?

9. Smith Medical Clinic has 2 telephones. The total basic monthly charge for both telephones is $31.65. The charge includes 160 outgoing local calls. The cost of additional local calls is $0.09 each. Smith Medical Clinic made 196 outgoing local calls this month. What is the total cost of telephone service, including tax, for the month?

For problems 10–11, refer to the chart at the right.

10. Charles Young, Inc., brokers for stocks and bonds, have 5 telephones for a total basic monthly charge of $88.60. The charge includes 500 local outgoing calls. The cost of additional local calls is $0.08 each. Last month, 863 outgoing local calls were made from the office. In addition, long-distance calls totaling $488.32 were made. Charles Young also has call forwarding and line-backer special services. The federal excise tax is applied to all charges. What is the total cost for the month?

11. Prime Realty Company has 3 telephone lines. The monthly charge of $47.85 includes 160 outgoing local calls. Each additional local call costs $0.11. Prime made 180 outgoing local calls this month. Prime also has call waiting and line-backer plus special services. What is the total cost of telephone service, including tax, for the month?

MONTHLY CHARGE PER SPECIAL SERVICE	
Call waiting	$9.60
Call forwarding	3.25
3-way calling	3.25
Line-backer	1.25
Line-backer plus	2.50

MAINTAINING YOUR SKILLS Look up the skills in parentheses if you need help or more practice.

Add. **(Skills 5, 6, 8, and 30)**

12. $26.15 + $1.60 + $0.83

13. $32.95 + $2.79 + $1.07

14. $26.15 + [(103 − 73) × $0.08]

15. $37.45 + $2.45 + $3.25 + (3% of $43.15)

Utilities Costs— Electricity

The monthly cost of electricity for your business depends on the demand charge and the energy charge. The demand charge is based on the peak load during the month. The peak load is the greatest number of kilowatts (kW) that your business uses at one time during the month. The energy charge is based on the total number of kilowatt-hours (kW·h) that your business uses during the month. Meters installed at your business record the kilowatt demand and the number of kilowatt-hours. The electric company may add a fuel adjustment charge to your monthly bill to help cover increases in the cost of fuel needed to produce your electricity.

$$\frac{\text{Total Cost}}{\text{for Month}} = \frac{\text{Demand}}{\text{Charge}} + \frac{\text{Energy}}{\text{Charge}} + \frac{\text{Fuel Adjustment}}{\text{Charge}}$$

EXAMPLE   *Skill* 8   *Application* A   *Term* Fuel adjustment charge

The Acme Manufacturing Company had a peak load of 100 kilowatts of electricity during April. The demand charge is $6.54 per kilowatt. Acme used a total of 30,000 kilowatt-hours of electricity during the month. The energy charge for the first 1000 kilowatt-hours is $0.076 per kilowatt-hour. The cost of the remaining kilowatt-hours is $0.058 per kilowatt-hour. The fuel adjustment charge for April is $0.0155 per kilowatt-hour. What is the total cost of the electricity that Acme used in April?

SOLUTION

A. Find the **demand charge.**
100 kW × $6.54 = $654.00 demand charge

B. Find the **energy charge.**
(1) 1000 kW·h × $0.076 = $76.00
(2) (30,000 − 1000) kW·h × $0.058 = $1682.00
$76.00 + $1682.00 = $1758.00 energy charge

C. Find the **fuel adjustment charge.**
30,000 kW·h × $0.0155 = $465.00 fuel adjustment charge

D. Find the **total cost for month.**
Demand Charge + Energy Charge + Fuel Adjustment Charge
$654.00   + $1758.00   + $465.00 = $2877.00
total cost

100 ⊠ 6.54 M+ 654 1000 ⊠ .076 = 76 30000 − 1000 = 29000 ⊠ .058 =
1682 76 + 1682 M+ 1758 30000 ⊠ .0155 M+ 465 RM 2877

1. Find the total cost for month.

Demand Charge   Energy Charge   Fuel Adjustment Charge

$(100 \times \$6.60) + [(1000 \times \$0.098) + (9000 \times \$0.067)] + (10{,}000 \times \$0.016)$

PROBLEMS

2. Flagg Mattress Company.
Used 17,000 kW·h in May.
Peak load of 100 kW.
Demand charge: $6.45 per kW.
Energy charge: $0.07 per kW·h.
Fuel adjustment: $0.018 per kW·h.
What is the total cost for May?

3. Town and Country Mall.
Used 18,700 kW·h in April.
Peak load of 90 kW.
Demand charge: $5.90 per kW.
Energy charge: $0.08 per kW·h.
Fuel adjustment: $0.02 per kW·h.
What is the total cost for April?

F.Y.I.
Supplies of natural gas and heating oil are ample and shortages or sharp price rises are unlikely (*Kiplinger*, December 1991, p. 17).

4. City Deli used 8700 kilowatt-hours of electricity with a peak load of 100 kilowatts in June. The demand charge is $7.25 per kilowatt. The energy charge is $0.08 per kilowatt-hour for the first 1000 kilowatt-hours and $0.06 per kilowatt-hour for all kilowatt-hours over 1000. The fuel adjustment charge is $0.03 per kilowatt-hour. What is the total cost of electricity for City Deli for June?

5. Pantry Supermarket used 21,400 kilowatt-hours of electricity last month. The peak load for the month was 120 kilowatts. The demand charge is $5.91 per kilowatt. The energy charge per kilowatt-hour is $0.0675 for the first 10,000 kilowatt-hours and $0.0455 for all kilowatt-hours over 10,000. The fuel adjustment charge is $0.015 per kilowatt-hour. What is the total cost of electricity for Pantry Supermarket for last month?

6. *The City Journal* used 12,417 kilowatt-hours of electricity in August. The peak load during the month was 78 kilowatts. The demand charge is $6.96 per kilowatt. The energy charge per kilowatt-hour is $0.0872 for the first 10,000 kilowatt-hours and $0.0685 for all kilowatt-hours over 10,000. There is no fuel adjustment charge. What is the total cost of electricity for *The City Journal* in August?

For problems 7–9, refer to the chart.

7. Riechle Manufacturing used 15,400 kilowatt-hours of electricity this month; the peak load was 80 kilowatts. What is the total cost of electricity for the month?

8. Prompt Answering Service, Inc., used 7400 kilowatt-hours of electricity for July. Its peak load was 75 kilowatts. What is the total cost of electricity for the month?

9. Combustible, Inc., used a total of 21,760 kilowatt-hours of electricity for the month. The peak load during the month was 170 kilowatts. What is the total cost of electricity for the month?

NORTHWEST POWER

Demand Charge

First 50 kW	$6.54/kW
Over 50 kW	$5.91/kW

Energy Charge

First 250 kW·h	$0.1055/kW·h
Next 750 kW·h	$0.0954/kW·h
Next 2000 kW·h	$0.0644/kW·h
Next 2000 kW·h	$0.0578/kW·h
Next 5000 kW·h	$0.0488/kW·h
Over 10,000 kW·h	$0.0455/kW·h
Fuel Adjustment	$0.017/kW·h

Multiply. **(Skill 8)**

10. 100 × $6.47

11. 1000 × $0.087

12. 8700 × $0.063

13. 9700 × $0.015

Professional Services

OBJECTIVE

Compute the total cost of professional services.

You may hire professional consultants to advise your business on a particular problem. The method of determining each consultant's fee may vary. Some consultants charge a flat fee, some charge a percent of the cost of the project, and some charge by the hour.

Total Cost = Sum of Consultants' Fees

EXAMPLE *Skills* 5, 8, 30 *Application* A *Term* Consultant's fee

Appleton Wholesale Grocers, Inc., plans to expand its use of computers to include billing, payroll, and inventory control. The company plans to construct a new building at a cost of $875,000 to house the new computer installation. Appleton hires an architect, a systems analyst, and a computer programmer. The architect charges 7% of the total cost of the building. The systems analyst charges a flat fee of $6000. The computer programmer charges $30 an hour and works 150 hours. What is the total cost of the professional services?

SOLUTION Find the **sum of consultants' fees.**

Architect: $875,000 × 7% $61,250
Systems analyst: Flat fee 6000
Computer programmer: $30 × 150........................ 4500
                                        Total Cost $71,750

875000 ⨉ 7 % 61250 M+ 6000 M+ 30 ⨉ 150 = 4500 M+ RM 71750

✔ SELF-CHECK Complete the problems, then check your answers in the back of the book.

Find the total cost.

1. Architect: 6% of $450,000.
Lawyer: 20 hours at $75
per hour.
Surveyor: $2500 flat fee.

2. Engineer: 25 hours at $42.50
per hour.
Computer programmer: $1000 flat fee.
Computer time: $450 flat fee.

PROBLEMS

	Professional Service	Fee Structure	Project Information	Total Fee
3.	Patent attorney	$75.00 per hour	Worked 20 hours	?
4.	Industrial nurse	$10.25 per hour	Worked 40 hours	?
5.	Productivity consultant	$5000 flat fee	Worked 25 hours	?
6.	Planning specialist	5% of project cost	$500,000 project	?

7. Bond broker.
Paid $1\frac{1}{2}\%$ of total bonds sold.
$10,000,000 sold.
What is the total fee?

8. Real estate broker.
Paid $3\frac{1}{2}\%$ of total sale.
Sale totals $615,500.
What is the total fee?

9. The June Co. hired a staff development specialist to conduct a workshop for 83 of its employees. The specialist was paid $110 per person. What was the total cost of the specialist's services?

10. Lakeside Hospital hired an industrial engineering firm to conduct a work sampling on the average nurse's day. Time Study, Inc., did the work sampling and charged $67.50 per hour. It took Time Study, Inc., 32 hours to complete the task. What did the work sampling cost Lakeside Hospital?

11. The Central Farmers Co-Op plans to issue $20,000,000 in bonds to pay for an extensive expansion. One bond broker will sell the bonds for a fee of 3.25% of the $20,000,000 face value of the bonds. What will the broker's services cost Central?

12. Modern Service Stations plans to sell a large piece of commercial property. The selling price of the property is $975,000. Roula Mathis will handle the entire transaction for 6.5% of the selling price. What will it cost Modern Service Stations to have Roula handle the transaction?

13. Liberty Bank and Trust plans to build a new branch office. Liberty hired Ed Baker, an architectural consultant. Ed's fee is 7% of the cost of the project. Liberty hired Tina Pike as project engineer for the new branch. Tina's fee is 4.5% of the cost of the project. Liberty hired Sara Charles to take care of the legal aspects of the project. Sara charges $75 per hour. It took her 10.5 hours to prepare the legal documents. The cost of the branch office is estimated to be $4,575,000. What amount should Liberty plan to pay for the 3 professional services?

14. Super Food Mart has installed automated checkout equipment in its 6 stores. It hired Robert Case to consult on its installation. Case charged 6% of the installation cost. It also hired Marilyn Lee to instruct its employees on the use of the equipment. Marilyn conducted six 4-hour sessions at $125 per session. The cost of the installation totaled $68,970. What was the total cost for professional services?

MAINTAINING YOUR SKILLS   Look up the skills in parentheses if you need help or more practice.

Solve. **(Skills 5, 8, 30)**

15. $175.00 + $493.75 + $216.80

16. $14,150.00 + $7185.90 + $417.95

17. (5% of $1,750,000) + $14,750

18. (7.5% of $863,900) + (25 × $9.50)

19. ($85 × 24.5) + $5000 + (2% of $2,500,000)

20. (6% of $3,500,000) + ($75 × 6.5) + $1475

Reviewing the Basics

(Skill 2)

Round answers to the nearest dollar.

1. $42.74   **2.** $117.40   **3.** $4171.60   **4.** $12,147.83

Solve. Round answers to the nearest hundredth.

(Skill 3)

5. $179 + $464 + $958   **6.** $201 + $743 + $6198

7. $17,946 + $9217

(Skill 4)

8. 743 − 295   **9.** 15,000 − 11,763   **10.** 74,000 − 7193

(Skill 5)

11. 916.87 + 1714.31   **12.** 716.47 + 176.178   **13.** 7.418 + 15.79

(Skill 6)

14. 8417.92 − 147.74   **15.** 4684.75 − 619.87   **16.** 219.47 − 156.26

(Skill 7)

17. 144 × 50   **18.** 36 × 74   **19.** 48 × 150

(Skill 8)

20. 26.12 × 615   **21.** 4217.64 × 0.655   **22.** 34,157 × 0.48

(Skill 10)

23. $4178 ÷ 12   **24.** $22,196 ÷ 24   **25.** $31,965 ÷ 36

(Skill 30)

26. $417,800 × 35%   **27.** $125,750 × 6.5%   **28.** $3,450,000 × 1.5%

(Application X)

Find the area of each rental space.

29. 30 ft by 40 ft   **30.** 25 ft by 45 ft   **31.** 120 ft by 150 ft

32. 15 ft by 15 ft   **33.** 10.5 ft by 48.5 ft   **34.** 115.5 ft by 42.75 ft

Use each term in one of the sentences.

35. Rent

36. Labor charge

37. Basic monthly charge

38. Fuel adjustment charge

39. Consultant's fee

a. Your __?__ for telephone service may include a certain number of outgoing local calls.

b. You will pay a __?__ when you hire a professional person to advise your business on a particular project.

c. A __?__ may be added to your electricity bill.

d. Your business may __?__ rather than buy office space or equipment.

e. The __?__ is the greatest number of kilowatts that your business uses at one time during the month.

f. The cost of having a maintenance firm do a particular job may include a __?__ and a charge for materials.

Refer to your reference files in the back of the book if you need help.

Unit Test

Lesson 18-1

1. Moore's Nursery rents warehouse space at an annual rate of $4.50 per square foot. Moore's warehouse measures 60 feet by 150 feet. What is the monthly rental charge?

Lesson 18-2

2. Town Mall hired 4 high school students to spring-clean the grounds. Each student worked for 2.5 hours at an hourly rate of $4.50. The materials charge was $46.75. What was the total charge for this service?

Lesson 18-3

3. Mary Taylor is opening a branch tax office for 3 months. She plans to rent this furniture at a monthly charge of 10% of the list price plus a 6% tax.

Item	List Price
2 desks	$279.79 each
2 desk chairs	97.50 each
4 guest chairs	74.75 each
1 file cabinet	112.45
1 bookcase	147.50

What is the total rental cost for the furniture?

Lesson 18-4

4. The Greco Import Store pays a basic monthly charge of $26.15 for telephone service. The charge includes 73 outgoing local calls. Each additional outgoing local call costs $0.08. Last month a total of 95 outgoing local calls were made from Greco's. A 3% federal tax was added to the bill for the month. What was the total cost of telephone service for the month?

Lesson 18-5

5. The Norris Company used 11,650 kilowatt-hours of electricity with a peak load of 120 kilowatts in April. The demand charge is $6.54 per kilowatt. The energy charge per kilowatt-hour is $0.076 for the first 10,000 kilowatt-hours and $0.058 for all kilowatt-hours over 10,000. The fuel adjustment charge is $0.0165 per kilowatt-hour. What is the total cost of electricity for The Norris Company for April?

Lesson 18-6

6. Innovative Automotive, Inc., hired a consultant from Engineering Consortium to help develop a new engine. The consultant's fee was 7.3% of the cost of the project. The project cost $843,500. What was the total cost of the consultant's services?

A SPREADSHEET APPLICATION

Services

To complete this spreadsheet application, you will need the template diskette for *Mathematics with Business Applications*. Follow the directions in the *User's Guide* to complete this activity.

Input the information in the following problems to find the monthly rent for square footage, the monthly share of gross sales, and the total monthly rent.

	Store	Rental Space Dimensions	Rent per Sq Ft per Yr	Monthly Sq Ft Rent	Plus % of Gross Sales	Annual Gross Sales	Monthly Share Gross Sales	Total Monthly Rent
1.	Video	45' × 60'	$ 7.50	?	5%	$124,600	?	?
2.	Clothes	90' × 120'	2.75	?	4%	215,780	?	?
3.	Food	20' × 30'	10.25	?	0	—	0	?
4.	Arcade	60' × 75'	9.80	?	0	—	0	?
5.	Photo	50' × 50'	6.75	?	3%	156,860	?	?
6.	Shoe	40' × 65'	8.25	?	2.5%	453,780	?	?
7.	Lunch	35' × 50'	12.50	?	0	—	0	?
8.	Music	50' × 75'	9.50	?	4.5%	267,750	?	?
9.	Flower	25' × 30'	17.50	?	2%	98,574	?	?
10.	Deli	35' × 40'	13.60	?	0	—	0	?

11. In #1, what is the monthly rent for square footage?

12. In #1, what is the monthly share of gross sales?

13. In #1, what is the total monthly rent?

14. In #2, what is the monthly rent for square footage?

15. In #2, what is the monthly share of gross sales?

16. In #2, what is the total monthly rent?

17. In #1, the rent per square foot per year is almost $5 more than in #2, yet the total monthly rent is less in #1. State two reasons why this is the case.

18. In #3, what is the total monthly rent?

19. In #4, what is the total monthly rent?

20. In #5, what is the monthly rent for square footage?

21. In #5, what is the monthly share of gross sales?

22. In #5, what is the total monthly rent?

23. In #6, what is the *annual* rent for square footage?

24. In #6, what is the *annual* share of gross sales?

25. In #6, what is the total *annual* rent?

26. In #7, what is the total monthly rent?

27. In #8, what is the monthly rent for square footage?

28. In #8, what is the monthly share of gross sales?

29. In #8, what is the total monthly rent?

30. In #9, what is the total monthly rent?

31. In #10, what is the total *annual* rent?

32. In #3, #4, #7, and #10, why does the monthly rent for square footage equal the total monthly rent?

Accounting

The accounting department prepares the company *payroll register*, which is a record of all your employees' income. The department also keeps records of all your business expenses and the value of your equipment. The value of your equipment goes down, or *depreciates*, yearly

because of age or wear and tear. There are four ways to determine depreciation: *straight-line method, sum-of-the-years'-digits method, double-declining-balance method,* and *modified accelerated cost recovery system.*

O U T L I N E

Instead of writing payroll checks, many companies now use computers to have wages automatically deposited to employees' bank accounts.

19-1

Payroll Register

OBJECTIVE
Complete a payroll register.

A payroll register is a record of the gross income, deductions, and net income of your company's employees. You may use a computer to prepare your payroll register. If you prepare the register by hand, you will probably refer to tables to determine the amount of income tax to withhold from each employee's pay.

EXAMPLE   *Skills* 5, 6, 8   *Application* C   *Term* Payroll register

Natural Foods Center pays its employees weekly. Mary Clark prepares the payroll register for the center's 5 employees from the following information.

Name	Regular Pay	Overtime Pay	Income Tax Information	Health Insurance Coverage
T.L. Wright	$4.50/hr.	Time and a half	Married, 2 allow.	Family
J.A. Bruss	$6.45/hr.	Time and a half	Single, 1 allow.	—
H.T. Perkins	$325/week	—	Single, 2 allow.	Single
N.J. Nystrand	$435/week	—	Married, 3 allow.	Family
W.K. Fine	$280/week plus 5% commission	—	Married, 4 allow.	Family

SOLUTION   Mary has prepared the payroll register for the week of March 2. What is the net pay for the week?

$280 plus 5% of $1860 in sales | From federal withholding tables | 7.65% of gross pay

Payroll Register for Week of: *March 2, 19–*

Name	Hours Worked Reg.	Hours Worked OT	Hourly Rate	Gross Pay	FIT	FICA	Hosp. Ins.	Total	Net Pay
T.L. Wright	40	3	$4.50	$200.25	$ 8.00	$ 15.32	$41.54	$64.86	$135.39
J.A. Bruss	40	—	6.45	258.00	28.00	19.74	20.77	68.51	189.49
H.T. Perkins	—	—	—	325.00	33.00	24.86	20.77	78.63	246.37
N.J. Nystrand	—	—	—	435.00	36.00	33.28	41.54	110.82	324.18
W.K. Fine	—	—	—	373.00	21.00	28.53	41.54	91.07	281.93
Total				$1591.25	$126.00	$121.73	$166.16	$413.89	$1177.36

$8 + 15.32 + 41.54 = 64.86$ M+ $200.25 −$ RM $64.86 = 135.39$

✔ **SELF-CHECK**   Complete the problem, then check your answer in the back of the book.

1. Use this information to find the net pay: gross weekly pay, $396; FIT, $37; FICA, $30.29; health insurance, $42.75.

488 ◆ Unit 19 Accounting

	Name	Gross Pay	FIT	FICA	Total Deductions	Net Pay
2.	Abbot, J.	$485.00	$ 56.00	$37.10	$93.10	?
3.	Eberly, T.	$136.00	$ 11.00	$10.40	$21.40	?
4.	Jackson, J.	$795.00	$107.00	$60.82	?	?
5.	Lewton, K.	$975.00	$157.00	$74.59	?	?
6.	Timmons, W.	$643.00	$100.00	$49.19	?	?
7.	Young, K.	$519.00	$ 63.00	$39.70	?	?

8. Pre-Fab Manufacturing Company.
Payroll for week.
Total gross pay: $16,478.43.
Total deductions: $3914.75.
What is the total net pay?

9. Lou's Variety Store.
Payroll for week.
Total gross pay: $416.74.
Total deductions: $97.82.
What is the total net pay?

Use the tables on pages 642–643 for federal withholding tax (FIT). Use the social security (FICA) tax rate of 7.65%.

10. The Goldstone Swimming Complex employs students in the summer. Goldstone pays a standard hourly rate of $5.00. The only deductions are federal withholding, social security, and city income tax (CIT). Complete the payroll register for the week.

Payroll Register for Goldstone Swimming Complex						Date	June 8, 19 –	
Employee	Income Tax Information	Hours Worked	Gross Pay	FIT	FICA	CIT	Total Deductions	Net Pay
Banks, Chuck	Single, 1 allow.	24	?	?	?	$2.52	?	?
Drake, Heather	Single, 0 allow.	34	?	?	?	3.57	?	?
Faust, Henry	Single, 1 allow.	36	?	?	?	3.78	?	?
Harakis, Pete	Single, 0 allow.	38	?	?	?	3.99	?	?
Kendrick, Peg	Single, 0 allow.	36	?	?	?	3.78	?	?
Reese, Tina	Single, 1 allow.	38	?	?	?	3.99	?	?
Singleton, Marie	Single, 0 allow.	30	?	?	?	3.15	?	?
Total			?	?	?	?	?	?

11. The Consortium, consulting engineers, pays its employees weekly salaries. Deductions include state and city income taxes. Complete the payroll register for the week.

| Name | FIT Information | Weekly Salary | Deductions | | | | | | | Net Pay |
			FIT	FICA (7.65%)	SIT	CIT	Hosp. Ins.	Total Deductions	
Cole	M, 3 allow.	$600.00	?	?	$16.50	$10.50	$42.75	?	?
Dobbs	M, 2 allow.	585.00	?	?	15.75	10.24	35.00	?	?
Haddad	S, 1 allow.	415.00	?	?	11.41	7.26	17.50	?	?
Micelli	M, 1 allow.	564.80	?	?	15.53	9.88	35.00	?	?
Nowakowski	S, 2 allow.	537.50	?	?	14.78	9.41	17.50	?	?
Presser	M, 4 allow.	619.20	?	?	17.03	10.84	42.75	?	?
Zatcoff	M, 0 allow.	612.44	?	?	16.84	10.72	17.50	?	?
Total		?	?	?	?	?	?	?	?

12. The Silver Wholesale Clothes Co. hires extra sales personnel during its warehouse sale. The extra help earns $3.50 an hour plus 6% commission on all sales. Deductions include $4.00 per week for health insurance. Complete the payroll register.

| Name | FIT Information | Hours Worked | Total Sales | Gross Pay | Deductions | | | | Net Pay |
					FIT	FICA (7.65%)	Insurance	Total	
Cramer	S, 1 allow.	30	$4200	?	?	?	$4.00	?	?
Grant	S, 0 allow.	30	2786	?	?	?	$4.00	?	?
Iskersky	M, 2 allow.	36	3865	?	?	?	$4.00	?	?
Miller	S, 1 allow.	40	4500	?	?	?	$4.00	?	?
Quinn	S, 0 allow.	25	1430	?	?	?	$4.00	?	?
Talbot	M, 3 allow.	40	5184	?	?	?	$4.00	?	?
Xerxes	S, 1 allow,	34	3146	?	?	?	$4.00	?	?
Total			?	?	?	?	?	?	?

13. King Sporting Goods, Inc., pays its employees weekly. Deductions include federal withholding, state income tax, social security (7.65%), and 1% of gross pay for unemployment compensation. Use this information to prepare the payroll for the week of November 30.

Name	FIT Information	Position	Pay Plan	Total Sales or Hours Worked
Adams	M, 2 allow.	Sales	$7\frac{1}{2}$% straight commission	$6740
Lightner	S, 1 allow.	Sales Manager	$7\frac{1}{2}$% straight commission	$7152
Rose	M, 3 allow.	Sales Manager	$420 + 6% commission	$3298
Ulrich	S, 0 allow.	Sales Trainee	5% straight commission	$4165
Goode	M, 1 allow,	Maintenance	$5.95 per hour	44 hours

KING SPORTING GOODS, INC.   Payroll Register   Week of: November 30, 19 –

Name	Gross Pay	FIT	SIT	FICA	Tax for Unemployment Compensation	Total Deductions	Net Pay
Adams	?	?	$20.22	?	?	?	?
Lightner	?	?	21.46	?	?	?	?
Rose	?	?	32.59	?	?	?	?
Ulrich	?	?	8.33	?	?	?	?
Goode	?	?	10.47	?	?	?	?
Total	?	?	?	?	?	?	?

Add. **(Skill 5)**

14. $17.96 + $8.44   **15.** $149.74 + $97.89   **16.** $212.47 + $197.83

Subtract. **(Skill 6)**

17. $412.76 − $26.40   **18.** $676.83 − $212.91   **19.** $1864.75 − $417.93

19-2

Business Expenses

OBJECTIVE

Compute the percent that a particular business expense is of the total expenses.

Your business must keep accurate records of all its expenses. You will use the information when you prepare income tax forms and when you calculate your company's profits. You may determine your total expenses monthly, quarterly, or annually. To plan for future spending, you may calculate the percent that each expense is of the total.

$$\text{Percent of Total} \ = \ \frac{\text{Particular Expense}}{\text{Total Expenses}}$$

EXAMPLE *Skills* 5, 31 *Application* A

Molded Plastic Products, Inc., manufactures plastic buckets, containers, and other products. Records of Molded Plastic's expenses for the first quarter of the year show:

Payroll	$42,171.84
Advertising	2500.00
Raw materials	12,417.83
Factory and showroom rental	4500.00
Office supplies	216.90
Insurance	517.85
Utilities	6417.93
Total	$68,742.35

What percent of total expenses did Molded Plastic spend on advertising during the quarter?

SOLUTION
Find the **percent of total.**
Particular Expense ÷ Total Expenses
$2500.00 ÷ $68,742.35 = 0.0363 = 3.6% of total was spent on advertising

2500 ÷ 68742.35 = .03636768

✔ SELF-CHECK Complete the problems, then check your answers in the back of the book.

1. Particular expenses: filing, $300; programming, $1200; payroll, $2500.
   Filing is what percent of the total?

2. Particular expenses: telephone, $175; electricity, $450; natural gas, $375; water, $200.
   Water is what percent of the total?

Find percents to the nearest tenth percent.

	Expenses for Supplies	÷	Total Expenses	=	Percent of Total
3.	$ 450	÷	$ 4500	=	?
4.	$ 1400	÷	$ 28,000	=	?
5.	$ 15,700	÷	$ 210,000	=	?
6.	$132,685	÷	$5,415,750	=	?

7. Trapper Financial Consultants.
Expenses total $72,650 for quarter.
Payroll for quarter is $48,435.
Payroll expense is what percent
of the total?

8. Moonlight Motel.
Annual expenses total $214,500.
Annual utilities cost $38,475.
Utility cost is what percent
of the total?

9. The Clear Pool Company had the following business expenses last year:

What are the total business expenses?
Payroll is what percent of the total?

Payroll	$84,565
Advertising	2000
Equipment	21,495
Supplies	17,194
Insurance	2000
Utilities	1450
Rent	3000

10. Parson's Traditional Clothiers, Inc.,
had the following business expenses
last month:

Store rental is what percent of the total
business expenses for the month?

Payroll	$ 1645.80
Advertising	325.00
Cost of goods sold	18,175.75
Store rental	900.00
Supplies	45.00
Insurance	120.00
Utilities	217.89

MAINTAINING YOUR SKILLS   Look up the skills in parentheses if you need help or more practice.

Add. **(Skill 5)**

11.	**12.**	**13.**	**14.**
$114,570	$2171.89	$212.43	$714,789.43
31,720	419.43	171.80	24,561.78
+ 56,493	+ 114.65	+ 64.95	+ 7147.93

Find the rate. Round answers to the nearest tenth of a percent. **(Skill 31)**

15. $35,000 is what percent of $140,000?

16. $212 is what percent of $1060?   **17.** $78 is what percent of $471?

18. $7500 is what percent of $171,432?

19. $17,893.45 is what percent of $914,719.40?

Apportioning Expenses

OBJECTIVE

Compute a department's share of the total business expense.

Your business may apportion, or distribute, certain expenses among its departments. Each department is charged a certain amount of the particular expense. Often, the amount that each department is charged depends on the space that it occupies.

$$\text{Amount Paid} = \frac{\text{Square Feet Occupied}}{\text{Total Square Footage}} \times \text{Total Expense}$$

EXAMPLE   *Skills* 10, 8   *Application* X   *Term* Apportion

Mike's Discount Stores, Inc., occupies a building that contains 200,000 square feet. The cut-rate clothing department occupies an area that measures 100 feet by 150 feet. The annual cost of security for the entire building is $28,400. Mike's apportions the cost based on square feet occupied. What annual amount for security does Mike's charge to the cut-rate clothing department?

SOLUTION

A. Find the **square feet occupied.**
100 ft × 150 ft = 15,000 square feet occupied

B. Find the **amount paid.**

$$\begin{pmatrix} \text{Square Feet} \\ \text{Occupied} \end{pmatrix} \div \begin{matrix} \text{Total Square} \\ \text{Footage} \end{matrix} \times \begin{matrix} \text{Total} \\ \text{Expense} \end{matrix}$$

15,000 ÷ 200,000 × $28,400

0.075 × $28,400 = $2130 amount paid

100 × 150 = 15000 ÷ 200000 = .075 × 28400 = 2130

✔ SELF-CHECK   Complete the problems, then check your answers in the back of the book.

1. The Law-for-All Clinic occupies 40 feet by 50 feet in a building with 50,000 square feet. Annual cost of janitorial service is $9600. What does The Law-for-All Clinic pay?

2. Family Shoes occupies 50 feet by 80 feet in a mall with 40,000 square feet. Monthly mall expenses are $12,500. What does Family Shoes pay?

PROBLEMS

	(Square Feet Occupied	÷	Total Square Footage)	×	Total Expense	=	Amount Paid
3.	(3200	÷	160,000)	×	$62,500	=	?
4.	(15,000	÷	160,000)	×	$62,500	=	?
5.	(50,000	÷	160,000)	×	$62,500	=	?

Round the intermediate answers to the nearest thousandth and the final answers to the nearest cent.

6. Building area: 980,000 sq ft.
Annual maintenance cost: $135,000.
Billing dept. is 40 ft by 60 ft.
What does the billing department pay annually for maintenance?

7. Building area: 450,000 sq ft.
Annual insurance charge: $5750.
Sales dept. is 30 ft by 40 ft.
What does the sales department pay annually for insurance?

8. The Dunberry Company apportions the annual cost of utilities among its departments. The total cost for the year was $186,470. The total area of the building is 750,000 square feet. The company's receiving department occupies an area that is 20 feet by 650 feet. What did the receiving department pay for utilities for the year?

9. Tubular Assemblies, Inc., pays a total of $960,000 per year to rent its building. The total area of the building is 480,000 square feet. Tubular Assemblies apportions the rental charge among its departments. How much does each of these departments pay for rent?
  a. Research and development, 45 ft by 60 ft
  b. Shipping, 50 ft by 120 ft

10. The Barteli Corporation owns and maintains a 5-store shopping mall. Barteli paid $12,000 for advertising last year. It apportions the cost based on square footage. How much does Barteli charge each store?

Shop	Dimensions
Ken's Carry-Out	40 ft by 60 ft
The Jean Place	30 ft by 50 ft
Pan's Pizza	25 ft by 60 ft
Universal Travel	30 ft by 30 ft
The Jewel Store	25 ft by 30 ft

11. Some businesses apportion costs among their departments on the basis of gross sales. The gross sales for Tent Mart totaled $3,750,000 last year. It distributed these annual expenses:

Maintenance	Utilities	Security
$6000	$30,000	$4800

  a. The women's wear department had $900,000 in gross sales last year. How much did it pay for each of the annual expenses?

  b. The shoe department had $650,000 in gross sales last year. How much did it pay for each of the annual expenses?

  c. The toy department had $475,650 in gross sales last year. How much did it pay for each of the annual expenses?

MAINTAINING YOUR SKILLS   Look up the skills in parentheses if you need help or more practice.

Divide. Round answers to the nearest thousandth. (Skill 10)

12. $900 \div 12,450$

13. $2000 \div 96,500$

14. $1200 \div 50,000$

15. $2475 \div 475,000$

19-4

Depreciation— Straight-Line Method

OBJECTIVE

Use the straight-line method to compute the annual depreciation of an item.

For tax purposes, the Internal Revenue Service allows you to recognize the depreciation of many of the items that your business owns. Depreciation is a decrease in the value of an item because of its age or condition. The straight-line method is one way of determining the annual depreciation of an item. This method assumes that the depreciation is the same from year to year. To calculate the depreciation, you must know the original cost, the estimated life, and the resale value of the item. The estimated life of an item is the length of time, usually in years, that it is expected to last. The resale value is the estimated trade-in, salvage, or scrap value at the end of the item's expected life.

$$\text{Annual Depreciation} = \frac{\text{Original Cost} - \text{Resale Value}}{\text{Estimated Life}}$$

EXAMPLE   *Skills* 4, 10   *Application* A   *Term* Straight-line method

The law firm of Charles A. Adams, Inc., purchased a new copier that cost $1745. The life of the copier is estimated to be 5 years. The total resale value after 5 years of use is estimated to be $245. Using the straight-line method, find the annual depreciation of the copier.

SOLUTION

Find the **annual depreciation.**
(Original Cost − Resale Value) ÷ Estimated Life
    ($1745   − $245)   ÷   5
             $1500   ÷   5   = $300 annual depreciation

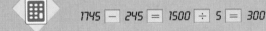 1745 − 245 = 1500 ÷ 5 = 300

✔ SELF-CHECK   Complete the problems, then check your answers in the back of the book.

Find the annual depreciation.

1. $6000 original cost; 10 years estimated life; $1000 resale value.

2. $475 original cost; 3 years estimated life; $25 resale value.

PROBLEMS

(Original Cost	−	Resale Value)	÷	Estimated Life	=	Annual Depreciation
3. ($ 1435	−	$ 235)	÷	3 years	=	?
4. ($ 7186	−	$ 1000)	÷	5 years	=	?
5. ($ 14,750	−	$ 2000)	÷	10 years	=	?
6. ($115,476	−	$10,000)	÷	15 years	=	?

7. New radio tower.
Cost is $175,000.
Estimated life of 20 years.
Salvage value estimated at $5000.
What is the annual depreciation?

8. New taxicab.
Cost is $16,250.
Estimated life is 3 years.
Salvage value estimated at $1250.
What is the annual depreciation?

9. Central Dental Clinic recently purchased a new computer system for a total cost of $64,735. The estimated life of the system is 5 years. The trade-in value of the system after 5 years is estimated to be $10,000. What is the annual depreciation?

10. Tina Cole is a chartered financial advisor. She purchased equipment for her office for $7843. The trade-in value of the equipment is estimated to be $500 after 7 years of use. What is the annual depreciation?

11. The Star Trucking Company purchased a new tractor unit for $125,000. After 4 years of useful life, the estimated salvage value is $10,000. What is the annual depreciation?

12. Tom Nichols is a representative for a landscape company. Tom uses his pickup truck entirely for business. He trades in his truck every 3 years. Tom paid $8500 for his present truck. He expects to receive $2000 for the truck when he trades it in after 3 years. What is the annual depreciation?

13. Third National Bank recently purchased 3 new typewriters. Each typewriter cost $845. The estimated life of each typewriter is 6 years. The trade-in value of each typewriter is expected to be $125 at the end of the 6 years. What is the annual depreciation for all 3 typewriters?

14. Annie's Laundromat recently purchased 6 new washing machines at a cost of $325 each and 5 new clothes dryers at a cost of $268 each. The salvage value of the washers is expected to be $50 each after 4 years of use. The salvage value of the dryers is expected to be $75 each after 5 years of use. What is the total annual depreciation?

MAINTAINING YOUR SKILLS   Look up the skills in parentheses if you need help or more practice.

Subtract. **(Skill 4)**

15. $1475
     − 1189

16. $743
     − 95

17. $725
     − 56

18. $16,417
     − 750

Divide. Round answers to the nearest cent. **(Skill 10)**

19. $14,718 ÷ 20

20. $9171.45 ÷ 15

21. $768.43 ÷ 5

22. $4171.60 ÷ 3

Multiply. **(Skill 8)**

23. $1470.00
     × 0.055

24. $963.49
     × 0.05

25. $24.49
     × 0.06

26. $11,419.90
     × 0.075

19-5

Depreciation— Book Value

OBJECTIVE

Use the straight-line method to compute the book value of an item.

Book value is the approximate value of an item after you have owned it and depreciated it for a period of time. The book value is the original cost minus the accumulated depreciation. The accumulated depreciation is the total depreciation to date. At the end of an item's life, its book value and resale value should be equal.

Book Value = Original Cost − Accumulated Depreciation

EXAMPLE   *Skill 7   Application A   Term* Book value

The law firm of Charles A. Adams, Inc., purchased a new copier for $1745. The total resale value after 5 years is estimated to be $245. Using the straight-line method, Charles A. Adams, Inc., determined that the copier will depreciate $300 per year. What will the book value be at the end of 4 years?

SOLUTION   A. Find the **accumulated depreciation** for fourth year.
$300 per year × 4 years = $1200 accumulated depreciation

B. Find the **book value.**
Original Cost − Accumulated Depreciation
$1745   − $1200   = $545 book value

✔ SELF-CHECK   Complete the problems, then check your answers in the back of the book.

A computer system costs $96,000. Its estimated life is 4 years. The annual depreciation is $20,000. What will the book value be:

1. At the end of 3 years?   **2.** At the end of 4 years?

PROBLEMS

	Annual Depreciation	×	Number of Years	=	Accumlated Depreciation	Original Cost	−	Accumulated Depreciation	=	Book Value
3.	$ 400	×	6 years	=	?	$3500	−	?	=	?
4.	$16,000	×	3 years	=	?	$86,500	−	?	=	?
5.	$ 50	×	2 years	=	?	$280	−	?	=	?

Use the straight-line method of depreciation. Round answers to the nearest dollar.

6. $4800 for office furniture. Depreciates $430 per year. What is the book value after 3 years?

7. $9165 for communications equipment. Depreciates $900 per year. What is the book value after 5 years?

8. New facsimile transmitting equipment.
Cost is $11,748.
Estimated life is 5 years.
Resale value of $3000.
What is the book value
after 3 years?

9. New automobile.
Cost is $18,417.
Estimated life of 3 years.
Resale value of $5000.
What is the book value
after 2 years?

10. Tina Cole purchased new office equipment for $7843. The trade-in value of the equipment is estimated to be $500 after 7 years of use. Find the book value after each year of use.

Year	Annual Depreciation	Accumulated Depreciation	Book Value
1	$1049	$1049	$6749
2	1049	2098	?
3	1049	?	?
4	?	?	?
5	?	?	?
6	?	?	?
7	?	?	?

11. Wayne Guest purchased a new over-the-road tractor for $114,760. The tractor is expected to have a trade-in value of $20,000 after 4 years. What is the book value at the end of each of the 4 years?

Year	Annual Depreciation	Accumulated Depreciation	Book Value
1	?	?	?
2	?	?	?
3	?	?	?
4	?	?	?

12. A Brief Case The **units-of-production method** is another method of computing depreciation. This method bases depreciation on the number of units produced rather than the years of estimated life.

$$\text{Depreciation} = \frac{\text{Number of Units Produced}}{\text{Estimated Units of Production}} \times \left(\begin{matrix} \text{Original} \\ \text{Cost} \end{matrix} - \begin{matrix} \text{Salvage} \\ \text{Value} \end{matrix} \right)$$

a. Ross Manufacturing purchased a new automatic punchpress for $216,500. The press is expected to be worth $20,000 after 200,000 units are produced. Use the units-of-production method to calculate the book value of the press after 50,000 units are produced.

b. Sustom Plastics purchased a new folding machine for $51,615. The salvage value of the machine is estimated to be $5000 after 750,000 units are produced. Use the units-of-production method to calculate the book value after 150,000 units are produced.

MAINTAINING YOUR SKILLS   Look up the skills in parentheses if you need help or more practice.

Multiply. **(Skills 7, 8)**

13. $550 × 3

14. $1240 × 5

15. $17.60 × 4

16. $2500 × 7

17. $11,400 × 9

18. $8175 × 6

19. $24,650 × 8

20. $17.47 × 2

21. $718.65 × 10

◆ **19-6** ◆

Depreciation—
Sum-of-the-Years'-Digits Method

<table>
<tr>
<td>

OBJECTIVE

Use the sum-of-the-year's-digits method to compute the yearly depreciation.

</td>
<td>

The **sum-of-the-years'-digits method** is another way to determine the annual depreciation of an item. This method assigns a fraction of the total depreciation to each year of the item's life. More depreciation is allowed in the early years of the item's life than in the later years. For example, if the life of a machine is estimated to be 4 years, the years' digits are 4, 3, 2, and 1. The sum of the years' digits is 4 + 3 + 2 + 1 = 10. The fractions of the total depreciation are $\frac{4}{10}$ for the first year, $\frac{3}{10}$ for the second year, $\frac{2}{10}$ for the third year, and $\frac{1}{10}$ for the fourth year. To use the sum-of-the-years'-digits method, you must know the original cost, the estimated life, and the resale value.

</td>
</tr>
</table>

$$\begin{array}{c} \text{Depreciation} \\ \text{for Year} \end{array} = \begin{array}{c} \text{Fraction of} \\ \text{Total Depreciation} \end{array} \times \left(\begin{array}{c} \text{Original} \\ \text{Cost} \end{array} - \begin{array}{c} \text{Resale} \\ \text{Value} \end{array} \right)$$

EXAMPLE   *Skills* 20, 2   *Application* A   *Term* Sum-of-the-years'-digits method

The General Assembly Company purchased a new numerically controlled riveter for $195,000. The expected life of the riveter is 5 years. At the end of the 5 years, the resale value of the machine is expected to be $15,000. General Assembly uses the sum-of-the-years'-digits method to calculate depreciation. To the nearest dollar, what is the depreciation for the first year of the life of the riveter?

S O L U T I O N

A. Find the **fraction of total depreciation.**

$$\frac{\text{Assigned Digit}}{\text{Sum of the Years' Digits}} = \frac{5}{5 + 4 + 3 + 2 + 1} = \frac{5}{15}$$

B. Find the **depreciation for year.**

$$\begin{array}{c} \text{Fraction of} \\ \text{Total Depreciation} \end{array} \times \left(\begin{array}{c} \text{Original} \\ \text{Cost} \end{array} - \begin{array}{c} \text{Resale} \\ \text{Value} \end{array} \right)$$

$$\frac{5}{15} \quad \times \ (\$195,000 - \$15,000)$$

$$\frac{5}{15} \quad \times \ \$180,000 \qquad = \$60,000$$
depreciation for first year

| 195000 | − | 15000 | = | 180000 | × | 5 | ÷ | 15 | = | 60000 |

✔ **SELF-CHECK**   Complete the problems, then check your answers in the back of the book.

1. $\frac{4}{15}$ × $180,000 = ? depreciation for the second year.

2. $\frac{3}{15}$ × $180,000 = ? depreciation for the third year.

Use the sum-of-the-years'-digits method to calculate depreciation. Round answers to the nearest dollar.

	Original Cost	Resale Value	Estimated Life	Depreciation 1st Year Fraction	Amount	2nd Year Fraction	Amount
3.	$ 9000	$ 1200	3 years	$\frac{3}{6}$	$3900	$\frac{2}{6}$	?
4.	$ 14,700	$ 2100	6 years	$\frac{6}{21}$	$3600	?	?
5.	$125,000	$15,000	10 years	?	?	?	?
6.	$ 9500	$ 500	5 years	?	?	?	?

F.Y.I.

The Internal Revenue Service requires that the salvage value equal the book value at the end of the useful life.

7. Robotic delivery unit.
Estimated life of 2 years.
Original cost is $12,500.
Resale value is $3500.
What is the depreciation for the first year?

8. Portable generator.
Estimated life of 3 years.
Original cost is $10,000.
Resale value is $400.
What is the depreciation for the first year?

9. The original cost of a small delivery truck is $7800. After 4 years, the resale value is estimated to be $1000. What is the depreciation for each of the 4 years?

10. The Office Temp. Service recently purchased 2 new typewriters for a total cost of $2480. The typewriters have an estimated life of 5 years. After 5 years, the resale value of each typewriter is expected to be $250. What is the depreciation for each typewriter for each of the 5 years?

11. The Perfect Plastic Co. recently purchased a plastic-molding machine for $80,000. The estimated life of the machine is 4 years. The resale value after 4 years is expected to be $8000. What is the book value of the machine after each year of use?

Year	Depreciation for Year	Accumulated Depreciation	Book Value
First	?	?	?
Second	?	?	?
Third	?	?	?
Fourth	?	?	?

Solve. Round answers to the nearest dollar. **(Skills 4, 20)**

12. $\frac{4}{15} \times$ $9000   **13.** $\frac{2}{3} \times$ $12,630   **14.** $\frac{3}{6} \times$ $4840   **15.** $\frac{3}{10} \times$ $17,800

16. $\frac{6}{21} \times$ ($147,500 $-$ $15,000)

17. $\frac{9}{55} \times$ ($7850 $-$ $900)

18. $\frac{7}{36} \times$ ($475 $-$ $100)

Depreciation—
Double-Declining-Balance Method

OBJECTIVE

Use the double-declining-balance method to compute the annual depreciation and book value.

Your business may use the double-declining-balance method to find the annual depreciation of an item. This method allows the highest possible depreciation in the first year of the item's life. The depreciation declines as the item approaches the end of its life. To use this method, you must know the item's original cost and estimated life.

$$\text{Annual Depreciation Rate} = 2 \times \frac{100\%}{\text{Estimated Life}}$$

$$\begin{array}{c} \text{Depreciation} \\ \text{for Year} \end{array} = \begin{array}{c} \text{Previous} \\ \text{Declining Balance} \end{array} \times \begin{array}{c} \text{Annual} \\ \text{Depreciation Rate} \end{array}$$

EXAMPLE   *Skill* 30   *Application* A   *Term* Double-declining-balance method

Major Media Productions purchased production equipment for $240,000. Major Media estimates that the equipment will become obsolete in 5 years. What are the depreciation and book value for each of the 5 years?

SOLUTION

A. Find the **annual depreciation rate.**
2 × (100% ÷   Estimated Life)
2 × (100% ÷   5)   = 40% annual rate

B. Find the **depreciation for year.**
Previous Declining Balance × Annual Depreciation Rate

40% of original cost

			Previous Declining Balance − Depreciation for Year

Year	Depreciation for Year	Accumulated Depreciation	Declining Balance (Book Value)
First	$96,000.00	$96,000.00	$144,000.00
Second	57,600.00	153,600.00	86,400.00
Third	34,560.00	188,160.00	51,840.00
Fourth	20,736.00	208,896.00	31,104.00
Fifth	12,441.60	221,337.60	18,662.40 ◄— Resale value

40% of previous declining balance of $144,000

✔ SELF-CHECK   Complete the problems, then check your answers in the back of the book.

A machine that cost $80,000 will be obsolete in 4 years.

1. Find the depreciation and book value for the first year.

2. Find the depreciation and book value for the second year.

	Estimated Life	Annual Depreciation Rate	×	Previous Declining Balance	=	Depreciation for Year
3.	10 years	20%	×	$ 16,450	=	?
4.	8 years	?	×	$ 7400	=	?
5.	6 years	?	×	$ 42,000	=	?
6.	3 years	?	×	$120,000	=	?

7. New forklift cost $7200. Estimated life of 6 years. What is the depreciation for the first year? The book value?

8. New incinerator cost $135,000. Estimated life of 4 years. What is the depreciation for the first year? The book value?

9. The Packer Company purchased new inventory control equipment for $50,000. The life of the equipment is estimated to be 5 years. The resale value is expected to be $3888. Complete the depreciation table for each of the 5 years.

Year	Depreciation for Year	Accumulated Depreciation	New Declining Balance
First	$20,000	?	?
Second	12,000	?	?
Third	?	?	?
Fourth	?	?	?
Fifth	?	?	?

10. A Brief Case When the double-declining-balance method is used to calculate depreciation, the declining balance in the last year of an item's life might not equal the expected resale value. When this happens, the depreciation for the last year is adjusted so that the final declining balance and the resale value are the same.

Samsen's purchased office equipment for $10,000. The resale value is expected to be $500 after a 4-year life. Complete the depreciation table.

Year	Depreciation for Year	Accumulated Depreciation	New Declining Balance
1	$5000	?	?
2	?	?	?
3	?	?	?
4	?	?	$500

This amount is computed to be $625. It must be increased so that the book value is $500.

MAINTAINING YOUR SKILLS Look up the skill in parentheses if you need help or more practice.

Find the percentage. Round answers to the nearest dollar. **(Skill 30)**

11. 40% of $1350

12. 50% of $5000

13. 20% of $14,850

14. $66\frac{2}{3}$% of $9600

15. 25% of $415,600

16. $33\frac{1}{3}$% of $1242

17. 28.6% of $1490

18. 22.2% of $12,560

Modified Accelerated Cost Recovery System (MACRS)

The modified accelerated cost recovery system (MACRS) is another method of computing depreciation. Introduced by the Tax Reform Act of 1986 and further modified by the Tax Bill of 1989, MACRS takes the place of the accelerated cost recovery system (ACRS) introduced by the Economic Recovery Tax Act of 1981. MACRS allows businesses to depreciate assets fully over a set period of time. This method encourages businesses to replace equipment earlier than they would if they used other depreciation methods. Under MACRS, assets can be depreciated fully over recovery periods of 4, 6, 8, 11, 16, or 21 years according to fixed percents.

Annual Depreciation = Original Cost × Fixed Percent

Book Value = Original Cost − Accumulated Depreciation

EXAMPLE   *Skills* 30, 3   *Term* Modified accelerated cost recovery system

Prince Pizza purchased a new delivery van for $12,400 to use in delivering pizzas. MACRS allows delivery vans to be depreciated fully in 6 years according to 6 fixed percents: 20% the first year, 32% the second year, 19.2% the third year, 11.52% the fourth and fifth years, and 5.76% the sixth year. What are the annual depreciation and book value for each of the six years?

SOLUTION   **A.** Find the **annual depreciation.**

Original Cost	×	Fixed Percent		
$12,400.00	×	20.00%	=	$2480.00
$12,400.00	×	32.00%	=	$3968.00
$12,400.00	×	19.20%	=	$2380.80
$12,400.00	×	11.52%	=	$1428.48
$12,400.00	×	11.52%	=	$1428.48
$12,400.00	×	5.76%	=	$ 714.24

B. Find the **book value.**

Original Cost	−	Accumulated Depreciation			
$12,400.00	−	$ 2480.00	=	$9920.00	$2480.00
$12,400.00	−	$ 6448.00	=	$5952.00	+ $3968.00
$12,400.00	−	$ 8828.80	=	$3571.20	$6448.00
$12,400.00	−	$10,257.28	=	$2142.72	
$12,400.00	−	$11,685.76	=	$ 714.24	$6448.00
$12,400.00	−	$12,400.00	=	0	+ $2380.80
					$8828.80

12400 M+ × 20 % 2480 RM 12400 × 32 % 3968 RM 12400 × 19.2 %
2380.8 RM 12400 × 11.52 % 1428.48 RM 12400 × 5.76 % 714.24

1. MACRS depreciates automobiles in 6 years according to the same 6 percents used for a delivery van in the example above. Find the annual depreciation for each of the 6 years for an automobile costing $17,000.

PROBLEMS

F.Y.I.
Using the MACRS method, the salvage value is not used since 100% of the cost is depreciated.

Use the modified accelerated cost recovery system (MACRS) to find the annual depreciation. Round answers to the nearest cent.

		2.	3.	4.	5.	6.
	Cost	Taxi $14,600	Car $9240	Truck $24,700	Manufacturing Equipment $34,840	TeleComp $147,500
Year	Percent	Annual Dep.	Annual Dep.	Annual Dep.	Annual Dep.	Annual Dep.
1	20.00%	?	?	?	?	?
2	32.00%	?	?	?	?	?
3	19.20%	?	?	?	?	?
4	11.52%	?	?	?	?	?
5	11.52%	?	?	?	?	?
6	5.76%	?	?	?	?	?

7. Office computer system. Original cost is $47,600. Fully depreciated in 8 years. Fixed percents are 14.28%, 24.49%, 17.49%, 12.49%, 8.93%, 8.93%, 8.93%, and 4.46%. What is the depreciation for each of the 8 years?

8. Three-year-old racehorse. Original cost is $164,500. Fully depreciated in 4 years. Fixed percents are 33%, 45%, 15%, and 7%. What is the depreciation for each of the 4 years?

9. The Extended Care Center purchased a van for $48,260 to transport residents to and from a shopping center. The van is fully depreciated in 6 years. Fixed percents are 20%, 32%, 19.2%, 11.52%, 11.52%, and 5.76%. What are the depreciation and book value for each year?

Year	Percent	Cost	Depreciation	Accumulated Depreciation	Book Value
1	20.00%	?	?	?	?
2	32.00%	?	?	?	?
3	19.20%	?	?	?	?
4	11.52%	?	?	?	?
5	11.52%	?	?	?	?
6	5.76%	?	?	?	?

10. Wastewater treatment plants are depreciated fully in 16 years. Complete the depreciation table for a plant costing $1,400,000.

Year	Percent	Original Cost	Depreciation	Accumulated Depreciation	Book Value
1	5.00%	?	?	?	?
2	9.50%	?	?	?	?
3	8.55%	?	?	?	?
4	7.69%	?	?	?	?
5	6.93%	?	?	?	?
6	6.23%	?	?	?	?
7	5.90%	?	?	?	?
8	5.90%	?	?	?	?
9	5.90%	?	?	?	?
10	5.90%	?	?	?	?
11	5.90%	?	?	?	?
12	5.90%	?	?	?	?
13	5.90%	?	?	?	?
14	5.90%	?	?	?	?
15	5.90%	?	?	?	?
16	3.00%	?	?	?	?

MAINTAINING YOUR SKILLS   Look up the skill in parentheses if you need help or more practice.

Find the percentage. Round answers to the nearest dollar. **(Skill 30)**

11. 38% of $74,500   **12.** 22% of $18,500 **13.** 8% of $1475 **14.** 12% of $840

Reviewing the Basics

Skills

(Skill 2)

Round answers to the nearest dollar.

1. $7151.444   **2.** $71,417.20   **3.** $143.51

Solve.

(Skill 4)

4. $24,600 − $6150   **5.** $214,650 − $17,172   **6.** $9170 − $917

(Skill 5)

7. $47.50 + $16.75 + $4.76   **8.** $4718.71 + $516.48

9. $2.49 + $1.79

(Skill 6)

10. $5176.47 − $3247.98   **11.** $22,417.87 − $5147.97

(Skill 7)

12. $512 × 6   **13.** $925 × 80   **14.** $4217 × 4

(Skill 8)

15. 0.16 × $42,500   **16.** 0.12 × $718   **17.** 0.37 × $3240

(Skill 20)

18. $\frac{4}{21}$ × $1344   **19.** $\frac{3}{10}$ × $185,750   **20.** $\frac{4}{15}$ × $23,480

Round answers to the nearest thousandth.

(Skill 10)

21. 2400 ÷ 37,000   **22.** 2000 ÷ 180,000   **23.** 1500 ÷ 50,000

Find the percentage. Round answers to the nearest cent.

(Skill 30)

24. 40% of $42,500   **25.** 12.5% of $8640   **26.** 16% of $970

27. 125% of $12,425   **28.** 10% of $115,475   **29.** 3% of $179.79

Round answers to the nearest tenth of a percent.

(Skill 31)

30. $648 is what percent of $19,719?

31. $7845 is what percent of $49,417?

32. $13,415 is what percent of $871,980?

33. $20,450 is what percent of $147,850?

Applications
(Application X)

Find the area.

34. 120 ft by 50 ft   **35.** 40 ft by 60 ft

Terms

Write your own definition for each term.

36. Payroll register   **37.** Straight-line method

38. Apportion   **39.** Sum-of-the-years'-digits method

40. Book value   **41.** Double-declining-balance method

42. Modified accelerated cost recovery system

Refer to your reference files in the back of the book if you need help.

Unit Test

Lesson 19-1

1. Complete the payroll register. Use the FICA tax rate of 7.65%.

Name	Gross Income	FIT	FICA	SIT	Total Deductions	Net Pay
Cole	$415.00	$52	?	$8.30	?	?
Miller	442.00	57	?	8.84	?	?
Peters	455.00	46	?	9.10	?	?
Thomas	438.00	36	?	8.76	?	?
Total	?	?	?	?	?	?

Lesson 19-2

2. Trace Realty had these expenses during the last quarter:

   Payroll $17,945
   Advertising   1750
   Office rental   2500
   Office supplies   547
   Utilities   723

   What are the total expenses? To the nearest tenth of a percent, payroll is what percent of the total?

Lesson 19-3

3. The Mallard Aerodynamics Company pays $247,500 per year for security. The company apportions the cost among its departments based on space. The research department occupies an area that measures 40 feet by 60 feet. The building contains 1,500,000 square feet. How much does the research department pay for security?

Lesson 19-4

4. The Temp. Company purchased a new copy machine for $7860. The machine has an estimated life of 4 years. The resale value is expected to be $500. Use the straight-line method to find the annual depreciation.

Lesson 19-5

5. Find the book value after each year of use for the copy machine in problem 4.

Lesson19-6

6. The All-Purpose Card Co. purchased new store fixtures for a new branch store. The cost of the fixtures was $9860. The salvage value after 5 years is estimated to be $860. Use the sum-of-the-years'-digits method to find the depreciation for each of the 5 years.

Lesson 19-7

7. The original price of a proof and transit machine is $54,000. The machine is expected to have a life of 3 years. Use the double-declining-balance method to find the annual depreciation and the book value for the first 2 years. Round answers to the nearest dollar.

Lesson19-8

8. Brown Construction purchased special tools for $8700. Brown will use the MACRS method to fully depreciate the tools over 4 years. The fixed percents for the 4 years are 33%, 45%, 15%, and 7%. Use the MACRS method to find the annual depreciation and book value for each of the 4 years.

A SPREADSHEET APPLICATION

Accounting

To complete this spreadsheet application, you will need the template diskette for *Mathematics with Business Applications*. Follow the directions in the *User's Guide* to complete this activity.

Select option 19, Accounting, from the menu. Input the information in the following problems and the computer will automatically compute the depreciation schedule for these methods: straight-line, sum-of-the-year's-digits, double-declining-balance, and modified accelerated cost recovery system (MACRS). Then answer the questions that follow.

	Original Cost	Estimated Life	Resale Value
1.	$70,000	5 years	$20,000
2.	14,500	5 years	2500
3.	53,758	5 years	8758

4. In #1, using the straight-line method:
  a. What is the first year's depreciation?
  b. Is it the same each year?
  c. What is the book value after the fifth year?
  d. Will it always equal the resale value?

5. In #2, using the straight-line method:
  a. What is the annual depreciation?
  b. What is the book value after the fifth year?

6. In #3, using the straight-line method:
  a. What is the accumulated depreciation after the third year?
  b. What is the book value after the fourth year?

7. Using the sum-of-the-year's-digits method:
  a. What is the first year's depreciation in #1? in #2? in #3?
  b. Is it the same each year in each case?
  c. What is the second year's depreciation in #1? in #2? in #3?
  d. What is the book value after the fifth year in #1? in #2? into #3?
  e. Does the book value after the fifth year equal the resale value for each of the problems?

8. Using the double-declining-balance method:
  a. What is the first year's depreciation in #1? in #2? in #3?
  b. What is the second year's depreciation in #1? in #2? in #3?
  c. Does the book value after the fifth year automatically equal the resale value in any of the problems?
  d. What must be done to cause the book value after the fifth year to equal the resale value?

9. Using the modified accelerated cost recovery system method:
  a. What is the first year's depreciation in #1? in #2? in #3?
  b. How many years in the recovery period?

▼ CAREER WISE

Accountant

Darryl Turner is an accountant for the Ace Technology Company. Prior to coming to Ace, he worked for an accounting firm that monitored expenses incurred by a small company that makes documentaries and music videos.

At Ace, Darryl's primary responsibility is handling the depreciation of the company's durable goods. This involves keeping purchase and use records. It also involves completing the income tax forms with regard to depreciation.

Recently, Darryl analyzed the straight-line (SL) and double-declining-balance (DDB) book values for a new $90,000 computer system with a useful life of 10 years and a resale value after that of $8500. His table shows the years of life, the SL book value, and the DDB book value. He also visualized the data in a graph.

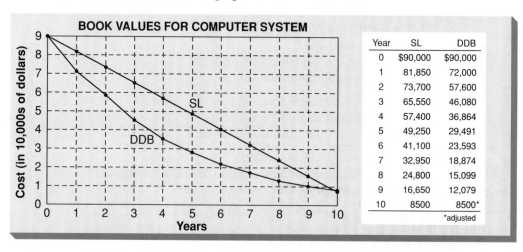

BOOK VALUES FOR COMPUTER SYSTEM

Year	SL	DDB
0	$90,000	$90,000
1	81,850	72,000
2	73,700	57,600
3	65,550	46,080
4	57,400	36,864
5	49,250	29,491
6	41,100	23,593
7	32,950	18,874
8	24,800	15,099
9	16,650	12,079
10	8500	8500*

*adjusted

1. The table and graph show a value of $90,000 at year 0. Explain why this makes sense.

2. Use the method in Lesson 19-4 to compute the book value after 2 years.

3. Use the method in Lesson 19-7 to compute the book value after 2 years.

4. Copy the table. Add a fourth column to show the difference between the SL book value and the DDB book value for each year. When is the difference the least? the greatest? Do the graphs confirm your answers?

20

Accounting Records

The accounting department also keeps other records that show the income, expenses, and value of your business. A *balance sheet* shows your cash, or *assets,* and the money you owe others, or your *liabilities.* An *income statement* details your income and operating expenses.

O U T L I N E

20-1 Assets, Liabilities, and Equity
20-2 Balance Sheet
20-3 Cost of Goods Sold
20-4 Income Statement
20-5 Vertical Analysis
20-6 Horizontal Analysis

◆ Reviewing the Basics
◆ Unit Test
◆ A Spreadsheet Application: Accounting Records

Assets must equal liabilities plus equity in order for a balance sheet to balance.

Assets, Liabilities, and Equity

OBJECTIVE

Compute the total assets, liabilities, and owner's equity.

When you start a business, you will need to buy either merchandise to sell or materials with which to make your products. You may need to purchase office supplies, equipment, buildings, or land. In addition, you must have cash to make change, pay bills, and meet other expenses. Assets are the total of your cash, the items that you have purchased, and any money that your customers owe you.

You may borrow money to start your business, or you may purchase merchandise on credit. Liabilities are the total amount of money that you owe to creditors. Owner's equity, net worth, or capital is the total value of assets that you own. Owner's equity plus liabilities equal assets.

Assets = Liabilities + Owner's Equity

Owner's Equity = Assets − Liabilities

Liabilities = Assets − Owner's Equity

EXAMPLE   *Skills* 5, 6   *Application* A   *Term* Owner's equity

Tina and John Agee recently opened The Clothing Store. They used $60,000 of their own money and a bank loan of $25,000. From the $85,000, Tina and John paid $15,000 for merchandise and $10,000 for supplies. This left a cash balance of $60,000. They received another shipment of $8000 worth of merchandise. They did not pay for this merchandise immediately. What was their owner's equity?

SOLUTION

A. Find the **assets.**

Cash: $85,000 − ($15,000 + $10,000)..................$60,000
Merchandise: $15,000 + $8000 23,000
Supplies.. 10,000
Total Assets $93,000

B. Find the **liabilities.**

Bank loan $25,000
Unpaid merchandise 8,000
Total Liabilities $33,000

C. Find the **owner's equity.**

Assets   −   Liabilities
$93,000   − $33,000 = $60,000 owner's equity

✔ SELF-CHECK   Complete the problems, then check your answers in the back of the book.

1. Find the assets.
Liabilities: $82,000.
Owner's equity: $50,000.

2. Find the liabilities.
Assets: $18,000 cash, $49,000 merchandise.
Owner's equity: $25,000.

	Liabilities	+	Owner's Equity	=	Assets
3.	$27,000	+	$ 50,000	=	?
4.	$14,750	+	$ 37,500	=	?
5.	$38,750	+	?	=	$ 75,000
6.	$54,690	+	?	=	$143,650
7.	?	+	$ 25,000	=	$ 42,450
8.	?	+	$147,470	=	$475,000

9. $17,740 in cash.
$74,800 in merchandise.
$11,475 in supplies.
$22,480 owed to bank.
$917.80 owed in taxes.
What are the total assets?
What are the total liabilities?
What is the owner's equity?

10. $11,420 in cash.
$17,590 in supplies.
$43,470 in equipment.
$7890 owed for supplies.
$12,740 owed to bank.
What are the total assets?
What are the total liabilities?
What is the owner's equity?

11. Family Express Pharmacy has these assets and liabilities.

Cash: $4187
Inventory: $17,450
Equipment: $36,475

Supplies: $7185
Building: $125,000
Land: $31,500

Unpaid merchandise: $11,410
Taxes owed: $847
Real estate loan: $130,000

What are the total assets? What are the total liabilities? What is the owner's equity?

12. Howard's Jewelers has these assets and liabilities.

Cash on hand: $3417
Customers owe: $6214
Inventory: $13,419
Supplies: $417.50

Store fixtures: $1250
Building: $35,750
Land: $20,000
Bank loan: $12,214

Unpaid merchandise: $6470
Taxes owed: $714.85
Wages owed: $274.35
Mortgage loan: $31,340

What are the total assets? What are the total liabilities? What is the owner's equity?

MAINTAINING YOUR SKILLS   Look up the skills in parentheses if you need help or more practice.

Add. **(Skill 5)**

13. $31,475.00
     10,719.50
+   563.85

14. $9187.40
     7341.85
+   1174.97

15. $12,471.80
     7,518.75
+   431.83

16. $14,817.48
     3,247.55
+   7,916.37

Subtract. **(Skill 6)**

17. $74,850.00
 − 35,798.00

18. $147,875.00
 −   74,917.00

19. $45,371.80
 − 23,747.75

20. $12,147.85
 −   7,591.97

Balance Sheet

A balance sheet shows the financial position of your company on a certain date. You may prepare a balance sheet monthly, quarterly, or annually. The balance sheet shows your total assets, total liabilities, and owner's equity. The balance sheet is designed so that the assets appear on the left. The liabilities and owner's equity appear on the right. The sum of the assets must equal the sum of the liabilities and owner's equity.

Assets = Liabilities + Owner's Equity

EXAMPLE   *Skill* 3   *Term* Balance sheet

The Agees used $60,000 of their own money plus a $25,000 loan to start their clothing store. They received two shipments of merchandise. They paid cash for the $15,000 shipment. They did not pay immediately for the $8000 shipment. They bought supplies for $10,000. The Agees have $60,000 left in cash. What does their balance sheet show?

SOLUTION

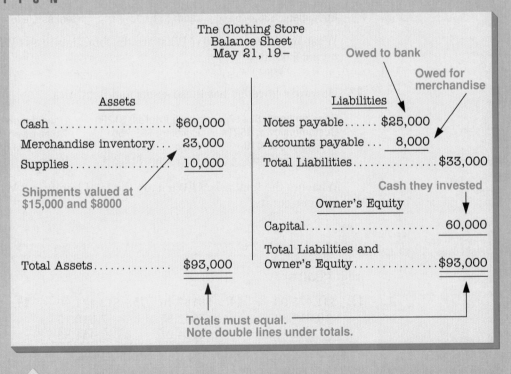

The Clothing Store
Balance Sheet
May 21, 19–

Owed to bank

Owed for merchandise

Assets		Liabilities	
Cash	$60,000	Notes payable......	$25,000
Merchandise inventory...	23,000	Accounts payable ...	8,000
Supplies.................	10,000	Total Liabilities..............	$33,000

Shipments valued at $15,000 and $8000

Cash they invested

Owner's Equity

Capital...................... 60,000

		Total Liabilities and	
Total Assets..............	$93,000	Owner's Equity..............	$93,000

Totals must equal.
Note double lines under totals.

60000 + 23000 + 10000 = 93000 25000 + 8000 = 33000 + 60000 = 93000

Assets		Liabilities	
Cash	$25,000	Notes payable	$10,000
Inventory	30,000	Accounts payable	7,000
Supplies	12,000	Owner's equity	50,000

1. Total assets . . . . .   ?

2. Total liabilities and
     owner's equity . . . . . .   ?

	3.	4.	5.	6.
Cash	$18,000	$ 45,000	$10,500	$125,400
Inventory	$11,500	$255,700	$41,200	$196,700
Supplies	$ 3,500	$ 11,400	$ 800	$ 12,800
Total Assets	?	?	?	?
Bank Loan	$10,500	$200,000	?	$175,000
Taxes Owed	$ 1,800	$ 8,800	$ 5,000	?
Total Liabilities	?	?	?	?
Owner's Equity	?	?	$25,000	$150,000

7. $25,000 in cash.
     $74,800 in merchandise.
     $65,000 owed to bank.
     What are the total assets?
     What are the total liabilities?
     What is the owner's equity?

8. $8450 in cash.
     $11,170 in equipment.
     $10,000 owed to bank.
     What are the total assets?
     What are the total liabilities?
     What is the owner's equity?

9. Complete the balance sheet for Howard's Jewelers.

Howard's Jewelers				
Balance Sheet				
June 30, 19–				
Assets		**Liabilities**		
Cash on hand	3 4 1 7 00	Bank loan	1 2 2 1 4 00	
Accts. receivable	6 2 1 4 00	Accts. payable	6 4 7 0 00	
Inventory	1 3 4 1 9 00	Taxes owed	7 1 4 85	
Supplies	4 1 7 50	Wages owed	2 7 4 35	
Store fixtures	1 2 5 0 00	Mortgage loan	3 1 3 4 0 00	
Building	3 5 7 5 0 00	Total Liabilities	?	
Land	2 0 0 0 0 00			
		Owner's Equity	?	
		Capital		
		Total Liabilities		
Total Assets	?	and Owner's Equity	?	

10. Complete the balance sheet for Pet Supplies, Inc.

Pet Supplies, Inc. Balance Sheet Dec. 31, 19–				
Assets		**Liabilities**		
Cash on hand	3 1 2 8 00	Bank loan	1 1 7 1 5 00	
Accts. receivable	2 1 4 85	Accts. payable	4 7 4 84	
Inventory	8 4 2 6 00	Taxes owed	4 6 1 72	
Supplies	5 6 1 7 40	Wages owed	2 3 4 85	
Store fixtures	2 4 0 5 75	Mortgage loan	5 4 1 7 5 00	
Building	4 3 4 5 0 00	Total Liabilities	?	
Land	2 4 5 0 0 00			
		Owner's Equity		
		Capital	?	
		Total Liabilities		
Total Assets	?	and Owner's Equity	?	

Prepare a balance sheet for each business.

11. Wholesale Grocer Supply Co. had these assets and liabilities on December 31:

Assets		**Liabilities**	
Cash	$ 2,417,600	Accounts payable	$84,640,000
Accounts receivable	53,591,500	Notes payable	56,119,400
Inventory	48,478,600	Income taxes	975,450
Property	75,750,000	Other liabilities	864,560
Investments	475,000		
Other assets	18,791,500		

12. A Brief Case   Metal Abrasives, Inc., shows these assets and liabilities as of June 30:

Assets (in millions of dollars)		**Liabilities (in millions of dollars)**	
Cash	$ 1.1	Notes payable	$14.9
Accounts receivable	9.8	Accounts payable	4.3
Inventories	11.4	Income taxes	1.2
Property	19.7	Stock	10.3
Foreign investments	1.3	Other liabilities	4.1
Other assets	4.2		

MAINTAINING YOUR SKILLS   Look up the skills in parentheses if you need help or more practice.

Add. **(Skills 3, 5)**

13. $14,780 + $13,190

14. $147,560 + $93,480

15. $4,175,000 + $897,400

16. $416,750 + $318,430 + $84,970

17. $1.7 + $2.3 + $4.5 + $1.4

20-3

Cost of Goods Sold

OBJECTIVE

Compute the cost of goods sold.

The balance sheet shows your total assets, total liabilities, and owner's equity at a given point in time. You also need to know whether the company is operating at a profit or loss. To determine if you are making money or losing money, you must know sales figures, expenses, and the cost of goods sold. The cost of goods sold is equal to the value of the beginning inventory plus the cost of any goods received (receipts) minus the value of the ending inventory.

Cost of Goods Sold = (Beginning Inventory + Receipts) − Ending Inventory

EXAMPLE   *Skills* 5, 6, 8   *Application* A   *Term* Cost of goods sold

The Clothing Store began the month with an inventory valued at $14,750. During the quarter, it received 100 belts that cost $6.49 each, 50 scarves at $8.24 each, 25 sweaters at $19.72 each, and 144 plastic raincoats at $2.50 each. The ending inventory was valued at $12,847. What was the cost of goods sold?

SOLUTION

A. Find the **receipts.**

100 belts	×	$ 6.49 each	=	$ 649.00
50 scarves	×	$ 8.24 each	=	412.00
25 sweaters	×	$19.72 each	=	493.00
144 raincoats	×	$ 2.50 each	=	360.00
		Total Receipts		$1914.00

B. Find the **cost of goods sold.**

(Beginning Inventory + Receipts) − Ending Inventory
   ($14,750.00   + $1914.00) − $12,847.00 = $3817.00
                                           cost of goods sold

100 × 6.49 = 649 M+ 50 × 8.24 = 412 M+ 25 × 19.72 = 493 M+ 144 × 2.5 = 360 M+ RM 1914 + 14750 − 12847 = 3817

✔ SELF-CHECK   Complete the problems, then check your answers in the back of the book.

Find the cost of goods sold.

	1.	**2.**
Beginning inventory	$156,470	$43,656
Receipts	21,960	11,712
Ending inventory	161,510	42,964

PROBLEMS

3. Beginning inventory: $417,600.
Receipts: $75,800.
Ending inventory: $396,800.
Find the cost of goods sold.

4. Beginning inventory: $125,400.
Receipts: $31,200.
Ending inventory: $131,400.
Find the cost of goods sold.

5. Beginning inventory: $75,470.
Receipts: $14,650.
Ending inventory: $72,170.
Find the cost of goods sold.

6. Beginning inventory: $19,560.
Receipts: $3780.
Ending inventory: $20,450.
Find the cost of goods sold.

7. Beginning inventory: $112,475.
Receipts: $42,164.
Ending inventory: $96,512.
Find the cost of goods sold.

8. Beginning inventory: $11,186.
Receipts: $2871.
Ending inventory: $12,074.
Find the cost of goods sold.

9. Miller Lamp Shop had a beginning inventory valued at $48,748.75. During the quarter, it had receipts of $9164.86. The value of the ending inventory was $50,041.93. Find the cost of goods sold.

10. The Popcorn Castle had a beginning inventory valued at $2146.73. During the quarter, receipts totaled $516.65. The value of the ending inventory was $1989.83. Find the cost of goods sold.

11. Allison's Dress Shop began the quarter with an inventory valued at $21,647. During the quarter, 4 shipments were received valued at $2248.60, $1874.55, $2516.43, and $2050.74. The ending inventory for the quarter was valued at $20,416. What was the cost of goods sold?

12. Central Auto Parts began the month with an inventory valued at $34,767.80. During the month, Central received 5 shipments valued at $1274.74, $4756.44, $983.45, $2465.39, and $416.93. Central's month-end inventory was valued at $36,193.48. Find the cost of goods sold.

13. The Shoehorn Repair Shop started the quarter with an inventory valued at $2178.43. During the quarter, it received 5 boxes of replacement heels at $4.35 a box, 3 boxes of replacement soles at $6.15 a box, 4 cards of shoelaces at $7.65 a card, 6 spools of nylon thread at $12.13 a spool, and 2 large bottles of industrial strength adhesive at $9.87 a bottle. The ending inventory for the quarter was valued at $2241.83. Find the cost of goods sold.

14. Leather Limited started the month with an inventory valued at $18,719.45. During the month, it received 9 men's jackets at $47.53 each, 15 belts at $2.79 each, 4 ladies' long coats at $84.37 each, 5 men's hats at $8.71 each, and 5 ladies' jackets at $38.84 each. The month-end inventory was valued at $16,478.54. Find the cost of goods sold.

MAINTAINING YOUR SKILLS Look up the skills in parentheses if you need help or more practice.

Add. **(Skill 5)**

15. $14,178.90 + $2417.36

16. $8174.32 + $2461.89

17. $56,147.83 + $9171.81

Subtract. **(Skill 6)**

18. $16,596.80 − $13,743.34

19. $9771.83 − $5421.95

20. $147,178.66 − $93,147.02

Multiply. **(Skill 8)**

21. 12 × $7.98 **22.** 5 × $46.73 **23.** 10 × $1.74 **24.** 144 × $11.43

Income Statement

An income statement, or profit-and-loss statement, shows in detail your income and operating expenses. If your gross profit is greater than your total operating expenses, your income statement will show a net income, or net profit.

Gross Profit = Net Sales − Cost of Goods Sold

Net Income = Gross Profit − Total Operating Expenses

EXAMPLE   *Skills* 3, 4   *Application* A   *Term* Income statement

Three months after opening The Clothing Store, Tina and John Agee prepare an income statement. Sales for the first three months totaled $12,174. Merchandise totaling $173 was returned to them. The Agees' inventory records show that the goods they sold cost them $3817. Records show that their operating expenses totaled $8047. What is the net income?

SOLUTION

The Clothing Store
Income Statement
For the Quarter Ended June 30, 19—

Income:		
Sales	$12,174	
Less: Sales returns and allowances	173	
Net Sale		$12,001
Cost of goods sold		3,817
Gross profit on sales		$8,184
Operating Expenses:		
Salaries and wages	$6,400	
Delivery expenses	100	
Rent	900	
Advertising	75	
Utilities	120	
Supplies	50	
Depreciation	150	
Insurance	210	
Miscellaneous	42	
Total operating expenses		$8,047
Net Income		$137

Total Sales − Returns

Net Sales − Cost of Goods Sold

Gross Profit − Total Operating Expenses

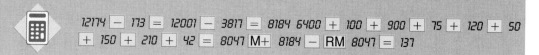

12174 − 173 = 12001 − 3817 = 8184 6400 + 100 + 900 + 75 + 120 + 50 + 150 + 210 + 42 = 8047 M+ 8184 − RM 8047 = 137

1. Sales: $95,000.
Returned merchandise: $3500.
Cost of goods sold: $30,000.
Total operating expenses: $34,700.
Find the gross profit on sales.
Find the net income.

2. Sales: $475,000.
Returned merchandise: $7500.
Cost of goods sold: $178,500.
Total operating expenses: $105,400.
Find the gross profit on sales.
Find the net income.

PROBLEMS

	Total Sales	Returns	Net Sales	Cost of Goods Sold	Gross Profit	Operating Expenses	Net Income
3.	$ 14,700	$ 540	$14,160	$ 8,750	$5410	$ 725	?
4.	$ 38,900	$4120	?	$ 11,175	?	$18,900	?
5.	$ 3,750	$ 75	?	$ 1,740	?	$ 828	?
6.	$174,945	0	?	$ 56,750	?	$42,193	?
7.	$674,916	$1274	?	$417,916	?	$96,419	?

Complete the income statements.

8. Income:

Sales.....................................$47,890
Less: Sales returns and allowances .... __976__
Net sales .... ?
Cost of goods sold .... __$21,742__
Gross profit on sales .... ?
Operating Expenses:
Salaries and wages $8,500
Taxes 497
Utilities 235
Depreciation............................... 975
Total operating expenses .... __?__
Net Income .... ?

9. Income:

Sales $463,575
Less: Sales returns and allowances .... __75,450__
Net sales .... ?
Cost of goods sold .... __$231,000__
Gross profit on sales .... ?
Operating Expenses:
Total operating expenses .... __$ 71,916__
Net Income .... ?

Prepare an income statement for each business in problems 10–13.

10. Last month Hurry-N-Go Service Station had total sales of $8961. Merchandise totaling $85 was returned. The goods that were sold cost Hurry-N-Go $5617. Operating expenses for the month were $718.

11. Advantage Housekeeping Service prepares an income statement monthly. For this past month, it collected $3150 from homeowners for services provided (sales). It had $70 in allowances for jobs that had to be redone. Salaries and wages totaled $1800, supplies cost $315, and other operating expenses totaled $94.

12. During the past quarter, Henri's Clothes Company had total sales of $27,418 and returns of $220. The cost of goods sold amounted to $9193. Operating expenses for the quarter included: salaries and wages of $2000, real estate loan payment of $1210, advertising at $190, utilities and supplies of $195, bank loan payment of $350, and other operating expenses of $375.

13. Hi-Tech Aluminum Co. prepares an annual income statement for distribution to its stockholders. This past year, net sales totaled $121.4 million. Cost of goods sold totaled $63.1 million. Operating expenses included: wages and salaries of $23.4 million, depreciation and amortization of $3.3 million, general taxes of $1.1 million, interest paid totaling $1.0 million, income taxes of $8.7 million, and miscellaneous operating expenses of $1.2 million.

14. A Brief Case If your total operating expenses are greater than your gross profit, your income statement will show a net loss.

Net Loss = Total Operating Expenses − Gross Profit

Find the net loss for this quarter.

Gross profit			1 9 2 3 80
Operating Expenses:			
Wages		1 6 7 5 65	
Supplies		4 8 3 00	
Miscellaneous		1 0 7 00	
Total			?
Net Loss			?

Add. **(Skill 5)**

15. $567.56
             74.92
         + 123.74

16. $18,516.50
              8,743.91
         + 191,434.84

17. $9871.47
             219.50
         + 4618.49

18. $231.42
             71.85
         + 63.79

Subtract. **(Skill 6)**

19. $174.93
          − 84.79

20. $746.90
          − 291.84

21. $7415
          − 1700

22. $135,463
          − 96,394

Vertical Analysis

OBJECTIVE

Analyze balance sheets and income statements by finding what percent a part is of the whole and computing the current ratio and the quick ratio.

Your business may analyze its income statement by finding what percent any given item is of the net sales.

Your business may analyze its balance sheet by finding certain defined ratios. The current ratio is the ratio of total assets to total liabilities. The quick ratio, sometimes called the acid-test ratio, is the ratio of total assets minus inventory to total liabilities.

$$\text{Percent of Net Sales} = \frac{\text{Amount for Item}}{\text{Net Sales}}$$

$$\text{Current Ratio} = \frac{\text{Total Assets}}{\text{Total Liabilities}}$$

$$\text{Quick Ratio} = \frac{\text{Total Assets} - \text{Inventory}}{\text{Total Liabilities}}$$

EXAMPLE   *Skill* 31   *Application* A   *Term* Net sales

Tina and John Agee analyze their income statement for the quarter. The statement shows:

Income:		
Sales	$12,174	
Less: Sales returns and allowances	173	
Net sales		$12,001
Cost of goods sold		3,817
Gross profit on sales		$8,184

What is the gross profit as a percent of net sales?

SOLUTION

Find the **percent of net sales.**

Amount for Item ÷ Net Sales

$8184 ÷ $12,001 = 0.6819 = 68.2% of net sales

8184 ÷ 12001 = .681943

✓ SELF-CHECK   Complete the problems, then check your answers in the back of the book.

1. Sales returns: $173.
Net sales: $12,001.
What are the sales returns as a percent of net sales?

2. Cost of goods sold: $3817.
Net sales: $12,001.
What is the cost of goods sold as a percent of net sales?

The Agee's analyze their May 21 balance sheet. The balance sheet shows:

Assets		Liabilities	
Cash	$60,000	Notes payable	$25,000
Merchandise inventory	23,000	Accounts payable	8,000
Supplies	10,000	Total Liabilities	$33,000
Total Assets	$93,000		

Owner's Equity

Capital 60,000

Total Liabilities and
Owner's Equity $93,000

What are the current ratio and the quick ratio?

SOLUTION

A. Find the **current ratio.**

Total Assets ÷ Total Liabilities

$93,000 ÷ $33,000 = 2.81 *or* 2.8 to 1 *or* 2.8:1 current ratio

This means that the Agees' total assets are 2.8 times their total liabilities. This is a good fiscal position. A current ratio of at least 2 to 1 is considered good.

B. Find the **quick ratio.**

(Total Assets − Inventory) ÷ Total Liabilities

($93,000 − $23,000) ÷ $33,000 = 2.12 *or* 2.1 to 1 *or* 2.1:1 quick ratio

This means that the Agees' quick assets (liquid or quickly converted to cash) are 2.1 times their total liabilities. This is a very good fiscal position. A quick ratio of at least 1 to 1 is considered good.

✔ SELF-CHECK   Complete the problems, then check your answers in the back of the book.

3. Total assets: $50,000.
Total liabilities: $30,000.
Find the current ratio.

4. Total assets: $56,500.
Inventory: $16,500.
Total liabilities: $30,000.
Find the quick ratio.

PROBLEMS

Round answers to the nearest tenth of a percent.

	Item	Amount for Item	÷	Net Sales	=	Percent of Net Sales
5.	Cost of goods sold	$20,000	÷	$40,000	=	?
6.	Operating expenses	$12,500	÷	$40,000	=	?
7.	Net income	$17,800	÷	$40,000	=	?

Find the ratios to the nearest tenth.

	Total Assets	Inventory	Total Liabilities	Current Ratio	Quick Ratio
8.	$ 78,500	$ 37,250	$ 40,000	?	?
9.	$325,800	$134,600	$163,200	?	?
10.	$ 4,897	$ 995	$ 2,750	?	?

11. Speedie-Quick's income statement for one month showed these figures.

a.
> Net sales $7690
> Cost of goods sold 4137
> Gross profit on sales .... ?

b. What is the gross profit as a percent of net sales?

12. The income statement for one quarter for Henri's Clothes Company showed these figures.

a.
> Net sales $27,198
> Cost of goods sold 13,100
> Gross profit on sales .... ?

b.
> Total operating expenses $12,195
> Net Income .... ?

c. What is the cost of goods sold as a percent of net sales?
d. What is the gross profit on sales as a percent of net sales?
e. What are the operating expenses as a percent of net sales?
f. What is the net income as a percent of net sales?

13. Use the balance sheet for Howard's Jewelers, dated June 30, from problem 9 on page 515.
   a. Find the current ratio.
   b. Find the quick ratio.

14. A Brief Case   Use the balance sheet for the Wholesale Grocer Supply Co., dated December 31, from problem 11 on page 516.
   a. Find the current ratio.
   b. Comment on the fiscal position based on the current ratio.
   c. Find the quick ratio.
   d. Comment on the fiscal position based on the quick ratio.

MAINTAINING YOUR SKILLS   Look up the skills in parentheses if you need help or more practice.

Find the rate. Round answers to the nearest tenth of a percent. **(Skill 31)**

15. n% of 495 = 45   **16.** n% of 120 = 30   **17.** n% of $74 = $58

Write as ratios with denominator of 1. Round answers to the nearest tenth **(Skill 22)**

18. $174:$94   **19.** $74.90:$76.40   **20.** $74,910:$36,120

Horizontal Analysis

OBJECTIVE

Compare two income statements using horizontal analysis and compute the percentage of change.

Horizontal analysis is the comparison of two or more income statements for different periods. The comparison is done by computing percent changes from one income statement to another. When computing percent change, the dollar amount on the earlier statement is the base figure. The amount of change is the difference between the base figure and the corresponding figure on the current statement. If the amount for an item decreases from one income statement to the next, both the amount of change and the percent change are negative.

$$\text{Percent Change} = \frac{\text{Amount of Change}}{\text{Base Figure}}$$

EXAMPLE   *Skill* 31   *Application* A   *Term* Percent change

The Agees prepare a second income statement and compare it to the one for the previous quarter. To the nearest tenth of a percent, what is the percent change for each item?

SOLUTION

Current Figure − Base Figure

	Last Quarter (Base)	This Quarter	Amount of Change	Percent Change
Net sales	$12,001	$13,174	$1,173	9.8%
Cost of goods sold	3817	4190	373	9.8%
Gross profit on sales	$ 8184	$ 8,984	$ 800	9.8%
Operating expenses	8047	7995	− 52	− 0.6%
Net income	$ 137	$ 989	$ 852	621.9%

Operating expenses decreased. Change is negative.

The Agee's net income increased by 621.9%

✔ **SELF-CHECK**   Complete the problems, then check your answers in the back of the book.

Find the percent change from last quarter to this quarter.

1. Last quarter: $12,000.
This quarter: $15,000.

2. Last quarter: $70,000.
This quarter: $50,000.

PROBLEMS

Round to the nearest tenth of a percent.

	Last Year (Base)	This Year	Amount of Change	Percent Change
3.	$720,000	$830,000	$110,000	?
4.	$ 45,000	$ 36,000	− $ 9,000	?
5.	$114,750	$137,840	? ?	

6. Last year net sales were $150,000. **7.** Last week cost of goods sold
This year net sales are $210,000.   was $4650.
What is the amount of change?   This week cost of goods sold
What is the percent change?   is $3875.
       What is the amount of change?
       What is the percent change?

8. Last year net sales were $400,000. **9.** Last month net income was
This year net sales are $600,000. $42,476.
What is the amount of change?   This month net income
What is the percent change?   is $51,419.
       What is the amount of change?
       What is the percent change?

10. Income statements for Hurry-N-Go Service Station showed these figures for March and April.

	March	April	Amount of Change	Percent Change
a. Net sales	$8876	$9172	?	?
b. Cost of goods sold	5617	5904	?	?
c. Gross profit on sales	3259	3268	?	?
d. Operating expenses	718	700	?	?
e. Net income	2541	2568	?	?

11. Income statements for Wholesale Grocer Supply Co. showed the following figures.

	Last Year (in thousands)	This Year (in thousands)	Amount of Change (in thousands)	Percent Change
a. Net sales	$117.4	$109.9	?	?
b. Cost of goods sold	56.9	54.7	?	?
c. Gross profit on sales	60.5	55.2	?	?
d. Operating expenses	35.7	36.2	?	?
e. Net income	24.8	19.0	?	?

12. Two annual income statements for Hi-Tech Aluminum Co. showed
these figures.

	Last Year (in millions)	This Year (in millions)	Amount of Change (in millions)	Percent Change
a. Net sales	$121.4	$123.5	?	?
b. Cost of goods sold	63.1	64.6	?	?
c. Gross profit on sales	58.3	58.9	?	?
d. Operating expenses	38.7	36.8	?	?
e. Net income	19.6	22.1	?	?

MAINTAINING YOUR SKILLS Look up the skill in parentheses if you need help or more practice.

Find the rate. Round answers to the nearest tenth of a percent. **(Skill 31)**

13. $n\%$ of 95 $=$ 19   **14.** $n\%$ of 74 $=$ 37   **15.** $n\%$ of 1450 $=$ 974

16. $n\%$ of $14,176 $=$ $4193   **17.** $n\%$ of 748 $=$ 1050

Reviewing the Basics

Skills

(Skill 3)

Solve.

1. $174,914 + $637,540 + $461,532 + $71,916

2. $14,179,831 + $24,816,121 + $7,816,550 + $31,147,641

(Skill 4)

3. $17,416 − $8154

4. $21,520 − $6473

5. $54,318 − $18,433

6. $41,117 − $24,823

(Skill 5)

7. $4718.75 + $176.94 + $71.93 + $3491.89

8. $4214.91 + $15,714.70 + $871.85 + $15,716.95

(Skill 6)

9. $8640.71 − $916.75

10. $41,917.62 − $2147.66

11. $74,917.52 − $2147.73

12. $516,531.78 − $391,478.91

Round answers to the nearest tenth of a percent.

(Skill 31)

13. $17,800 is what percent of $91,400?

14. $7142 is what percent of $27,172?

15. n% of 91 = 27

16. n% of 35 = 147

Applications

Use the following formula to solve problems 17–20.

Gross Profit = Net Sales − Cost of Goods Sold

(Application A)

17. Net sales are $41,741. Cost of goods sold is $22,416. What is the gross profit?

18. Net sales are $28,419. Cost of goods sold is $17,816. What is the gross profit?

19. Net sales are $4,419,500. Cost of goods sold is $3,516,000. What is the gross profit?

20. Net sales are $8175. Cost of goods sold is $4395. What is the gross profit?

Terms

Match each term with its definition on the right.

21. Owner's equity

22. Balance sheet

23. Income statement

24. Percent change

a. the total amount of money that you owe to creditors

b. shows your sales, operating expenses, and net profit or loss

c. the difference between your assets and liabilities

d. is negative if the amount decreases from one income statement to the next

e. shows your assets, liabilities, and owner's equity

Refer to your reference files in the back of the book if you need help.

Unit Test

Lesson 20-1

1. The Shoe Box had these assets and liabilities on June 30.

Cash: $4750
Inventory: $12,500
Supplies: $735
Store furnishings: $7215

Unpaid merchandise: $6800
Taxes owed: $517
Bank loan: $5000

What are the total assets? What are the total liabilities? What is the owner's equity?

Lesson 20-2

2. Prepare a balance sheet for The Shoe Box. Use the information in problem 1.

Lesson 20-3

3. For one period, The Shoe Box had a beginning inventory of $11,450, receipts of $4360, and an ending inventory of $10,716. Find the cost of goods sold.

Lesson 20-4

4. Global Manufacturing, Inc., prepares an annual income statement. This past year, Global had net sales of $789,640. The cost of goods sold totaled $341,620. Operating expenses included: wages and salaries totaling $107,119, depreciation and amortization totaling $5447, interest paid totaling $9179, product recall totaling $4172, and miscellaneous operating expenses of $3178. Prepare an income statement for Global Manufacturing.

Lesson 20-5

5. Use the income statement for Global Manufacturing, Inc., in problem 4. To the nearest tenth of a percent, find the cost of goods sold as a percent of net sales.

Lesson 20-5

6. Use the balance sheet for The Shoe Box in problem 2.

  a. Find the current ratio.
  b. Find the quick ratio.

Lesson 20-6

7. Income statements for B.J. Benson's showed the following figures for two quarters. Round answers to the nearest tenth of a percent.

	First Quarter	Second Quarter	Amount of Change	Percent Change
a. Net sales	$8500	$9100	?	?
b. Cost of goods sold	4300	4500	?	?
c. Gross profit on sales	4200	4600	?	?
d. Operating expenses	2900	2800	?	?
e. Net income	1300	1800	?	?

A SPREADSHEET APPLICATION

Accounting Records

To complete this spreadsheet application, you will need the template diskette for *Mathematics with Business Applications.* Follow the directions in the *User's Guide* to complete this activity.

Input the information in the problem to create a balance sheet and an income statement for The Home Furniture Mart. After you have finished entering the data, answer the questions that follow.

The Home Furniture Mart
Balance Sheet
March 31, 19–

Assets (thousands)		Liabilities (thousands)	
Cash	$125.6	Accounts payable	$35.4
Accounts receivable	214.9	Notes payable	28.5
Inventory	56.3	Other liabilities	10.7
Supplies	8.6	Total Liabilities	?
Other Assets	24.6	Owner's Equity (thousands)	
Total Assets	?	Capital	?
		Total Liabilities and Owner's Equity	?

1. What are the total assets?

2. What are the total liabilities?

3. What is the owner's equity (capital)?

4. What are the total liabilities and owner's equity?

5. What is the current ratio?

6. What is the quick ratio?

The Home Furniture Mart
Income Statement
For the Quarter Ended March 31,19–

Income:

   Sales $34,576

   Less: Returns 678

   Net sales .... $?

Cost of goods sold 16,589

Gross profit on sales .... $?

Operating Expenses:

   Salaries and wages .... $ 9862

   Utilities 1423

   Supplies 856

   Total operating expenses .... $?

Net Income .... $?

7. What are the net sales?

8. What is the gross profit on sales?

9. What are the total operating expenses?

10. What is the net income?

11. What is the cost of goods sold as a percent of net sales?

12. What are the total operating expenses as a percent of net sales?

13. What is the net income as a percent of net sales?

14. Salaries and wages are what percent of the total operating expenses?

Financial Management

Financial management of your business includes your investments, loans, and taxes. You may borrow money by taking a *commercial loan,* or business loan, from a bank. If you want to raise money, you can sell shares of your business, or stocks and bonds. Money that you don't need for day-to-day expenses can be invested to earn additional interest. Your business may invest its surplus cash in *United States Treasury Bills* or in other businesses with a high credit rating.

O U T L I N E

Today, corporations own more than 90 percent of all stocks and bonds traded on a stock exchange.

Corporate Income Taxes

Your business must pay federal income taxes. The tax rates vary depending on the size and type of your business. Corporations, businesses owned by stockholders, are subject to federal tax rates ranging from 15% to 39% of taxable income. Taxable income is the portion of your company's gross income that remains after normal business expenses are deducted. Normal business expenses include wages, rent, utilities, interest paid on loans, property taxes, depreciation, and so on. The structure for federal corporate income taxes is graduated.

If taxable income is:		Tax is:	
Over—	But not over—		Of the amount over—
0	50,000	15%	0
$ 50,000	$ 75,000	$ 7500 + 25%	$ 50,000
75,000	100,000	13,250 + 34%	75,000
100,000	—	21,750 + 39%	100,000

Taxable Income = Annual Gross Income − Deductions

EXAMPLE

Skill 30   *Application* C   *Term* Corporate income taxes

The World Corporation had a gross income of $623,145 for the year. It had these business expenses during the year:

| Wages | $412,120 | Utilities | $11,219 | Depreciation | $14,720 |
| Rent | $18,750 | Interest | $7195 | Other deductions | $19,745 |

What federal corporate income tax must the World Corporation pay?

SOLUTION

A. Find the **deductions.**
$412,210 + $18,750 + $11,219 + $7195 + $14,720 + $19,745 = $483,749 in deductions

B. Find the **taxable income.**
Annual Gross Income − Deductions
$623,145 − $483,749 = $139,396 taxable income

C. Find the **federal corporate income tax.** (Refer to the table above.)
$21,750 + 39% of ($139,396 − $100,000)
$21,750 + (0.39 × $39,396) =
$21,750 + $15,364.44 = $37,114.44 federal corporate income tax

412120 + 18750 + 11219 + 7195 + 14720 + 19745 = 483749 M+ 623145 −
RM 483749 = 139396 − 100000 = 39396 × 39 % 15364.44 + 21750 =
37114.44

✔ SELF-CHECK

Complete the problems, then check your answers in the back of the book.

Use the tax table above to find the federal corporate income tax.

1. Gross income: $145,000.
Deductions: $60,000

2. Gross income: $219,000.
Deductions: $154,000.

Refer to the table on page 532 for federal corporate income taxes.

	Annual Gross Income	–	Deductions	=	Taxable Income	Total Tax
3.	$ 74,918	–	$ 38,172	=	$36,746	?
4.	$212,971	–	$125,539	=	?	?
5.	$279,434	–	$214,611	=	?	?
6.	$720,338	–	$621,913	=	?	?
7.	$916,418	–	$721,534	=	?	?

8. Johnson-Etna Mfg. Corp.
Annual gross income is $316,921.
Deductions total $234,847.
What is the federal corporate
income tax?

9. The Towncraft Company.
Annual gross income is $721,173.
Deductions total $670,224.
What is the federal corporate
income tax?

F.Y.I.
In 1989, the Internal Revenue Service collected $117 billion in corporate income taxes.

10. The Magno-Met Company had these business expenses for the year:

Wages	$816,147.80	Property taxes	$97,493.70
Rent	$60,000.00	Depreciation	$26,916.00
Utilities	$74,617.46	Other deductions	$34,124.76
Interest	$5119.54		

The Magno-Met Company had a gross income of $1,516,749 for the year.
a. What are the total business expenses?
b. What is the taxable income?
c. What is the federal corporate income tax for the year?

11. Gayle Sturgeon, D.V.M., formed a corporation. The corporation had these business expenses for the year:

Wages	$24,000.00	Property taxes	$7413.00
Utilities	$15,774.35	Depreciation	$3849.74
Interest	$1491.84	Other deductions	$4916.47

Dr. Sturgeon's gross income for the year was $134,176. What is Dr. Sturgeon's federal corporate income tax for the year?

Find the percentage. Round answers to the nearest hundredth. **(Skill 30)**

12. 6% of 74   **13.** 7.65% of $80.77   **14.** $\frac{7}{8}$% of 160   **15.** 145% of 350

16. 35% of 4700   **17.** 34% of 816   **18.** $16\frac{1}{2}$% of 470   **19.** 2.64% of 1700

Subtract. **(Skill 6)**

20. $87,434 − $75,000   **21.** $127,816 − $100,000   **22.** 74.718 − 16.634

Issuing Stocks and Bonds

OBJECTIVE

Compute the selling expenses and the net proceeds from an issue of stocks and bonds.

Your business may raise money by issuing stocks or bonds. When you issue stocks, the buyer becomes a part owner of your business. When you issue bonds, the buyer is lending money to your business.

When you issue stocks or bonds, you must pay certain expenses. One expense is an underwriting commission, a commission to the investment banker who helps you distribute the stocks or bonds. Other expenses include accounting costs, legal fees, and printing costs. The amount that your business actually receives from the sale of the stocks or bonds after paying these expenses is the net proceeds.

Net Proceeds = Value of Issue − Total Selling Expenses

EXAMPLE *Skill* 30 *Application* A *Term* Net proceeds

The Landover Company is planning a major expansion program. To finance the program, Landover plans to sell an issue of 300,000 shares of stocks at $41.50 per share. The underwriting commission will be 6.5% of the value of the stocks. Accounting fees, legal fees, printing costs, and other expenses are estimated to be 0.9% of the value of the stocks. If all the shares of stock are sold, what net proceeds will Landover Company receive?

SOLUTION

A. Find the **value of issue.**
$41.50 × 300,000 = $12,450,000

B. Find the **total selling expenses.**
Underwriting commission:
$12,450,000 × 6.5%............................$809,250
Other expenses:
$12,450,000 × 0.9%............................$112,050
Total Selling Expenses $921,300

C. Find the **net proceeds.**
Value of Issue   −   Total Selling Expenses
$12,450,000   − $921,300   = $11,528,700 net proceeds

41.5 ☒ 300000 ═ 12450000 M+ ☒ 6.5 % 809250 RM 12450000 ☒ .9 %
112050 ＋ 809250 M− 921300 RM 11528700

✔ SELF-CHECK   Complete the problems, then check your answers in the back of the book.

Find the net proceeds.

1. Value of stock: $8,000,000.
Total selling expenses: 5.4% of value of stocks.

2. Value of stock: $37,450,000.
Underwriting commission: $372,500.
Total selling expenses: 0.3% of value of stock.

Underwriting Commission					Total	
(Value of Issue	× Percent =	Dollar Amount)	+ Other Expenses	=	Selling Expenses	Net Proceeds
3. ($6,800,000 × 6%		= $408,000)	+ $47,500	=	?	?
4. ($950,000 × 9%		= ?)	+ $26,500	=	?	?
5. ($21,000,000 × 5%		= ?)	+ $54,650	=	?	?

6. Value of stocks is $750,000. Underwriting commission is 10%. Other expenses total $21,640. What are the net proceeds?

7. Value of bonds is $7,435,750. Underwriting commission is 5.5%. Other expenses total $48,650. What are the net proceeds?

8. Universal Waste Disposal, Inc., sold 1,350,000 shares of stock at $24.625 per share. The investment banker's commission was 5% of the value of the stock. The other expenses were 1% of the value of the stock. What net proceeds did UWD, Inc., receive?

9. The Mercury Electric Company issued 20,000 shares of stocks at $62.50 per share. Find the net proceeds after these selling expenses are deducted.

Underwriting Expenses		Other Expenses	
Commissions	$75,000	Printing costs	$4500
Legal fees	2500	Legal fees	2500
Advertising	2500	Accounting fees	3500
Miscellaneous	1250	Miscellaneous	2000

Use the table to solve problems 10 and 11.

10. The Hawthorne Company sold 25,000 bonds at $65 per bond. What are the net proceeds?

11. The Stock Paper Corporation sold 75,000 bonds for $35.75 per bond. What are the net proceeds?

Size of Issue (in millions)	Underwriting Commission	Other Expenses
Under $0.5	6.5%	1.5%
$0.5–$0.9	6%	1.3%
$1.0–$1.9	5.5%	1.0%
$2.0–$4.9	5%	0.8%

12. A Brief Case   The Athens City Council needs to raise $2.8 million to continue required services for the remainder of the year. The council plans to sell bonds at $37.50 each to raise the money. Selling expenses are expected to be 5.8% of the value of the bonds. How many bonds must the council sell to obtain net proceeds of $2.8 million?

MAINTAINING YOUR SKILLS   Look up the skill in parentheses if you need help or more practice.

Find the percentage. Round answers to the nearest hundredth. **(Skill 30)**

13. 6.5% of $980,000

14. 5% of $3,150,000

15. 6% of $743,000

16. 5.5% of $1.5 million

21-3

Borrowing

OBJECTIVE
Compute the maturity value of a commerical loan.

Your business may borrow money to buy raw materials, products, or equipment. **Commercial loans,** or business loans, are similar to personal loans. The interest rates are generally one to two percentage points higher than the **prime rate.** The prime rate is the lowest rate of interest available to commercial customers at a given time.

The **maturity value** of your loan is the total amount you repay. The maturity value includes both the principal borrowed and the interest owed on the loan. Commercial loans usually charge ordinary interest at exact time. That is, the length of time of the loan is calculated by dividing the exact number of days of the loan by 360 days.

Interest = Principal × Rate × Time

Maturity Value = Principal + Interest Owed

EXAMPLE   *Skills* 28, 10   *Application* A   *Term* Maturity value

Harm's Drugstore borrowed $80,000 from First National Bank to pay for remodeling costs. The bank loaned the money at 10.5% ordinary interest for 60 days. What is the maturity value of the loan?

SOLUTION

A. Find the **interest owed.**

Principal × Rate × Time
$80,000 × 10.5% × $\frac{60}{360}$
($80,000 × 0.105 × 60) ÷ 360
$504,000   ÷ 360 = $1400 interest owed

B. Find the **maturity value.**

Principal + Interest Owed
$80,000 + $1400 = $81,400 maturity value

80000 [M+] [×] 10.5 [%] 8400 [×] 60 [÷] 360 [=] 1400 [+] [RM] [=] 81400

✔ SELF-CHECK   Complete the problems, then check your answers in the back of the book.

Find the maturity value of the loan.

1. $100,000 borrowed at 11.25% ordinary interest for 90 days.

2. $150,000 borrowed at 9.5% ordinary interest for 140 days.

Use ordinary interest at exact time to solve.

	Principal	×	Rate	×	Time	=	Interest	Maturity Value
3.	$70,000	×	10%	×	$\frac{90}{360}$	=	?	?
4.	$95,000	×	11%	×	$\frac{100}{360}$	=	?	?
5.	$37,500	×	10.5%	×	$\frac{120}{360}$	=	?	?

6. $64,000 is borrowed for 60 days. **7.** $37,650 is borrowed for 180 days.
Interest rate is 11.5%.   Interest rate is 9.75%.
What is the interest owed?   What is the interest owed?
What is the maturity value?   What is the maturity value?

8. Trust Bank, Inc., loaned $90,000 to Fernandez Home Builders, Inc. The term of the loan was 270 days. The interest rate was 11.75%. What was the maturity value of the loan?

9. To take advantage of a bicycle manufacturer's closeout special on touring bikes, Wheels, Inc., borrowed $50,000 from the Union Trust Company. Union Trust charged 10.75% interest on the loan. The term of the loan was 175 days. What was the maturity value of the loan?

10. The Gibraltar Construction Company was granted a 1-year construction loan of $650,000 to finance the construction of an apartment complex. Gibraltar borrowed the money from Citizens Trust Company at an interest rate of 10.25%. What was the maturity value of the loan?

11. The Solar Panel Manufacturing Company needs $1,450,000 for 270 days to help finance the production of an experimental solar hot water heater. The financial manager has arranged financing from 3 sources. Each loan charges ordinary interest at exact time.

Swancreek Trust Company
$500,000 loan for 270 days
Interest rate: 10.5%

Universal Investment Company
$450,000 loan for 270 days
Interest rate: 10.9%

Investment Bankers, Inc.
$500,000 loan for 270 days
Interest rate: 11.25% for first 90 days
11.00% for next 180 days

a. What is the total interest for the 3 loans?
b. What is the total maturity value?

Write the percents as decimals. **(Skill 28)**

12. 12.7%   **13.** 340%   **14.** 0.6%   **15.** $\frac{7}{10}$%   **16.** 0.05%

Investments

OBJECTIVE

Compute the cost of a Treasury Bill.

Your business may invest surplus cash that is not needed for day-to-day operations. One way to invest your money is to purchase United States Treasury Bills. You may purchase the bills through a bank. Some banks charge a service fee for the paperwork involved in obtaining the bills.

When you purchase a Treasury Bill, you are actually lending money to the government. In return, you receive interest at the rate that is in effect at the time you purchase the bill. The interest is ordinary interest at exact time. Treasury Bills are issued on a discount basis. That is, the interest is computed and then subtracted from the face value of the bill to determine the cost of the bill. The face value of the Treasury Bill is the amount of money you will receive on the maturity date of the bill. Maturity dates for Treasury Bills range from 91 days to a year.

Cost of Treasury Bill = (Face Value of Bill − Interest) + Service Fee

---

EXAMPLE | *Skills* 8, 10 *Application* A *Term* United States Treasury Bill

The financial manager of the Osinski Manufacturing Company has decided to invest the company's surplus cash in a $50,000 United States Treasury Bill for 120 days. The interest rate is 9.25%. The bank charges a service fee of $25 to obtain the Treasury Bill. What is the cost of the Treasury Bill?

SOLUTION

A. Find the **interest**.

Principal	×	Rate	×	Time
$50,000.00	×	9.25%	×	$\frac{120}{360}$

($50,000.00 × 0.0925 × 120) ÷ 360

$555,000.00 ÷ 360 = $1541.666 = $1541.67

B. Find the **cost of Treasury Bill**.

(Face Value of Bill − Interest) + Service Fee

($50,000.00 − $1541.67) + $25.00

$48,458.33 + $25.00 = $48,483.33 cost of Treasury Bill

50000 M+ × 9.25 % 4625 × 120 ÷ 360 = 1541.666 M− RM 48458.333 + 25 = 48483.33

✔ SELF-CHECK Complete the problem, then check your answer in the back of the book.

1. $100,000 Treasury Bill purchased at 8.9% interest for 180 days. Service fee is $50. Find the cost of the Treasury Bill.

Use ordinary interest at exact time to solve.

	Face Value of Treasury Bill	Interest Rate	Time in Days	Interest	Bank Service Fee	Cost of Treasury Bill
2.	$ 70,000	9%	100	?	$25	?
3.	$100,000	8.7%	120	?	$35	?
4.	$150,000	9.35%	180	?	No fee	?
5.	$ 85,000	10.213%	91	?	$30	?

6. Face value of bill is $60,000.
Interest rate is 9.4%.
Bill matures in 120 days.
Bank service fee is $30.
What is the interest?
What is the cost of the bill?

7. Face value of bill is $97,000.
Interest rate is 8.78%.
Bill matures in 100 days.
Bank service fee is $40.
What is the interest?
What is the cost of the bill?

8. The J. P. Cotton Company purchased a $40,000 Treasury Bill at 10% ordinary interest for 120 days. The bank charged a service fee of $20. What was the interest? What was the cost of the Treasury Bill?

9. The Pacer Investment Company purchased a $95,000 United States Treasury Bill at 9.24% interest. The Treasury Bill matures in 91 days. The bank service fee is $35. What is the cost of the Treasury Bill?

10. The Masters Printing Company had $115,000 in surplus cash. Sue Masters decided to invest in United States Treasury Bills at an interest rate of 10.45%. She purchased a 110-day bill with a face value of $115,000. The bank charged a $35 service fee. What was the cost of the Treasury Bill?

11. Robert Tompkins sold his apartment house for $240,000. After paying various expenses, he had $210,000 to invest. He purchased a $100,000 United States Treasury Bill for 120 days at an interest rate of 9.672%. He purchased a second Treasury Bill with a face value of $110,000. The second bill was for 91 days at an interest rate of 9.842%. Robert's bank does not charge a service fee on Treasury Bills of $100,000 or over. What was the total cost of the 2 bills?

Multiply. Round answers to the nearest thousandth. **(Skill 8)**

12. 17.49 × 8 **13.** 7.475 × 8.1 **14.** 0.57 × 0.41 **15.** 21.43 × 0.214

Divide. Round answers to the nearest thousandth. **(Skill 10)**

16. 74 ÷ 6 **17.** 7.47 ÷ 1000 **18.** 971 ÷ 40 **19.** 700 ÷ 32

Investments— Net Interest

Your business may invest extra cash in **commercial paper**. Commercial paper is a 30-day to 270-day loan issued by a company with a very high credit rating. When your business invests in commercial paper, you are actually lending money to another company. Commercial paper earns ordinary interest at exact time. Usually, your business obtains commercial paper through a bank. The bank may charge a service fee. Your **net interest** is the total interest that you earn after any bank service fees have been deducted.

$$\text{Net Interest} = \text{Total Interest Earned} - \text{Service Fee}$$

EXAMPLE | *Skills* 8, 10 | *Application* A | *Term* Commercial paper

Boat Works has a $75,000 cash surplus. The financial manager used the cash to make the following investments in commercial paper:

Hastings Corporation	$20,000 at 10.9% for 30 days
A.B.C. Financial Corporation	$40,000 at 10.5% for 60 days
Lake Erie Fish Company	$15,000 at 9.9% for 120 days

The bank charges a service fee of $35. What net interest will Boat Works earn?

SOLUTION

A. Find the **total interest earned.**

(1) Hastings Corporation
$20,000.00 \times 10.9\% \times \frac{30}{360} = \$181.666 = \qquad \$\ 181.67$

(2) A.B.C. Financial Corporation
$40,000.00 \times 10.5\% \times \frac{60}{360} = \qquad\qquad \$\ 700.00$

(3) Lake Erie Fish Company
$15,000.00 \times 9.9\% \times \frac{120}{360} = \qquad\qquad \underline{\$\ 495.00}$

$\qquad\qquad\qquad\qquad\qquad$ Total Interest Earned $\quad \$1376.67$

B. Find the **net interest.**

Total Interest Earned $-$ Service Fee
$\$1376.67 \qquad - \qquad \$35.00 \ = \$1341.67$ net interest

$20000\ \boxed{\times}\ 10.9\ \boxed{\%}\ 2180\ \boxed{\times}\ 30\ \boxed{\div}\ 360\ \boxed{=}\ 181.666\ \boxed{M+}\ 40000\ \boxed{\times}\ 10.5\ \boxed{\%}\ 4200\ \boxed{\times}$
$60\ \boxed{\div}\ 360\ \boxed{=}\ 700\ \boxed{M+}\ 15000\ \boxed{\times}\ 9.9\ \boxed{\%}\ 1485\ \boxed{\times}\ 120\ \boxed{\div}\ 360\ \boxed{=}\ 495\ \boxed{M+}\ \boxed{RM}$
$1376.666\ \boxed{-}\ 35\ \boxed{=}\ 1341.666$

✔ SELF-CHECK | Complete the problem, then check your answer in the back of the book.

1. $100,000 invested at 10.4% interest for 30 days; no service fee. Find the net interest.

Use ordinary interest at exact time to solve.

	Borrower	Value of Note	Interest Rate	Time in Days	Interest	Bank Service Fee	Net Interest
2.	Warehouse Corp.	$ 50,000	9.7%	60	?	$20	?
3.	Tel-Com Inc.	$120,000	10.3%	120	?	$25	?
4.	Furniture Factory	$250,000	10.7%	270	?	No fee	?

5. $75,000 note for 90 days.
Purchased from The Kord
    Corporation.
Interest rate is 9.9%.
Bank service fee is $20.
What is the net interest?

6. $50,000 note for 75 days.
Purchased from Amos
    Packing Co.
Interest rate is 10.2%.
Bank service fee is $25.
What is the net interest?

7. The Carmine Coal Company has a $45,000 cash surplus. The financial manager made the following investments in commercial paper:

Rural Farm Electric $25,000 at 10.5% for 270 days
Farmers' Grain Supply $20,000 at 10.65% for 180 days

The bank charges a service fee of $15 for each note. What net interest will the Carmine Coal Company receive?

8. Judy Myers, the financial manager of Myers Shoe Mfg. Co., made the following investments in commercial paper:

Piedmont Paper Co. $75,000 at 10.75% for 270 days
Ace Development Co. $80,000 at 10.5% for 180 days
Investment Properties Inc. $95,000 at 11.5% for 30 days

The bank charges a service fee of $8 per note. What net interest will Myers Shoe Mfg. Co. receive?

9. A Brief Case   On June 1, Adelphia Investments, Inc., received $272,500 for the sale of an industrial park. Harry Pappas, the financial manager for the company, invested the money in the following commercial paper:

Donovan Corporation $75,000 at 10.4% until June 16
Commercial Telephone $97,500 at 10.9% until September 25
Central Power Co. $50,000 at 11.25% until October 20
General Land Development Co. $50,000 at 11.4% until November 4

The bank service fee is $25 for each note under $51,000 and $15 for each note over $51,000. What net interest will Adelphia Investments, Inc., receive from these investments?

Multiply. Round answers to the nearest hundredth. **(Skill 8)**

10. 52.74 × 1000 **11.** 0.478 × 0.49 **12.** 4173 × 0.005 **13.** 7.165 × 0.047

Growth Expenses

OBJECTIVE

Compute the total cost of expanding a business.

You may expand your business in several different ways. You may purchase or build an addition to your present building. You may purchase another business to become part of your business. Your business may merge, or combine, with another business to form a new business. Growth expenses for your business may include construction fees, consultation fees, legal fees, and so on.

Total Cost of Expansion = Sum of Individual Costs

EXAMPLE   *Skill* 5   *Application* A   *Term* Growth Expenses

The owner of Posner's Deli plans to expand the business by opening a new store. Growth expenses include:

Marketing survey: $3500 plus $3.75 per person for 150 people interviewed
Land: $60,000
Building construction: $750,000
Architect's fee: 5% of the cost of construction
Surfacing parking lot: $0.90 per square foot for 6000 square feet
Legal fees: $975
Equipment and fixtures: $67,500
Additional stock: $13,000
Miscellaneous expenses: $4000

What is the total cost for the expansion of Posner's Deli?

SOLUTION

Find the **sum of individual costs.**

Marketing survey: $3500.00 + ($3.75 × 150) ....$	4062.50
Land	60,000.00
Building construction	750,000.00
Architect's fee: $750,000.00 × 5%	37,500.00
Surfacing parking lot: $0.90 × 6000	5400.00
Legal fees	975.00
Equipment and fixtures	67,500.00
Additional stock	13,000.00
Miscellaneous expenses	4000.00

Total Cost of Expansion $942,437.50

3.75 ✕ 150 = 562.5 + 3500 = 4062.5 M+ 60000 M+ 750000 M+ ✕ 5 %
37500 M+ .9 ✕ 6000 = 5400 M+ 975 M+ 67500 M+ 13000 M+ 4000 M+
RM 942437.5

Find the total cost of expansion.

1. Expansion costs for Wyatt Jewelry store: legal fees, $1500; display cases, $25,000; safe, $10,150; miscellaneous, $3200.

2. Expansion for Harriet's Fashions: new construction, $74,800; display racks, $9450; clothing stock, $57,500; advertisements, $500.

PROBLEMS

3. Expand housecleaning calls. Solicit new customers for $125. Purchase new van for $12,470. What is the total cost of expansion?

4. Expand lawn-care business. Purchase new truck for $10,540. Purchase supplies for $875. What is the total cost of expansion?

5. Grandview Farms is opening a new beef stick outlet in the Western Mall. Grandview pays rent in advance for 3 months. The rental charge is $1150 per month. Grandview makes a 20% down payment on refrigeration equipment that costs a total of $13,980. Grandview also purchased additional supplies for $5795. What is the total cost of expansion?

6. Outdoors, Inc., is adding a new department that will specialize in hunting equipment. Outdoors pays $775 for redecoration of an area of the store. Outdoors also makes a 30% down payment on new stock that costs a total of $17,785. What is the total of these growth expenses?

7. The Antique Mart is expanding by adding 6 new stalls available for lease. The Mart is converting a storage area of 1800 square feet into these stalls. The costs of the expansions are:

Description	Cost
Construction permit	$120
Removal of two walls	$800
New lighting fixtures and installation	$2785
Carpet and installation	$18.90 per square yard
Down payment on fixtures	25% of $4260

What is the total cost for The Antique Mart to expand?

8. Central National Bank plans to open a new branch office. Central purchased property for $34,500. Construction costs for a new building totaled $875,980. In addition, Central paid an architect's fee of 7.5% of the cost of construction. Legal fees for the expansion totaled $6000. New equipment and fixtures cost $24,675. Other expenses came to $7215. What was the total cost of Central's expansion?

9. JP Industries, Inc., is searching for a new business to buy. Finders, Inc., a company that specializes in locating firms for sale, has located a small machine plant. If JP Industries purchases the firm, it must pay Finders, Inc., a finder's fee of 4.5% of the total worth of the machine plant. To acquire the plant, JP Industries must pay the plant its total worth of $2.45 million. JP Industries must pay legal fees amounting to 1.25% of the total worth of the machine plant. In addition, JP Industries must pay the debts of the machine plant. The debts amount to $165,850 plus 12.4% interest for 1 year. What will be the total cost of the expansion?

MAINTAINING YOUR SKILLS   Look up the skill in parentheses if you need help or more practice.

Add. **(Skill 5)**

10.	**11.**	**12.**	**13.**	**14.**
78.14	571.8	747.249	4.718	8456.87
+ 234.854	+ 28.912	+ 221.84	+ 0.643	+ 4318.93

Reviewing the Basics

(Skill 5)

Solve.

1. $6718.75 + $54,193.84 + $715.66

2. $76,000.00 + $4218.74 + $42,193.75 + $21,424.95

(Skill 8)

3. $86,000 × 0.095 × 120   4. $35,000 × 0.105 × 180

5. $17,500 × 0.0925 × 90   6. $62,550 × 0.0975 × 270

Round answers to the nearest cent.

(Skill 10)

7. $980,400 ÷ 360   8. $661,500 ÷ 360   9. $1,650,000 ÷ 360

Write as a decimal.

(Skill 28)

10. 10.75%   11. $9\frac{1}{2}$%   12. 9.475%   13. 175%

Find the percentage. Round answers to the nearest cent.

(Skill 30)

14. 11% of $41,750   15. 35% of $24,760   16. 2.1% of $186

17. 9.875% of $18,125   18. $6\frac{1}{2}$% of $3250   19. $\frac{1}{4}$% of $86

(Application C)

Find the tax.

20. Taxable income is $41,650.

21. Taxable income is $93,460.

If taxable income is:		Tax is:		
Over—	But not over—			Of the amount over—
0	50,000		15%	0
$ 50,000	$ 75,000	$ 7,500 +	25%	$ 50,000
75,000	100,000	13,250 +	34%	75,000
100,000	—	21,750 +	39%	100,000

Write your own definition for each term.

22. Corporate income taxes   23. Net proceeds

24. Maturity value   25. United States Treasury Bill

26. Commercial paper   27. Growth expenses

Refer to your reference files in the back of the book if you need help.

Unit Test

Lesson 21-1

1. The Bellview Corporation had an annual gross income of $834,718 for the year. Business deductions totaled $697,497. What federal corporate income tax must Bellview pay for the year?

If taxable income is:		Tax is:	
Over—	But not over—		Of the amount over—
0	50,000	15%	0
$ 50,000	$ 75,000	$ 7500 + 25%	$ 50,000
75,000	100,000	13,250 + 34%	75,000
100,000	—	21,750 + 39%	100,000

Lesson 21-2

2. Core Control Inc. is issuing 500,000 shares of stock. Each share will be sold at $36.75. The underwriting commission is 3% of the value of the stock. The other expenses are expected to be 0.75% of the value of the stock. If all the shares of stock are sold, what net proceeds will Core Control receive?

Lesson 21-3

3. Mill Grove Fruit Farms borrowed $62,500 for 150 days at the prime rate of interest. When Mill Grove borrowed the money, the prime rate was 9.5%. The bank charged ordinary interest at exact time. What was the maturity value of the loan?

Lesson 21-4

4. Horton Manufacturing had $250,000 in surplus cash. The financial manager decided to invest in a United States Treasury Bill with a face value of $250,000. The bill matured in 126 days. The interest rate was 10.875% ordinary interest at exact time. The bank service fee was $35. What was the cost of the Treasury Bill?

Lesson 21-5

5. The Tom Brown Kitchen Supply Company sold its entire stock of kitchen cupboards for installation in an apartment complex. Tom Brown invested the $132,500 income from the sale in the following commercial paper:

Machinko Construction Co.	$90,000 at 9.95% for 30 days
Tobar Excavations, Inc.	$42,500 at 10.25% for 60 days

Both investments earn ordinary interest at exact time. The bank service fee is $45 for both. What net interest will the Tom Brown Kitchen Supply Company receive?

Lesson 21-6

6. The Campus Bookstore opened a new branch store, which it rents for $1250 per month. The Campus Bookstore paid the rent in advance for 3 months. The Campus Bookstore purchased new furniture and store fixtures for $34,175. The Campus Bookstore also purchased additional merchandise for $7185. Miscellaneous expenses totaled $2465. What was the total cost of expansion?

A SPREADSHEET APPLICATION

Financial Management

To complete this spreadsheet application, you will need the template diskette for *Mathematics with Business Applications.* Follow the directions in the *User's Guide* to complete this activity.

Input information in the following problem to find the value of the issue, the total selling expenses, and the net proceeds. Then answer the questions that follow.

Stock Symbol	# of Shares	$ per Share	Value of Issue	Under. Comm. %	Selling Expense	Net Proceeds
AMR	500,000	$ 56.75	?	4.1%	?	?
CSX	250,000	36	?	6.8%	?	?
EDO	1,000,000	5.875	?	5%	?	?
GM	300,000	19.625	?	4.5%	?	?
IBM	200,000	127.875	?	3.75%	?	?
KFC	100,000	72.5	?	6%	?	?
MGM	750,000	11.375	?	5.5%	?	?
OEA	350,000	40.625	?	6.5%	?	?
QMS	600,000	19.25	?	5.875%	?	?
SBX	450,000	2.125	?	6.25%	?	?
URS	750,000	9.625	?	5.75%	?	?
WB	150,000	108.75	?	6.125%	?	?

1. What is the value of the issue for each of these stocks: AMR, EDO, IBM, MGM, QMS, and URS?

2. What is the selling expense for each of these stocks: CSX, GM, KFC, OEA, SBX, and WB?

3. What are the net proceeds for each of these stocks: AMR, CSX, EDO, GM, IBM, KFC, MGM, OEA, QMS, SBX, URS, and WB?

4. How many shares of stock must KFC sell at $72.50 per share to raise $10 million if the underwriter commission is 6%?

5. How many shares of stock must WB sell at $108.75 per share to raise $10 million if the underwriter commission is 6.125% and other expenses total $25,000?

6. How many shares of stock must EDO sell at $5.875 per share to raise $7.5 million if the underwriter commission is 5% and other expenses total 1.5% of the value of the stock?

7. If OEA could reduce the underwriter commission to 5.5%, how much more would the net proceeds be equal to?

8. If MGM's stock went up $0.50, how much more would the net proceeds be equal to?

▼ CAREER WISE

Stockbroker

Vida Kasic is a stockbroker for a large investment firm. As a broker, she acts as an agent for someone who wants to buy or sell part ownership in a company.

Not only does Vida handle transactions for her clients, she also watches market trends. It is her hope that she can help people buy shares when the price is low and sell shares when the price increases.

Mergers and acquisitions are extremely important to Vida, since they can greatly affect the price of stock. When she finds that the health of the economy is changing, she might advise a client to alter his or her portfolio, the collection of investments the client has.

The line graph below shows the price of one share of Xylex Company stock for each business day in the month of July. Prices are closing prices.

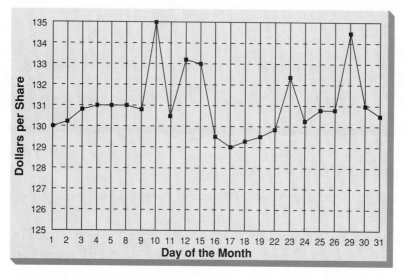

1. Over which 3-day period did the price remain unchanged?

2. Over which two days was the greatest price increase?

3. Did the price rise or fall from July 23 to July 24? By how much?

4. If 100 shares were bought on July 1 and sold on July 29 (at closing prices), was there a profit or loss? Of how much?

5. What were the minimum and maximum closing prices during July?

6. What was the average closing price for the stock in July?

▼
A SIMULATION

Stock Market

Your employer has given you a $2000 bonus. You decide to invest the money in stocks. To help you decide which stocks to buy, you study the past performance of many stocks. You also read financial newspapers and magazines, and talk to people who are well-informed about the stock market.

52 Week High	Low	Stock	Dividend	Yield (%)	P.E. Ratio	Volume (000)	High	Low	Close	Net Change
26 1/2	23	Cablvsn	—	—	58	265	26 1/2	23 1/2	23	−3 1/2
22 3/8	20 1/2	LaZBoy	—	—	67	257	20	20 7/8	20	0
59	55 7/8	QuakrO	$0.25	6.8%	19	7446	57 3/8	55 1/8	56 3/8	1/2
8 1/2	5 3/8	TCBY	—	—	11	120	7 1/2	6	6 1/8	−1 1/2
44 3/8	34 1/2	Upjohn	1.16	7.4%	38	1456	44 1/2	41 7/8	42 1/4	3 3/8

1. You purchase 30 shares of LaZBoy stock at today's closing price. What is the cost of the stock?

2. Your stockbroker charges a 1.5% commission each time you buy or sell stock. What is the commission on this purchase?

3. What is the total amount you pay for the stock?

4. Of the $2000 that you want to invest in stock, how much is now available to be invested?

5. LaZBoy declares a dividend of $1.25 per share. What is the annual yield?

6. What is the total dividend you receive on your 30 shares?

7. Including the dividend, how much money is now available to be invested in stocks?

Buying and Selling Stock

During the next year, you continue to study the stock market. Sometimes you buy stocks, starting with the money you have available after buying the LaZBoy stock. Sometimes you sell stock, either because the company is not doing well or because you want the money to buy stocks in a different company. You sell all your stock at the end of the year. While you owned some of the stocks, you received the dividends shown. You add the dividend to the money left to invest.

	Name of Stock	Number of Shares	Bought (B) Sold (S)	Price per Share	Amount Paid	Amount Received	Commission 1.5%	Total Cost	Divi- dend	Dividend Income	Total Income	Money Left to Invest
	LaZBoy	30	B	$20.000	$600.00	—	$9.00	$609.00	$1.25	$37.50	$37.50	$1428.50
8.	QuakrO	15	B	57.375	?	—	?	?	?	?	?	?
9.	TCBY	25	B	6.125	?	—	?	?	—	—	—	?
10.	Upjohn	8	B	43.000	?	—	?	?	—	—	—	?
11.	QuakrO	10	S	68.500	—	?	?	—	—	—	?	?
12.	TCBY	25	S	7.875	—	?	?	—	—	—	?	?
13.	Upjohn	16	B	40.125	?	—	?	?	?	?	?	?
14.	CtyCm	60	B	3.500	?	—	?	?	—	—	—	?
15.	QuakrO	5	S	62.125	—	?	?	—	—	—	?	?
16.	LaZBoy	30	S	21.000	—	?	?	—	—	—	?	?
17.	CtyCm	80	B	3.875	?	—	?	?	—	—	—	?
18.	Upjohn	24	S	44.875	—	?	?	—	—	—	?	?
19.	CtyCm	140	S	3.125	—	?	?	—	—	—	?	?

20. What is the total dividend you received?

21. Have you made a net profit or a net loss? How much?

22. What is the percent profit or loss on your $2000 investment?

A SIMULATION
(CONTINUED)

Comparing Mutual Funds

You realize at the end of the year that you do not have time to study the stock market. You decide to invest $2100 in three mutual finds. The following table shows the net asset value per share for each fund at the end of each month of the previous year.

23. On graph paper, set up a graph like the one at the bottom of the page. Make a line graph using the values in the table. Use a different color for each fund.

NET ASSET VALUE PER SHARE												
Month	Jan	Feb	Mar	Apr	May	June	July	Aug	Sep	Oct	Nov	Dec
South Bay Fund	$6.42	$6.55	$6.72	$6.89	$6.75	$6.81	$6.97	$7.14	$7.29	$7.50	$7.68	$7.90
Salinas Valley Fund	6.59	6.74	6.89	7.03	6.82	6.85	6.92	7.25	7.41	7.78	7.95	7.83
Central Section Fund	6.54	6.62	6.73	6.84	6.78	6.86	6.95	7.04	7.15	7.24	7.32	7.40

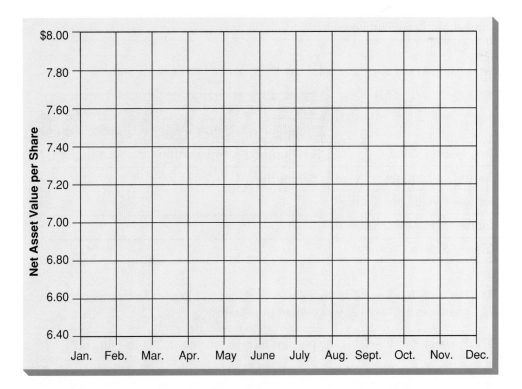

A SIMULATION
(CONTINUED)

Investing in Mutual Funds

You decide to invest $700 in each of the three funds. South Bay and Salinas Valley have loading charges. Central Section is a no-load fund.

   The table shows the total market value of each fund's portfolio and the number of shares outstanding on the day you make your investments. Calculate the number of shares of each fund you can purchase.

		FUND	
	South Bay	**Salinas Valley**	**Central Section**
Total Market Value	$49.0 million	$30.6 million	$44.1 million
Shares Outstanding	5.6 million	3.6 million	4.9 million
24. Net Asset Value per Share	?	?	?
Amount Invested	$700	$700	$700
Loading Rate	10%	8.5%	0
25. Loading Charge	?	?	?
26. Amount Invested Minus Loading Charge	?	?	?
27. Number of Shares Purchased	?	?	?

At the end of the year, you calculate the value of your investments.

		FUND	
	South Bay	**Salinas Valley**	**Central Section**
Total Market Value	$63.4 million	$40.9 million	$52.2 million
Shares Outstanding	6.0 million	4.1 million	5.2 million
28. Net Asset Value per Share	?	?	?
29. Number of Shares Owned	?	?	?
30. Value of Shares Owned	?	?	?

31. For each fund, what is the percent increase in the net asset value per share during the year you owned the shares?

32. For each fund, what is the percent increase in your $700 investment?

22

Corporate Planning

Corporate planning takes place for strategic short-term planning as well as extensive long-range planning. Economic considerations that play an important role in corporate planning include the *inflation rate* at the time, the size of the real and per capita *gross national product,* and the measure of the *consumer price index* for particular items. All these considerations, and more, go into planning and implementing a corporate *budget.*

Business executives meet regularly to discuss and plan strategies that enable them to deal successfully with ever-changing economic conditions.

22-1

Inflation

OBJECTIVE

Compute the inflation rate, current price, and original price.

Inflation is an economic condition during which there are price increases in the cost of goods and services. At the corporate level, inflation is observed as increases in (1) wholesale prices, (2) cost of utilities, (3) cost of production and shipping, and (4) demands for scarcer materials. Some of the causes of inflation are (1) heavy spending, resulting in high demand, (2) increased production costs while producers try to maintain profit levels, and (3) lack of competition.

The inflation rate is found by subtracting the original price from the current price and dividing the result by the original price. The inflation rate is expressed as a percent increase over a specified time period, usually to the nearest tenth of a percent.

The current price is found by adding the original price to the product of the original price times the inflation rate.

The original price is found by dividing the current price by 1 plus the inflation rate.

$$\text{Inflation Rate} = \frac{(\text{Current Price} - \text{Original Price})}{\text{Original Price}}$$

$$\text{Current Price} = \text{Original Price} + (\text{Original Price} \times \text{Inflation Rate})$$

$$\text{Original Price} = \frac{\text{Current Price}}{(1 + \text{Inflation Rate})}$$

EXAMPLE   *Skills* 5, 10, 30, 31, 32   *Application* A   *Term* Inflation rate

Liza Turner works in the summer as a junior staff reporter for the local newspaper. Liza was asked to write a story about the changes in the economy. One section was on inflation. She had the following data:

- New automobile costs $15,000 today. Same model sold for $12,500 a year ago.

- Home sold for $130,000 2 years ago. Inflation rate over the 2-year period for homes is 10%.

- Current price for a lawn mower is $395. Inflation rate over the last year for lawn mowers is 5%.

  A. What is the inflation rate for the automobile?

  B. What is the current price of the home?

  C. What was the original price of the lawn mower last year?

A. Find the **inflation rate** for the automobile.
(Current Price − Original Price) ÷ Original Price
($15,000 − $12,500) ÷ $12,500
$2500 ÷ $12,500 = 0.2 = 20%
inflation rate

B. Find the **current price** of the home.
Original Price + (Original Price × Inflation Rate)
$130,000 + ($130,000 × 10%)
$130,000 + $13,000 = $143,000 current price

C. Find the **original price** of the lawn mower.
Current Price ÷ (1 + Inflation Rate)
$395.00 ÷ (1 + 5%)
$395.00 ÷ 1.05 = $376.1904 = $376.19
original price

`15000 ⊟ 12500 ⊟ 2500 ÷ 12500 ⊟ 0.2`

✔ SELF-CHECK Complete the problems, then check your answers in the back of the book.

1. Original price was $75. Current price is $80. Find the inflation rate.

2. Original price was $12,400. Inflation rate is 8.2%. Find the current price.

3. Current price is $28.05. Inflation rate is 10%. Find the original price.

PROBLEMS

	Inflation Rate	Current Price	Original Price
4.	?	$ 4670.00	$ 4389.80
5.	?	$ 79.99	$ 78.79
6.	5.0%	?	$ 147.00
7.	3.6%	?	$17,458.00
8.	0.7%	$ 1.29	?
9.	12.4%	$243,750.00	?

10. Original price: $16.95.
Current price: $18.49.
Find the inflation rate.

11. Original price: $3.50.
Inflation rate: 12.4%.
Find the current price.

12. Current price: $4370.
Inflation rate: 6.3%.
Find the original price.

13. Original price: $0.988.
Current price: $1.249.
Find the inflation rate.

14. Original price: $134,980.
Inflation rate: 4.2%
Find the current price.

15. Current price: $27,495.
Inflation rate: 14.61%.
Find the original price.

16. Original price: $3680.
Current price: $11,850.
Find the inflation rate.

17. Original price: $0.49.
Inflation rate: 19.8%.
Find the current price.

18. The inflation rate for grocery products is 5.7%. How much would a cart of groceries cost today if it cost $72.70 last year?

19. A month ago, a friend purchased a mini van for $16,865. The inflation rate for mini vans over the past month is 0.4%. What is the current price of a mini van?

20. At the grand opening of the Bayview Market in 1938, a 10-ounce box of cereal cost 10¢. Today the box costs $1.99. What is the inflation rate for cereal over that time period?

21. At Bayview Market, a quart of milk could be purchased for $0.11 in 1938. The inflation rate for milk since that time is 600%. What does a quart of milk cost today?

22. At Bayview Market, 2 pounds of coffee cost $5.99 today. The rate of inflation for coffee since 1938 is 1715.2%. What did 2 pounds of coffee cost in 1938?

23. A Brief Case   A hand-held calculator that had an original price of $79.95 currently sells for $19.95.

   a. What is the inflation rate?

   b. What is true of the inflation rate when the current price is less than the original price?

   c. Instead of inflation, what might this be called?

MAINTAINING YOUR SKILLS   Look up the skills in parentheses if you need help or more practice.

Find the percentage. Round answers to the nearest cent. **(Skill 30)**

24. 3.7% of $436.80   **25.** 11.5% of $48.98   **26.** 0.4% of $1.79

Find the rate. Round answers to the nearest tenth of a percent. **(Skill 31)**

27. $12.85 is what percent of $257?

28. $3.79 is what percent of $199.99?

29. $2500 is what percent of $22,387.65?

Gross National Product

The gross national product is a measure of nation's economic performance. The GNP is the estimated total value of all goods and services produced by a nation during a year. Only goods, such as automobiles, machinery, food, and clothing, and services, such as haircuts, appliance repairs, and inventory costs, that add to the national income are included. The government uses the GNP to monitor the health of the country's economy.

Because of inflation, a nation's GNP can appear to be growing faster than it actually is. For this reason, the real GNP, or adjusted GNP, is corrected for inflation.

There per capita GNP provides an indication of the nation's standard of living. It is the GNP distributed over the population.

$$\text{Real GNP} = \text{GNP} - (\text{GNP} \times \text{Inflation Rate})$$

$$\text{Per Capita GNP} = \frac{\text{GNP}}{\text{Population}}$$

EXAMPLE   *Skills* 1, 6, 11, 30   *Application* A   *Term* Gross national product

The United States had a population of 248,700,000 in 1990, an inflation rate of 4.6%, and a GNP of $5200.8 billion. What is the real GNP? What is the per capita GNP?

SOLUTION

A. Find the **real GNP.**   Rounded to the nearest tenth of a billion

GNP   − (GNP   × Inflation Rate)
$5200.8 billion − ($5200.8 billion ×   4.6%)
$5200.8 billion − $239.2 billion   = $4961.6 billion
real GNP

B. Find the **per capita GNP.**   248,700,000 ÷ 1,000,000,000 = 0.2487 billion

GNP   ÷   Population
$5200.8 billion ÷ $248,700,000, or 0.2487 billion
= $20,911.942 = $20,911.94 per capita GNP

5200.8 [M+] [×] 4.6 [%] 239.2368 [M−] [RM] 4961.56

---

1. GNP is $43,800,000.
Inflation rate is 6.4%.
Find the real GNP.

2. GNP is $92.8 million
(or $92,800,000).
Population is 78,650.
Find the per capita GNP.

	GNP	Inflation Rate	Population	Real GNP	Per Capita GNP
3.	$123.6 million	4.3%	0.4 million	?	?
4.	$57,459,650	0.4%	5,365,760	?	?
5.	$478.6 billion	6.2%	74.2 million	?	?
6.	$756,500,000	2.1%	2,586,450	?	?
7.	$2,876.7 billion	7.8%	2,456,980,000	?	?
8.	$98,760,000,000	12.6%	31,478,000	?	?

9. GNP is $768,000,000.
Inflation rate is 4.6%.
Find the real GNP.

10. GNP is $45,678,000,000.
Population is 9,670,000.
Find the per capita GNP.

11. Australia has a population of 16,646,000 and a GNP of $220 billion. The inflation rate in Australia is 7.6%. Find the real GNP and the per capita GNP.

12. Greece has a GNP of $43,500,000,000. Its population is 10,066,000, and the inflation rate is 13.7%. Find the real GNP and the per capita GNP.

13. Canada has a GNP of $486,000,000,000. It has a population of 26,527,000 and an inflation rate of 5.0%. Find the real GNP and the per capita GNP.

14. The GNP for Israel is $36 billion. Its population is 4.371 million. The rate of inflation in Israel is 20.2%. Find the real GNP and the per capita GNP.

15. The GNP for Belgium is $153 billion. Its population is 9.898 million. The rate of inflation in Belgium is 3.1%. Find the real GNP and the per capita GNP.

16. The population of China is 1,130,065,000. China's GNP is $350,000,000,000, and the inflation rate is 16.3%. Find the real GNP and the per capita GNP.

17. A Brief Case

Nation	GNP	Population	Inflation Rate
France	$943 billion	56,184,000	3.5%
Italy	$825 billion	57,657,000	6.2%
U.K.	$758 billion	57,121,000	7.8%

Using the information above, determine:
a. Which nation has the highest real GNP.
b. Which nation has the lowest real GNP.
c. Which nation has the highest per capita GNP.
d. Which nation has the lowest per capita GNP.

18. A Brief Case Argentina has a GNP of $74.3 billion and a reported inflation rate of 3079%.
a. What real GNP do you get using the formula?
b. Why is this impossible?

22-3

Consumer Price Index

OBJECTIVE

Compute the consumer price index (CPI), current cost, and cost in 1967 of any given commodity.

The **consumer price index** is a measure of the average change in prices of a certain number of goods and services. The year 1967 is used as the base year and the CPI for 1967 is set at 100. This means that a commodity that cost $100 in 1967 and that has a CPI of 159 today would cost $159 today. To find the CPI for any given commodity, divide the current cost by the cost in 1967 and then multiply by 100.

If you know the CPI for a given commodity and its cost in 1967, you can find the current cost by multiplying the cost in 1967 by the CPI and then dividing by 100.

If you know the CPI for a given commodity and its current cost, you can find the cost in 1967 by dividing the current cost by the CPI and then multiplying by 100.

$$\text{CPI} = \frac{\text{Current Cost}}{\text{Cost in 1967}} \times 100$$

$$\text{Current Cost} = \frac{(\text{Cost in 1967} \times \text{CPI})}{100}$$

$$\text{Cost in 1967} = \frac{\text{Current Cost}}{\text{CPI}} \times 100$$

EXAMPLE   *Skills* 7, 8, 10   *Application* A   *Term* Consumer price index

Liza Turner was asked to write a story about the consumer price index. She had the following data:

• The current cost of a briefcase is $48.50. The cost in 1967 was $19.95.

• The cost in 1967 of a good pair of sweat socks was $3.75. The CPI for sweat socks is 170.

• The current cost of a classical cassette is $8.00. The CPI for classical cassettes is 200.

A. What is the CPI for the briefcase?

B. What is the current cost of the sweat socks?

C. What was the cost of the cassette in 1967?

A. Find the **CPI.**

(Current Cost ÷ Cost in 1967) × 100

($48.50 ÷ $19.95) × 100

2.4310 × 100 = 243.1 CPI

B. Find the **current cost.**

(Cost in 1967 × CPI) ÷ 100

($3.75 × 170) ÷ 100

$637.50 ÷ 100 = $6.375 = $6.38 current cost

C. Find the **cost in 1967.**

(Current Cost ÷ CPI) × 100

($8.00 ÷ 200) × 100

$0.04 × 100 = $4.00 cost in 1967

48.5 ⌷÷⌷ 19.95 ⌷=⌷ 2.4310776 ⌷×⌷ 100 ⌷=⌷ 243.10776

✔ SELF-CHECK Complete the problems, then check your answers in the back of the book.

1. The current cost is $350. The cost in 1967 was $140. What is the CPI?

2. The CPI is 350. The cost in 1967 was $765.95. What is the current cost?

3. The CPI is 140. The current cost is $24.50. Find the cost in 1967.

PROBLEMS

Round answers to the nearest tenth of a percent or to the nearest cent.

	Item	CPI	Current Cost	Cost in 1967
4.	Sport coat	?	$139.50	$ 74.95
5.	Dinner out for 2	?	$ 45.75	$ 21.50
6.	Apartment rent	110.6	?	$325.00
7.	Lawn mower	156.7	?	$425.50
8.	Bicycle	315.8	$658.78	?
9.	Book	875.4	$ 49.97	?

10. Current cost is $48.
Cost in 1967 was $40.
What is the CPI?

11. Cost in 1967 was $78.
The CPI is 210.
What is the current cost?

12. The current cost is $540.
The CPI is 225.
What was the cost in 1967?

13. Current cost is $125,500.
Cost in 1967 was $37,500.
What is the CPI?

Use the table for problems 14–20.

14. In 1967, Josh paid $4577.13 for a new car. What could he expect to pay today for a similar car?

15. In 1967, the Johnson family paid $48,650 for their home. What would a similar home sell for today?

16. In 1967, the Johnson family paid $34.70 for their weekly groceries. How much could a similar family expect to pay for their weekly groceries today?

INDICATORS FOR JUNE	
Commodity	**CPI**
Fuel oil	504.8
Gasoline	364.6
Rent	238.9
Home	243.5
Groceries	315.6
Dining out	328.2
New car	312.7
Clothing	186.3

F.Y.I.

Over the last 45 years, the consumer price index averaged 4.2% a year (Better Investing, December 1991, p. 36).

17. Last month's fuel oil bill for the Crowleys was $256.36. What would the Crowleys have paid for the same amount of fuel oil in 1967?

18. Bill and Mary Ellis spent $87.95 to have dinner out last Friday. How much would they have spent in 1967 to have a similar dinner out?

19. Alice McCarty pays $450 per month rent for her efficiency apartment. What would Alice have paid in 1967 for a similar apartment?

20. Tammy Spencer just purchased a new Treble four-door sedan for $18,457.50. What would a similar new car have cost Tammy in 1967?

21. James Manos paid $11 for a haircut and styling that would have cost him $6.50 in 1967. What is the CPI?

22. Tonia Curtis paid $6.75 for a movie ticket last weekend. In 1967, she would have paid $3.75 for the same ticket. What is the CPI?

23. A Brief Case   Generally, in comparing prices from 1967 to the present:

a. If the price of a commodity was more in 1967 than it is now (for example, hand-held calculators, VCRs, kilowatts of electricity), would the CPI be more or less than 100 today?

b. If something you purchased in 1967 was given away free today, what would the CPI equal?

c. If something you would pay for today had been given away free in 1967, would it be possible to determine the CPI?

MAINTAINING YOUR SKILLS   Look up the skills in parentheses if you need help or more practice.

Divide. Round answers to the nearest tenth. **(Skill 11)**

24. $234.50 ÷ $78.25

25. $24.79 ÷ $18.45

26. $234,750.00 ÷ $489,980.00

27. $2.49 ÷ $5.79

Multiply. Round answers to the nearest cent. **(Skill 8)**

28. 234.3 × $45.75   **29.** 156.8 × $378.89   **30.** 112.4 × $7.89

◆ 22-4

Budget

OBJECTIVE

Allocate revenue and expenses and analyze a budget.

A budget is a financial plan that enables a business, a governmental agency, or an individual to identify what sources are expected to produce revenues (earn money) and what amounts are allocated to various departments or categories for expenses. Revenues and expenses are allocated as a percent of the total income. The main purpose of a budget is to monitor revenues and expenses. The actual amount spent must be compared with the budget allocation.

Budget Allocation = Total Income × Percent

Difference = Actual Amount − Budget Allocation

EXAMPLE | *Skills* 6, 30 *Application* A *Term* Budget

The Kimberly Auto Company wanted to earn $2,457,890 in revenues for the year. The company expected to earn its revenues by the following sources: 75% from sales, 15% from services, and 10% from return on investments. At the end of the year, a budget analysis was sent to each manager showing the budgeted amounts and actual amounts of each area. Sales showed revenues of $1,755,965; services showed $369,240; and return on investments showed $245,790. Did the company reach its goal in revenues?

SOLUTION

A. Find the amount each source was budgeted, the **budget allocation.**

	Total Income	×	Percent		
Sales	$2,457,890	×	75%	=	$1,843,417.50
Services	$2,457,890	×	15%	=	$ 368,683.50
Return on investments	$2,457,890	×	10%	=	$ 245,789.00

B. Find the **difference** between the actual amount and budget allocation for each source.

	Actual Amount	−	Budget Allocation	=	Difference
Sales	$1,755,965.00	−	$1,843,417.50	=	−$87,452.50
Services	369,240.00	−	368,683.50	=	+556.50
Return on investments	245,790.00	−	245,789.00	=	+1.00
Total	$2,370,995.00	−	$2,457,890.00	=	−$86,895.00

The company was $86,895 short of its goal for the year.

2457890 M+ × 75 % 1843417.5 RM 2457890 × 15 % 368683.5 RM
2457890 × 10 % 245789

Wagner Enterprises had a $400,000 budget. For each category, determine:

1. Amount budgeted   **2.** Difference

	Expected Percent	Actual Amount
Salaries	83%	$360,000
Supplies	6%	$ 20,000
New equipment . . .	5%	$ 26,000
Maintenance	4%	$ 9,000
Misc.	2%	$ 3,000

PROBLEMS

	Total Revenue	Expected Percent	Actual Amount	Budget Allocation	Difference
3.	$ 80,000	25%	$ 15,000	?	?
4.	$ 120,000	60%	$ 74,000	?	?
5.	$ 860,000	7%	$ 64,000	?	?
6.	$ 987,500	80%	$ 760,000	?	?
7.	$5,880,000	64%	$3,562,000	?	?
8.	$7,644,000	8%	$ 756,253	?	?

9. Allocate $150,000 as shown:
Sales: 70%.
Services: 20%.
Investments: 10%.

10. Allocate $600,000 as shown:
Interest: 15%.
Stocks: 65%.
Bonds: 20%.

11. Find the difference.

	Actual	Budget
Sales	$460,000	$500,000
Services	150,000	180,000
Investments	60,000	50,000
Total	$670,000	$730,000

12. Find the difference.

	Actual	Budget
Supplies	$155,400	$144,000
Equipment	450,600	390,000
Services	90,300	155,620
Total	$696,300	$689,620

13. The Downey Corporation is budgeting total revenues of $15,219,000 next year. Out of the total, 96% is expected to come from sales, 2% is expected to come from trading profits, and 2% is expected to come from other sources. How many dollars has Downey budgeted in expected revenues in each category?

14. The Downey Corporation from problem 13 had actual revenues of $14,700,000 from sales; $240,000 from trading profits; and $120,000 from other sources. Did the company reach its revenue goals?

15. The Mason Plastics Company wanted to earn $1,500,000 in revenues for the year. The company expected to earn its revenues by the following sources: 85% of the total from sales, 10% from services, and 5% from return on investments. At the end of the year, budget analysis was sent to each manager showing the budgeted amounts and actual amounts of each area. Sales showed revenues of $1,590,000; services showed $120,000; and return on investments showed $34,000. Did the company reach its goal in revenues?

16. The Posy Fruit Farm had planned earnings of $680,000 for the year. The farm expected its revenues to come from the following sources: 60% from apple sales, 20% from strawberry sales, 15% from blueberry sales, and 5% from cherry sales. At the end of the season, budget figures showed sales revenues of: apples, $420,000; strawberries, $91,320; blueberries, $91,340; and cherries, $39,900. Did the Posy Fruit Farm reach its goal in revenues?

17. A Brief Case Suppose you have a small lawn service and nursery business. Your expected revenues of $45,000 are to come from the following sources: 60% from lawn care, 20% from nursery stock sales, 10% from fall cleanup, 8% from landscaping supplies, and 2% from other sales. At the end of the year, your budget analysis indicated this income data: $24,500 from lawn care, $8300 from nursery stock sales, $4100 from fall cleanup, $2400 from landscaping supplies, and $1200 from other sales. Determine the amount budgeted for each source and find the difference between the actual amount and budget allocation.

MAINTAINING YOUR SKILLS Look up the skill in parentheses if you need help or more practice.

Find the percentage. Round answers to the nearest cent. (Skill 30)

18. 78% of $348,970

19. 43% of $7,385,450

20. 4.6% of $23,684,780

21. 14.7% of $836,712

22. 83.7% of $3,769,628

23. 1.6% of $598,684

Reviewing the Basics

Skills

(Skill 2)

Round to the nearest cent.

1. $48.9858   **2.** $0.04179   **3.** $2841.9691   **4.** $1.625

Round to the nearest tenth percent.

5. 24.97%   **6.** 8.7801%   **7.** 17.405%   **8.** 7.65%

Solve

(Skill 5)   **9.** $3.95 + $2.49   **10.** $14.97 + $29.79 + $134.99

(Skill 6)   **11.** $212.95 − $168.75   **12.** $147,875.50 − $14,787.55

(Skill 8)   **13.** 1.124 × $475.75   **14.** 1.065 × $9568.85

(Skill 11)   **15.** $24.79 ÷ 1.045   **16.** $147,690 ÷ 1.105

Find the percentage. Round answers to the nearest cent.

(Skill 30)   **17.** 9.8% of $378.45   **18.** 15.48% of $37.79

19. 109.6% of $24.95   **20.** 112.5% of $32,759.85

Find the rate. Round answers to the nearest tenth percent.

(Skill 31)   **21.** n% of $4.50 = $.13   **22.** n% of $379.89 = $402.64

Find the base. Round answers to the nearest cent.

(Skill 32)   **23.** 105.5% of n = $54.98   **24.** 5.3% of n = $12.85

Applications

(Application C)   **25.** $16,890 in 1986.

26. $2186 in 1941.

27. $37,490 in 1990.

28. $17,852 in 1981.

Find the amount withheld for social security.

Years	Tax Rate
1937–49	1.00%
1981	6.65%
1982–83	6.70%
1984	7.00%
1985	7.05%
1986–87	7.15%
1988–89	7.51%
1990–97	7.65%

Terms

Write your own definintion for each term.

29. Inflation   **30.** Inflation rate

31. Gross national product   **32.** Real GNP

33. Per Capita GNP   **34.** Consumer Price Index

35. Budget

Refer to your reference files in the back of the book if you need help.

Unit Test

Lesson 22-1

1. Last year a 2-liter bottle of grapefruit juice cost $0.99. Today it costs $1.09. What is the inflation rate?

Lesson 22-1

2. Ten years ago, a ballpoint pen cost $14.95. The inflation rate for the pen since that time is 16.8%. What does that pen cost today?

Lesson 22-1

3. The inflation rate for groceries over the past year has been 5.7%. What would you have paid a year ago for a bag of groceries that cost $34.80 today?

Lesson 22-2

4. The nation of Angola has a GNP of $4.7 billion and a population of 8,802,000. The inflation rate for Angola is 8.4%.
a. What is the real GNP?
b. What is the per capita GNP?

Lesson 22-3

5. The CPI for recreational services such as theater tickets is 180. Theater tickets for a Broadway show in New York City, which cost $52 today, would have cost how much in 1967?

Lesson 22-3

6. A new automobile purchased by the Jones family in 1967 for $8780 would cost $27,455 today. What is the consumer price index for automobiles?

Lesson 22-3

7. The CPI for new homes is 243.5. A new home costing $54,800 in 1967 would cost how much today?

Lesson 22-4

8. The Able Manufacturing Company has budgeted total revenues of $3,800,000. Able allocates its annual budget expenditures according to the following percents:

Salaries	78.6%
Research	16.4%
Equipment	2.8%
Supplies	2.2%

End-of-the-year data showed these actual expenditures: salaries, $3,000,000; research, $643,000; equipment, $145,000; and supplies, $74,900.
a. How much is allocated to each of the categories?
b. Find the difference between the actual amount and the budget allocation.

Corporate Planning

To complete this spreadsheet application, you will need the template diskette for *Mathematics with Business Applications.* Follow the directions in the *User's Guide* to complete this activity.

Input information in the following problems to find the original price, current price, inflation rate, real GNP, and CPI; and then, answer the questions that follow.

	Original Price	Current Price	Inflation Rate	Gross National Product (GNP)	Real GNP	Consumer Price Index (CPI)*
a.	$ 18.75	$ 21.45	?	$45,800,000,000	?	?
b.	$ 219.97	$299.99	?	$ 235.7 billion	?	?
c.	$ 47.79	?	6.2%	$ 562,000,000	?	?
d.	$8,750.00	?	10.4%	$ 73.9 million	?	?
e.	?	$ 4.75	7.2%	$ 7,650,000,000	?	?
f.	?	$847.50	1.6%	$1,750.6 million	?	?
g.	$ 649.98	?	?	$ 2,625.8 billion	?	215.6
h.	$ 98.98	?	?	$ 95,750,000	?	168.4
i.	?	$ 75.75	?	$ 1.4 billion	?	345.2
j.	?	$549.99	?	$97,748,251,000	?	116.4

*To work with the CPI, assume the original price occurred in 1967.

1. What is the inflation rate in part **a**?

2. What is the real GNP in part **a**?

3. What is the CPI in part **a**?

4. What is the current price in part **c**?

5. What is the real GNP in part **c**?

6. What is the CPI in part **c**?

7. What is the original price in part **e**?

8. What is the real GNP in part **e**?

9. What is the CPI in part **e**?

10. What is the current price in part **g**?

11. What is the inflation rate in part **g**?

12. What is the real GNP in part **g**?

13. What is the original price in part **i**?

14. What is the inflation rate in part **i**?

15. What is the real GNP in part **i**?

16. Explain the relationship between the inflation rate and the consumer price index. Knowing one, how can you find the other?

17. If the current price is less than the original price:
    a. What will be true about the inflation rate?
    b. What will be true about the CPI?
    c. Can the CPI ever be negative? Why or why not?
    d. How is the GNP related to the real GNP?

18. If the current price is equal to the original price:
    a. What is the inflation rate?
    b. What is the CPI?
    c. How is the GNP related to the real GNP?

19. If the inflation rate is greater than 100%:
    a. What happens to the real GNP? Is that possible?
    b. What happens to the CPI?

Cumulative Review
(Skills/Applications) Units 17–22

Skills

(Skill 2)

Round answers to the nearest dollar.

1. $178.51   **2.** $6.444   **3.** $4218.17   **4.** $14.99

Solve. Round answers to the nearest cent.

(Skill 3)

5. 48 + 87   **6.** 724 + 57   **7.** $694 + $375

8. $2186 + $3125   **9.** $85 + $144   **10.** $1596 + $15,895 + $48

(Skill 4)

11. $85 − $36   **12.** $337 − $98   **13.** $400 − $216

14. $1345 − $1155   **15.** $8740 − $984   **16.** $1946 − $1473

(Skill 5)

17. $14.95 + $4.75   **18.** $1.19 + $0.45   **19.** $22.25 + $62.87

20. $1500 + $19.56 + $3.52   **21.** $3187.48 + $751.94 + $100.53

(Skill 6)

22. $30.39 − $17.91   **23.** $500.99 − $46.84

24. $1543.60 − $9.56   **25.** $531.25 − $18.79   **26.** $1327.81 − $871.93

(Skill 7)

27. $43 × 8   **28.** $51 × 6   **29.** $1500 × 5

30. $579 × 91   **31.** $585 × 55   **32.** $3591 × 72

(Skill 8)

33. $140 × 0.06   **34.** $2.85 × 0.12   **35.** $26.24 × 0.86

36. $793.02 × 4.70   **37.** $42.24 × 75   **38.** $1.59 × 57

(Skill 10)

39. $75 ÷ 5   **40.** $427 ÷ 24   **41.** $5512 ÷ 52

42. $329 ÷ 184   **43.** $5275 ÷ 35   **44.** $43,615 ÷ 360

(Skill 11)

45. $30.65 ÷ 12   **46.** $479.30 ÷ 24   **47.** $5191.20 ÷ 365

48. $8921.70 ÷ 4.12   **49.** $12,291.60 ÷ 21.7

(Skill 20)

50. $85 × $\frac{4}{5}$ **51.** $96 × $\frac{3}{8}$ **52.** $32 × $\frac{3}{4}$

53. $40,817 × $\frac{9}{10}$ **54.** $\frac{6}{15}$ × $526   **55.** $2274 × $\frac{2}{3}$

Write as a decimal.

(Skill 28)

56. 7.65%   **57.** 8%   **58.** 10.4%   **59.** 74.35%

Solve. Round answers to the nearest cent.

(Skill 30)

60. 10% of $475   **61.** 58% of $856   **62.** 40% of $8748

63. $5\frac{1}{4}$% of $506   **64.** $9\frac{1}{2}$% of $14,612   **65.** $3\frac{1}{2}$% of $415,758

Solve. Round answers to the nearest tenth of a percent.

(Skill 31)

66. $35 is what percent of $84?   **67.** $150 is what percent of $315?

68. $15,000 is what percent of $70,000?

69. $43,470 is what percent of $210,000?

Applications

(Application C)

Find the tax bracket.

70. Taxable income is $45,690. **71.** Taxable income is $114,750.

72. Taxable income is $91,495. **73.** Taxable income is $18,710.

Taxable Income		Your tax is:		
Over—	But not over—			Of the amount over—
0	$ 50,000		15%	0
$ 50,000	75,000	$ 7,500 +	25%	$ 50,000
75,000	100,000	13,250 +	34%	75,000
100,000	—	21,750 +	39%	100,000

Find the charge per 100 pounds.

74. Weight: 2490 pounds
Distance: 623 miles

75. Weight: 6540 pounds
Distance: 1300 miles

BASIC RATES PER 100 POUNDS		
Weight Group (in pounds)	Distance (in miles)	
	500-1000	1001-1500
2001-5000	$40.25	$66.20
5001-10,000	38.00	53.00

Find the average cost to the nearest cent.

(Application Q)

76. 2 units at $378 each and
3 units at $398 each.

77. 30 units at $2.35 each and
21 units at $2.25 each.

78. 75 units at $3.12 each and
14 units at $2.98 each.

79. 423 units at $0.56 each and
180 units at $0.60 each.

Find the area.

(Application X)

80. 29 m by 12 m   **81.** 15 ft by 12 ft   **82.** 580 ft by 75 ft

83. 727 in by 60 in   **84.** 7.5 ft by 14.5 ft   **85.** 110 cm by 15 cm

Find the volume.

(Application Z)

86. 7 ft by 3 ft by 10 ft   **87.** 2.7 m by 8 m by 11.4 m

88. 9 in by $13\frac{1}{2}$ in by 6 in   **89.** $1\frac{3}{4}$ ft by 4 ft by $5\frac{1}{2}$ ft

Terms

Write your own definition for each term.

90. Inventory   **91.** U.S. Treasury Bill   **92.** Income statement

93. Inventory card   **94.** Apportion   **95.** Gross national product

96. Owner's equity   **97.** Book value   **98.** Net proceeds

99. Maturity value   **100.** Balance sheet   **101.** Consumer price index

Refer to your reference files in the back of the book if you need help.

Cumulative Review Test
Units 17–22

Lesson 17-1

1. The Office Equipment Company manufactures some 3-drawer file cabinets. Each cabinet is stored in a box measuring 3.8 feet high, 1.5 feet wide, and 2.5 feet long. How many cubic feet of space does Office Equipment need to store 1200 file cabinets?

Lesson 17-2

2. How many RunAway models does Clarion Motor Coach have on its lot on October 1?

```
Item: RunAway
Stock Number: KT 4403 G

                Opening                            Inventory at
Month of        Balance      Receipts    Issues    End of Month
August            75            12          8           ?
September         79            15         35           ?
October            ?
```

Lesson 17-3

3. In problem 2, the 75 in the opening balance were valued at $12,000 each, the 12 received in August were valued at $12,400 each, and the 15 received in September were valued at $12,600 each. Using the average-cost method, what is the value of the inventory?

Lesson 18-2

4. Motel Seven hired 4 high school students to clean its swimming pool. Each student worked for 2 hours at an hourly rate of $7.45. The materials charge was $43.70. What was the total charge for this service?

Lesson 18-5

5. The Huston Corporation used 15,400 kilowatt-hours of electricity with a peak load of 120 kilowatts in April. The demand charge is $6.40 per kilowatt. The energy charge per kilowatt-hour is $0.08 for the first 10,000 kilowatt-hours and $0.06 for all kilowatt-hours over 10,000. The fuel adjustment charge is $0.04 per kilowatt-hour. What is the total cost of electricity for the Huston Corporation for April?

Lesson 19-1

6. Complete the payroll register. Use the FICA tax rate of 7.65%.

Name	Gross Income	FIT	FICA	SIT	Total Deductions	Net Pay
Bisset	$300.00	$41.00	?	$4.50	?	?
Heller	435.70	64.00	?	6.54	?	?
Molnar	405.00	45.00	?	6.08	?	?
Stein	425.65	47.00	?	6.38	?	?
Total	?	?	?	?	?	?

Lesson 19-3

7. The Foster Machine Company pays $125,000 per year for security. The company apportions the cost among its departments based on space. The research department occupies an area that measures 60 feet by 60 feet. The building contains 1,200,000 square feet. How much does the research department pay for security?

Lessons 19-4, 19-5

8. Central Copy purchased a new copy machine for $12,400. The machine has an estimated life of 4 years. The resale value is expected to be $1400. Use the straight-line method to find the annual depreciation. Find the book value after each year of use.

Lessons 20-1

9. The Running Shop had these assets and liabilities on May 31.

Cash: $4500
Inventory: $28,000
Supplies: $356
Store furnishings: $17,860

Unpaid merchandise: $13,470
Taxes owed: $975
Bank loan: $11,600

What are the total assets? What are the total liabilities? What is the owner's equity?

Lesson 20-4

10. Globe Manufacturing, Inc., prepares an annual income statement. This past year, Globe had net sales of $800,000. The cost of goods sold totaled $375,750. Operating expenses included: wages and salaries totaling $165,970; depreciation and amortization totaling $19,590; interest paid totaling $15,540; product recall totaling $3596; and miscellaneous operating expenses of $6570. Prepare an income statement for Globe Manufacturing.

Lesson 21-2

11. Solar Energy, Inc., is issuing 500,000 shares of stock. Each share will be sold at $45. The underwriting commission is 3% of the value of the stock. The other expenses are expected to be 0.6% of the value of the stock. If all the shares of stock are sold, what net proceeds will Solar Energy receive?

Lesson 21-3

12. Hereford Ranch borrowed $48,500 for 120 days at the prime rate of interest. When Hereford Ranch borrowed the money, the prime rate was 10.8%. The bank charged ordinary interest at exact time. What was the maturity value of the loan?

Lesson 21-4

13. Carlton Manufacturing had $80,000 in surplus cash. The financial manager decided to invest in a United States Treasury Bill with a face value of $80,000. The bill matured in 126 days. The interest rate was 9.8% ordinary interest at exact time. The bank service fee was $25. What was the cost of the Treasury Bill?

Lesson 22-3

14. The CPI for clothing is 186.3. A sweater that cost $24.95 in 1967 would cost how much today?

▼ A SIMULATION

Copyfax Center

You are the manager of Copyfax Center, a small business that does copying, printing, and word processing. Copyfax Center has 3 employees. Here is a chart of the hours they work each day, Monday through Friday.

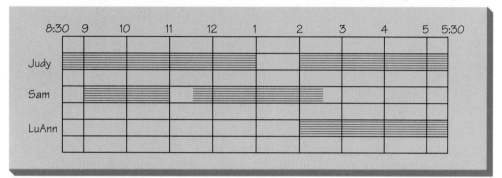

Each week you make up the weekly payroll. Each employee must pay 7.65% for FICA and 5% for state tax. You pay employees time and a half for overtime.

Use these work sheets to calculate this week's net pay for each employee. Use the tax tables on pages 642–643 to determine the federal income tax (FIT). This week Judy worked 3 hours overtime. Sam and LuAnn did not work overtime.

	Employee	Hourly Pay	Hours Worked Regular	Hours Worked Overtime	Gross Pay
1.	Judy Olson	$7.60	40	3	?
2.	Sam Borden	$6.50	25	0	?
3.	LuAnn Tozzi	$4.50	17.5	0	?

	Employee	Tax Status	FIT	FICA	State Tax	Health Insur.	Total Deduct.	Net Pay
4.	Judy Olson	Married 3 allow.	?	?	?	$8.50	?	?
5.	Sam Borden	Single 2 allow.	?	?	?	$6.50	?	?
6.	LuAnn Tozzi	Single 1 allow.	?	?	?	0	?	?

A SIMULATION
(CONTINUED)

Pricing

The price that Copyfax Center charges for copies depends on the number of copies made. This chart shows the current prices.

Number of Copies	Price per Page*
1–24	$0.15
25–49	$0.135
50–99	$0.13
100–199	$0.125
200–499	$0.12
over 500	$0.115

*plus 6% sales tax

Regular customers get discounts. A few receive a trade discount of 20%. Others get a cash discount of 2% if they pay within 10 days. Both discounts are deducted before the sales tax is calculated.

Use these invoices to calculate the prices charged these customers.

7.

COPYFAX CENTER	Invoice no. **2257**
	Date *June 15, 19—*

Customer	*Christine's Hardware*

	Number of copies	800
a.	Price per page	$?
b.	Total price	$?
c.	Discount: __20__ % ☒trade ☐cash	$?
d.	Net price	$?
e.	Sales tax	$?
f.	Invoice price	$?

9. What is the price per page?

8.

COPYFAX CENTER	Invoice no. **2289**
	Date *June 17, 19—*

Customer	*Mike's Travel Service*

	Number of copies	150
a.	Price per page	$?
b.	Total price	$?
c.	Discount: __2__ % ☐trade ☒cash	$?
d.	Net price	$?
e.	Sales tax	$?
f.	Invoice price	$?

10. What is the price per page?

A SIMULATION
(CONTINUED)

Depreciation

Depreciation of Copyfax Center's equipment is considered a business expense.

Use this work sheet to calculate the depreciation of these items, using the straight-line method.

Item	11. High-Speed Copier	12. Word Processor	13. Computer Typesetter	14. Delivery Van
Year Purchased	1992	1989	1990	1992
Original Cost	$18,500	$4500	$9750	$14,248
Salvage Value	$ 1,000	$ 500	$ 750	$ 1,648
Total Depreciation	?	?	?	?
Estimated Life	5	5	6	6
Annual Depreciation	?	?	?	?

As part of your long-range planning, you decide to calculate how much each item will be worth each year.

Use this work sheet to calculate the accumulated depreciation and the book value (B/V) of each item listed. A zero has been put in for any years that do not apply.

	15. High-Speed Copier		16. Word Processor		17. Computer Typesetter		18. Delivery Van	
Year	**A/D**	**B/V**	**A/D**	**B/V**	**A/D**	**B/V**	**A/D**	**B/V**
1989	0	0	?	?	0	0	0	0
1990	0	0	?	?	?	?	0	0
1991	0	0	?	?	?	?	0	0
1992	?	?	?	?	?	?	?	?
1993	?	?	?	?	?	?	?	?
1994	?	?	0	0	?	?	?	?
1995	?	?	0	0	?	?	?	?
1996	?	?	0	0	0	0	?	?
1997	0	0	0	0	0	0	?	?

Balance Sheet

Each month you prepare a balance sheet. The balance sheet lists Copyfax Center's assets (what it owns), its liabilities (what it owes), and the owner's equity (assets minus liabilities).

You start by listing your assets and liabilities in a running account:

Copyfax Center	December 31, 19—
Cash on hand	$ 3000
Accounts receivable	$ 3500
Accounts payable	$10,500
Equipment (less accumulated depreciation)	$25,798
Inventory (paper, ink, etc.)	$ 2100
Prepaid insurance	$ 1800
Office supplies	$ 520
Taxes owed	$ 1500
Notes payable	$ 9500

19. Use this form to complete Copyfax Center's balance sheet. Enter the assets and liabilities from your running account.

COPYFAX CENTER Balance Sheet Dec. 31, 19—					
Assets		**Liabilities**			
a.			$		h.
b.					i.
c.					j.
d.					k.
e.		Total Liabilities	$		l.
f.		Owner's Equity	$		m.
g. Total Assets		Total Liabilities and Equity	$		n.

A SIMULATION
(CONTINUED)

Income Statement

Each month you also prepare Copyfax Center's income statement. Here are the sales and expenses you have recorded during the month.

Copyfax Center			December 19—
Wages	$2510	Insurance	$ 600
Advertising	$ 135	Taxes	$ 500
Delivery	$ 160	Depreciation	$ 660
Postage	$ 300	Supplies	$ 170
Rent	$ 750	Gross sales	$8002
Utilities	$ 325	Sales discounts	$ 880

20. Use this form to prepare an income statement.

	Copyfax Center Income Statement for Month Ended December 31, 19—		
a.	**Income:** Gross sales	?	
b.	Less sales discounts	?	
c.	Net sales		?
d.	**Expenses:**	?	
e.		?	
f.		?	
g.		?	
h.		?	
i.		?	
j.		?	
k.		?	
l.		?	
m.		?	
n.	**Total operating expenses**		?
o.	**Net Income**		?

A SIMULATION
(CONTINUED)

Annual Report and Comparative Analysis

At the end of each year, you prepare an annual income statement. You can then compare Copyfax Center's finances for two years, in order to plan for next year.

This chart shows Copyfax Center's income and expenses for last year and this year. Fill in the missing numbers. Then, for each amount, calculate the percent increase from last year to this year.

21.	Copyfax Center Annual Income Statement		Last Year	This Year	Percent Increase
a.	**Income:**	Gross sales	$89,560	$99,480	?
b.		Less sales discounts	9985	10,750	?
c.		Net sales	?	?	?
d.	**Expenses:**	Wages	28,650	30,075	?
e.		Advertising	1350	1550	?
f.		Delivery	1520	1840	?
g.		Postage	3260	3758	?
h.		Rent	8400	8800	?
i.		Utilities	3315	3900	?
j.		Insurance	4100	4200	?
k.		Taxes	4456	4826	?
l.		Depreciation	2100	7900	?
m.		Supplies	2136	2254	?
n.	**Total operating expenses**		?	?	?
o.	**Net Income**		?	?	?

Complete the following statements comparing last year's and this year's figures.

22. The expense with the largest percent increase was _____.

23. The next largest was _____.

24. Net sales increased by _____%.

25. Total operating expenses increased by _____%.

26. Net income changed by _____%.

The Bottom Line

What would you do differently next year to improve Copyfax Center's financial picture?

YOUR REFERENCE FILES

Skills Thirty-two computational skills, from whole number and decimal operations through rates and percents, provide the basics you will need as a consumer and future business person.

Applications Twenty-six mathematical applications, covering money, time, measurement, graphs, and more, provide the tools you will need for consumer and business use.

Terms Key terms with their definitions from all lessons provide the background you will need in consumer and business situations.

Numbers

Place Value

In 4532.869, give the place and value of each digit.

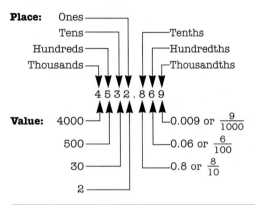

Give the place and value of the underlined digit.

1. 6<u>5</u>   **2.** 9<u>6</u>   **3.** 4<u>7</u>2

4. <u>2</u>3   **5.** 10<u>8</u>   **6.** <u>5</u>36

7. 15<u>0</u>6   **8.** 28<u>2</u>1   **9.** 7<u>7</u>84

10. <u>1</u>492   **11.** 100<u>9</u>   **12.** 1<u>9</u>26

13. 0.<u>3</u>7   **14.** 1.6<u>1</u>   **15.** 2.73<u>9</u>

16. <u>6</u>.3   **17.** 9.4<u>2</u>   **18.** 4.3<u>7</u>

19. 24.<u>0</u>4   **20.** 37.3<u>2</u>9   **21.** 1.8<u>2</u>4

22. <u>4</u>93.89   **23.** 90.25<u>7</u>   **24.** 23.<u>5</u>72

25. 12.76<u>3</u>   **26.** 0.<u>0</u>78   **27.** 5.46<u>1</u>

Numbers as Words

Write each number in words.

Number	Words
105 ➤	One hundred five
26 ➤	Twenty-six
17 ➤	Seventeen
$98.09 ➤	Ninety-eight and $\frac{9}{100}$ dollars
$33.13 ➤	Thirty-three and $\frac{13}{100}$ dollars

Write in words.

28. 18   **29.** 34   **30.** 159   **31.** 78

32. 103   **33.** 842   **34.** 207   **35.** 5012

36. 6005   **37.** 119   **38.** 72   **39.** 1240

40. 5102   **41.** 194   **42.** 6590   **43.** $25.00

44. $6.24   **45.** $17.09   **46.** $112.35   **47.** $120.17

48. $4.37   **49.** $7749   **50.** $65.90   **51.** $0.50

Numbers as Decimals

Rewrite these large numbers as decimals.

$14,700,000 ➤ $14.7 million

$245,600 ➤ $245.6 thousand

Rewrite these decimals as numbers.

$1.2 billion = $1,200,000,000

$43.6 thousand = $43,600

Write in millions.
52. $1,700,000 **53.** $71,640,000   **54.** $618,700,000

Write in thousands.
55. 18,400   **56.** $9,640   **57.** 171,600

Write in billions.
58. $16,450,000,000   **59.** 2,135,000,000
60. $171,200,000,000

Write completely in numbers.
61. $3.4 million **62.** 16.2 thousand **63.** $11.2 billion
64. 17.2 million **65.** 7.3 billion   **66.** $74.21 thousand
67. 0.4 million **68.** $0.5 thousand **69.** $0.72 billion

Comparing Whole Numbers

Which number is greater: 5428 or 5431?

5428 = 5000 + 400 + 20 + 8

5431 = 5000 + 400 + 30 + 1

Same   Same   30 is greater than 20.

So 5431 is greater than 5428.

Which number is greater?

70. 23 or 32

71. 54 or 45

72. 459 or 462

73. 741 or 835

74. 810 or 735

75. 125 or 211

76. 3450 or 6450

77. 5763 or 925

78. 1000 or 999

79. 444 or 4444

80. 3002 or 4000

81. 1236 or 820

82. 150 or 149

83. 493 or 650

84. $2000 or $1997

85. $101 or $99

86. $482 or $600

87. $39 or $93

88. $1686 or $1668

89. $568 or $742

90. $86,432 or $101,000

91. $791,000 or $768,000.

Comparing Decimals

Which number is greater: 24.93 or 24.86?

24.93 = 20 + 4 + 0.9 + 0.03

24.86 = 20 + 4 + 0.8 + 0.06

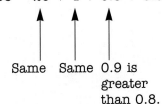

Same   Same   0.9 is greater than 0.8.

So 24.93 is greater than 24.86

Which number is greater?

92. 3.1 or 1.3

93. 1.2 or 2.0

94. 4.50 or 4.05

95. 25.1 or 20.8

96. 18.43 or 17.88

97. 56.84 or 58

98. 0.4 or 0.6

99. 0.01 or 0.1

100. 0.82 or 0.28

101. 0.5 or 0.06

102. 8.739 or 10

103. 0.002 or 0.020

104. $5.99 or $5.00

105. $10 or $8.50

106. $23.85 or $19.84

107. $11.19 or $19

108. $6.98 or $7.50

109. $83.59 or $600

110. $4327.75 or $6297.86

111. $8391.34 or $9521.39

112. $4640.66 or $4646.40

113. $2000.00 or $1997.98

Round Numbers

Whole Numbers

Round 7863 to the nearest hundred.

7863 ——— Hundreds place

↑——— 5 or more? Yes.

7900 ——— Add 1 to hundreds place.

↑——— Change to zeros.

Round answers to the place value shown.

Nearest ten: **1.** 26

2. 37   **3.** 68   **4.** 195

5. 217   **6.** 302   **7.** 8099

Nearest hundred: **8.** 119

9. 649   **10.** 2175   **11.** 6042

Nearest thousand: **12.** 7423

13. 15,602   **14.** 22,094   **15.** 750

Decimals

Round 0.6843 to the nearest thousandth.

0.6843 ——— Thousandths place

↑——— 5 or more? No.

0.684 ——— Do not change.

↑——— Drop the final digit.

Round answers to the place value shown.

Nearest tenth: **16.** 0.63

17. 0.091   **18.** 0.407   **19.** 0.452

Nearest hundredth: **20.** 0.652

21. 0.474   **22.** 0.168   **23.** 0.355

Nearest thousandth: **24.** 0.4291

25. 0.6007   **26.** 0.0097   **27.** 0.2126

28. 6.3942   **29.** 137.4920   **30.** 9.9999

Mixed Practice

Round answers to the place value shown.

Nearest thousand:	**31.** 37,874	**32.** 19,266	**33.** 48,092
Nearest hundred:	**34.** 751	**35.** 919	**36.** 6771
Nearest ten:	**37.** 26	**38.** 6533	**39.** 575
Nearest one:	**40.** 6.2	**41.** 35.73	**42.** 17.392
Nearest tenth:	**43.** 189.673	**44.** 10.009	**45.** 0.07
Nearest hundredth:	**46.** 0.392	**47.** 152.430	**48.** 0.6974
Nearest thousandth:	**49.** 0.1791	**50.** 16.0005	**51.** 108.437

Add Whole Numbers

Without Carrying

Add.

```
 723        723
 154    ▶   154
+212       +212
           1089
```

1.   65
       +41

2.   76
       +32

3.   97
       +40

4.   32
       +25

5. 352
       +837

6. 361
       +834

7. 448
       +351

8. 125
       +604

9. 864
       + 33

10. 721
        + 77

11. 423
        + 65

12. 108
        + 91

13. 9037
        +1841

14. 9520
        +1379

15. 3924
        +5063

16. 2840
        +1152

With Carrying

Add.

```
          1 11
 8679     8 679
+9748  ▶ +9 748
         18,427
```

17.   32
        +39

18.   54
        +48

19. 187
        +23

20.   49
        +86

21. 728
        +169

22. 527
        +284

23. 845
        +697

24. 697
        +546

25. 3046
        +1592

26. 7801
        +3564

27. 5246
        +6978

28. 8347
        +1528

29. 8448
        +3753

30. 108
        +7665

31. 9179
        +3608

32. 982
        +2165

Mixed Practice

33. 1481
        +2317

34. 8495
        +1417

35. 5783
        +6535

36. 3950
        +1615

37. 6259
        +1893

38. 8347
        +1528

39. 6845
        +2639

40. 5692
        +1204

41. 2642
        +4135

42. 7921
        +2639

43. 7884
        +7069

44. 46,234
        +11,325

45. 17,694
        +15,893

46. 37,491
        +21,308

47. 59,641
        +27,840

48. 9100
         536
        +2413

49. 7749
         1240
        +6010

50. 6590
         2408
        +5001

51. 5783
         6535
        +2132

52. 6259
         503
        +1893

Subtract Whole Numbers

Without Borrowing

Subtract.

$$
\begin{array}{r} 9876 \\ -7545 \\ \hline \end{array}
\quad\blacktriangleright\quad
\begin{array}{r} 9876 \\ -7545 \\ \hline 2331 \end{array}
$$

1. $\begin{array}{r} 784 \\ -453 \\ \hline \end{array}$ **2.** $\begin{array}{r} 985 \\ -734 \\ \hline \end{array}$ **3.** $\begin{array}{r} 693 \\ -542 \\ \hline \end{array}$ **4.** $\begin{array}{r} 199 \\ -158 \\ \hline \end{array}$

5. $\begin{array}{r} 7659 \\ -4217 \\ \hline \end{array}$ **6.** $\begin{array}{r} 8436 \\ -6223 \\ \hline \end{array}$ **7.** $\begin{array}{r} 5792 \\ -2481 \\ \hline \end{array}$ **8.** $\begin{array}{r} 4877 \\ -3614 \\ \hline \end{array}$

9. $\begin{array}{r} 6754 \\ -5643 \\ \hline \end{array}$ **10.** $\begin{array}{r} 1866 \\ -853 \\ \hline \end{array}$ **11.** $\begin{array}{r} 8191 \\ -171 \\ \hline \end{array}$ **12.** $\begin{array}{r} 1187 \\ -145 \\ \hline \end{array}$

13. $\begin{array}{r} 479 \\ -473 \\ \hline \end{array}$ **14.** $\begin{array}{r} 3987 \\ -3085 \\ \hline \end{array}$ **15.** $\begin{array}{r} 6358 \\ -127 \\ \hline \end{array}$ **16.** $\begin{array}{r} 1721 \\ -720 \\ \hline \end{array}$

With Borrowing

Subtract.

$$
\begin{array}{r} 9672 \\ -4136 \\ \hline \end{array}
\quad\blacktriangleright\quad
\begin{array}{r} 9672 \\ -4136 \\ \hline 5536 \end{array}
$$

$$
\begin{array}{r} 8352 \\ -4584 \\ \hline \end{array}
\quad\blacktriangleright\quad
\begin{array}{r} 8352 \\ -4584 \\ \hline 3768 \end{array}
$$

17. $\begin{array}{r} 100 \\ -36 \\ \hline \end{array}$ **18.** $\begin{array}{r} 512 \\ -43 \\ \hline \end{array}$ **19.** $\begin{array}{r} 602 \\ -503 \\ \hline \end{array}$ **20.** $\begin{array}{r} 250 \\ -162 \\ \hline \end{array}$

21. $\begin{array}{r} 6932 \\ -4674 \\ \hline \end{array}$ **22.** $\begin{array}{r} 8724 \\ -2932 \\ \hline \end{array}$ **23.** $\begin{array}{r} 4329 \\ -3163 \\ \hline \end{array}$ **24.** $\begin{array}{r} 9721 \\ -6842 \\ \hline \end{array}$

25. $\begin{array}{r} 6123 \\ -4214 \\ \hline \end{array}$ **26.** $\begin{array}{r} 9231 \\ -6453 \\ \hline \end{array}$ **27.** $\begin{array}{r} 7450 \\ -3783 \\ \hline \end{array}$ **28.** $\begin{array}{r} 7734 \\ -5935 \\ \hline \end{array}$

29. $\begin{array}{r} 8121 \\ -6846 \\ \hline \end{array}$ **30.** $\begin{array}{r} 9000 \\ -7997 \\ \hline \end{array}$ **31.** $\begin{array}{r} 9107 \\ -8248 \\ \hline \end{array}$ **32.** $\begin{array}{r} 7734 \\ -5935 \\ \hline \end{array}$

Mixed Practice

33. $\begin{array}{r} 6140 \\ -3157 \\ \hline \end{array}$ **34.** $\begin{array}{r} 8005 \\ -6246 \\ \hline \end{array}$ **35.** $\begin{array}{r} 7000 \\ -5432 \\ \hline \end{array}$ **36.** $\begin{array}{r} 9297 \\ -9286 \\ \hline \end{array}$ **37.** $\begin{array}{r} 9811 \\ -700 \\ \hline \end{array}$

38. $\begin{array}{r} 9148 \\ -954 \\ \hline \end{array}$ **39.** $\begin{array}{r} 2625 \\ -763 \\ \hline \end{array}$ **40.** $\begin{array}{r} 1850 \\ -975 \\ \hline \end{array}$ **41.** $\begin{array}{r} 7469 \\ -5231 \\ \hline \end{array}$ **42.** $\begin{array}{r} 6342 \\ -5793 \\ \hline \end{array}$

43. $\begin{array}{r} 10{,}743 \\ -7{,}842 \\ \hline \end{array}$ **44.** $\begin{array}{r} 16{,}947 \\ -14{,}523 \\ \hline \end{array}$ **45.** $\begin{array}{r} 22{,}493 \\ -5{,}967 \\ \hline \end{array}$ **46.** $\begin{array}{r} 64{,}654 \\ -57{,}312 \\ \hline \end{array}$ **47.** $\begin{array}{r} 79{,}850 \\ -42{,}347 \\ \hline \end{array}$

48. $\begin{array}{r} 172{,}493 \\ -67{,}254 \\ \hline \end{array}$ **49.** $\begin{array}{r} 249{,}657 \\ -123{,}254 \\ \hline \end{array}$ **50.** $\begin{array}{r} 300{,}692 \\ -147{,}593 \\ \hline \end{array}$ **51.** $\begin{array}{r} 647{,}593 \\ -546{,}972 \\ \hline \end{array}$ **52.** $\begin{array}{r} 800{,}000 \\ -627{,}351 \\ \hline \end{array}$

Add Decimals

Same Number of Places

Add.

```
  178.79        178.79
  596.24   ▶    596.24
 +631.43       +631.43
               1406.46
```

1.   317.83
      +161.16

2.   821.76
      +178.23

3.   504.76
      +296.36

4.   536.67
      +197.47

5.   189.71
      +601.48

6.   27.492
      +31.608

7.   148.810
      221.097
      +173.206

8.   18.009
      149.910
      +251.295

9.   272.005
      42.111
      +502.543

Different Number of Places

Add.

```
   0.91         0.910
   6.647   ▶    6.647
   5.3          5.300
  +16          +16.000
                28.857
         Placeholders
```

10.   10
       + 6.3

11.   42.9
       +21.64

12.   63.21
       +42.327

13.   141.21
       4.136
       +32.003

14.   3.64
       4.378
       +0.213

15.   19.7
       0.261
       + 1.94

16.   18.09
       49.91
       +51.295

17.   0.161
       18.24
       + 6.059

18.   2.19
       0.09
       +0.7

19.   16.74
       39.1
       +26.492

20.   1.943
       27.21
       + 3.691

21.   2.5
       1.692
       +5.93

Mixed Practice

22.   60.148
       +16.623

23.   271.195
       +189.714

24.   163.2
       + 37.915

25.   427.9
       + 74.275

26.   693.642
       +193.871

27.   973.573
       81.91
       + 9.20

28.   69
       27.198
       +178.26

29.   0.64
       0.378
       +0.223

30.   0.007
       0.986
       +0.034

31.   19.7
       0.261
       + 1.94

32.   17.92
       23.81
       42.63
       +15.27

33.   103.69
       27.4
       83.621
       + 5.9

34.   327.219
       27.543
       111.621
       +257.619

35.   471.6
       2.937
       57.11
       +241.005

36.   327.915
       1.002
       40.07
       +154.6

6

Subtract Decimals

Same Number of Places

Subtract.

$$
\begin{array}{r}
597.18 \\
-392.35 \\
\end{array}
\blacktriangleright
\begin{array}{r}
{}^{6\ \ 11}\\
59\cancel{7}.\cancel{1}8 \\
-392.35 \\
\hline
204.83 \\
\end{array}
$$

1. $\begin{array}{r} 65.46 \\ -14.31 \\ \hline \end{array}$ **2.** $\begin{array}{r} 48.58 \\ -15.47 \\ \hline \end{array}$ **3.** $\begin{array}{r} 151.02 \\ -16.13 \\ \hline \end{array}$

4. $\begin{array}{r} 36.25 \\ -13.67 \\ \hline \end{array}$ **5.** $\begin{array}{r} 87.56 \\ -82.47 \\ \hline \end{array}$ **6.** $\begin{array}{r} 51.634 \\ -27.849 \\ \hline \end{array}$

7. $\begin{array}{r} 69.37 \\ -43.86 \\ \hline \end{array}$ **8.** $\begin{array}{r} 89.63 \\ -\ \ 7.99 \\ \hline \end{array}$ **9.** $\begin{array}{r} 109.46 \\ -\ 29.78 \\ \hline \end{array}$

10. $\begin{array}{r} 521.52 \\ -\ 38.56 \\ \hline \end{array}$ **11.** $\begin{array}{r} 321.02 \\ -117.18 \\ \hline \end{array}$ **12.** $\begin{array}{r} 572.24 \\ -283.35 \\ \hline \end{array}$

Different Number of Places

Subtract.

$$
\begin{array}{r}
86.9 \\
-\ 3.84 \\
\end{array}
\blacktriangleright
\begin{array}{r}
{}^{8\ 10}\\
86.9\cancel{0} \\
-\ 3.84 \\
\hline
83.06 \\
\end{array}
$$

Placeholder

13. $\begin{array}{r} 79.6 \\ -\ 8.75 \\ \hline \end{array}$ **14.** $\begin{array}{r} 95.1 \\ -\ 9.87 \\ \hline \end{array}$ **15.** $\begin{array}{r} 100.1 \\ -\ 15.78 \\ \hline \end{array}$

16. $\begin{array}{r} 16.8 \\ -\ 5.91 \\ \hline \end{array}$ **17.** $\begin{array}{r} 36 \\ -16.4 \\ \hline \end{array}$ **18.** $\begin{array}{r} 42 \\ -12.94 \\ \hline \end{array}$

19. $\begin{array}{r} 17.9 \\ -\ 9.83 \\ \hline \end{array}$ **20.** $\begin{array}{r} 21 \\ -19.7 \\ \hline \end{array}$ **21.** $\begin{array}{r} 67.2 \\ -\ 9.76 \\ \hline \end{array}$

22. $\begin{array}{r} 136.1 \\ -\ 69.542 \\ \hline \end{array}$ **23.** $\begin{array}{r} 771.9 \\ -394.27 \\ \hline \end{array}$ **24.** $\begin{array}{r} 4578 \\ -\ 878.127 \\ \hline \end{array}$

Mixed Practice

25. $\begin{array}{r} 87.56 \\ -82.47 \\ \hline \end{array}$ **26.** $\begin{array}{r} 39.27 \\ -18.38 \\ \hline \end{array}$ **27.** $\begin{array}{r} 36.1 \\ -16.117 \\ \hline \end{array}$ **28.** $\begin{array}{r} 4.546 \\ -2.558 \\ \hline \end{array}$ **29.** $\begin{array}{r} 653.05 \\ -327.19 \\ \hline \end{array}$

30. $\begin{array}{r} 198.20 \\ -\ 64.897 \\ \hline \end{array}$ **31.** $\begin{array}{r} 854.01 \\ -649.656 \\ \hline \end{array}$ **32.** $\begin{array}{r} 316.07 \\ -118.29 \\ \hline \end{array}$ **33.** $\begin{array}{r} 800.04 \\ -242.17 \\ \hline \end{array}$ **34.** $\begin{array}{r} 985.93 \\ -\ 99.794 \\ \hline \end{array}$

35. $\begin{array}{r} 6194.9 \\ -\ 978.954 \\ \hline \end{array}$ **36.** $\begin{array}{r} 719.3 \\ -\ 47.832 \\ \hline \end{array}$ **37.** $\begin{array}{r} 5.9871 \\ -4.8693 \\ \hline \end{array}$ **38.** $\begin{array}{r} 17.9328 \\ -\ 6.2973 \\ \hline \end{array}$ **39.** $\begin{array}{r} 843.002 \\ -\ 64.973 \\ \hline \end{array}$

40. $\begin{array}{r} 87.69 \\ -86.9975 \\ \hline \end{array}$ **41.** $\begin{array}{r} 4.97652 \\ -1.37846 \\ \hline \end{array}$ **42.** $\begin{array}{r} 3.29131 \\ -2.19378 \\ \hline \end{array}$ **43.** $\begin{array}{r} 6.962 \\ -4.21698 \\ \hline \end{array}$ **44.** $\begin{array}{r} 9.7 \\ -8.65947 \\ \hline \end{array}$

Multiply Whole Numbers

Without Carrying

Multiply.

$$
\begin{array}{r}
442 \\
\times 211 \\
\end{array}
\quad
\begin{array}{r}
442 \\
\times 211 \\
\hline
442 \longleftarrow 1 \times 442 \\
4\,420 \longleftarrow 10 \times 442 \\
88\,400 \longleftarrow 200 \times 442 \\
\hline
93{,}262 \longleftarrow 211 \times 442 \\
\end{array}
$$

1. $\begin{array}{r} 73 \\ \times 21 \\ \hline \end{array}$ **2.** $\begin{array}{r} 42 \\ \times 22 \\ \hline \end{array}$ **3.** $\begin{array}{r} 212 \\ \times 412 \\ \hline \end{array}$

4. $\begin{array}{r} 311 \\ \times 232 \\ \hline \end{array}$ **5.** $\begin{array}{r} 321 \\ \times 312 \\ \hline \end{array}$ **6.** $\begin{array}{r} 223 \\ \times 323 \\ \hline \end{array}$

7. $\begin{array}{r} 232 \\ \times 333 \\ \hline \end{array}$ **8.** $\begin{array}{r} 7143 \\ \times 102 \\ \hline \end{array}$ **9.** $\begin{array}{r} 8643 \\ \times 111 \\ \hline \end{array}$

With Carrying

Multiply.

$$
\begin{array}{r}
6524 \\
\times 273 \\
\end{array}
\quad
\begin{array}{r}
6524 \\
\times 273 \\
\hline
19\,572 \longleftarrow 3 \times 6524 \\
456\,680 \longleftarrow 70 \times 6524 \\
1\,304\,800 \longleftarrow 200 \times 6524 \\
\hline
1{,}781{,}052 \longleftarrow 273 \times 6524 \\
\end{array}
$$

10. $\begin{array}{r} 61 \\ \times 76 \\ \hline \end{array}$ **11.** $\begin{array}{r} 78 \\ \times 36 \\ \hline \end{array}$ **12.** $\begin{array}{r} 437 \\ \times 571 \\ \hline \end{array}$

13. $\begin{array}{r} 465 \\ \times 541 \\ \hline \end{array}$ **14.** $\begin{array}{r} 542 \\ \times 168 \\ \hline \end{array}$ **15.** $\begin{array}{r} 8023 \\ \times 532 \\ \hline \end{array}$

16. $\begin{array}{r} 64 \\ \times 27 \\ \hline \end{array}$ **17.** $\begin{array}{r} 37 \\ \times 45 \\ \hline \end{array}$ **18.** $\begin{array}{r} 68 \\ \times 71 \\ \hline \end{array}$

19. $\begin{array}{r} 836 \\ \times 372 \\ \hline \end{array}$ **20.** $\begin{array}{r} 7501 \\ \times 447 \\ \hline \end{array}$ **21.** $\begin{array}{r} 5327 \\ \times 312 \\ \hline \end{array}$

Mixed Practice

22. $\begin{array}{r} 480 \\ \times 10 \\ \hline \end{array}$ **23.** $\begin{array}{r} 230 \\ \times 300 \\ \hline \end{array}$ **24.** $\begin{array}{r} 641 \\ \times 237 \\ \hline \end{array}$ **25.** $\begin{array}{r} 231 \\ \times 122 \\ \hline \end{array}$ **26.** $\begin{array}{r} 122 \\ \times 40 \\ \hline \end{array}$

27. $\begin{array}{r} 510 \\ \times 700 \\ \hline \end{array}$ **28.** $\begin{array}{r} 8233 \\ \times 2584 \\ \hline \end{array}$ **29.** $\begin{array}{r} 6010 \\ \times 6000 \\ \hline \end{array}$ **30.** $\begin{array}{r} 9000 \\ \times 7011 \\ \hline \end{array}$ **31.** $\begin{array}{r} 2793 \\ \times 1504 \\ \hline \end{array}$

32. $\begin{array}{r} 19{,}008 \\ \times 8000 \\ \hline \end{array}$ **33.** $\begin{array}{r} 8791 \\ \times 5000 \\ \hline \end{array}$ **34.** $\begin{array}{r} 6743 \\ \times 27 \\ \hline \end{array}$ **35.** $\begin{array}{r} 4231 \\ \times 253 \\ \hline \end{array}$ **36.** $\begin{array}{r} 8427 \\ \times 19 \\ \hline \end{array}$

37. $\begin{array}{r} 13{,}010 \\ \times 13 \\ \hline \end{array}$ **38.** $\begin{array}{r} 14{,}231 \\ \times 12 \\ \hline \end{array}$ **39.** $\begin{array}{r} 17{,}822 \\ \times 35 \\ \hline \end{array}$ **40.** $\begin{array}{r} 22{,}300 \\ \times 15 \\ \hline \end{array}$ **41.** $\begin{array}{r} 31{,}942 \\ \times 41 \\ \hline \end{array}$

42. $\begin{array}{r} 27{,}642 \\ \times 321 \\ \hline \end{array}$ **43.** $\begin{array}{r} 13{,}231 \\ \times 212 \\ \hline \end{array}$ **44.** $\begin{array}{r} 14{,}402 \\ \times 121 \\ \hline \end{array}$ **45.** $\begin{array}{r} 49{,}237 \\ \times 321 \\ \hline \end{array}$ **46.** $\begin{array}{r} 64{,}159 \\ \times 347 \\ \hline \end{array}$

Multiply Decimals

Decimals Greater Than One

Multiply.

$$17.45 \quad \blacktriangleright \quad 17.45 \longleftarrow \text{2 places}$$
$$\times 2.7 \qquad \times 2.7 \longleftarrow \text{+1 place}$$
$$\qquad\qquad 47.115 \longleftarrow \text{3 places}$$

1.   2.5
    ×1.8

2.   8.36
    ×1.5

3.   10.2
    ×8.61

4.   15.36
    ×5.3

5.   25.14
    ×7.5

6.   19.36
    ×7.12

7.   27.06
    ×8.53

8.   4.367
    ×8.5

9.   5.564
    ×7.9

10.   32.63
    ×9.2

11.   31.20
    ×9.21

12.   6.715
    ×9.03

Decimals Less Than One

Multiply.

$$0.08 \quad \blacktriangleright \quad 0.08 \qquad \text{2 places}$$
$$\times 0.4 \qquad \times 0.4 \qquad \text{+1 place}$$
$$\qquad\qquad 0.032 \qquad \text{3 places}$$

13.   0.144
    ×0.7

14.   0.86
    ×0.5

15.   0.96
    ×0.1

16.   0.56
    ×0.07

17.   0.73
    ×0.8

18.   0.05
    ×0.9

19.   0.81
    ×0.76

20.   0.47
    ×0.84

21.   0.63
    ×0.09

22.   0.57
    ×0.03

23.   1.23
    ×0.07

24.   0.01
    ×0.05

Mixed Practice

25.   41.16
    ×100

26.   0.923
    ×0.49

27.   0.12
    ×300

28.   67.32
    ×10

29.   7.243
    ×121

30.   557.4
    ×100

31.   327.8
    ×3.7

32.   14.923
    ×0.76

33.   1.125
    ×100

34.   0.009
    ×1000

35.   2.014
    ×40.7

36.   2.854
    ×0.04

37.   6243.78
    ×25.9

38.   5.9312
    ×5.62

39.   0.534
    ×0.293

40.   16.4591
    ×51.23

41.   96.00
    ×0.875

42.   0.3172
    ×0.2008

43.   0.1543
    ×0.4931

44.   0.7984
    ×0.0003

SKILL 9

Divide (Fractional Remainder)

Two-Digit Divisor

Divide. $46\overline{)703}$

$$\begin{array}{r} 1 \\ 46\overline{)703} \\ -46 \\ \hline 24 \end{array} \blacktriangleright \begin{array}{r} 15 \\ 46\overline{)703} \\ -46\downarrow \\ \hline 243 \\ -230 \\ \hline 13 \end{array} \blacktriangleright \begin{array}{r} 15\frac{13}{46} \\ 46\overline{)703} \\ -46\downarrow \\ \hline 243 \\ -230 \\ \hline 13 \end{array}$$

1. $27\overline{)63}$ 2. $38\overline{)89}$
3. $41\overline{)84}$ 4. $46\overline{)613}$
5. $53\overline{)754}$ 6. $61\overline{)685}$
7. $21\overline{)1781}$ 8. $55\overline{)4273}$
9. $43\overline{)2900}$ 10. $73\overline{)3956}$
11. $23\overline{)1369}$ 12. $34\overline{)2167}$
13. $81\overline{)5793}$ 14. $93\overline{)78}$

Three-Digit Divisor

Divide. $472\overline{)9463}$

$$\begin{array}{r} 2 \\ 472\overline{)9463} \\ -944 \\ \hline 2 \end{array} \blacktriangleright \begin{array}{r} 20\frac{23}{472} \\ 472\overline{)9463} \\ -944\downarrow \\ \hline 23 \\ -0 \\ \hline 23 \end{array}$$

15. $114\overline{)1837}$ 16. $216\overline{)5417}$
17. $321\overline{)4832}$ 18. $429\overline{)9038}$
19. $892\overline{)7928}$ 20. $910\overline{)11,849}$
21. $409\overline{)9429}$ 22. $900\overline{)7621}$
23. $625\overline{)8742}$ 24. $710\overline{)15,694}$
25. $843\overline{)17,691}$ 26. $937\overline{)22,474}$

Mixed Practice

27. $19\overline{)342}$ 28. $46\overline{)782}$ 29. $71\overline{)1136}$ 30. $279\overline{)9778}$
31. $35\overline{)2520}$ 32. $509\overline{)5089}$ 33. $621\overline{)8716}$ 34. $49\overline{)4959}$
35. $549\overline{)6937}$ 36. $953\overline{)9597}$ 37. $842\overline{)4975}$ 38. $87\overline{)2139}$
39. $473\overline{)85,642}$ 40. $192\overline{)91,845}$ 41. $622\overline{)54,641}$ 42. $812\overline{)55,692}$
43. $51\overline{)81,603}$ 44. $23\overline{)44,364}$ 45. $88\overline{)22,222}$ 46. $34\overline{)27,945}$
47. $417\overline{)86,902}$ 48. $71\overline{)103,205}$ 49. $514\overline{)691,507}$ 50. $64\overline{)237,605}$

Divide (Decimal Remainder)

Exact Quotient

Divide. 28)378

```
      13                13.5
28 )378      ➧    28 )378.0
    -28                -28
     98                 98
    -84                -84
     14                140
                      -140
```

1. 8)89

2. 10)123

3. 15)318

4. 16)264

5. 20)102

6. 50)530

7. 12)237

8. 25)1310

9. 32)1104

10. 96)2988

Rounded Quotient

Divide. 39)818

Round to the nearest hundredth.

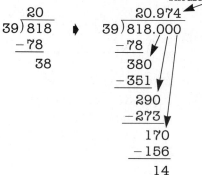

20.974 rounded to the nearest hundredth is 20.97 ◄——— Skill 2

Round answers to the place value shown.

Nearest tenth:

11. 14)319

12. 26)347

13. 23)371

14. 46)9415

15. 47)9719

Nearest hundredth:

16. 19)427

17. 83)168

18. 47)432

Nearest thousandth:

19. 37)402

20. 24)643

21. 21)452

Mixed Practice

Round answers to the nearest hundredth.

22. 28)1022

23. 24)209

24. 72)1665

25. 29)303

26. 85)1802

27. 24)747

28. 67)701

29. 71)400

30. 23)273

31. 36)630

32. 44)858

33. 59)852

34. 37)673

35. 41)9432

36. 73)1079

37. 24)994

38. 27)3365

39. 35)894

40. 42)5264

41. 110)4345

Divide Decimals

Divisor Greater Than One

Divide. $7.2\overline{)16.92}$

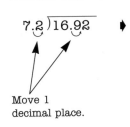

$$\begin{array}{r} 2.35 \\ 72\overline{)169.20} \\ -144 \\ \hline 252 \\ -216 \\ \hline 360 \\ -360 \end{array}$$

Move 1 decimal place.

Round answers to the nearest hundredth.

1. $2.5\overline{)11.5}$
2. $3.8\overline{)4.56}$
3. $3.2\overline{)18.272}$
4. $7.5\overline{)27.823}$
5. $3.15\overline{)53.55}$
6. $24.12\overline{)369.036}$
7. $4.08\overline{)26.52}$
8. $3.02\overline{)10.57}$
9. $4.23\overline{)181.5}$
10. $6.67\overline{)25.963}$

Divisor Less Than One

Divide. $0.032\overline{)14.400}$

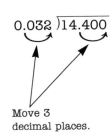

$$\begin{array}{r} 450 \\ 32\overline{)14400} \\ -128 \\ \hline 160 \\ -160 \\ \hline 00 \\ -00 \end{array}$$

Move 3 decimal places.

Round answers to the nearest hundredth.

11. $0.24\overline{)6.24}$
12. $0.372\overline{)6.324}$
13. $0.154\overline{)4.774}$
14. $0.48\overline{)2.938}$
15. $0.37\overline{)9.62}$
16. $0.21\overline{)1.374}$
17. $0.51\overline{)4.569}$
18. $0.67\overline{)2.693}$
19. $0.73\overline{)9.641}$
20. $0.81\overline{)11.632}$

Mixed Practice

Round answers to the nearest hundredth.

21. $3.12\overline{)4.386}$
22. $0.73\overline{)9.48}$
23. $0.136\overline{)33.32}$
24. $0.21\overline{)130.2}$
25. $6.94\overline{)8.378}$
26. $1.23\overline{)0.3813}$
27. $0.065\overline{)16.64}$
28. $8.34\overline{)7.416}$
29. $0.63\overline{)42.51}$
30. $2.91\overline{)5.932}$
31. $1.07\overline{)24.153}$
32. $1.1\overline{)29.9}$
33. $15.93\overline{)27.931}$
34. $12.12\overline{)36.422}$
35. $0.05\overline{)1.925}$
36. $2.03\overline{)21.249}$
37. $0.007\overline{)0.692}$
38. $1.05\overline{)25.421}$
39. $0.31\overline{)0.00354}$
40. $15.42\overline{)113.005}$
41. $5.02\overline{)86}$
42. $0.03\overline{)29}$
43. $5.4\overline{)0.062}$
44. $0.068\overline{)0.009}$

Equivalent Fractions

Higher Terms

Complete the equivalent fraction.

$$\frac{3}{4} = \frac{?}{20}$$

$$\frac{3}{4} = \frac{?}{20} \quad \blacktriangleright \quad \frac{3}{4} = \frac{15}{20}$$

with $\times 5$ applied to top and bottom

Solve.

1. $\frac{5}{6} = \frac{?}{12}$ **2.** $\frac{4}{9} = \frac{?}{27}$ **3.** $\frac{7}{10} = \frac{35}{?}$

4. $\frac{11}{12} = \frac{22}{?}$ **5.** $\frac{8}{17} = \frac{?}{51}$ **6.** $\frac{15}{17} = \frac{90}{?}$

7. $\frac{11}{18} = \frac{?}{36}$ **8.** $\frac{4}{19} = \frac{12}{?}$ **9.** $\frac{16}{17} = \frac{48}{?}$

10. $\frac{11}{13} = \frac{?}{39}$ **11.** $\frac{3}{4} = \frac{?}{32}$ **12.** $\frac{15}{17} = \frac{60}{?}$

13. $\frac{7}{9} = \frac{?}{45}$ **14.** $\frac{11}{12} = \frac{?}{72}$ **15.** $\frac{5}{8} = \frac{45}{?}$

16. $\frac{11}{8} = \frac{33}{?}$ **17.** $\frac{5}{24} = \frac{?}{96}$ **18.** $\frac{23}{73} = \frac{?}{365}$

Lowest Terms

Reduce $\frac{12}{28}$ to lowest terms

$$\frac{12}{28} = \frac{3}{7} \quad \blacktriangleright \quad \frac{12}{28} = \frac{3}{7}$$

with $\div 4$ applied to top and bottom

Reduce to lowest terms.

19. $\frac{6}{9}$ **20.** $\frac{4}{16}$ **21.** $\frac{9}{18}$ **22.** $\frac{20}{22}$

23. $\frac{18}{27}$ **24.** $\frac{24}{32}$ **25.** $\frac{16}{48}$ **26.** $\frac{18}{42}$

27. $\frac{55}{66}$ **28.** $\frac{30}{50}$ **29.** $\frac{20}{54}$ **30.** $\frac{21}{27}$

31. $\frac{8}{14}$ **32.** $\frac{19}{57}$ **33.** $\frac{14}{28}$ **34.** $\frac{15}{25}$

35. $\frac{21}{28}$ **36.** $\frac{200}{365}$ **37.** $\frac{188}{366}$ **38.** $\frac{150}{365}$

39. $\frac{180}{360}$ **40.** $\frac{225}{365}$ **41.** $\frac{190}{360}$ **42.** $\frac{65}{75}$

43. $\frac{72}{468}$ **44.** $\frac{183}{366}$ **45.** $\frac{232}{1450}$ **46.** $\frac{1792}{5120}$

Change Mixed Numbers/ Improper Fractions

Mixed Numbers to Improper Fractions

Write $3\frac{7}{8}$ as an improper fraction.

$3\frac{7}{8} = \dfrac{(3 \times 8) + 7}{8}$

$3\frac{7}{8} = \dfrac{24 + 7}{8}$

$3\frac{7}{8} = \dfrac{31}{8}$

Write as an improper fraction.

1. $6\frac{1}{4}$ 2. $4\frac{3}{4}$ 3. $7\frac{3}{8}$ 4. $9\frac{1}{2}$

5. $6\frac{2}{3}$ 6. $3\frac{5}{6}$ 7. $7\frac{4}{5}$ 8. $3\frac{1}{8}$

9. $4\frac{3}{10}$ 10. $1\frac{1}{16}$ 11. $4\frac{17}{32}$ 12. $4\frac{1}{5}$

13. $5\frac{1}{3}$ 14. $2\frac{1}{6}$ 15. $2\frac{9}{10}$ 16. $3\frac{1}{10}$

17. $3\frac{5}{16}$ 18. $4\frac{9}{16}$ 19. $5\frac{5}{32}$ 20. $6\frac{3}{5}$

21. $7\frac{2}{11}$ 22. $8\frac{5}{6}$ 23. $11\frac{2}{5}$ 24. $13\frac{1}{3}$

Improper Fractions to Mixed Numbers

Write $\frac{17}{3}$ as a mixed number.

$\frac{17}{3}$ ◆ $\begin{array}{r} 5\frac{2}{3} \\ 3\overline{)17} \\ -15 \\ \hline 2 \end{array}$ ◀── Skill 9

$\frac{17}{3} = 5\frac{2}{3}$

Write as a mixed number. Reduce any fractional parts to lowest terms.

25. $\frac{13}{2}$ 26. $\frac{18}{4}$ 27. $\frac{46}{8}$ 28. $\frac{19}{4}$

29. $\frac{33}{6}$ 30. $\frac{30}{7}$ 31. $\frac{21}{9}$ 32. $\frac{28}{8}$

33. $\frac{45}{18}$ 34. $\frac{49}{21}$ 35. $\frac{62}{21}$ 36. $\frac{23}{3}$

37. $\frac{35}{6}$ 38. $\frac{47}{7}$ 39. $\frac{57}{18}$ 40. $\frac{37}{9}$

41. $\frac{49}{8}$ 42. $\frac{63}{2}$ 43. $\frac{45}{7}$ 44. $\frac{71}{4}$

Change Fractions/Decimals

Fraction to Decimal

Write $\frac{5}{12}$ as a decimal. Round to the nearest hundredth.

$$\frac{5}{12} \quad \blacklozenge \quad 12\overline{)5.000} \xrightarrow{\text{0.416}} \text{Skill 10}$$

0.416 rounded to the nearest hundredth is 0.42. ◄——— Skill 2

$$\frac{5}{12} = 0.42$$

Write as a decimal. Round answers to the nearest hundredth.

1. $\frac{4}{5}$ 2. $\frac{7}{20}$ 3. $1\frac{1}{8}$ 4. $\frac{3}{4}$

5. $2\frac{3}{7}$ 6. $\frac{7}{12}$ 7. $1\frac{11}{30}$ 8. $\frac{2}{5}$

9. $\frac{3}{10}$ 10. $3\frac{4}{25}$ 11. $4\frac{3}{8}$ 12. $\frac{7}{10}$

13. $3\frac{1}{12}$ 14. $\frac{1}{15}$ 15. $\frac{1}{30}$ 16. $2\frac{9}{10}$

17. $5\frac{3}{20}$ 18. $\frac{9}{20}$ 19. $2\frac{3}{25}$ 20. $\frac{1}{8}$

21. $7\frac{1}{7}$ 22. $2\frac{5}{8}$ 23. $\frac{4}{9}$ 24. $1\frac{7}{8}$

Decimal to Fraction

Write 0.42 as a fraction in lowest terms.

$$0.42 = \frac{42}{100} \xleftarrow{\text{Skill 1}} = \frac{21}{50} \xleftarrow{} \text{Skill 12}$$

$$0.42 = \frac{21}{50}$$

Write as a fraction in lowest terms.

25. 0.1   26. 3.7   27. 0.30

28. 2.25   29. 0.03   30. 0.09

31. 0.53   32. 1.75   33. 0.003

34. 0.010   35. 0.064   36. 4.206

37. 4.444   38. 0.732   39. 0.469

40. 2.9   41. 7.5   42. 0.32

43. 1.08   44. 0.06   45. 2.83

46. 1.039   47. 0.105   48. 2.422

49. 0.0005   50. 0.2482   51. 1.6432

52. 0.0058   53. 0.0002   54. 6.66

SKILL 15

Add Fractions, Like Denominators

Fractions

Add.

Like denominators

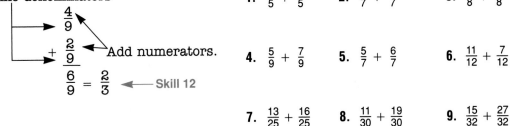

$$\frac{4}{9}$$
$$+\frac{2}{9}$$
$$\frac{6}{9} = \frac{2}{3} \longleftarrow \text{Skill 12}$$

Add numerators.

Express answers in lowest terms.

1. $\frac{4}{5} + \frac{3}{5}$ **2.** $\frac{4}{7} + \frac{2}{7}$ **3.** $\frac{1}{8} + \frac{5}{8}$

4. $\frac{5}{9} + \frac{7}{9}$ **5.** $\frac{5}{7} + \frac{6}{7}$ **6.** $\frac{11}{12} + \frac{7}{12}$

7. $\frac{13}{25} + \frac{16}{25}$ **8.** $\frac{11}{30} + \frac{19}{30}$ **9.** $\frac{15}{32} + \frac{27}{32}$

10. $\frac{21}{40} + \frac{31}{40}$ **11.** $\frac{22}{45} + \frac{31}{45}$ **12.** $\frac{11}{50} + \frac{19}{50}$

Mixed Numbers

Add.

$$4\frac{5}{7}$$
$$+8\frac{6}{7}$$
$$\frac{11}{7} = 1\frac{4}{7}$$

Skill 12

$$1$$
$$4\frac{5}{7}$$
$$+8\frac{6}{7}$$
$$13\frac{4}{7}$$

Express answers in lowest terms.

13. $4\frac{1}{3} + 7\frac{1}{3}$ **14.** $13\frac{4}{5} + 8\frac{3}{5}$

15. $5\frac{4}{7} + 8\frac{5}{7}$ **16.** $12\frac{3}{8} + 14\frac{1}{8}$

17. $6\frac{11}{12} + 5\frac{5}{12}$ **18.** $2\frac{12}{13} + 7\frac{2}{13}$

19. $8\frac{9}{16} + 8\frac{11}{16}$ **20.** $14\frac{5}{24} + 15\frac{7}{24}$

21. $4\frac{13}{32} + 5\frac{15}{32}$ **22.** $15\frac{19}{45} + 6\frac{28}{45}$

23. $7\frac{8}{35} + 4\frac{6}{35}$ **24.** $9\frac{5}{32} + 14\frac{7}{32}$

Add Fractions, Unlike Denominators

Fractions

Add.

Express answers in lowest terms.

Unlike denominators   Like denominators

Skill 12

$$\frac{5}{8} \rightarrow \frac{15}{24} \leftarrow \frac{15}{24}$$
$$+\frac{2}{3} \quad +\frac{16}{24} \leftarrow +\frac{16}{24}$$
$$\frac{31}{24} = 1\frac{7}{24}$$

Skill 15

Skill 13

1. $\frac{1}{2} + \frac{3}{5}$ 2. $\frac{3}{4} + \frac{1}{6}$ 3. $\frac{2}{7} + \frac{2}{3}$

4. $\frac{3}{8} + \frac{1}{5}$ 5. $\frac{5}{6} + \frac{1}{3}$ 6. $\frac{7}{12} + \frac{3}{7}$

7. $\frac{9}{11} + \frac{3}{10}$ 8. $\frac{13}{16} + \frac{9}{8}$ 9. $\frac{7}{20} + \frac{19}{30}$

10. $\frac{5}{18} + \frac{19}{24}$ 11. $\frac{1}{25} + \frac{13}{30}$ 12. $\frac{2}{15} + \frac{29}{30}$

Mixed Numbers

Add.

Express answers in lowest terms.

Skill 12

$$12\frac{2}{5} \rightarrow 12\frac{8}{20} \leftarrow 12\frac{8}{20}$$
$$+9\frac{3}{4} \quad +9\frac{15}{20} \leftarrow +9\frac{15}{20}$$
$$\frac{23}{20} = 1\frac{3}{20} \quad 22\frac{3}{20}$$

Skill 15

13. $5\frac{1}{2} + 3\frac{2}{3}$ 14. $4\frac{3}{8} + 9\frac{3}{4}$

15. $7\frac{5}{6} + 8\frac{1}{2}$ 16. $9\frac{7}{8} + 10\frac{7}{16}$

17. $11\frac{2}{13} + 9\frac{15}{26}$ 18. $18\frac{2}{3} + 5\frac{7}{11}$

19. $12\frac{4}{7} + 15\frac{7}{9}$ 20. $16\frac{5}{6} + 10\frac{6}{7}$

21. $16\frac{5}{8} + 12\frac{1}{7}$ 22. $11\frac{4}{9} + 13\frac{3}{8}$

23. $4\frac{2}{3} + 24\frac{7}{16}$ 24. $15\frac{5}{8} + 9\frac{4}{7}$

SKILL

17
Subtract Fractions, Like Denominators

Fractions

Subtract.

Like denominators

$$\frac{11}{12}$$

$$-\frac{7}{12}$$

→ $\frac{11}{12}$ ← Subtract numerators.

$-\frac{7}{12}$

$\frac{4}{12} = \frac{1}{3}$ ← Skill 13

Express answers in lowest terms.

1. $\frac{4}{9} - \frac{2}{9}$ 2. $\frac{9}{8} - \frac{3}{8}$ 3. $\frac{11}{12} - \frac{1}{12}$

4. $\frac{5}{6} - \frac{1}{6}$ 5. $\frac{5}{7} - \frac{1}{7}$ 6. $\frac{19}{27} - \frac{1}{27}$

7. $\frac{11}{16} - \frac{9}{16}$ 8. $\frac{16}{25} - \frac{6}{25}$ 9. $\frac{37}{40} - \frac{29}{40}$

10. $\frac{29}{36} - \frac{11}{36}$ 11. $\frac{7}{10} - \frac{3}{10}$ 12. $\frac{19}{36} - \frac{5}{36}$

13. $\frac{19}{24} - \frac{13}{24}$ 14. $\frac{13}{28} - \frac{7}{28}$ 15. $\frac{29}{60} - \frac{8}{60}$

Mixed Numbers

Subtract.

$8\frac{11}{16}$

$-5\frac{9}{16}$

→ $8\frac{11}{16}$

$-5\frac{9}{16}$

$\frac{2}{16} = \frac{1}{8}$

→ $8\frac{11}{16}$

$-5\frac{9}{16}$

$3\frac{1}{8}$

Skill 12

Express answers in lowest terms.

16. $7\frac{3}{4} - 6\frac{1}{4}$ 17. $8\frac{6}{7} - 7\frac{5}{7}$

18. $10\frac{4}{5} - 6\frac{1}{5}$ 19. $12\frac{5}{8} - 9\frac{3}{8}$

20. $2\frac{16}{21} - 1\frac{5}{21}$ 21. $7\frac{11}{24} - 5\frac{7}{24}$

22. $14\frac{17}{20} - 8\frac{9}{20}$ 23. $15\frac{15}{32} - 10\frac{9}{32}$

24. $14\frac{7}{8} - 9\frac{3}{8}$ 25. $12\frac{11}{13} - 10\frac{4}{13}$

26. $18\frac{17}{20} - 15\frac{9}{20}$ 27. $35\frac{39}{40} - 27\frac{19}{40}$

28. $74\frac{41}{75} - 29\frac{16}{75}$ 29. $103\frac{11}{35} - 78\frac{1}{35}$

18

Subtract Fractions, Unlike Denominators

Fractions

Subtract.

Express answers in lowest terms.

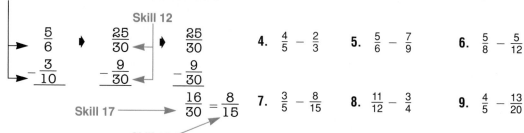

Unlike denominators

Like denominators

Skill 12

$\frac{5}{6}$

$\frac{25}{30}$

$\frac{25}{30}$

$-\frac{3}{10}$

$-\frac{9}{30}$

$-\frac{9}{30}$

$\frac{16}{30} = \frac{8}{15}$

Skill 17

Skill 12

1. $\frac{3}{4} - \frac{3}{8}$ **2.** $\frac{1}{3} - \frac{2}{9}$ **3.** $\frac{9}{10} - \frac{3}{4}$

4. $\frac{4}{5} - \frac{2}{3}$ **5.** $\frac{5}{6} - \frac{7}{9}$ **6.** $\frac{5}{8} - \frac{5}{12}$

7. $\frac{3}{5} - \frac{8}{15}$ **8.** $\frac{11}{12} - \frac{3}{4}$ **9.** $\frac{4}{5} - \frac{13}{20}$

10. $\frac{6}{7} - \frac{10}{21}$ **11.** $\frac{1}{2} - \frac{2}{5}$ **12.** $\frac{3}{4} - \frac{11}{16}$

Mixed Numbers

Subtract.

Express answers in lowest terms.

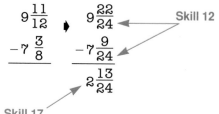

$9\frac{11}{12}$ $9\frac{22}{24}$ Skill 12

$-7\frac{3}{8}$ $-7\frac{9}{24}$

$2\frac{13}{24}$

Skill 17

13. $9\frac{3}{4} - 5\frac{1}{2}$ **14.** $7\frac{5}{6} - 4\frac{2}{3}$

15. $14\frac{4}{9} - 8\frac{1}{6}$ **16.** $15\frac{3}{7} - 10\frac{1}{3}$

17. $34\frac{11}{12} - 29\frac{4}{5}$ **18.** $48\frac{4}{7} - 37\frac{5}{9}$

19. $12\frac{5}{8} - 9\frac{1}{2}$ **20.** $2\frac{16}{21} - 1\frac{2}{3}$

21. $13\frac{3}{4} - 11\frac{3}{5}$ **22.** $21\frac{5}{6} - 19\frac{5}{8}$

23. $7\frac{11}{24} - 5\frac{1}{3}$ **24.** $20\frac{15}{38} - 20\frac{1}{19}$

SKILL 19

Subtract Mixed Numbers, Borrowing

Whole Number and a Mixed Number

Subtract.

$$8 \rightarrow 7\frac{5}{5} \quad \leftarrow \text{Like denominators}$$
$$-4\frac{3}{5} \quad -4\frac{3}{5}$$
$$\overline{} \quad \overline{3\frac{2}{5}}$$

Express answers in lowest terms.

1. $3 - 1\frac{1}{2}$ **2.** $4 - 2\frac{3}{4}$ **3.** $10 - 7\frac{9}{10}$

4. $6 - 4\frac{3}{8}$ **5.** $12 - 9\frac{5}{11}$ **6.** $15 - 12\frac{7}{16}$

Mixed Numbers with Unlike Denominators

Subtract.

Skill 12

$$6\frac{2}{3} \rightarrow 6\frac{10}{15} \leftarrow \rightarrow 5\frac{25}{15}$$
$$-2\frac{4}{5} \quad -2\frac{12}{15} \leftarrow \quad -2\frac{12}{15}$$
$$\overline{} \qquad \qquad \qquad \overline{3\frac{13}{15}}$$

Express answers in lowest terms.

7. $5\frac{1}{2} - 2\frac{4}{9}$ **8.** $8\frac{1}{2} - 3\frac{5}{8}$

9. $18\frac{3}{4} - 5\frac{5}{6}$ **10.** $27\frac{2}{7} - 13\frac{3}{4}$

11. $28\frac{3}{7} - 26\frac{9}{10}$ **12.** $40\frac{19}{30} - 39\frac{3}{4}$

13. $6\frac{3}{4} - 3\frac{7}{8}$ **14.** $16\frac{1}{12} - 9\frac{3}{10}$

Mixed Practice

15. $4 - 2\frac{5}{8}$ **16.** $9 - 5\frac{11}{12}$ **17.** $12 - 11\frac{7}{8}$ **18.** $16 - 7\frac{5}{11}$

19. $8\frac{1}{3} - 4\frac{5}{9}$ **20.** $13\frac{1}{2} - 7\frac{3}{5}$ **21.** $14\frac{1}{5} - 13\frac{3}{4}$ **22.** $13 - 6\frac{5}{8}$

23. $27\frac{1}{4} - 15\frac{15}{16}$ **24.** $38\frac{3}{7} - 29\frac{9}{10}$ **25.** $18\frac{1}{12} - 3\frac{11}{36}$ **26.** $17\frac{8}{9} - 5\frac{3}{8}$

27. $19 - 5\frac{2}{7}$ **28.** $16\frac{4}{5} - 6\frac{3}{4}$ **29.** $19\frac{11}{32} - 7\frac{3}{16}$ **30.** $25 - 8\frac{7}{10}$

Multiply Fractions/Mixed Numbers

Fractions

Multiply.

$$\frac{5}{8} \times \frac{2}{3} = \frac{5 \times 2}{8 \times 3} = \frac{10}{24} = \frac{5}{12}$$

Skill 12

Express answers in lowest terms.

1. $\frac{1}{2} \times \frac{2}{3}$ **2.** $\frac{1}{4} \times \frac{3}{5}$ **3.** $\frac{5}{6} \times \frac{3}{4}$

4. $\frac{8}{9} \times \frac{2}{7}$ **5.** $\frac{11}{12} \times \frac{5}{11}$ **6.** $\frac{4}{13} \times \frac{8}{9}$

7. $4 \times \frac{1}{2}$ **8.** $17 \times \frac{2}{5}$ **9.** $21 \times \frac{3}{7}$

Mixed Numbers

Multiply.

$$4\frac{1}{3} \times 2\frac{1}{4} = \frac{13}{3} \times \frac{9}{4} = \frac{117}{12} = 9\frac{3}{4}$$

Skill 13

Express answers in lowest terms.

10. $1\frac{1}{2} \times 1\frac{1}{3}$ **11.** $3\frac{2}{3} \times 4\frac{2}{5}$

12. $\frac{1}{8} \times 4\frac{4}{5}$ **13.** $7 \times 8\frac{1}{3}$

14. $12\frac{1}{2} \times 1\frac{1}{2}$ **15.** $20\frac{1}{2} \times 2\frac{1}{4}$

Mixed Practice

Express answers in lowest terms.

16. $\frac{2}{5} \times \frac{4}{7}$ **17.** $\frac{3}{8} \times \frac{4}{9}$ **18.** $\frac{2}{3} \times 1\frac{1}{8}$ **19.** $3\frac{2}{3} \times \frac{1}{4}$ **20.** $6 \times \frac{4}{5}$

21. $\frac{3}{4} \times 8$ **22.** $3\frac{1}{3} \times 4\frac{2}{5}$ **23.** $2\frac{1}{4} \times 1\frac{2}{3}$ **24.** $\frac{3}{4} \times 2\frac{5}{6}$ **25.** $6 \times 3\frac{1}{4}$

26. $18\frac{1}{2} \times 2\frac{2}{3}$ **27.** $1\frac{3}{5} \times 5\frac{2}{6}$ **28.** $5 \times 15\frac{1}{2}$ **29.** $10 \times 6\frac{2}{5}$ **30.** $2\frac{2}{7} \times 1\frac{1}{9}$

31. $3\frac{4}{9} \times 7\frac{3}{8}$ **32.** $\frac{7}{10} \times \frac{5}{9}$ **33.** $11\frac{3}{4} \times 8\frac{1}{3}$ **34.** $\frac{11}{12} \times \frac{4}{33}$ **35.** $6\frac{7}{8} \times 10\frac{1}{3}$

Divide Fractions/Mixed Numbers

Fractions

Divide.

$$\frac{2}{3} \div \frac{3}{8} = \frac{2}{3} \times \frac{8}{3} = 1\frac{7}{9}$$

Skill 20

Express answers in lowest terms.

1. $\frac{1}{4} \div \frac{1}{8}$ 2. $\frac{3}{8} \div \frac{1}{4}$ 3. $\frac{3}{4} \div \frac{2}{5}$

4. $5 \div \frac{5}{6}$ 5. $7 \div \frac{14}{15}$ 6. $8 \div \frac{4}{11}$

7. $\frac{3}{4} \div \frac{1}{8}$ 8. $\frac{2}{5} \div \frac{5}{6}$ 9. $\frac{4}{7} \div \frac{1}{28}$

Mixed Numbers

Divide.

Skill 13

$$5\frac{1}{6} \div 1\frac{1}{4} = \frac{31}{6} \div \frac{5}{4}$$
$$= \frac{31}{6} \times \frac{4}{5}$$
$$= \frac{124}{30}$$
$$= 4\frac{2}{15}$$

Skill 20

Express answers in lowest terms.

10. $3\frac{2}{3} \div 1\frac{1}{2}$ 11. $3\frac{5}{6} \div 1\frac{1}{5}$

12. $2\frac{1}{6} \div 3\frac{1}{3}$ 13. $4\frac{2}{3} \div 1\frac{3}{5}$

14. $\frac{3}{4} \div 3\frac{1}{2}$ 15. $\frac{5}{8} \div 2\frac{1}{2}$

16. $16 \div 1\frac{1}{8}$ 17. $11\frac{1}{3} \div 2\frac{1}{5}$

18. $12 \div 2\frac{12}{17}$ 19. $13\frac{1}{7} \div 3\frac{1}{6}$

Mixed Practice

Express answers in lowest terms.

20. $7\frac{1}{2} \div 1\frac{1}{4}$ 21. $8\frac{1}{4} \div 5\frac{1}{2}$ 22. $\frac{5}{11} \div 2\frac{1}{5}$ 23. $\frac{9}{10} \div \frac{3}{5}$ 24. $\frac{7}{12} \div \frac{1}{6}$

25. $1\frac{3}{8} \div 2\frac{1}{16}$ 26. $13\frac{1}{3} \div \frac{1}{10}$ 27. $8\frac{1}{3} \div \frac{5}{9}$ 28. $\frac{19}{21} \div 3\frac{1}{2}$ 29. $3\frac{5}{7} \div \frac{13}{21}$

30. $8 \div 1\frac{3}{5}$ 31. $20 \div \frac{2}{3}$ 32. $2\frac{1}{16} \div 11$ 33. $5\frac{3}{8} \div 22$ 34. $22 \div \frac{3}{5}$

35. $\frac{11}{15} \div 1\frac{12}{13}$ 36. $\frac{11}{12} \div \frac{11}{12}$ 37. $2\frac{11}{14} \div \frac{3}{7}$ 38. $\frac{15}{16} \div \frac{5}{32}$ 39. $19\frac{1}{2} \div 19\frac{1}{2}$

Write Ratios

Compare Two Numbers

What is the ratio of desks to computers?

Write the ratio.

OFFICE INVENTORY		
Computers	Calculators	Desks
6	5	15

Ratio of desks to computers:

15 to 6 or 15:6 or $\frac{15}{6}$

DEPARTMENT SALES		
Meat	Produce	Dairy
$1500	$600	$1200

1. Ratio of meat to produce

2. Ratio of meat to dairy

3. Ratio of produce to dairy

4. Ratio of dairy to produce

5. Ratio of produce to meat

6. Ratio of dairy to meat

Ratios as Fractions

Write the ratio of tables to chairs as a fraction in lowest terms.

A restaurant has 40 chairs and 16 tables.

Ratio of tables to chairs is $\frac{16}{40}$.

$\frac{16}{40} = \frac{2}{5}$ ◄────── Skill 12

Write the ratio as a fraction in lowest terms.

7. Nurse-to-patient ratio in a hospital is 6 nurses to 30 patients.

8. Teacher-to-student ratio in a school is 8 teachers to 160 students.

9. Door-to-window ratio in a house is 3 doors to 45 windows.

10. Width-to-length ratio of a box is 22 cm to 30 cm.

11. Room-to-desk ratio in a school is 40 rooms to 1000 desks.

12. Land area-to-people ratio in a county is 6 km^2 to 138 people.

13. Car-to-people ratio in a town is 4000 cars to 10,000 people.

14. Hit-to-strikeout ratio of a batter is 15 hits to 25 strikeouts.

Proportions

Checking Proportions

Is this proportion true?

$$\frac{5}{9} \overset{?}{=} \frac{35}{63}$$

Cross multiply.

$$\frac{5}{9} \overset{?}{=} \frac{35}{63} \quad \blacktriangleright \quad 5 \times 63 \overset{?}{=} 9 \times 35$$

$$315 = 315$$

True. The products are equal.

Tell whether the proportions are true.

1. $\frac{1}{3} \overset{?}{=} \frac{4}{12}$ 2. $\frac{2}{5} \overset{?}{=} \frac{4}{10}$ 3. $\frac{4}{5} \overset{?}{=} \frac{16}{19}$

4. $\frac{5}{6} \overset{?}{=} \frac{25}{29}$ 5. $\frac{3}{4} \overset{?}{=} \frac{6}{9}$ 6. $\frac{1}{2} \overset{?}{=} \frac{11}{24}$

7. $\frac{3}{8} \overset{?}{=} \frac{9}{24}$ 8. $\frac{2}{7} \overset{?}{=} \frac{14}{49}$ 9. $\frac{16}{32} \overset{?}{=} \frac{1}{2}$

10. $\frac{13}{23} \overset{?}{=} \frac{1}{2}$ 11. $\frac{10}{15} \overset{?}{=} \frac{5}{8}$ 12. $\frac{22}{30} \overset{?}{=} \frac{11}{15}$

Solving Proportions

Solve for the number that makes the proportion true.

$$\frac{7}{8} = \frac{28}{a}$$

Cross multiply.

$$\frac{7}{8} = \frac{28}{a} \quad \blacktriangleright \quad 7 \times a = 28 \times 8$$

$$7 \times a = 224$$

$$a = \frac{224}{7}$$

$$a = 32$$

Solve.

13. $\frac{1}{2} = \frac{9}{h}$ 14. $\frac{2}{3} = \frac{12}{y}$ 15. $\frac{6}{a} = \frac{9}{11}$

16. $\frac{8}{y} = \frac{12}{15}$ 17. $\frac{t}{6} = \frac{1}{11}$ 18. $\frac{a}{7} = \frac{2}{19}$

19. $\frac{5}{9} = \frac{n}{10}$ 20. $\frac{4}{11} = \frac{x}{33}$ 21. $\frac{8}{13} = \frac{y}{20}$

22. $\frac{42}{66} = \frac{y}{11}$ 23. $\frac{11}{33} = \frac{c}{60}$ 24. $\frac{8}{40} = \frac{5}{n}$

25. $\frac{n}{7} = \frac{15}{21}$ 26. $\frac{2}{3} = \frac{18}{n}$ 27. $\frac{96}{16} = \frac{n}{4}$

28. $\frac{13}{42} = \frac{65}{c}$ 29. $\frac{16}{3} = \frac{t}{12}$ 30. $\frac{h}{25} = \frac{15}{125}$

31. $\frac{27}{y} = \frac{81}{45}$ 32. $\frac{a}{21} = \frac{9}{27}$ 33. $\frac{121}{11} = \frac{550}{h}$

Solve a Rate Problem

Equal Rates as a Proportion

Write the proportion.

300 words typed in 5 minutes.
How many words in 3 minutes?

$$\frac{300 \text{ words}}{5 \text{ minutes}} = \frac{n \text{ words}}{3 \text{ minutes}}$$

$$\frac{300}{5} = \frac{n}{3}$$

Write the proportion.

1. 12 oranges cost 99¢. How much for 7 oranges?

2. A car travels 84 km on 7 L of gas. How many kilometers on 13 L of gas?

3. On a map, 5 cm represents 40 km. How many kilometers does 17 cm represent?

4. A machine uses 50 kW•h in 6 hours. How many kilowatt-hours in 45 hours?

Solve a Rate Problem

Find the number of boxes.

5000 envelopes in 10 boxes.
3000 envelopes in how many boxes?

$$\frac{5000 \text{ envelopes}}{10 \text{ boxes}} = \frac{3000 \text{ envelopes}}{n \text{ boxes}}$$

$$\frac{5000}{10} = \frac{3000}{n}$$

$$n = 6 \longleftarrow \text{Skill 23}$$

Solve.

5. 5 apples cost 79¢. How much for 22 apples?

6. A machine produces 121 bolts in 3 hours. How many bolts in 15 hours?

7. 36 pages typed in 4 hours. How many pages in 13 hours?

8. A car travels 19 km on 2 L of gas. How many kilometers on 76 L of gas?

Mixed Practice

Write the proportion and solve.

9. A machine produces 660 wheels in 4 hours. How many wheels are made in 9 hours?

10. Fuel costs 93¢ for 3 L. How much would it cost to fill a car with an 84-L tank?

11. The telephone rate to Europe is $6 for 3 minutes. How much for 14 minutes?

12. Sally can run 11 km in 60 minutes. How far can she run in 105 minutes?

13. Juan can read 6 pages in 5 minutes. How long will it take him to read 40 pages?

Compare Rates

Find the Unit Rate

Find the number of books.

4200 books in 200 boxes.
How many books in 1 box?

$$\frac{4200 \text{ books}}{200 \text{ boxes}} = \frac{n \text{ books}}{1 \text{ box}}$$

$$n = 21 \longleftarrow \text{Skill 24}$$

The unit rate is 21 books per box.

Find the unit rate.

1. 3456 oranges in 12 crates. How many oranges in 1 crate?

2. Ida prints 4950 words in 9 minutes. How many words in 1 minute?

3. Lou drives 560 km in 7 hours. How many kilometers in 1 hour?

4. $14.25 for 15 dozen eggs. How much for 1 dozen?

5. Sam drives 384 km in 6 hours. How far does he drive in 1 hour? How far in 10 hours?

6. 2500 sheets of bond paper weigh 100 pounds. How many sheets weigh 1 pound? What will 10,000 sheets weigh?

Compare Unit Rates

Which is faster?

Car A: 280 km in 4 hours
Car B: 455 km in 7 hours

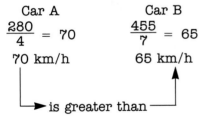

Car A is faster.

Compare the unit rates.

7. Which uses less fuel per kilometer?
Car A: 52 km on 6 L
Car B: 59 km on 7 L

8. Which costs less per kilogram?
Carton A: $89.25 for 15 kg
Carton B: $117.00 for 20 kg

9. Which shipment holds more per crate?
Shipment A: 1584 apples in 11 crates
Shipment B: 1937 apples in 13 crates

10. Which makes more cogs per hour?
Machine A: 288 cogs in 4 hours
Machine B: 414 cogs in 6 hours

11. Which gets more km per liter?
Car A: 162 km on 18 L of fuel
Car B: 220 km on 20 L of fuel

Write Decimals as Percents

Two or More Decimal Places

Write 0.037 as a percent.

0.037 = ?%

0.037 ◆ 0.037

        Move 2 places to the right.

0.037 = 3.7%

Write as a percent.

1. 0.10	**2.** 0.15	**3.** 0.25
4. 0.74	**5.** 0.82	**6.** 0.93
7. 2.13	**8.** 4.212	**9.** 5.753
10. 0.267	**11.** 0.391	**12.** 0.914
13. 12.104	**14.** 0.625	**15.** 10.82
16. 0.007	**17.** 0.106	**18.** 0.008
19. 0.04	**20.** 0.001	**21.** 0.503

Fewer Than Two Decimal Places

Write 0.5 as a percent.

0.5 = ?%

       Use zero as placeholder.

0.5 ◆ 0.50%

        Move 2 places to the right.

0.5 = 50%

Write as a percent.

22. 0.4	**23.** 0.7	**24.** 0.9
25. 7.1	**26.** 9.3	**27.** 10.5
28. 11.0	**29.** 12.6	**30.** 17.1
31. 22.5	**32.** 29.0	**33.** 37.2
34. 0.1	**35.** 7.5	**36.** 1.8
37. 16.3	**38.** 0.3	**39.** 6.2

Mixed Practice

Write as a percent.

40. 0.57	**41.** 3.80	**42.** 0.001	**43.** 0.67	**44.** 20.7
45. 8.8	**46.** 0.2915	**47.** 0.32	**48.** 17.4	**49.** 3.003
50. 139.25	**51.** 0.2187	**52.** 9.1	**53.** 25.00	**54.** 12.004
55. 0.14	**56.** 2.185	**57.** 0.51	**58.** 9.3	**59.** 1.868
60. 8.554	**61.** 0.003	**62.** 3.246	**63.** 0.26	**64.** 0.07
65. 0.0032	**66.** 3.642	**67.** 0.29	**68.** 0.052	**69.** 3.42
70. 0.30	**71.** 2.1	**72.** 0.0001	**73.** 1.67	**74.** 1.0
75. 2.549	**76.** 0.005	**77.** 2.077	**78.** 1.5	**79.** 0.19

27 Write Fractions/ Mixed Numbers as Percents

Fractions

Write $\frac{3}{5}$ as a percent.

$\frac{3}{5} = ?\%$

$\frac{3}{5} = 0.6 = 60\%$

Skill 14   Skill 26

Write as a percent.

1. $\frac{1}{4}$ 2. $\frac{2}{5}$ 3. $\frac{3}{4}$ 4. $\frac{7}{10}$

5. $\frac{1}{8}$ 6. $\frac{3}{20}$ 7. $\frac{7}{20}$ 8. $\frac{12}{25}$

9. $\frac{7}{50}$ 10. $\frac{7}{40}$ 11. $\frac{5}{8}$ 12. $\frac{11}{16}$

Mixed Numbers

Write $2\frac{3}{8}$ as a percent.

$2\frac{3}{8} = ?\%$

$2\frac{3}{8} = \frac{19}{8} = 2.375 = 237.5\%$

Skill 13   Skill 14   Skill 26

Write as a percent.

13. $2\frac{3}{4}$ 14. $4\frac{5}{8}$ 15. $6\frac{2}{3}$

16. $7\frac{1}{6}$ 17. $8\frac{5}{9}$ 18. $5\frac{4}{15}$

19. $8\frac{7}{40}$ 20. $14\frac{3}{16}$ 21. $7\frac{11}{50}$

Mixed Practice

Write as a percent.

22. $\frac{3}{5}$ 23. $5\frac{3}{8}$ 24. $12\frac{5}{6}$ 25. $\frac{7}{12}$ 26. $\frac{1}{9}$

27. $15\frac{3}{10}$ 28. $\frac{9}{40}$ 29. $20\frac{4}{5}$ 30. $10\frac{6}{25}$ 31. $18\frac{5}{8}$

32. $9\frac{3}{40}$ 33. $\frac{15}{16}$ 34. $8\frac{19}{50}$ 35. $11\frac{5}{14}$ 36. $16\frac{9}{20}$

37. $\frac{7}{9}$ 38. $7\frac{3}{16}$ 39. $5\frac{11}{15}$ 40. $\frac{11}{30}$ 41. $17\frac{7}{20}$

Write Percents as Decimals

28

Percent in Decimal Form

Write 8.75% as a decimal.

Drop % sign.

8.75% ▸ 0 08.75

Move 2 places to the left.

8.75% = 0.0875

Write as a decimal.

1. 10.5%	**2.** 15.7%	**3.** 40%
4. 85%	**5.** 120%	**6.** 137%
7. 6.7%	**8.** 7.1%	**9.** 8.9%
10. 95%	**11.** 119%	**12.** 7.9%
13. 17.2%	**14.** 85.6%	**15.** 100%
16. 1.35%	**17.** 5.3%	**18.** 142%

Percent in Fractional Form

Write $\frac{3}{20}$% as a decimal.

$\frac{3}{20}$% = 0.15% ▸ 0 00.15

Skill 14

$\frac{3}{20}$% = 0.0015

Write as decimals.

19. $\frac{7}{10}$%	**20.** $\frac{11}{20}$%	**21.** $\frac{5}{8}$%
22. $\frac{2}{25}$%	**23.** $5\frac{3}{5}$%	**24.** $7\frac{1}{10}$%
25. $6\frac{4}{5}$%	**26.** $22\frac{3}{4}$%	**27.** $30\frac{2}{5}$%
28. $89\frac{13}{20}$%	**29.** $57\frac{21}{25}$%	**30.** $12\frac{3}{4}$%

Mixed Practice

Write as a decimal.

31. 74.25%	**32.** $6\frac{1}{4}$%	**33.** $18\frac{4}{5}$%	**34.** 0.97%	**35.** $14\frac{9}{20}$%
36. $28\frac{3}{40}$%	**37.** 0.125%	**38.** 6.25%	**39.** $\frac{7}{50}$%	**40.** 6.63%
41. 8.79%	**42.** 100%	**43.** $\frac{19}{20}$%	**44.** 127%	**45.** $\frac{1}{40}$%
46. $20\frac{1}{4}$%	**47.** $22\frac{1}{8}$%	**48.** 14.6%	**49.** $25\frac{1}{5}$%	**50.** 18.9%

Write Percents as Fractions

Percent in Decimal Form

Write 37.5% as a fraction.

$$37.5\% = 0.375 = \frac{3}{8}$$

Skill 28   Skill 14

Write as a fraction in lowest terms.

1. 45%	**2.** 80%	**3.** 175%
4. 200%	**5.** 11.7%	**6.** 0.1%
7. 0.15%	**8.** 67.3%	**9.** 10.6%
10. 78.55%	**11.** 37.63%	**12.** 51.42%
13. 50%	**14.** 75%	**15.** 10.1%
16. 0.12%	**17.** 80.9%	**18.** 42.1%

Percent in Fractional Form

Write $16\frac{2}{3}\%$ as a fraction.

$$16\frac{2}{3}\% = 16\frac{2}{3} \div 100$$

$$= \frac{50}{3} \div 100$$

Skill 13

$$= \frac{50}{300} \longleftarrow \text{Skill 21}$$

$$= \frac{1}{6} \longleftarrow \text{Skill 12}$$

Write as a fraction in lowest terms.

19. $6\frac{1}{4}\%$	**20.** $37\frac{1}{2}\%$	**21.** $43\frac{3}{4}\%$
22. $3\frac{1}{2}\%$	**23.** $9\frac{1}{4}\%$	**24.** $10\frac{1}{3}\%$
25. $15\frac{1}{2}\%$	**26.** $20\frac{1}{2}\%$	**27.** $25\frac{1}{2}\%$
28. $15\frac{3}{4}\%$	**29.** $16\frac{2}{3}\%$	**30.** $33\frac{1}{3}\%$

Mixed Practice

Write as a fraction in lowest terms.

31. 30%	**32.** 57.2%	**33.** $31\frac{1}{4}\%$	**34.** 63.5%	**35.** $12\frac{1}{2}\%$
36. $13\frac{1}{3}\%$	**37.** $26\frac{2}{3}\%$	**38.** 64.75%	**39.** $11\frac{1}{9}\%$	**40.** 78.55%
41. 50%	**42.** 75%	**43.** $1\frac{1}{4}\%$	**44.** 100%	**45.** $12\frac{3}{4}\%$
46. $4\frac{3}{4}\%$	**47.** $5\frac{3}{8}\%$	**48.** 75.91%	**49.** $6\frac{2}{3}\%$	**50.** 121%

Find the Percentage

Decimal Percents

Find 15.5% of 36.

$$15.5\% \text{ of } 36 = n$$

Skill 28

$$0.155 \times 36 = n$$

$$5.58 = n$$

$$15.5\% \text{ of } 36 = 5.58$$

Find the percentage.

1. 15% of 60   **2.** 30% of 72

3. 4% of 96   **4.** 9% of 122

5. 6.5% of 120   **6.** 8.3% of 150

7. 17.8% of 80   **8.** 31.2% of 140

9. 5.81% of 60   **10.** 7.32% of 45

11. 67.7% of 67   **12.** 8.92% of 35

Fractional Percents

Find $\frac{3}{4}$% of 1600.

$$\frac{3}{4}\% \text{ of } 1600 = n$$

Skill 29

$$\frac{3}{400} \times 1600 = n$$

$$12 = n$$

$$\frac{3}{4}\% \text{ of } 1600 = 12$$

Find the percentage.

13. $2\frac{1}{2}$% of 400   **14.** $4\frac{1}{2}$% of 200

15. $33\frac{1}{3}$% of 120   **16.** $3\frac{5}{6}$% of 600

17. $\frac{3}{4}$% of 800   **18.** $\frac{3}{5}$% of 50

19. $\frac{3}{8}$% of 600   **20.** $\frac{7}{8}$% of 80

Mixed Practice

Find the percentage.

21. 80% of 160   **22.** $8\frac{1}{3}$% of 72   **23.** 9.1% of 90   **24.** $15\frac{3}{5}$% of 90

25. $16\frac{2}{3}$% of 90   **26.** 45% of 72   **27.** $\frac{1}{4}$% of 800   **28.** 0.75% of 1000

29. 125% of 64   **30.** $12\frac{1}{4}$% of 65   **31.** 7.2% of 127   **32.** $6\frac{1}{4}$% of 1600

33. $4\frac{1}{6}$% of 600   **34.** 16.5% of 84   **35.** $8\frac{1}{2}$% of 75   **36.** 24.7% of 80

SKILL

31

Find the Rate

Percentage Less Than Base

What percent of 75 is 60?

$n\%$ of 75 = 60

Write as a proportion and solve.

$$\frac{n}{100} = \frac{60}{75}$$

$75 \times n = 6000$

$n = 80 \longleftarrow$ Skill 23

80% of 75 = 60

Solve. Round answers to the nearest tenth of a percent.

1. $n\%$ of 40 = 20
2. $n\%$ of 60 = 15
3. $n\%$ of 90 = 18
4. $n\%$ of 60 = 9
5. $n\%$ of 70 = 25
6. $n\%$ of 90 = 65
7. $n\%$ of 65 = 58
8. $n\%$ of 84 = 75
9. $n\%$ of 113 = 79
10. $n\%$ of 410 = 295.5
11. $n\%$ of 94 = 82
12. $n\%$ of 296 = 239.76

Percentage Greater Than Base

What percent of 90 is 162?

$n\%$ of 90 = 162

Write as a proportion and solve.

$$\frac{n}{100} = \frac{162}{90}$$

$90 \times n = 16{,}200$

$n = 180$

180% of 90 = 162

Solve. Round answers to the nearest tenth of a percent.

13. $n\%$ of 20 = 30
14. $n\%$ of 36 = 45
15. $n\%$ of 50 = 60
16. $n\%$ of 70 = 91
17. $n\%$ of 60 = 80
18. $n\%$ of 72 = 96
19. $n\%$ of 24 = 31
20. $n\%$ of 56 = 77
21. $n\%$ of 37 = 148
22. $n\%$ of 60 = 90
23. $n\%$ of 81 = 415.75
24. $n\%$ of 110 = 130

Mixed Practice

Solve. Round answers to the nearest tenth of a percent.

25. $n\%$ of 125 = 100
26. $n\%$ of 100 = 115
27. $n\%$ of 130 = 60
28. $n\%$ of 56 = 96
29. $n\%$ of 89 = 49
30. $n\%$ of 84 = 108
31. $n\%$ of 115 = 92
32. $n\%$ of 42 = 52.5
33. $n\%$ of 64 = 76.8
34. $n\%$ of 64 = 24
35. $n\%$ of 204 = 51
36. $n\%$ of 42 = 6.72
37. $n\%$ of 36 = 50
38. $n\%$ of 173 = 136.25
39. $n\%$ of 18 = 36
40. $n\%$ of 21 = 63
41. $n\%$ of 120 = 80
42. $n\%$ of 15.5 = 62
43. $n\%$ of 90 = 4.5
44. $n\%$ of 125 = 18.75
45. $n\%$ of 75 = 110

Find the Base

Decimal Percents

42 is 37.5% of what number?   **Find the number.**

37.5% of n = 42

0.375 × n = 42

Skill 28

n = 42 ÷ 0.375

n = 112

37.5% of 112 = 42

1. 12.5% of n = 9   **2.** 62.5% of n = 55

3. 8.25% of n = 664   **4.** 7% of n = 3.5

5. 5.75% of n = 92   **6.** 9% of n = 8.2

7. 11% of n = 37   **8.** 3% of n = 8.7

9. 5% of n = 3   **10.** 12.5% of n = 21

11. 37.3% of n = 50   **12.** 81% of n = 14

Fractional Percents

3 is $6\frac{1}{4}$% of what number?   **Find the number.**

$6\frac{1}{4}$% of n= 3

$\frac{1}{16}$ × n = 3

Skill 29

n = 3 ÷ $\frac{1}{16}$

n = 48

$6\frac{1}{4}$% of 48 = 3

13. $2\frac{1}{2}$% of n = 4   **14.** $3\frac{1}{8}$% of n = 2

15. $33\frac{1}{3}$% of n = 20   **16.** $\frac{1}{4}$% of n = 2

17. $37\frac{1}{2}$% of n = 34   **18.** $\frac{3}{5}$% of n = 6

19. $45\frac{1}{2}$% of n = 25   **20.** $\frac{3}{4}$% of n = 7

Mixed Practice
Find the number.

21. 80% of n = 60   **22.** 12.5% of n = 39   **23.** 116% of n = 2.9

24. 10% of n = 2.9   **25.** 8% of n = 4.2   **26.** 40% of n = 6.8

27. $16\frac{2}{3}$% of n = 13   **28.** 145% of n = 6.38   **29.** $66\frac{2}{3}$% of n = 120

30. 25% of n = 36   **31.** $2\frac{2}{3}$% of n = 6   **32.** 90% of n = 72

33. 6% of n = 3.3   **34.** 7.8% of n = 9   **35.** $16\frac{3}{4}$% of n = 33

Substituting in a Formula

To use a formula to solve a problem, substitute known values in the formula. Then solve. Remember to perform any computations within parentheses first.

▶ An airplane travels at a rate of 960 km/h. How far can the plane travel in 7 hours?

Formula: Distance = Rate × Time

Substitute: Distance = 960 × 7

Solve: Distance = 6720

The plane travels 6720 km in 7 hours.

Find the distance.

	Distance = Rate × Time		
	Rate	Time	Distance
1.	100 mi/h	6 h	?
2.	55 km/h	4 h	?
3.	14 m/s	32 s	?
4.	30 mi/h	6 h	?
5.	42 m/s	5 s	?

Find the area of the triangle.

	Area of Triangle $= \frac{1}{2} \times$ Base × Height		
	Base	Height	Area
6.	1 m	0.5 m	?
7.	64 ft	38 ft	?
8.	18 in	5 in	?
9.	30 km	24 km	?

Find the percent of sales made.

	% of Sales $= \dfrac{\text{Sales Made}}{\text{Possible Sales}} \times 100$		
	Possible Sales	Sales Made	Percent
10.	$ 4,600	$ 1,900	?
11.	$19,100	$ 6,000	?
12.	$23,000	$13,000	?
13.	$64,000	$15,000	?
14.	$ 8,200	$ 3,000	?

Find the price of the stereo system.

	Price = Amp + CD Player + Speakers			
	Amp	CD Player	Speakers	Price
15.	$199.00	$ 99.95	$139.00	?
16.	$395.00	$119.99	$259.00	?
17.	$685.00	$249.00	$495.00	?
18.	$449.00	$188.00	$350.00	?

Find the long-distance telephone charge.

	Charge = Cost per min × Minutes		
	Minutes	Cost per min	Charge
19.	15	$0.35	?
20.	9	$1.12	?
21.	31	$1.49	?
22.	22	$0.54	?
23.	38	$1.32	?

Multiplying or Dividing to Solve Formulas

To solve some formulas after substituting, you may have to multiply or divide both sides of the equation by the same number.

▶ An airplane cruising at a rate of 575 km/h has traveled 2645 km. How long did it take to make this trip?

Formula: Distance = Rate × Time

Substitute: 2645 = 575 × Time

Divide: $\dfrac{2645}{575} = \dfrac{575 \times \text{Time}}{575}$

Solve: 4.6 h = Time

Find the missing amount.

	Distance = Rate × Time		
	Rate	Time	Distance
1.	88 in/min	3.5 min	?
2.	60 km/h	?	195 km
3.	675 yd/h	13 h	?
4.	?	1.75 h	78.75 km
5.	24 km/h	?	192 km
6.	?	$4\frac{1}{2}$ h	$247\frac{1}{2}$ ft
7.	3.2 m/s	?	96 m

Adding or Subtracting to Solve Formulas

To solve some formulas after substituting, you may have to add the same number to, or subtract it from, both sides of the equation.

▶ The temperature of a solution is 212°F (degrees Fahrenheit). What is this in degrees Celsius (°C)?

Formula: °F = (1.8 × °C) + 32

Substitute: 212 = (1.8 × °C) + 32

Subtract: − 32   − 32
                 ─────────────────────
                 180 = (1.8 × °C)

Divide: $\dfrac{180}{1.8} = \dfrac{1.8 \times °C}{1.8}$

Solve: 100 = °C

Find the missing amount.

	°F = (1.8 × °C) + 32	
	Degrees Celsius	Degrees Fahrenheit
8.	50	?
9.	?	86
10.	?	32
11.	120	?
12.	?	95
13.	?	68
14.	144	?
15.	10	?
16.	?	5

Tables and Charts

Reading Tables and Charts

To read a table or chart, find the row containing one of the conditions of the information you seek. Run down the column containing the other condition until it crosses that row. Read the answer.

▶ How many sales were made at the Lincoln office?

The table shows that there were 537 sales made at the Lincoln office.

IMAG CORP. SALES REPORT			
Office	Agents	Sales	Income
Bronx	252	640	$12,000,000
Calgary	204	307	1,800,000
Columbus	92	778	1,200,000
Dallas	50	761	1,300,000
Denver	76	398	1,500,000
Lincoln	35	537	1,500,000
San Diego	128	813	14,300,000
Toronto	710	390	7,900,000

Find the income for each of the following offices.

1. Columbus   **2.** Denver   **3.** Toronto   **4.** Bronx

5. Which offices made more than 500 sales?

6. Which offices have less than 200 agents?

Using Tables and Charts

To classify an item, find the row or column that contains the given data. Then read the classification from the head of the row or column.

▶ An order weighing 5 lb 5 oz cost $9.25 in shipping charges. To which zone was it delivered?

The table shows that the order was delivered within Zone 2.

DELIVERY CHART	Zone 1	Zone 2	Custom Delivery			
			Express		Express Plus	
			Ordered	Delivered	Ordered	Delivered
Find Your State at Right ▶ Shipping Weight ▼	AL, AR, DC, DE, IA, IL, IN, KS, KY, MD, MI, MN, MO, MS, NC, NE, ND, OH, OK, PA, SC, SD, TN, VA, WI, WV	AK, AZ, CA, CO, CT, FL, GA, HI, ID, LA, MA, ME, MT, NH, NJ, NM, NV, NY, OR, RI, TX UT, VT, WA, WY	Mon. Tue. Wed. Thur. Fri. Sat. Sun.	Thur. Fri. Mon. Tue. Wed. Wed. Wed.	Mon. Tue. Wed. Thur. Fri. Sat. Sun.	Wed. Thur. Fri. Mon. Tue. Tue. Tue.
Minimum Charge	$ 4.00	$ 4.50	$ 9.00		$19.00	
2 lb 1 oz to 4 lb	6.50	6.75	11.00		21.00	
4 lb 1 oz to 8 lb	9.00	9.25	15.75		25.75	
8 lb 1 oz to 14 lb	11.25	11.75	22.25		32.25	
14 lb 1 oz to 20 lb	13.00	15.00	28.25		38.25	

Find the missing information.

7. An order weighing 16 lb cost $38.25 in shipping charges. How was the order sent?

8. An order weighing 9 lb 12 oz cost $11.25 in shipping charges. To what zone was it delivered?

9. An order weighing 10 lb cost $22.25 in shipping charges.
   a. How was the order sent?
   b. When would the order be delivered if it was ordered on a Wednesday?

Making Change: Building Up

To make change, build up to the next nickel, dime, quarter, and so on, until the amount presented is reached.

▶ Compute the change from $5.00 for a purchase of $3.67.

			$3.67
	3 pennies	⟶	+ .03
C			3.70
H	1 nickel	⟶	+ .05
A			3.75
N	1 quarter	⟶	+ .25
G			4.00
E	1 dollar	⟶	+1.00
	Amount presented	⟶	$5.00

Compute the change from $10. Use the fewest coins and bills.

1. $3.75
2. $7.78
3. $1.99
4. $3.47
5. $5.63
6. $9.19

Compute the change. Use the fewest coins and bills.

7. Amount spent: $15.05
   Amount presented: $20

8. Amount spent: $3.37
   Amount presented: $20

9. Amount spent: $2.29
   Amount presented: $3

Making Change: Loose Coins

The amount presented may contain loose coins so that fewer coins and bills are returned in the change.

▶ Compute the change from $5.02 for a purchase of $3.67.

		Amount Presented	Purchase
a.	Subtract the loose change.	$5.02 − .02	$3.67 − .02
		$5.00	$3.65

b. Now compute the change from $5.00 for a purchase of $3.65.

			$3.65
C	1 dime	⟶	+ .10
H			3.75
A	1 quarter	⟶	+ .25
N			4.00
G	1 dollar	⟶	+1.00
E			$5.00

Compute the change using the fewest coins and bills.

10. Amount spent: $5.26
    Amount presented: $6.01

11. Amount spent: $7.79
    Amount presented: $10.04

12. Amount spent: $1.07
    Amount presented: $5.07

13. Amount spent: $3.92
    Amount presented: $10.02

14. Amount spent: $15.21
    Amount presented: $20.01

15. Amount spent: $7.81
    Amount presented: $10.06

16. Amount spent: $6.03
    Amount presented: $21.03

17. Amount spent: $9.62
    Amount presented: $21.12

Rounding Time: Nearest Quarter Hour

To round to the nearest quarter hour, find the quarter-hour interval that the given time is between. Subtract the earlier quarter from the given time. Subtract the given time from the later quarter. Round to whichever is closer.

▶ Round 6:19 to the nearest quarter hour.

6:19	6:30
− 6:15	− 6:19
4 minutes	11 minutes

4 is less than 11; round to 6:15

Round to the nearest quarter hour.

1. 8:39	**2.** 10:40	**3.** 6:55
4. 1:23	**5.** 11:17	**6.** 3:11
7. 12:09	**8.** 4:44	**9.** 3:32
10. 11:11	**11.** 8:51	**12.** 1:50
13. 7:35	**14.** 5:04	**15.** 9:40
16. 10:34	**17.** 12:39	**18.** 6:41
19. 6:01	**20.** 3:49	**21.** 11:36
22. 8:05	**23.** 10:10	**24.** 2:15

Application F **Elapsed Time (Hours)**

Finding Elapsed Time

To find the elapsed time, subtract the earlier time from the later time. If it is necessary to "borrow," rewrite the later time by subtracting 1 from the hours and adding 60 to the minutes.

▶ How much time has elapsed between 11:45 a.m. and 12:23 p.m.?

12:23 =	11:23 + :60 =	11:83
− 11:45 =	− 11:45	= − 11:45
		38 min

Find the elapsed time.

1. From 6:05 p.m. to 8:51 p.m.

2. From 4:18 a.m. to 6:52 a.m.

3. From 8:30 a.m. to 11:45 a.m.

4. From 6:45 p.m. to 11:59 p.m.

5. From 10:45 a.m. to 11:36 a.m.

6. From 3:45 p.m. to 5:30 p.m.

Finding Elapsed Time Spanning 1 O'Clock

To find the elapsed time when the period spans 1 o'clock, add 12 hours to the later time before subtracting.

▶ How much time has elapsed between 7:45 p.m. and 2:52 a.m.?

2:52 =	2:52 + 12:00 =	14:52
− 7:45 =	− 7:45	= − 7:45
		7h:7 min

Find the elapsed time.

7. From 8:15 a.m. to 1:30 p.m.

8. From 4:55 a.m. to 1:00 p.m.

9. From 11:30 p.m. to 7:50 a.m.

10. From 9:45 p.m. to 8:00 a.m.

11. From 3:30 a.m. to 2:15 p.m.

APPLICATIONS

Application G — Elapsed Time (Days)

To find the number of days **between** two dates, or **from** one date **to** the next, use the table on page 645. Find the position of each date in the year. When the dates run from one year to the next, add 365 to the later date. Then subtract the earlier date from the later.

Day No.	Jan.	Feb.	Mar.	Apr.	May	June	July	Aug.	Sept.
1	1	32	60	91	121	152	182	213	244
2	2	33	61	92	122	153	183	214	245
3	3	34	62	93	123	154	184	215	246
4	4	35	63	94	124	155	185	216	247
5	5	36	64	95	125	156	186	217	248
6	6	37	65	96	126	157	187	218	249

To find the number of days **from** one date **through** another, find the elapsed time between the dates and add 1 day.

▶ How many days have elapsed between April 4 and January 6?

Jan. 6:   (6 + 365)   =   371
Apr. 4:   – 94   = – 94
                               277 days

Find the elapsed time. Use the table on page 645.

1. Between August 10 and December 25
2. From March 21 through September 30
3. Between June 1 and December 30
4. From May 21 to July 2
5. From September 5 through May 20

Application H — Determining Leap Years

A year is a leap year if it is exactly divisible by 4. Exceptions are years that end in '00' (1900, 2000, and so on); they must be divisible by 400.

▶ Was 1848 a leap year?

1848 ÷ 4 = 462; 1848 was a leap year.

Find the leap years.

1. 1949	**2.** 1980	**3.** 2000
4. 1912	**5.** 2100	**6.** 1954
7. 1992	**8.** 1951	**9.** 2400
10. 1882	**11.** 1920	**12.** 2004

Application I — Elapsed Time in a Leap Year

When the dates span February 29 in a leap year, add 1 day to the later date.

▶ How many days elapsed between January 5, 1980, and September 3, 1980?

Sept. 3, 1980:   (246 + 1) =   247
Jan. 5, 1980: –   5   = –   5
                                 242 days

Find the elapsed time. Use the table on page 645.

1. January 12, 1978, to April 1, 1978
2. July 1, 1979, through March 21, 1980
3. February 5, 1984, to September 23, 1984
4. January 9, 1988, through July 2, 1988
5. February 21, 2000, to September 30, 2000

618 ◆ Applications

Application J ◆ Fractional Parts of a Year

Changing Months Into a Part of a Year

To change months into a part of a year, write as a fraction with denominator 12. Express answers in lowest terms.

▶ 6 months is what part of a year?

$$6 \text{ months} = \frac{6}{12} \text{ year} = \frac{1}{2} \text{ year}$$

Change each term to a part of a year.

1. 9 months   2. 7 months   3. 12 months

4. 20 months   5. 36 months   6. 18 months

7. 21 months   8. 8 months   9. 14 months

Changing Days Into a Part of a Year

To change days into a part of a year, write as a fraction with denominator 365 when using the exact year or 360 when using the ordinary year. Express answers in lowest terms.

▶ 240 days is what part of an exact year?

$$240 \text{ days} = \frac{240}{365} \text{ year} = \frac{48}{73} \text{ year}$$

Change each term to a part of a year.

10. 90 days as an exact year

11. 300 days as an ordinary year

12. 120 days as an exact year

13. 175 days as an exact year

14. 146 days as an ordinary year

15. 250 days as an exact year

Application K ◆ Chronological Expressions

COMMON CHRONOLOGICAL EXPRESSIONS (number of occurrences per year)		
	Weekly (52)	Biweekly (26)
Semimonthly (24)	Monthly (12)	Bimonthly (6)
	Quarterly (4)	
Semiannually (2)	Annually (1)	

Find the number of occurrences. Use the table at the left.

1. Weekly for 1 year

2. Biweekly for 2 years

3. Quarterly for 4 years

4. Semimonthly for 1 year

5. Bimonthly for 2 years

6. Annually for 25 years

7. Monthly for $\frac{1}{2}$ year

8. Weekly for 1.5 years

9. Semimonthly for $2\frac{1}{2}$ years

▶ Monthly payments for 30 years. How many payments in all?

Monthly: 12 times a year
In 30 years: $12 \times 30 = 360$ payments

The Complement of a Number

To find the complement of a percent or decimal less than 1, subtract the percent from 100% or the decimal from 1.

▶ What is the complement of 65%?

The complement of 65% = 100% − 65% = 35%

▶ What is the complement of 0.25?

The complement of 0.25 = 1 − 0.25 = 0.75

Find the complement.

1. 0.15	**2.** 0.72	**3.** 0.35
4. 0.50	**5.** 0.80	**6.** 0.34
7. 91%	**8.** 85%	**9.** 47%
10. 66%	**11.** 7%	**12.** 45%
13. 3.5%	**14.** 56.6%	**15.** 75.9%

Application M Reading Bar Graphs

To read a bar graph, find the bar that represents the information you seek. Trace an imaginary line from the top of the bar to the scale on the left. Read the value represented by the bar from the scale.

▶ In 1985, how much did it cost to paint the exterior of a nine-room house?

The bar graph shows that it cost about $800 to paint the exterior in 1985.

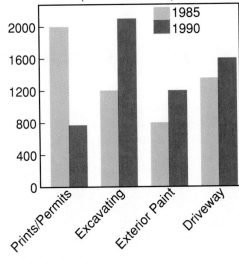

COMPARISON OF BUILDING COSTS
(nine-room house)

Answer the following. Use the bar graph shown.

1. In 1985, how much did it cost to construct the driveway?

2. In 1990, how much did it cost to obtain the prints and permits?

3. In 1990, how much did it cost to excavate?

4. What type of building costs decreased rather than increased?

5. In 1990, how much more did it cost to excavate than in 1985?

6. In 1985, how much more did it cost to obtain prints and permits than in 1990?

7. In 1990, how much more did it cost to paint the exterior of the house than in 1985?

Reading Line Graphs

To read a line graph, find the point that represents the information you seek. Trace an imaginary line from the point to the scale on the left. Read the value represented by the point from the scale.

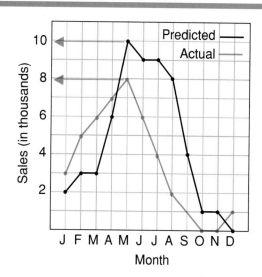

▶ What was the predicted and the actual amount of sales for the month of May?

The line graph shows that the predicted amount of sales was $10,000. The actual amount was $8000.

Answer the following. Use the line graph shown.

1. What was the predicted amount of sales for the month of August?

2. Which month(s) actually had the least amount of sales?

3. What was the actual amount of sales for the month of October?

4. In which month was the actual amount of sales equal to the amount predicted?

Reading Pictographs

To read a pictograph, find the scale to see how much each picture represents. Then multiply that amount times the number of pictures on each line to get the total amount.

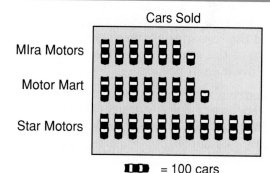

▶ How many cars did Motor Mart sell?

Each picture represents 100 cars sold. The Motor Mart row has 7.5 pictures.

$$7.5 \times 100 = 750 \text{ cars}$$

Answer the following. Use the pictograph shown.

1. How many cars did Mira Motors sell?

2. How many cars did Star Motor sell?

3. How many more cars did Star Motors sell than Motor Mart?

APPLICATIONS

Reading Circle Graphs

A circle graph is used to compare the parts to the whole. To find what part of the whole each section represents, multiply the amount, or percent, per section times the total amount.

▶ How much of the annual budget was spent on social services?

Social Services: 10¢ of the tax dollar

$$0.10 \times \$408,000 = \$40,800$$

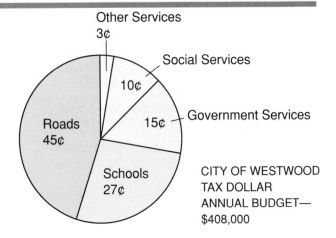

CITY OF WESTWOOD TAX DOLLAR ANNUAL BUDGET— $408,000

Answer the following. Use the circle graph shown.

1. How much of every tax dollar was spent to operate the schools?

2. How much more of the annual budget was spent on city roads than on schools?

3. How much of the annual budget was spent to operate the schools?

4. Does the city spend more on the schools than on all other services combined?

Constructing Circle Graphs

To construct a circle graph, find the percent of the total that each section represents. Multiply each percent times 360° to find the measure of the angle for each portion of the circle graph. Use a protractor to draw each portion.

▶ Construct that portion of a circle graph that represents the Bradford family spending 40¢ out of every dollar for food.

$$\frac{40}{100} = 40\%; \quad 0.40 \times 360° = 144°$$

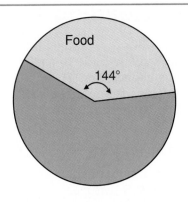

5. Complete the circle graph shown. Use the following information.

   The Bradford family budgets the rest of each dollar as follows: rent, 30¢; books, 15¢; clothing and personal items, 10¢; and all other expenses, 5¢.

6. Draw a circle graph. Use the following information.

   A factory budgets each dollar spent as follows: salaries, 35¢; raw materials, 30¢; utilities, 10¢; plant maintenance, 15¢; and research and development, 10¢.

Application **Q** Mean

To find the mean (average) of a group of numbers, find the sum of the group and divide it by the number of items in the group.

▶ What is the mean of the following: 454, 376, 416, 472?

$$\text{Mean} = \frac{454 + 376 + 416 + 472}{4}$$

$$= 429.5$$

Find the mean for each group.

1. 43, 19, 61, 72, 81, 50

2. 116, 147, 136, 151, 123, 117, 120

3. 4615, 5918, 7437, 8937

4. 4.0, 3.5, 4.0, 3.0, 3.5, 3.0, 2.0, 0.5, 4.6, 5.0

5. $580,000; $625,000; $105,583; $733,358; $5,750,000; $255,000; $600,000

Application **R** Median

To find the median of a group of numbers, arrange the items in order from smallest to largest. The median is the number in the middle. If there is an even number of items, find the mean of the two middle numbers.

▶ What is the median of the following: 454, 376, 416, 472?

Arrange in order: 376, <u>416</u>, <u>454</u>, 472

$$\text{Median} = \frac{416 + 454}{2} = 435$$

Find the median for each group.

1. 141, 136, 191, 187, 149, 148

2. 17, 21, 30, 35, 27, 25, 15

3. 91, 92, 85, 98, 100, 76, 80, 75

4. 4.2, 3.7, 3.1, 4.8, 2.4, 3.0, 2.9

5. 0.07, 0.05, 0.10, 0.12, 0.09, 0.17, 0.01

6. $121,500; $49,750; $72,175; $65,449

7. 2.06, 2.04, 2.00, 2.10, 2.08, 2.24, 1.55, 2.04, 2.13, 2.08, 2.09

Application **S** Mode

To find the mode of a group of numbers, look for the number that appears most often. A group may have no mode, or it may have more than one mode.

▶ What is the mode of the following: 92, 88, 76, 84, 92, 96, 1200?

The number that appears most often, the mode, is 92.

Find the mode for each group.

1. $51, $13, $24, $62, $55, $57, $24

2. 800, 600, 800, 500, 600, 700

3. 4.1, 4.7, 4.5, 4.3, 4.2, 4.4

4. 3, 3, 5, 3, 4, 2, 3, 2, 4, 4, 4

5. 0.01, 0.1, 0.01, 1.0, 0.01, 0.11, 1.00

6. $45, $63, $27, $91, $65, $8, $43, $90

The common units of length in the metric system are the millimeter, centimeter, meter, and the kilometer.

A millimeter(mm) is about the thickness of a U.S. dime.

A centimeter(cm) is about the thickness of a sugar cube.

A meter(m) is about the length of a baseball bat.

A kilometer(km) is 1000 m and is used to measure large distances.

Name the unit that is most commonly used to measure the object.

1. The length of a tennis court

2. The distance from Detroit to Montreal

3. The thickness of a magazine

4. The length of a standard paper clip

5. The length of your living room

6. The thickness of a soda straw

7. The length of a postage stamp

8. The width of your foot

Application **U** **Metric Mass and Volume**

The common units of mass and volume in the metric system are the gram, kilogram, metric ton, liter, and kiloliter.

A gram(g) is about the mass of a paper clip.

A kilogram(kg) is about the mass of a hammer.

A metric ton(t) is 1000 kg and is used to measure very heavy objects.

A liter(L) is about the amount of liquid in a can of motor oil.

A kiloliter(kL) is 1000 L and is used to measure large volumes.

Name the unit that is most commonly used to measure the object.

1. The mass of your best friend

2. The amount of gas in a car's tank

3. The mass of a postage stamp

4. The volume of helium in a blimp

5. The mass of an adult blue whale

6. The mass of a baseball

7. The volume of a large fuel tank

8. The mass of a dime

9. The mass of a gumdrop

10. The volume of air in a toy balloon

The distance around a shape is its
perimeter. To find the perimeter of a
shape, add the lengths of all its sides.

▶ What is the perimeter of the triangle at
the right?

Perimeter:
10 cm + 15 cm + 20 cm = 45 cm

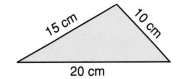

Find the perimeter.

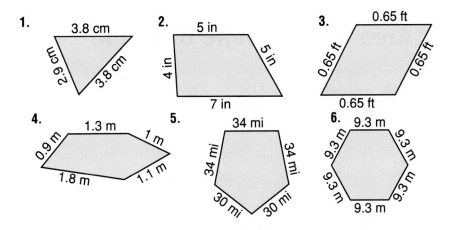

1. 3.8 cm, 2.6 cm, 3.8 cm

2. 5 in, 4 in, 5 in, 7 in

3. 0.65 ft, 0.65 ft, 0.65 ft, 0.65 ft

4. 0.9 m, 1.3 m, 1 m, 1.1 m, 1.8 m

5. 34 mi, 34 mi, 34 mi, 30 mi, 30 mi

6. 9.3 m, 9.3 m, 9.3 m, 9.3 m, 9.3 m, 9.3 m

The length of a diameter of a circle is
twice the length of its radius. The circumference is the distance around a
circle. The circumference is estimated
by multiplying 3.14 times the length of
a diameter.

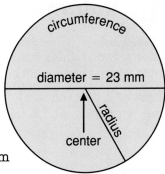

▶ The diameter of a circle is 23 mm.
What is its circumference?

Circumference = 3.14 × Diameter
Circumference = 3.14 × 23 mm = 72.22 mm

Find the circumference.

1. Diameter: 4 mi

2. Diameter: 3.5 cm

3. Diameter: 0.8 m

4. Radius: 35 mm

5. Radius: 0.75 ft

6. Radius: 13 in

7. Diameter: 18 ft

8. Radius: 120 km

9. Diameter: 12.5 mi

APPLICATIONS

Area of a Rectangle or Square

To find the area of a rectangle, multiply length times width. In a square, the length and width are equal.

▶ What is the area of a rectangular board that measures 3.5 m by 1.9 m?

$$\text{Area} = \text{Length} \times \text{Width}$$
$$\text{Area} = 3.5 \text{ m} \times 1.9 \text{ m} = 6.65 \text{ m}^2$$

Find the area.

1. Rectangle: 25 in long, 6 in wide
2. Square: 45.4 cm per side
3. Rectangle: 0.9 m long, 0.75 m wide
4. Square: $3\frac{1}{2}$ yd per side
5. Rectangle: $4\frac{2}{3}$ ft long, $2\frac{1}{6}$ ft wide

 Application Y

Area of a Triangle or Circle

To find the area of a triangle or circle, use the formulas below.

$$\text{Area of Triangle} = \frac{1}{2} \times \text{Base} \times \text{Height}$$
$$\text{Area of Circle} = 3.14 \times \text{Radius}^2$$

▶ What is the area of a circle with a radius of 27 m?

$$\text{Area of Circle} = 3.14 \times \text{Radius}^2$$
$$\text{Area of Circle} = 3.14 \times 27^2$$
$$\text{Area of Circle} = 2289.06 \text{ m}^2$$

Find the area.

1. Triangle: base is 3 in, height is 9 in
2. Circle: radius is 5 cm
3. Triangle: base is 16 m, height is 25 m
4. Triangle: base is 132 m, height is 9.6 m
5. Circle: radius is $2\frac{1}{2}$ yd
6. Triangle: base is $9\frac{1}{3}$ ft, height is $3\frac{1}{4}$ ft
7. Circle: radius is 86 mm

 Application Z

Volume of a Rectangular Solid

To find the volume of a rectangular solid, multiply length times width times height.

▶ What is the volume of a rectangular box that is 6 m long, 4 m wide, and 2 m deep?

$$\text{Volume} = \text{Length} \times \text{Width} \times \text{Height}$$
$$\text{Volume} = 6 \text{ m} \times 4 \text{ m} \times 2 \text{ m}$$
$$\text{Volume} = 48 \text{ m}^3$$

Finds the volume.

1. 36 m long, 44 m wide, 328 m high
2. 1.6 m long, 0.55 m wide, 0.8 m high
3. 11 cm long, 12 cm wide, 3 cm high
4. 68 in long, 18 in wide, 18 in high
5. $12\frac{1}{3}$ yd long, $4\frac{1}{4}$ yd wide, 3 yd high
6. $7\frac{1}{2}$ ft long, $3\frac{1}{2}$ ft wide, $5\frac{5}{8}$ ft high

TERMS

accumulated depreciation (p. 498) The total depreciation to date.

acid-test ratio (p. 522) The ratio of total assets minus inventory to total liabilities.

age group (p. 246) A category to which an automobile is assigned by an insurance company depending on its age. Age group affects the base premium.

air freight (p. 460) A way of shipping goods to customers.

amount (p. 157) The balance in a savings account at the end of an interest period, computed by adding the principal and interest.

amount financed (p. 220) The portion of an item's purchase price that you still owe after making the down payment.

amount of change (p. 525) The difference between the base figure and the corresponding figure on the current income statement.

annual expenses (p. 318) Amounts for real estate taxes, insurance payments, contributions, and so on, that occur once a year.

annual interest rate (p. 155) The percent of the principal that you earn as interest based on one year.

annual percentage rate (APR) (p. 222) An index used to compare the relative cost of borrowing money.

annual premium (p. 246) The amount you pay each year for insurance coverage.

annual yield (p. 300) The total interest or return for one year expressed as a percent of the principal or amount invested.

apportion (p. 494) To distribute among several departments.

assessed value (p. 275) The value of property for tax purposes, found by multiplying the market value by the rate of assessment.

assets (p. 512) The total of your cash, the items that you have purchased, and any money that your customers owe you.

average-cost method (p. 455) A way of calculating the value of an inventory based on the average cost of the goods you received.

average daily balance (p. 206) The average of the balances in a charge account at the end of each day in the billing period.

balance (p. 130) The amount of money in a bank account.

balance sheet (p. 514) A statement that shows your total assets, total liabilities, and owner's equity.

bar graph (p. 38) A picture that graphically displays and compares numerical facts in the form of vertical or horizontal bars.

base figure (p. 525) The dollar amount on the earlier income statement when you are computing percent change.

TERMS

base premium (p. 246) The portion of your annual automobile insurance premium determined by the amount of coverage you want, the age group of your car, and the insurance-rating group.

base price (p. 240) The price of the engine, chassis, and any other piece of standard equipment for a new automobile.

basic monthly charge (p. 476) The regular monthly charge for a utility before any additions are made for extra services.

basic plan (p. 290) A health insurance program that includes only hospital and surgical-medical insurance. The basic plan does not include major medical insurance.

basic rate (p. 462) The charge, usually per 100 pounds, to ship goods by truck. It depends on the type of goods, the weight, and the distance to be shipped.

beneficiary (p. 294) The person named in a policy to receive the insurance benefits if the insured dies.

bodily injury insurance (p. 246) Insurance that protects you against financial loss if your automobile causes personal injury to someone.

bond (pp. 308, 534) A promise to pay a specified amount of money, with interest, on a certain date for lending money to a corporation or government.

book value (p. 498) The original cost of an item minus the accumulated depreciation.

break-even analysis (p. 366) An investigation of or by a manufacturing business to find out how many units of a product must be made and sold to cover production expenses.

break-even point (p. 366) The point at which income from sales equals the cost of production.

budget (p. 562) A financial plan that enables a business, a governmental agency, or an individual to identify what sources are expected to produce revenue (earn money) and what amounts are allocated to various departments or categories for expenses.

budget sheet (p. 318) An outline of your total monthly expenses, which can help you plan future expenditures and savings.

canceled check (p. 133) A check that the bank has paid by deducting money from your account.

capital (p. 512) The difference between your assets and liabilities.

cash discount (p. 394) A discount from a supplier to encourage prompt payment of a bill.

cash value (p. 296) The amount of money you will receive if you cancel a whole life insurance policy.

catalog price (p. 384) The price at which a business generally sells an item to a consumer.

certificate of deposit (p. 298) A written record of a special kind of savings account that earns higher interest than a regular account. You must leave the money on deposit for a specified time or be penalized for early withdrawal.

chain discount (p. 392) A series of trade discounts on an item.

check (p. 128) A written order directing the bank to deduct money from a checking account to make a payment.

check register (p. 130) A record of the deposits made into a checking account, electronic transfers, and of the checks written.

closed-end lease (p. 253) A type of automobile lease where you make a specified number of payments, return the car at the end of the lease period, and owe nothing unless you damaged the car or exceeded the mileage limit.

closing costs (p. 270) Fees paid at the time documents are signed transferring ownership of a home.

coinsurance clause (p. 292) A clause in your major medical policy stating the percent of the bill you must pay after subtracting the deductible.

collision insurance (p. 246) Insurance to pay for repairs to your automobile if it is involved in an accident.

collision waiver (p. 255) An agreement by which you can obtain complete insurance coverage on a rented automobile by paying an additional charge per day.

commercial loan (p. 536) A business loan.

commercial paper (p. 540) A 30- to 270- day loan issued by a company with a very high credit rating.

commission (p. 89) An amount of money that you are paid for selling a product or service, based on the dollar value or number of sales.

commission rate (p. 89) A specified amount of money that you are paid for each sale or a percent of the total value of your sales.

complement method (p. 386) A way to find the net price when a trade discount is given.

compound interest (p. 157) Interest earned not only on the original principal but also on the interest earned during previous interest periods.

comprehensive insurance (p. 246) Insurance that protects you from loss of or damage to your automobile caused by fire, vandalism, theft, and so on.

comprehensive plan (p. 290) A health insurance program that includes hospital, surgical-medical, and major medical insurance.

consultant (p. 481) A person who gives technical or expert advice.

consultant's fee (p. 481) The payment to a consultant for professional services.

consumer price index (CPI) (p. 559) A measure of the average change in prices of a certain number of goods and services.

TERMS

TERMS

corporate income taxes (p. 532) Federal and state taxes on the earnings of a corporation.

corporation (p. 532) A business made up of a number of owners, called stockholders.

cost (p. 404) The amount that a business pays for a product, including expenses such as freight charges.

cost of goods sold (p. 517) The value of the beginning inventory plus the cost of any goods received minus the value of the ending inventory.

cost-of-living adjustment (p. 344) A salary raise to help an employee keep up with inflation.

coupon (p. 177) A certificate or ticket that offers the customer a special discount on an item or service.

current ratio (p. 522) The ratio of total assets to total liabilities.

dealer's cost (p. 242) The price that the automobile dealer pays for a new car.

deductible clause (p. 246) A clause in an insurance policy stating that the insured must pay a portion of any repair bill, or a portion of the amount not covered by hospital or surgical-medical insurance.

defective (p. 368) Incorrect, broken, or damaged.

demand charge (p. 478) The charge on an electric bill based on the peak load during the month.

deposit (pp. 126, 146) An amount of money put into a bank account.

depreciation (pp. 250, 496) A decrease in the value of an item because of its age or condition.

destination charge (p. 240) The cost of shipping a new automobile from the factory to the dealer.

direct labor cost (p. 364) The wages paid to the employees who manufacture an item.

direct material cost (p. 364) The cost of the goods that you use to manufacture an item.

disability insurance (p. 348) Pays benefits to an individual who must miss work due to an illness or injury.

discount 1. (pp. 179, 418) The difference between the regular selling price of an item and its sale price. 2. (p. 538) The difference between the face value and maturity value of a bond or United States Treasury Bill.

discount rate (p. 179) The discount on an item expressed as a percent of its regular selling price.

dividends (p. 304) Return on an investment in the stock of a corporation.

double-declining-balance method (p. 502) A method of computing depreciation that allows the highest possible depreciation in the first year of an item's life.

double time (p. 80) An overtime rate that is 2 times the regular hourly rate.

down payment (p. 220) The portion of an item's purchase price that you pay at the time of buying when you purchase the item with an installment loan.

driver-rating factor (p. 246) A number assigned by an insurance company to each driver of an automobile to reflect the driver's age, marital status, and so on. A higher driver-rating factor results in a higher annual premium.

emergency fund (p. 321) Money set aside for unpredictable expenses, such as medical bills or repair bills.

employee benefits (p. 346) Benefits an employee receives in addition to salary, such as health insurance, pension, or paid vacation.

end-of-month dating (p. 396) Terms such as "6/15 EOM" on an invoice. In this case, a 6% cash discount is offered if a bill is paid within 15 days after the end of the month in which the invoice is issued.

energy charge (p. 478) The charge on an electric bill based on the total number of kilowatt-hours used during the month.

escrow account (p. 280) An account often required by the lender of a mortgage loan to guarantee payment of the homeowner's real estate taxes and fire insurance premium.

estimated life (p. 496) The length of time, usually in years, that an item is expected to last.

exact interest (p. 218) Interest calculated by basing the length of time of a loan on a 365-day year.

expenditures (p. 316) Amounts of money paid out.

expense summary (p. 321) A totaling of actual monthly expenses used to compare the amounts spent to the amounts budgeted.

face value (pp. 294, 308, 538) The amount of money printed on an insurance policy, bond, or United States Treasury Bill.

factor (p. 434) A company's present market share.

factor method (p. 434) A way to project sales by using a company's present market share and the total estimated sales for the market in the future.

FIFO (first in, first out) method (p. 457) A method of valuing inventory which assumes that the first items received are the first items sold.

final payment (p. 228) The previous balance plus the current month's interest.

finance charge (p. 197) Interest that is charged to a customer for delaying payment of a bill.

fire protection class (p. 279) A number assigned to a home by an insurance company according to the quality of fire protection available in the area.

fixed costs 1. (p. 250) The costs of operating and maintaining an automobile that remain about the same regardless of the number of miles driven. 2. (p. 366) The costs of manufacturing that remain the same regardless of the number of units produced.

fixed expenses (p. 318) Amounts for rent or mortgage payments, installment payments, savings deposits, and so on, that do not vary from one month to the next.

fuel adjustment charge (p. 478) An additional charge on an electric bill to help cover increases in the cost of fuel needed to produce electricity.

graduated commission (p. 92) A method of payment in which the commission rate changes for each of several levels of sales, increasing as sales increase.

graduated income tax (p. 105) An income tax that involves a different tax rate for each of several levels of income.

gross national product (GNP) (p. 557) A measure of a nation's economic performance. It is the estimated total value of all goods and services produced by a nation during a year.

gross profit (p. 404) The difference between the selling price and the cost of a product when the selling price is greater than the cost.

group insurance (p. 109) Insurance offered to members of a group, such as the employees of a business, at a lower cost than individual insurance.

growth expenses (p. 542) Expenses from expanding a business, such as construction expenses, legal fees, or consultant's fees.

health insurance (p. 290) Insurance coverage against overwhelming medical expenses.

homeowner's insurance (p. 277) Insurance to protect yourself and your home from losses due to fire, theft of contents, or personal liability.

horizontal analysis (p. 525) The comparison of two or more income statements for different periods.

hospital insurance (p. 292) Insurance that pays most of the cost of hospitalization, including a semiprivate room and most laboratory tests.

hourly rate (p. 78) The amount of money earned per hour.

income statement (p. 519) A statement that shows in detail the sales, operating expenses, and net profit or loss of a business.

income tax (p. 100) A tax on the earnings, or income, of a person or business.

inflation (p. 554) An economic condition during which there are price increases in the cost of goods and services.

inflation rate (p. 554) Found by subtracting the original price from the current price and dividing the result by the original price.

installment loan (p. 220) A loan that is repaid in several equal payments over a specified period of time.

insurance-rating group (p. 246) A category to which an insurance company assigns an automobile depending on its size and value. The insurance-rating group affects the base premium.

Interest (pp. 155, 268) The amount of money paid for the use of money.

inventory (p. 452) The items that you have in stock.

inventory card (p. 452) A record of each item in stock.

invoice (p. 394) A written record of a purchase, listing the quantities and costs of the items purchased.

issues (p. 452) The number of outgoing items that are subtracted from the previous inventory.

kilowatt (kW) (p. 478) A unit of power equal to 1000 watts.

kilowatt-hour (kW·h) (p. 478) A unit to measure the consumption of electrical energy. It is equivalent to using one kilowatt of power constantly for one hour.

labor charge (p. 472) The cost of paying the people who do a job.

lease (p. 470) To agree to specific payments and other conditions in order to use property belonging to someone else for a period of time.

liabilities (p. 512) The total amount of money that you owe to creditors.

liability insurance (p. 246) Insurance that protects you against financial losses if your car is involved in an accident. Liability insurance includes bodily injury insurance and property damage insurance.

life insurance (p. 294) Financial protection that you purchase for your dependents in case of your death.

LIFO (last in, first out) method (p. 457) A method of valuing inventory which assumes that the last items received are the first items shipped.

limited-payment life insurance (p. 296) Lifetime insurance protection for which you pay premiums only for a specified number of years or until you reach a certain age.

line graph (p. 38) A picture used to compare facts over a period of time. An excellent way to show trends (increases or decreases).

list price (p. 384) The price at which a business generally sells an item to a consumer.

living expenses (p. 318) Amounts spent for food, clothing, utility bills, and so on, that vary from month to month.

loan value (p. 296)   The amount of money the insurance company will lend you if you have a whole life insurance policy.

loss (p. 306)   The difference between the cost and selling price of an item when the cost is greater than the selling price.

loss-of-use coverage (p. 277)   Insurance coverage to pay for some of the expenses of living away from home while damage to your home is being repaired.

major medical insurance (p. 292)   Insurance that pays for medical costs not covered by hospital or surgical-medical insurance.

markdown (pp. 179, 418)   The difference between the regular selling price and the sale price of an item.

markdown rate (pp. 179, 418)   The markdown on an item expressed as a percent of its regular selling price.

market (p. 428)   The total number of people who might purchase the type of product being made.

market share (p. 430)   The portion of the market that purchases a specific product.

market value (p. 275)   The price at which a house or other article can be bought or sold.

markup (p. 404)   The difference between the selling price and the cost of a product when the selling price is greater.

markup rate (p. 406)   The markup expressed as a percent of the selling price or cost of an item.

materials charge (p. 472)   The cost of the materials used to complete a job.

maturity date (p. 538)   The date on which a note, bill, or bond is due and payable.

maturity value (pp. 218, 536)   The total amount you repay on a loan, including both the principal borrowed and the interest owed.

merge (p. 542)   To combine with another business to form a new business.

merit increase (p. 344)   A raise in your salary to reward you for the quality of your work.

mill (p. 275)   A unit often used to express real estate tax rates, equal to $1.00 in tax for each $1000 of assessed value.

modified accelerated cost recovery system (MACRS) (p. 504)   A method of computing depreciation that allows depreciation according to fixed percents over a set period of time.

mortgage loan (p. 266)   A loan to finance the purchase of a home whereby the lender has the right to sell the property if the loan payments are not made.

net income (p. 519)   The difference between the gross profit and the total operating expenses if the gross profit is greater than the operating expenses.

TERMS

net interest (p. 540) The actual interest earned from investing in commercial paper after any bank service fees have been deducted.

net loss (p. 521) The difference between the markup, or gross profit, and the overhead, or operating expenses, when the overhead is greater than the markup.

net pay (p. 111) The amount of pay you have left after all tax withholdings and personal deductions have been subtracted.

net price (p. 384) The price a business actually pays for an item after discounts have been subtracted.

net-price rate (p. 392) The percent a business pays for an item, found by multiplying the complements of the chain discounts.

net proceeds (p. 534) The amount that a business actually receives from the sale of stocks or bonds after subtracting all selling expenses.

net profit (pp. 408, 519) The difference between the markup, or gross profit, and the overhead, or operating expenses, when the markup is greater than the overhead.

net-profit rate (p. 410) Net profit expressed as a percent of the selling price of an item.

net sales (p. 519) The total sales minus returns.

net worth (p. 512) The difference between your assets and liabilities.

nongroup insurance (p. 290) Insurance for people not enrolled in a group.

open-end lease (p. 253) A type of automobile lease where, at the end of the lease, you can buy the car for its expected value.

operating expenses (p. 408) Expenses of running a business, such as wages of employees, rent, utility charges, and taxes.

opinion-research firm (p. 426) A business that specializes in product testing and opinion surveys and the interpretation of the results.

opinion survey (p. 426) A series of questions answered by a group of volunteers. The results should approximate the opinions of a much larger group of people.

option (p. 240) An extra on a new automobile for convenience, safety, or appearance, such as a radio or air-conditioning.

ordinary dating (p. 394) Terms such as "2/10, net 30" on an invoice. In this case, a 2% cash discount is offered if the bill is paid within 10 days of date of invoice; the net price must be paid within 30 days.

ordinary interest (p. 218) Interest calculated by basing the length of time of a loan on a 360-day year.

outstanding check/deposit (p. 136) A check or deposit that appears in your check register but has not reached the bank in time to be listed on your statement.

overdraw (p. 128) To write checks in excess of the amount of money in a checking account.

overhead (p. 408) Expenses of running a

TERMS

business, such as wages of employees, rent, utility charges, and taxes.

overtime pay (p. 80)   The amount of money you earn when you work more than your regular hours. The overtime rate may be $1\frac{1}{2}$ or 2 times the regular hourly rate.

owner's equity (p. 512)   The difference between your assets and liabilities.

packaging (p. 375)   Placing a product in a container for shipment.

passbook (p. 150)   A book given to you in which the bank records all the deposits and withdrawals for your account and any interest earned.

payroll register (p. 488)   A record of the gross income, deductions and net income of a company's employees.

peak load (p. 478)   The greatest number of kilowatts used by a consumer at one time during the month.

per capita GNP (p. 557)   An indicator of the nation's standard of living. It is the GNP distributed over the population.

percent change (p. 525)   The amount of change from an earlier income statement to the current one, divided by the dollar amount on the earlier statement.

periodic rate (p. 200)   The monthly finance charge rate.

personal exemptions (p. 103)   Subtractions made from your gross pay for yourself and every other person you support. They are used to determine your taxable wages.

personal liability and medical coverage (p. 277)   Insurance coverage to protect you from financial losses if someone is injured on your property.

piecework (p. 85)   Work for which you are paid according to the number of items you complete.

point (p. 271)   One percent of the mortgage loan. Points are included in the closing costs.

premium 1. (p. 279)   The amount paid for insurance coverage. 2. (p. 308)   Bonds sold for more than face value.

previous balance (p. 200)   The amount owed by a credit card user on the closing date of the last statement.

prime cost (p. 364)   The total of the direct material cost and the direct labor cost.

prime rate (p. 536)   The lowest rate of interest available on business loans at a given time.

principal (pp. 155, 218)   The amount of money on which interest is paid.

product test (p. 426)   An attempt to determine how well a product is going to sell by asking a group of people to try it.

profit (p. 366)   The amount you receive after the costs of selling or production have been met.

profit-and-loss statement (p. 519)   A statement

that shows in detail your sales, operating expenses, and net profit or loss.

profit margin (p. 404) The difference between the selling price and the cost of a product.

promissory note (p. 218) A type of single-payment loan. It is a written promise to pay a certain sum of money at a specific future date.

property damage insurance (p. 246) Insurance that protects you against financial loss if your automobile damages the property of other people.

quality control (p. 368) Periodic inspection of the products you manufacture.

quality control chart (p. 368) A chart to show the percent of defective products that is allowable in a manufacturing business.

quick ratio (p. 522) The ratio of total assets minus inventory to total liabilities.

quoted price (p. 308) The actual cost of a bond, usually a percent of the face value.

rate of assessment (p. 275) The percent used to calculate the assessed value of a home, given the market value.

real estate taxes (p. 275) Fees collected from the owners of property and used to maintain government operations.

real GNP (p. 557) The gross national product corrected for inflation.

rebate (p. 177) A special incentive to purchase a particular item where the consumer must mail to the manufacturer a rebate coupon along with the sales slip and the universal product code label from the item.

rebate schedule (p. 233) A table of percent refunds used to determine the part of the finance charge refunded when an installment loan is repaid before the final due date.

receipts (p. 452) The number of incoming items that are added to the previous inventory.

reconcile (p. 136) To obtain agreement between your check register and the bank statement.

recruiting costs (p. 342) Expenses of finding a new employee, such as advertising fees, interviewing costs, and hiring expenses.

refund (p. 177) A special incentive to purchase a particular item where the manufacturer or store coupon is redeemed at the time of purchase.

release time (p. 354) Time spent away from your job in training programs for which you are still paid your regular wages or salary.

rent (pp. 255, 470, 474) To pay a regular fee for the use of a car, equipment, or space in a building.

rent escalation clause (p. 471) A statement in a lease that the rent may increase by a certain percent after a certain period of time.

repayment schedule (p. 225) A schedule that

TERMS

TERMS

shows the distribution of interest and principal over the life of a loan.

replacement value (p. 277) The amount of money required to reconstruct a home if it is destroyed.

resale value (p. 496) The estimated trade-in or salvage value of an item at the end of its expected life.

residual value (p. 253) The expected value of an automobile at the end of the lease period.

retail price (p. 404) The price that a business expects a consumer to pay for a product.

salary (p. 87) A fixed amount of money that you earn on a regular basis.

salary scale (p. 344) A table of wages or salaries used to compare various jobs in a business.

sale price (p. 179) A purchase price that is lower than the regular selling price.

sales potential (p. 428) An estimate of the number of products that you might sell during a specified period of time.

sales projection (p. 432) An estimate of the dollar volume or unit sales expected to occur during a future time period.

sales quota (p. 93) The amount of sales expected of a sales representative for a given period of time.

sales receipt (pp. 170, 194) A sales slip or cash register tape that shows the selling price of the items purchased, any sales tax, and the total purchase price. If a credit card is used, the receipt also shows the name and signature of the purchaser and the account number.

sales tax (p. 168) A tax on the selling price of an item or service that you purchase.

sample (p. 428) A selected group of people, representative of a much greater number of people, who are asked to evaluate a new product.

selling price (p. 404) The price that a business expects a consumer to pay for a product.

service charge (p. 133) A charge by a bank for handling a checking account.

simple interest (p. 155) Interest paid only on the original principal.

simple interest installment loan (p. 222) A loan repaid with equal monthly payments where part of each payment is used to pay the interest on the unpaid balance of the loan and the remaining part is used to reduce the balance.

single equivalent discount (SED) (p. 392) One discount that is equal to a chain discount.

single-payment loan (p. 218) A loan that you repay with one payment after a specified period of time.

social security (p. 107) A federal law (FICA) to provide hospitalization insurance for people over 65, retirement income, survivor's benefits, and disability benefits. The money for this program is provided by a tax paid by employers and employees.

statement (pp. 133, 152, 197) A record of an account sent by a bank or other business listing all transactions that have been recorded during a specified period of time.

sticker price (p. 240) The total of the base price, options, and destination charge for a new automobile.

stockbroker (p. 302) Someone who buys and sells stocks for other people.

stock certificate (p. 302) A printed form that proves part ownership of a corporation.

stocks (pp. 302, 534) Shares of ownership in the corporation issuing the stock.

storage space (p. 450) A place in which to keep products until they are needed.

straight commission (p. 89) A method of payment in which the only pay you receive is a commission on the sales you make.

straight-line method (p. 496) A method of computing depreciation which assumes that the depreciation is the same from year to year.

straight-time pay (p. 78) The total amount of money you earn for a pay period at your regular hourly rate.

sum-of-the-years'-digits method (p. 500) A method of computing depreciation that assigns a fraction of the total depreciation to each year of an item's life, with more depreciation allowed in the early years.

surgical-medical insurance (p. 292) Insurance that pays your doctor's fee, up to a certain amount, for surgery.

taxable income (p. 532) The portion of the gross income of a business that remains after normal business expenses are deducted.

tax rate 1. (p. 275) Expressed in mills and used to determine real estate taxes. 2. (p. 168) Expressed as a percent and used to determine sales taxes.

term (p. 218) The amount of time for which a loan is granted.

term life insurance (p. 294) A policy written for a specified amount of time, such as five years, or to a specified age. The policy expires at the end of the time unless renewed.

time and a half (p. 80) An overtime rate $1\frac{1}{2}$ times the regular hourly rate.

time study (p. 371) Determining how long a particular job should take.

trade discount (p. 384) A discount that a business receives when purchasing goods from a supplier.

trade-discount rate (p. 388) The trade discount on an item expressed as a percent of the list price.

travel expenses (p. 352) Money spent by an employee for expenses, such as transportation, lodging, and meals, while traveling on business.

TERMS

TERMS

underwriting commission   (p. 534)   The amount of money paid to the investment banker who helps a business distribute stocks or bonds.

unit   (p. 499)   One unit of insurance has a face value of $1000.

United States Treasury Bill   (p. 538)   A written promise by the federal government to repay the purchase price of the bill plus interest at a future date.

unit price   (p. 173)   The cost of an item expressed per unit of measure or count.

units-of-production method   (p. 499)   A method of computing depreciation based on the number of units produced.

universal life insurance   (p. 296)   Lifetime insurance protection, which is a combination of life insurance and a savings plan. You pay a minimum premium, with anything over the minimum going into a savings account.

unpaid balance   (p. 203)   The portion of the balance on your previous charge account or credit card statement that you have not yet paid.

used-car guide   (p. 244)   A monthly publication that gives price information on used cars purchased from dealers during the previous month.

utilities costs   (p. 281, 476)   Charges for public services, such as electricity or water.

variable costs   **1.** (p. 250)   The costs of operating and maintaining an automobile, which increase as the number of miles driven increases. **2.** (p. 366)   Manufacturing costs that increase as the number of units produced increases.

wage scale   (p. 344)   A table of wages or salaries used to compare various jobs in a business.

warehouse   (p. 450)   A building or part of one used to store raw materials or products until they are needed.

weekly time card   (p. 82)   A record of the times you report for work and the times you leave.

whole life insurance   (p. 296)   Life insurance that offers financial protection for your entire life.

withdrawal   (p. 148)   An amount of money taken out of a savings account.

withholding allowance   (p. 100)   A limitation on the amount of federal income tax withheld from your pay. You may claim one allowance for yourself, one for your spouse if married, and one for every other person you support.

TABLE OF MEASURES

TIME

60 seconds (s) = 1 minute (min)
60 minutes = 1 hour (h)
24 hours = 1 day
7 days = 1 week

52 weeks = 1 year
12 months = 1 year
100 years = 1 century

METRIC SYSTEM

Length

10 millimeters (mm) = 1 centimeter (cm)
100 centimeters = 1 meter (m)
1000 meters = 1 kilometer (km)

Volume

1000 milliliters (mL) = 1 liter (L)
1000 cubic centimeters (cm^3) = 1 liter
10 milliliters = 1 centiliter (cL)
10 deciliters (dL) = 1 liter

Area

100 square millimeters (mm^2) = 1 square centimeter (cm^2)
10,000 square centimeters = 1 square meter (m^2)
10,000 square meters = 1 hectare (ha)

Mass

1000 milligrams (mg) = 1 gram (g)
1000 grams = 1 kilogram (kg)
1000 kilograms = 1 metric ton (t)

CUSTOMARY SYSTEM

Length

12 inches (in) = 1 foot (ft)
3 feet = 1 yard (yd)
5280 feet = 1 mile (mi)

Volume

8 fluid ounces (oz) = 1 cup (c)
2 cups = 1 pint (pt)
2 pints = 1 quart (qt)
4 quarts = 1 gallon (gal)

Area

144 square inches (in^2) = 1 square foot (ft^2)
9 square feet = 1 square yard (yd^2)
4837 square yards = 1 acre (A)

Weight

16 ounces = 1 pound (lb)
2000 pounds = 1 ton (t)

SINGLE Persons—WEEKLY Payroll Period
(For Wages Paid After December)

And the wages are—		And the number of withholding allowances claimed is—										
At least	But Less than	0	1	2	3	4	5	6	7	8	9	10
		The amount of income tax to be withheld shall be—										
$0	$25	$0	$0	$0	$0	$0	$0	$0	$0	$0	$0	$0
25	30	1	0	0	0	0	0	0	0	0	0	0
30	35	1	0	0	0	0	0	0	0	0	0	0
35	40	2	0	0	0	0	0	0	0	0	0	0
40	45	3	0	0	0	0	0	0	0	0	0	0
45	50	4	0	0	0	0	0	0	0	0	0	0
50	55	4	0	0	0	0	0	0	0	0	0	0
55	60	5	0	0	0	0	0	0	0	0	0	0
60	65	6	0	0	0	0	0	0	0	0	0	0
65	70	7	0	0	0	0	0	0	0	0	0	0
70	75	7	1	0	0	0	0	0	0	0	0	0
75	80	8	2	0	0	0	0	0	0	0	0	0
80	85	9	3	0	0	0	0	0	0	0	0	0
85	90	10	3	0	0	0	0	0	0	0	0	0
90	95	10	4	0	0	0	0	0	0	0	0	0
95	100	11	5	0	0	0	0	0	0	0	0	0
100	105	12	6	0	0	0	0	0	0	0	0	0
105	110	13	6	0	0	0	0	0	0	0	0	0
110	115	13	7	1	0	0	0	0	0	0	0	0
115	120	14	8	2	0	0	0	0	0	0	0	0
120	125	15	9	2	0	0	0	0	0	0	0	0
125	130	16	9	3	0	0	0	0	0	0	0	0
130	135	16	10	4	0	0	0	0	0	0	0	0
135	140	17	11	5	0	0	0	0	0	0	0	0
140	145	18	12	5	0	0	0	0	0	0	0	0
145	150	19	12	6	0	0	0	0	0	0	0	0
150	155	19	13	7	1	0	0	0	0	0	0	0
155	160	20	14	8	1	0	0	0	0	0	0	0
160	165	21	15	8	2	0	0	0	0	0	0	0
165	170	22	15	9	3	0	0	0	0	0	0	0
170	175	22	16	10	4	0	0	0	0	0	0	0
175	180	23	17	11	4	0	0	0	0	0	0	0
180	185	24	18	11	5	0	0	0	0	0	0	0
185	190	25	18	12	6	0	0	0	0	0	0	0
190	195	25	19	13	7	0	0	0	0	0	0	0
195	200	26	20	14	7	1	0	0	0	0	0	0
200	210	27	21	15	9	2	0	0	0	0	0	0
210	220	29	22	16	10	4	0	0	0	0	0	0
220	230	30	24	18	12	5	0	0	0	0	0	0
230	240	32	25	19	13	7	1	0	0	0	0	0
240	250	33	27	21	15	8	2	0	0	0	0	0
250	260	35	28	22	16	10	4	0	0	0	0	0
260	270	36	30	24	18	11	5	0	0	0	0	0
270	280	38	31	25	19	13	7	0	0	0	0	0
280	290	39	33	27	21	14	8	2	0	0	0	0
290	300	41	34	28	22	16	10	3	0	0	0	0
300	310	42	36	30	24	17	11	5	0	0	0	0
310	320	44	37	31	25	19	13	6	0	0	0	0
320	330	45	39	33	27	20	14	8	2	0	0	0
330	340	47	40	34	28	22	16	9	3	0	0	0
340	350	48	42	36	30	23	17	11	5	0	0	0
350	360	50	43	37	31	25	19	12	6	0	0	0
360	370	51	45	39	33	26	20	14	8	2	0	0
370	380	53	46	40	34	28	22	15	9	3	0	0
380	390	54	48	42	36	29	23	17	11	5	0	0
390	400	56	49	43	37	31	25	18	12	6	0	0
400	410	57	51	45	39	32	26	20	14	8	1	0
410	420	59	52	46	40	34	28	21	15	9	3	0
420	430	61	54	48	42	35	29	23	17	11	4	0
430	440	64	55	49	43	37	31	24	18	12	6	0
440	450	67	57	51	45	38	32	26	20	14	7	1
450	460	70	58	52	46	40	34	27	21	15	9	3
460	470	73	61	54	48	41	35	29	23	17	10	4
470	480	75	64	55	49	43	37	30	24	18	12	6
480	490	78	67	57	51	44	38	32	26	20	13	7
490	500	81	69	58	52	46	40	33	27	21	15	9
500	510	84	72	61	54	47	41	35	29	23	16	10
510	520	87	75	63	55	49	43	36	30	24	18	12
520	530	89	78	66	57	50	44	38	32	26	19	13
530	540	92	81	69	58	52	46	39	33	27	21	15

MARRIED Persons—WEEKLY Payroll Period
(For Wages Paid After December)

And the wages are—		And the number of withholding allowances claimed is—										
At least	But Less than	0	1	2	3	4	5	6	7	8	9	10
		The amount of income tax to be withheld shall be—										
$0	$70	$0	$0	$0	$0	$0	$0	$0	$0	$0	$0	$0
70	75	1	0	0	0	0	0	0	0	0	0	0
75	80	1	0	0	0	0	0	0	0	0	0	0
80	85	2	0	0	0	0	0	0	0	0	0	0
85	90	3	0	0	0	0	0	0	0	0	0	0
90	95	4	0	0	0	0	0	0	0	0	0	0
95	100	4	0	0	0	0	0	0	0	0	0	0
100	105	5	0	0	0	0	0	0	0	0	0	0
105	110	6	0	0	0	0	0	0	0	0	0	0
110	115	7	0	0	0	0	0	0	0	0	0	0
115	120	7	1	0	0	0	0	0	0	0	0	0
120	125	8	2	0	0	0	0	0	0	0	0	0
125	130	9	3	0	0	0	0	0	0	0	0	0
130	135	10	3	0	0	0	0	0	0	0	0	0
135	140	10	4	0	0	0	0	0	0	0	0	0
140	145	11	5	0	0	0	0	0	0	0	0	0
145	150	12	6	0	0	0	0	0	0	0	0	0
150	155	13	6	0	0	0	0	0	0	0	0	0
155	160	13	7	1	0	0	0	0	0	0	0	0
160	165	14	8	2	0	0	0	0	0	0	0	0
165	170	15	9	2	0	0	0	0	0	0	0	0
170	175	16	9	3	0	0	0	0	0	0	0	0
175	180	16	10	4	0	0	0	0	0	0	0	0
180	185	17	11	5	0	0	0	0	0	0	0	0
185	190	18	12	5	0	0	0	0	0	0	0	0
190	195	19	12	6	0	0	0	0	0	0	0	0
195	200	19	13	7	1	0	0	0	0	0	0	0
200	210	21	14	8	2	0	0	0	0	0	0	0
210	220	22	16	10	3	0	0	0	0	0	0	0
220	230	24	17	11	5	0	0	0	0	0	0	0
230	240	25	19	13	6	0	0	0	0	0	0	0
240	250	27	20	14	8	2	0	0	0	0	0	0
250	260	28	22	16	9	3	0	0	0	0	0	0
260	270	30	23	17	11	5	0	0	0	0	0	0
270	280	31	25	19	12	6	0	0	0	0	0	0
280	290	33	26	20	14	8	2	0	0	0	0	0
290	300	34	28	22	15	9	3	0	0	0	0	0
300	310	36	29	23	17	11	5	0	0	0	0	0
310	320	37	31	25	18	12	6	0	0	0	0	0
320	330	39	32	26	20	14	8	1	0	0	0	0
330	340	40	34	28	21	15	9	3	0	0	0	0
340	350	42	35	29	23	17	11	4	0	0	0	0
350	360	43	37	31	24	18	12	6	1	0	0	0
360	370	45	38	32	26	20	14	7	1	0	0	0
370	380	46	40	34	27	21	15	9	3	0	0	0
380	390	48	41	35	29	23	17	10	4	0	0	0
390	400	49	43	37	30	24	18	12	6	0	0	0
400	410	51	44	38	32	26	20	13	7	1	0	0
410	420	52	46	40	33	27	21	15	9	2	0	0
420	430	54	47	41	35	29	23	16	10	4	0	0
430	440	55	49	43	36	30	24	18	12	5	0	0
440	450	57	50	44	38	32	26	19	13	7	1	0
450	460	58	52	46	39	33	27	21	15	8	2	0
460	470	60	53	47	41	35	29	22	16	10	4	0
470	480	61	55	49	42	36	30	24	18	11	5	0
480	490	63	56	50	44	38	32	25	19	13	7	0
490	500	64	58	52	45	39	33	27	21	14	8	2
500	510	66	59	53	47	41	35	28	22	16	10	3
510	520	67	61	55	48	42	36	30	24	17	11	5
520	530	69	62	56	50	44	38	31	25	19	13	6
530	540	70	64	58	51	45	39	33	27	20	14	8
540	550	72	65	59	53	47	41	34	28	22	16	9
550	560	73	67	61	54	48	42	36	30	23	17	11
560	570	75	68	62	56	50	44	37	31	25	19	12
570	580	76	70	64	57	51	45	39	33	26	20	14
580	590	78	71	65	59	53	47	40	34	28	22	15
590	600	79	73	67	60	54	48	42	36	29	23	17
600	610	81	74	68	62	56	50	43	37	31	25	18
610	620	82	76	70	63	57	51	45	39	32	26	20
620	630	84	77	71	65	59	53	46	40	34	28	21

AMOUNT OF $1.00

TOTAL INTEREST PERIOD	INTEREST RATE PER PERIOD						
	1.2500%	1.3750%	1.5000%	2.7500%	2.8750%	3.0000%	3.1250%
1	1.01250	1.01375	1.01500	1.02749	1.02875	1.03000	1.03125
2	1.02515	1.02768	1.03022	1.05575	1.05832	1.06090	1.06347
3	1.03797	1.04182	1.04567	1.08478	1.08875	1.09272	1.09671
4	1.05094	1.05614	1.06136	1.11462	1.12005	1.12550	1.13098
5	1.06408	1.07066	1.07728	1.14527	1.15225	1.15927	1.16632
6	1.07738	1.08538	1.09344	1.17676	1.18538	1.19405	1.20277
7	1.09085	1.10031	1.10984	1.20912	1.21946	1.22987	1.24036
8	1.10448	1.11544	1.12649	1.24237	1.25452	1.26677	1.27912
9	1.11829	1.13078	1.14339	1.27654	1.29059	1.30477	1.31909
10	1.13227	1.14632	1.16054	1.31165	1.32769	1.34391	1.36031
11	1.14642	1.16209	1.17795	1.34772	1.36586	1.38423	1.40282
12	1.16075	1.17806	1.19562	1.38478	1.40513	1.42576	1.44666
13	1.17526	1.19426	1.21355	1.42286	1.44553	1.46853	1.49187
14	1.18995	1.21068	1.23175	1.46199	1.48709	1.51258	1.53849
15	1.20482	1.22733	1.25023	1.50219	1.52984	1.55796	1.58657
16	1.21988	1.24421	1.26898	1.54350	1.57382	1.60470	1.63615
17	1.23513	1.26132	1.28802	1.58595	1.61907	1.65284	1.68728
18	1.25057	1.27866	1.30734	1.62956	1.66562	1.70243	1.74000
19	1.26620	1.29624	1.32695	1.67438	1.71351	1.75350	1.79438
20	1.28203	1.31406	1.34685	1.72042	1.76277	1.80611	1.85045
21	1.29806	1.33213	1.36706	1.76773	1.81345	1.86029	1.90828
22	1.31428	1.35045	1.38756	1.81635	1.86559	1.91610	1.96791
23	1.33071	1.36902	1.40838	1.86630	1.91922	1.97358	2.02941
24	1.34735	1.38784	1.42950	1.91762	1.97440	2.03279	2.09283

AMOUNT OF $1.00 AT 5.5%, COMPOUNDED DAILY, 365-DAY YEAR

DAY	AMOUNT	DAY	AMOUNT	DAY	AMOUNT	DAY	AMOUNT	DAY	AMOUNT
1	1.00015	11	1.00165	21	1.00316	31	1.00468	50	1.00755
2	1.00030	12	1.00180	22	1.00331	32	1.00483	60	1.00907
3	1.00045	13	1.00196	23	1.00347	33	1.00498	70	1.01059
4	1.00060	14	1.00211	24	1.00362	34	1.00513	80	1.01212
5	1.00075	15	1.00226	25	1.00377	35	1.00528	90	1.01364
6	1.00090	16	1.00241	26	1.00392	36	1.00543	100	1.01517
7	1.00105	17	1.00256	27	1.00407	37	1.00558	110	1.01670
8	1.00120	18	1.00271	28	1.00422	38	1.00574	120	1.01823
9	1.00135	19	1.00286	29	1.00437	39	1.00589	130	1.01977
10	1.00150	20	1.00301	30	1.00452	40	1.00604	140	1.02131

ELAPSED TIME TABLE

THE NUMBER OF EACH DAY OF THE YEAR

Day No.	Jan.	Feb.	Mar.	Apr.	May	June	July	Aug.	Sept.	Oct.	Nov.	Dec.
1	1	32	60	91	121	152	182	213	244	274	305	335
2	2	33	61	92	122	153	183	214	245	275	306	336
3	3	34	62	93	123	154	184	215	246	276	307	337
4	4	35	63	94	124	155	185	216	247	277	308	338
5	5	36	64	95	125	156	186	217	248	278	309	339
6	6	37	65	96	126	157	187	218	249	279	310	340
7	7	38	66	97	127	158	188	219	250	280	311	341
8	8	39	67	98	128	159	189	220	251	281	312	342
9	9	40	68	99	129	160	190	221	252	282	313	343
10	10	41	69	100	130	161	191	222	253	283	314	344
11	11	42	70	101	131	162	192	223	254	284	315	345
12	12	43	71	102	132	163	193	224	255	285	316	346
13	13	44	72	103	133	164	194	225	256	286	317	347
14	14	45	73	104	134	165	195	226	257	287	318	348
15	15	46	74	105	135	166	196	227	258	288	319	349
16	16	47	75	106	136	167	197	228	259	289	320	350
17	17	48	76	107	137	168	198	229	260	290	321	351
18	18	49	77	108	138	169	199	230	261	291	322	352
19	19	50	78	109	139	170	200	231	262	292	323	353
20	20	51	79	110	140	171	201	232	263	293	324	354
21	21	52	80	111	141	172	202	233	264	294	325	355
22	22	53	81	112	142	173	203	234	265	295	326	356
23	23	54	82	113	143	174	204	235	266	296	327	357
24	24	55	83	114	144	175	205	236	267	297	328	358
25	25	56	84	115	145	176	206	237	268	298	329	359
26	26	57	85	116	146	177	207	238	269	299	330	360
27	27	58	86	117	147	178	208	239	270	300	331	361
28	28	59	87	118	148	179	209	240	271	301	332	362
29	29	. . .	88	119	149	180	210	241	272	302	333	363
30	30	. . .	89	120	150	181	211	242	273	303	334	364
31	31	. . .	90	. . .	151	. . .	212	243	. . .	304	. . .	365

Monthly Payments of $1,000 Loan

Rate (Percent)	Number of Years of Loan							
	5	10	15	20	25	30	35	40
7	$19.81	$11.62	$8.99	$7.76	$7.07	$6.66	$6.39	$6.22
7 1/2	20.04	11.88	9.28	8.06	7.39	7.00	6.75	6.59
8	20.28	12.14	9.56	8.37	7.72	7.34	7.11	6.96
8 1/2	20.52	12.40	9.85	8.68	8.06	7.69	7.47	7.34
9	20.76	12.67	10.15	9.00	8.40	8.05	7.84	7.72
9 1/2	21.01	12.94	10.45	9.33	8.74	8.41	8.22	8.11
10	21.25	13.22	10.75	9.66	9.09	8.78	8.60	8.50
10 1/2	21.50	13.50	11.06	9.99	9.45	9.15	8.99	8.89
11	21.75	13.78	11.37	10.33	9.81	9.53	9.37	9.29
11 1/2	22.00	14.06	11.69	10.67	10.17	9.91	9.77	9.69
12	22.25	14.35	12.01	11.02	10.54	10.29	10.16	10.09
12 1/2	22.50	14.64	12.33	11.37	10.91	10.68	10.56	10.49
13	22.76	14.94	12.66	11.72	11.28	11.07	10.96	10.90
13 1/2	23.01	15.23	12.99	12.08	11.66	11.46	11.36	11.31
14	23.27	15.53	13.32	12.44	12.04	11.85	11.76	11.72
14 1/2	23.53	15.83	13.66	12.80	12.43	12.25	12.17	12.13
15	23.79	16.14	14.00	13.17	12.81	12.65	12.57	12.54
15 1/2	24.06	16.45	14.34	13.54	13.20	13.05	12.98	12.95
16	24.32	16.76	14.69	13.92	13.59	13.45	13.39	13.36
17	24.86	17.38	15.40	14.67	14.38	14.26	14.21	14.19
18	25.40	18.02	16.11	15.44	15.18	15.08	15.03	15.02

ANNUAL PERCENTAGE RATE TABLE FOR MONTHLY PAYMENT PLANS

# of Pmts.	Annual Percentage Rate (Finance Charge per $100 of Amount Financed)									
▼	2.00%	2.25%	2.50%	2.75%	3.00%	3.25%	3.50%	3.75%	4.00%	4.25%
6	$0.58	$0.66	$0.73	$0.80	$0.88	$0.95	$1.02	$1.10	$1.17	$1.24
12	1.09	1.22	1.36	1.50	1.63	1.77	1.91	2.04	2.18	2.32
18	1.59	1.79	1.99	2.19	2.39	2.59	2.79	2.99	3.20	3.40
24	2.10	2.36	2.62	2.89	3.15	3.42	3.69	3.95	4.22	4.49
30	2.60	2.93	3.26	3.59	3.92	4.25	4.58	4.92	5.25	5.58
36	3.11	3.51	3.90	4.30	4.69	5.09	5.49	5.89	6.29	6.69
42	3.62	4.08	4.54	5.00	5.47	5.93	6.40	6.86	7.33	7.80
48	4.14	4.66	5.19	5.72	6.24	6.78	7.31	7.84	8.38	8.92
54	4.65	5.24	5.83	6.43	7.03	7.63	8.23	8.83	9.44	10.04
60	5.17	5.82	6.48	7.15	7.81	8.48	9.15	9.82	10.50	11.18
	4.50%	4.75%	5.00%	5.25%	5.50%	5.75%	6.00%	6.25%	6.50%	6.75%
6	$1.32	$1.39	$1.46	$1.54	$1.61	$1.68	$1.76	$1.83	$1.90	$1.98
12	2.45	2.59	2.73	2.87	3.00	3.14	3.28	3.42	3.56	3.69
18	3.60	3.80	4.00	4.21	4.41	4.61	4.82	5.02	5.22	5.43
24	4.75	5.02	5.29	5.56	5.83	6.10	6.37	6.64	6.91	7.18
30	5.92	6.25	6.59	6.92	7.26	7.60	7.94	8.28	8.61	8.96
36	7.09	7.49	7.90	8.30	8.71	9.11	9.52	9.93	10.34	10.75
42	8.27	8.74	9.91	9.69	10.16	10.64	11.12	11.60	12.08	12.56
48	9.46	10.00	10.54	11.09	11.63	12.18	12.73	13.28	13.83	14.39
54	10.65	11.26	11.88	12.49	13.11	13.73	14.36	14.98	15.61	16.23
60	11.86	12.54	13.23	13.92	14.61	15.30	16.00	16.70	17.40	18.10
	7.00%	7.25%	7.50%	7.75%	8.00%	8.25%	8.50%	8.75%	9.00%	9.25%
6	$2.05	$2.13	$2.20	$2.27	$2.35	$2.42	$2.49	$2.57	$2.64	$2.72
12	3.83	3.97	4.11	4.25	4.39	4.52	4.66	4.80	4.94	5.08
18	5.63	5.84	6.04	6.25	6.45	6.66	6.86	7.07	7.28	7.48
24	7.45	7.73	8.00	8.27	8.55	8.82	9.09	9.37	9.64	9.92
30	9.30	9.64	9.98	10.32	10.66	11.01	11.35	11.70	12.04	12.39
36	11.16	11.57	11.98	12.40	12.81	13.23	13.64	14.06	14.48	14.90
42	13.04	13.52	14.01	14.50	14.98	15.47	15.96	16.45	16.95	17.44
48	14.94	15.50	16.06	16.62	17.18	17.75	18.31	18.88	19.45	20.02
54	16.86	17.50	18.13	18.77	19.41	20.05	20.69	21.34	21.98	22.63
60	18.81	19.52	20.23	20.94	21.66	22.38	23.10	23.82	24.55	25.28
	9.50%	9.75%	10.00%	10.25%	10.50%	10.75%	11.00%	11.25%	11.50%	11.75%
6	$2.79	$2.86	$2.94	$3.01	$3.08	$3.16	$3.23	$3.31	$3.38	$3.45
12	5.22	5.36	5.50	5.64	5.78	5.92	6.06	6.20	6.34	6.48
18	7.69	7.90	8.10	8.31	8.52	8.73	8.93	9.14	9.35	9.56
24	10.19	10.47	10.75	11.02	11.30	11.58	11.86	12.14	12.42	12.70
30	12.74	13.09	13.43	13.78	14.13	14.48	14.83	15.19	15.54	15.89
36	15.32	15.74	16.16	16.58	17.01	17.43	17.86	18.29	18.71	19.14
42	17.94	18.43	18.93	19.43	19.93	20.43	20.93	21.44	21.94	22.45
48	20.59	21.16	21.74	22.32	22.90	23.48	24.06	24.64	25.23	25.81
54	23.28	23.94	24.59	25.25	25.91	26.57	27.23	27.90	28.56	29.23
60	26.01	26.75	27.48	28.22	28.96	29.71	30.45	31.20	31.96	32.71

ANNUAL PERCENTAGE RATE TABLE FOR MONTHLY PAYMENT PLANS

# of Pmts.	Annual Percentage Rate (Finance Charge per $100 of Amount Financed)									
▼	12.00%	12.25%	12.50%	12.75%	13.00%	13.25%	13.50%	13.75%	14.00%	14.25%
6	$3.53	$3.60	$3.68	$3.75	$3.83	$3.90	$3.97	$4.05	$4.12	$4.20
12	6.62	6.76	6.90	7.04	7.18	7.32	7.46	7.60	7.74	7.89
18	9.77	9.98	10.19	10.40	10.61	10.82	11.03	11.24	11.45	11.66
24	12.98	13.26	13.54	13.82	14.10	14.38	14.66	14.95	15.23	15.51
30	16.24	16.60	16.95	17.31	17.66	18.02	18.38	18.74	19.10	19.45
36	19.57	20.00	20.43	20.87	21.30	21.73	22.17	22.60	23.04	23.48
42	22.96	23.47	23.98	24.49	25.00	25.51	26.03	26.55	27.06	27.58
48	26.40	26.99	27.58	28.18	28.77	29.37	29.97	30.57	31.17	31.77
54	29.91	30.58	31.25	31.93	32.61	33.29	33.98	34.66	35.35	36.04
60	33.47	34.23	34.99	35.75	36.52	37.29	38.06	38.83	39.61	40.39
	14.50%	14.75%	15.00%	15.25%	15.50%	15.75%	16.00%	16.25%	16.50%	16.75%
6	$4.27	$4.35	$4.42	$4.49	$4.57	$4.64	$4.72	$4.79	$4.87	$4.94
12	8.03	8.17	8.31	8.45	8.59	8.74	8.88	9.02	9.16	9.30
18	11.87	12.08	12.29	12.50	12.72	12.93	13.14	13.35	13.57	13.78
24	15.80	16.08	16.37	16.65	16.94	17.22	17.51	17.80	18.09	18.37
30	19.81	20.17	20.54	20.90	21.26	21.62	21.99	22.35	22.72	23.08
36	23.92	24.35	24.80	25.24	25.68	26.12	26.57	27.01	27.46	27.90
42	28.10	28.62	29.15	29.67	30.19	30.72	31.25	31.78	32.31	32.84
48	32.37	32.98	33.59	34.20	34.81	35.42	36.03	36.65	37.27	37.88
54	36.73	37.42	38.12	38.82	39.52	40.22	40.92	41.63	42.33	43.04
60	41.17	41.95	42.74	43.53	44.32	45.11	45.91	46.71	47.51	48.31
	17.00%	17.25%	17.50%	17.75%	18.00%	18.25%	18.50%	18.75%	19.00%	19.25%	19.50%
6	$5.02	$5.09	$5.17	$5.24	$5.32	$5.39	$5.46	$5.54	$5.61	$5.69	$5.76
12	9.45	9.59	9.73	9.87	10.02	10.16	10.30	10.44	10.59	10.73	10.87
18	13.99	14.21	14.42	14.64	14.85	15.07	15.28	15.49	15.71	15.93	16.14
24	18.66	18.95	19.24	19.53	19.82	20.11	20.40	20.69	20.98	21.27	21.56
30	23.45	23.81	24.18	24.55	24.92	25.29	25.66	26.03	26.40	26.77	27.14
36	28.35	28.80	29.25	29.70	30.15	30.60	31.05	31.51	31.96	32.42	32.87
42	33.37	33.90	34.44	34.97	35.51	36.05	36.59	37.13	37.67	38.21	38.76
48	38.50	39.13	39.75	40.37	41.00	41.63	42.26	42.89	43.52	44.15	44.79
54	43.75	44.47	45.18	45.90	46.62	47.34	48.06	48.79	49.51	50.24	50.97
60	49.12	49.92	50.73	51.55	52.36	53.18	54.00	54.82	55.64	56.47	57.30
	19.75%	20.00%	20.25%	20.50%	20.75%	21.00%	21.25%	21.50%	21.75%	22.00%	22.25%
6	$5.84	$5.91	$5.99	$6.06	$6.14	$6.21	$6.29	$6.36	$6.44	$6.51	$6.59
12	11.02	11.16	11.31	11.45	11.59	11.74	11.88	12.02	12.17	12.31	12.46
18	16.36	16.57	16.79	17.01	17.22	17.44	17.66	17.88	18.09	18.31	18.53
24	21.86	22.15	22.44	22.74	23.03	23.33	23.62	23.92	24.21	24.51	24.80
30	27.52	27.89	28.26	28.64	29.01	29.39	29.77	30.14	30.52	30.90	31.28
36	33.33	33.79	34.25	34.71	35.17	35.63	36.09	36.56	37.02	37.49	37.95
42	39.30	39.85	40.40	40.95	41.50	42.05	42.60	43.15	43.71	44.26	44.82
48	45.43	46.07	46.71	47.35	47.99	48.64	49.28	49.93	50.58	51.23	51.88
54	51.70	52.44	53.17	53.91	54.65	55.39	56.14	56.88	57.63	58.38	59.13
60	58.13	58.96	59.80	60.64	61.48	62.32	63.17	64.01	64.86	65.71	66.57

AMOUNT PER $1.00 INVESTED, MONTHLY, QUARTERLY, AND DAILY COMPOUNDING

Annual Rate	Interest Period: 1 year			Interest Period: 4 year		
	Monthly	Quarterly	Daily	Monthly	Quarterly	Daily
7.00%	1.072290	1.071859	1.072500	1.322053	1.319929	1.323094
7.25%	1.074958	1.074495	1.075185	1.335261	1.332961	1.336389
7.50%	1.077632	1.077135	1.077875	1.348599	1.346114	1.349817
7.75%	1.080312	1.079781	1.080573	1.362066	1.359388	1.363380
8.00%	1.082999	1.082432	1.083277	1.375666	1.372785	1.377079
8.25%	1.085692	1.085087	1.085988	1.389398	1.386306	1.390916
8.50%	1.088390	1.087747	1.088706	1.403264	1.399951	1.404891
8.75%	1.091095	1.090413	1.091430	1.417266	1.413723	1.419008
9.00%	1.093806	1.093083	1.094162	1.431405	1.427621	1.433265
9.25%	1.096524	1.095758	1.096900	1.445682	1.441647	1.447666
9.50%	1.099247	1.098438	1.099645	1.460098	1.455803	1.462212
9.75%	1.101977	1.101123	1.102397	1.474655	1.470088	1.476903
10.00%	1.104713	1.103812	1.105155	1.489354	1.484505	1.491742
10.25%	1.107455	1.106507	1.107921	1.504196	1.499055	1.506731
10.50%	1.110203	1.109207	1.110693	1.519183	1.513738	1.521869
10.75%	1.112958	1.111911	1.113473	1.534317	1.528555	1.537160
11.00%	1.115718	1.114621	1.116259	1.549598	1.543509	1.552604
11.25%	1.118485	1.117335	1.119052	1.565028	1.558600	1.568203
11.50%	1.121259	1.120055	1.121853	1.580608	1.573829	1.583959
11.75%	1.124039	1.122779	1.124660	1.596340	1.589197	1.599873
12.00%	1.126825	1.125508	1.127474	1.612226	1.604706	1.615946
12.25%	1.129617	1.128243	1.130295	1.628266	1.620357	1.632182
12.50%	1.132416	1.130982	1.133124	1.644462	1.636151	1.648580
12.75%	1.135221	1.133726	1.135959	1.660817	1.652089	1.665142
13.00%	1.138032	1,136475	1.138802	1.677330	1.668172	1.681871
13.25%	1.140850	1.139230	1.141651	1.694004	1.684402	1.698768
13.50%	1.143674	1.141989	1.144508	1.710841	1.700780	1.715835
13.75%	1.146505	1.144753	1.147371	1.727841	1.717308	1.733073
14.00%	1.149342	1.147523	1.150242	1.745006	1.733986	1.750484
14.25%	1.152185	1.150297	1.153121	1.762339	1.750815	1.768070
14.50%	1.155035	1.153076	1.156006	1.779840	1.767798	1.785832
14.75%	1.157891	1.155861	1.158898	1.797511	1.784935	1.803773
15.00%	1.160754	1.158650	1.161798	1.815354	1.802227	1.821894

INDEX

INDEX

INDEX

INDEX

SELECTED ANSWERS

Workshop 1 **Self-Check** **1.** two hundred forty-one
2. seven and three hundred seventeen thousandths
3. seven hundred sixty-one and thirteen one hundredths dollars or seven hundred sixty-one and $\frac{13}{100}$ dollars, or seven hundred sixty-one dollars and thirteen cents **Problems** **5.** $967.46 **7.** $647.56
9. $65.75 **11.** $95.07 **13.** fifteen and forty three one hundredths of a dollar or fifteen and $\frac{43}{100}$ dollars or fifteen dollars and 43 cents **15.** three hundred and five tenths **17.** 420,000 **19.** 960,000 **21.** 312,000
23. 868,000 **25.** 41,500 **27.** 17,800 **29.** 7290
31. 21,950 **33.** 3 **35.** 43 **37.** 21.2 **39.** 4718.2
41. 5.22 **43.** 31.80 **45.** 16,000,000 **47.** 15,749,000
49. 15,750,000 **51.** $14.72 **53.** $72.20 **55.** $119.90
57. $7.00 **59.** $152.00 **61.** $624.00 **63.** $3150
65. $520.00 **67.** $19,900 **69.** $4100 **71.** **a.** Seven and $\frac{17}{100}$ dollars **b.** Nine and $\frac{75}{100}$ dollars **c.** Forty-two and $\frac{91}{100}$ dollars **d.** Fifty-one and $\frac{80}{100}$ dollars **e.** Two hundred seventeen and $\frac{23}{100}$ dollars **f.** Five hundred thirty-one and $\frac{49}{100}$ dollars **g.** Nine thousand seven hundred fifty-one and $\frac{63}{100}$ dollars **h.** One thousand two hundred seventy and $\frac{82}{100}$ dollars **73.** **a.** $618,000
b. $719,000 **c.** $272,000 **d.** $915,000 **e.** $535,000
f. $420,000 **g.** $178,000 **h.** $572,000 **i.** $423,000

Workshop 2 **Self-Check** **1.** 8891 **2.** 759.80 **3.** 3079
4. 56.633 **5.** 284.4 **6.** 0.06 **7.** 62.313 **Problems**
9. 554 **11.** 3943 **13.** 81.2 **15.** 0.4 **17.** 0.9
19. 2.244 **21.** $329.20 **23.** 29 miles to the gallon
25. 1.36, 1.37, 1.39 **27.** 7.18, 7.38, 7.58 **29.** 15.231,
15.270, 15.321 **31.** 121.012, 121.021, 121.210
33. 0.1234, 0.1342, 0.1423 **35.** 10.100, 11.01, 11.11
37. 3.062, 3.09, 3.1 **39.** **a.** 513.12, 519.03, 532.626,
571.113, 587.41 **b.** 11.7, 32.615, 34.9, 67.192, 94.79
c. 15.04, 18.7, 26.311, 46.94, 71.21 **d.** 22.5, 38.9,
48.275, 67.21, 93.047 **41.** **a.** 873.15 and 877.142
b. 332.749 and 333.54 **43.** **a.** $2.45 **b.** $2.95
c. $2.45 **d.** $2.95 **e.** $2.95 **f.** $3.95 **g.** $4.45
h. $1.95 **i.** $2.95 **j.** $3.45 **45.** **a.** 2, 1, 3 **b.** 3, 1, 2
c. 1, 3, 2 **d.** 2, 1, 3 **e.** 3, 1, 2 **f.** 1, 3, 2

Workshop 3 **Self-Check** **1.** 41.76 **2.** 96.14
3. 174.70 **4.** 198.65 **5.** 198.208 **6.** 43.124
7. $44.06 **8.** $111.12 **Problems** **9.** 88.01
11. 1328.11 **13.** 86.007 **15.** 11.018 **17.** 205.2
19. 15.688 **21.** 463.491 **23.** 966.678 **25.** 645.356

27. 994.7351 **29.** $51.76 **31.** $130.40 **33.** 124.03
35. 303.1416 **37.** $123.96 **39.** $289.17 **41.** $148.87
43. $571.63 **45.** Subtotal $41.55, Total $44.87
47. **a.** 39.4 **b.** 48.8 **c.** 77.6 **d.** 75.5 **e.** 70.5

Workshop 4 **Self-Check** **1.** 53.61 **2.** 38.3 **3.** 610.08
4. 68.76 **5.** $16.78 **6.** $912.06 **7.** $332.15
8. $39.05 **Problems** **9.** 63.3 **11.** 11.33 **13.** 39.30
15. 4.639 **17.** 25.303 **19.** 5.9551 **21.** 55.617
23. 720.168 **25.** 407.22 **27.** 2.741 **29.** $67.69
31. $665.90 **33.** $264.03 **35.** $16,681.07 **37.** 117.5
39. 4.5 **41.** 22.9279 **43.** $28.91 **45.** $5.37
47. $863.15 **49.** **a.** $0.06 **b.** $0.88 **c.** $0.03
d. $2.00 **e.** $0.00 **f.** $1.89 **g.** $0.10 **h.** $2.58
i. $1.00 **j.** $0.16 **51.** **a.** Subtotal $132.55
b. Total deposit $112.55 **53.** **a.** Subtotal $96.59
b. Total deposit $56.59 **55.** **a.** Subtotal $577.13
b. Total deposit $502.13 **57.** **a.** $250.33 **b.** $102.60
c. $892.85 **d.** $378.72 **e.** $60.33 **f.** $1020.42
g. $500.60 **h.** $380.25 **i.** $210.80 **j.** $8.48

Workshop 5 **Self-Check** **1.** 8.151 **2.** 89.452
3. 0.0858 **4.** 0.0592 **5.** $35.275 = $35.28
6. 714 **7.** 4186.1 **8.** 3179.4 **Problems** **9.** 8.26
11. 21.2 **13.** 3.0601 **15.** 30.318 **17.** 0.0372
19. 0.0082 **21.** 0.0306 **23.** 0.003175 **25.** 0.16684
27. 2.18673 **29.** 0.000142 **31.** 1.866214 **33.** $45
35. $20.475 = $20.48 **37.** $43.102 = $43.10
39. 317 **41.** 65.17 **43.** 3285 **45.** 398.7 **47.** 72,716
49. 415 **51.** 69,178 **53.** 947,189 **55.** **a.** $3.87
b. $7.96 **c.** $2.98 **d.** $9.98 **e.** $8.76 **f.** $9.27
g. $7.49 **h.** $50.31 **57.** **a.** $314; $24.02; $78.50;
$14.13; $116.65; $197.35 **b.** $306; $23.41; $76.50;
$13.77; $113.68; $192.32 **c.** $350; $26.78; $87.50;
$15.75; $130.03; $219.97 **d.** $339.50; $25.97; $84.88;
$15.28; 126.13; $213.37 **e.** $388; $29.68; $97;
$17.46; $144.14; $243.86 **f.** $1697.50; $129.86;
$424.38; $76.39; $630.63; $1066.87

Workshop 6 **Self-Check** **1.** 6.2 **2.** 4.29 **3.** 29.4
4. 36 **5.** $1.36 **6.** 0.79 **7.** 1.389 **8.** 9.8628
Problems **9.** 1.8 **11.** 9.3 **13.** 7.4 **15.** 7.7 **17.** 8.5
19. 2.8 **21.** 73.2 **23.** 5.2 **25.** 1.83 **27.** 12.00
29. 53.39 **31.** 0.56 **33.** $2.66 **35.** $8.70
37. $4.31 **39.** $10.20 **41.** 0.93 **43.** 72.681
45. 4.298 **47.** 84.6265 **49.** 5.8969 **51.** 0.022098
53. 6.365218 **55.** 0.041549 **57.** 33 **59.** $4.09

61. a. 16.5 **b.** 18.7 **c.** 24.2 **d.** 29.6 **e.** 13.2
f. 6.6

Workshop 7 **Self-Check 1.** 0.875 **2.** 0.6 **3.** 0.778
4. 0.667 **5.** $\frac{4}{10} = \frac{2}{5}$ **6.** $\frac{12}{100} = \frac{3}{25}$ **7.** $1\frac{125}{1000} = 1\frac{1}{8}$
8. $7\frac{82}{100} = 7\frac{41}{50}$ **Problems 9.** 0.2 **11.** 0.714 **13.** 0.45
15. 0.444 **17.** 0.425 **19.** 0.7 **21.** 0.36 **23.** 0.417
25. 0.25 **27.** 0.13 **29.** 0.86 **31.** 0.3 **33.** 0.03
35. 0.93 **37.** 0.143 **39.** 0.583 **41.** 0.267 **43.** 0.567
45. 0.54 **47.** 0.982 **49.** $\frac{1}{4}$ **51.** $\frac{1}{2}$ **53.** $\frac{3}{10}$ **55.** $\frac{9}{20}$
57. $\frac{1}{8}$ **59.** $\frac{151}{200}$ **61.** $\frac{3}{16}$ **63.** $\frac{13}{40}$ **65.** $\frac{17}{50}$ **67.** $\frac{15}{16}$ **69.** $1\frac{1}{10}$
71. $7\frac{2}{25}$ **73.** $5\frac{117}{1000}$ **75.** $3\frac{7}{16}$ **77.** $21\frac{39}{40}$ **79.** $43\frac{11}{32}$
81. 30.25 **83.** 0.25 **85.** $\frac{4}{5}$ **Applications 87.** $52.50
b. $22.88 **c.** $12.63 **d.** $45.25 **e.** $8.38 **f.** $26.38
g. $39.50 **h.** $5.63 **89. a.** $222\frac{2}{25}$ **b.** $216\frac{8}{25}$
c. $216\frac{1}{100}$ **d.** $212\frac{29}{100}$ **e.** $210\frac{1}{20}$ **f.** $208\frac{21}{50}$ **c.** $208\frac{1}{4}$

Workshop 8 **Self-Check 1.** 0.25 **2.** 0.375 **3.** 1.42
4. 0.09 **5.** 85% **6.** 7% **7.** 30% **8.** 155% **9.** 0.4%
Problems 11. 0.34 **13.** 0.60 **15.** 0.87 **17.** 0.99
19. 0.976 **21.** 0.818 **23.** 0.344 **25.** 0.736 **27.** 8.17
29. 3.76 **31.** 5.08 **33.** 17.83 **35.** 0.03 **37.** 0.09
39. 0.047 **41.** 0.072 **43.** 0.0803 **45.** 0.050921
47. 13% **49.** 35% **51.** 24% **53.** 97% **55.** 5%
57. 1% **59.** 4.2% **61.** 9.65% **63.** 0.5% **65.** 0.61%
67. 937.1% **69.** 720% **71.** 298% **73.** 700%
75. 0.9% **77.** 4600% **79.** 10% **81.** 108,300%
83. 1.123 **85.** 0.30; 0.50 **87. a.** 1.6% **b.** 0.2%
c. 1.5% **d.** 4.5% **e.** 4.2% **f.** 1.3% **g.** 1.9%
89. a. 0.017 **b.** 0.013 **c.** 0.011 **d.** 0.009 **e.** 0.007
f. 0.005

Workshop 9 **Self-Check 1.** 28 **2.** 30 **3.** 1.5
4. $290 **Problems 5.** 24.5 **7.** 10.8 **9.** 140.7
11. 100.8 **13.** 209.5 **15.** 3.8 **17.** 6.51 **19.** 82.08
21. 0.764 **23.** 780 **25.** 1677 **27.** 3170.5
29. 6835.5 **31.** 37,308.7 **33.** 6.1875 **35.** 7.48
37. 26.28 **39.** 192.075 **41.** 623 **43.** 741.258
45. $4.20 **47.** $27.30 **49.** $19.98 **51.** $1.62
53. $1.26 **55.** $0.90 **57.** $435 **59. a.** $9, $20.99
b. $10.20, $23.79 **c.** $12, $27.99 **61. a.** 54 **b.** 40
c. 33 **d.** 34 **e.** 108 **63. a.** $30.00 **b.** $51.00
c. $117.60 **d.** $117.00 **e.** $61.88 **f.** $149.36
g. $251.69 **h.** $175.00 **i.** $121.48 **j.** Negotiated

Workshop 10 **Self-Check 1.** 5 **2.** 5.1 **3.** 130
4. $41.33 **Problems 5.** 10 **7.** 80 **9.** 140.3 **11.** 743
13. 4.38 **15.** 3.05 **17.** $14.20 **19.** $77.33 **21.** 6.7
23. 33.9 **25.** 40.04 **27.** $408,000 **29.** 78.0
31. $150.00 **33.** 93.5 **35.** 94 **37. a.** 3 **b.** 51
c. 8 **d.** 5 **e.** 16 **f.** 17; 15; 21; 15; 15 **39.** $1.31
41. 10.25; 27; 3321

Workshop 11 **Self-Check 1.** 7 h:15 min **2.** 5 h:15 min
3. 5 h:55 min **4.** 2 h:57 min **Problems 5.** 9 h:10 min
7. 7 h:43 min **9.** 4 h:15 min **11.** 6 h:15 min **13.** 6 h:

50 min **15.** 3 h:37 min **17.** 8 h:30 min **19.** 3 h:53 min
21. 4 h:45 min **23.** 7 h:45 min **25.** 7 h:15 min
27. 8 h:35 min **29.** 8 h **31.** 8 h:45 min **33.** 7 h:55
min **35.** 6 h:37 min **37.** 6 h:34 min **39.** 8 h:27
min **41. a.** 0 h:42 min **b.** 3 h:15 min **c.** 4 h:32
min **d.** 2 h:26 min **e.** 1 h:55 min **f.** 1 h:28 min
g. 2 h:55 min

Workshop 12 **Self-Check 1.** $1.25 **2.** $1.55 **3.** $2.34
4. $3.32 **5.** Size 38 or 39 **6.** Size 42 **Problems**
7. $1.87 **9.** $2.46 **11.** $3.01 **13.** $1.79 **15.** $2.58
17. $1.40 **19. a.** 2 lb **b.** 11 lb **c.** 13 lb **d.** 10 lb
e. 14 lb **f.** 8 lb **21.** 36 or 37 **23.** 33 **25.** 29
27. Bangor **29.** $54.00 **31.** $50.00 **33.** $56.00
35. $610; $620 **37.** $510; $520 **39.** $530; $540
41. $68 **43.** $70 **45.** $79

Workshop 13 **Self-Check**

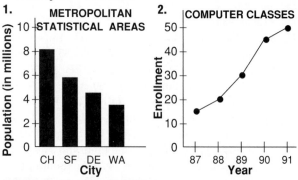

1. METROPOLITAN STATISTICAL AREAS (Population in millions by City: CH, SF, DE, WA)

2. COMPUTER CLASSES (Enrollment by Year: 87, 88, 89, 90, 91)

Problems 3. CITY GOVERNMENT EMPLOYMENT National Summary (Employment in thousands: SC, HO, HI, PO, FI, PR)

5. a. 1989; About $48 billion
b. 1982; About $12 billion
c. About $35 billion

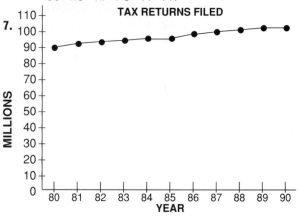

7. TAX RETURNS FILED (Millions by Year: 80–90)

654 ◆ Selected Answers

Workshop 14 **Self-Check 1.** 108 in **2.** 150 mL **3.** 8 yd **4.** 3.5 m **Problems 5.** 27 ft **7.** 112 oz **9.** 96 oz **11.** 3800 m **13.** 3200 g **15.** 2.25 gal **17.** 3.5 gal **19.** 2 kg **21.** 3.3 L **23.** 72 100 g **25.** 0.723 kg **27.** 180mm **29.** 40 in **31.** 11 pt **33.** 10 qt **35.** 29 pt **37.** 20 qt **39.** 8 c **41.** 240 in **43.** 3900 **45.** 2.13 L **47. a.** 36 ft **b.** 36 ft; 48 ft; 84 ft **c.** 21 ft; 18 ft 8 in; 39 ft 8 in **d.** 22 ft 6 in; 15 ft 4 in; 37 ft 10 in **e.** 22 ft 8 in; 26 ft 4 in; 49 ft **f.** 35 ft 4 in; 25 ft 6 in; 60 ft 10 in **g.** 16 ft 10 in; 15 ft 6 in; 32 ft 4 in; Total 381 ft 8 in

Workshop 15 **Self-Check 1.** $40 - 30 = 10$; 16.495 **2.** $50 \div 7 = 7$; 6.32 **3.** $60 \times 10 = 600$; 602.6713 **4.** $5 \times 5 = 25$; $23\frac{11}{12}$ **Problems 5.** 8000; 7789 **7.** 2; 2.14 **9.** 72; 73.9728 **11.** 7; 7.1 **13.** 6; $4\frac{1}{8}$ **15.** 3; $3\frac{1}{8}$ **17.** $35; $34.97 **19.** 80; 83.2 **21.** $10; $9.24 **23.** $9; $8.89 **25.** $67; $68.47 **27.** $26

Workshop 16 **Self-Check: 1.** 900; 1100; 1111 **2.** 15,000; 17,000; 16,784 **3.** $13.00; $15.00; $15.01 **Problems: 5.** 16,000; 16,846 **7.** 120; 120.4 **9.** 1200; 1195 **11.** $73; $74.96 **13.** $400; $462.81 **15.** $23.00; $23.39 **Applications: 17.** $12.00; $11.79 **19.** $70.00; $71.20 **21.** 450; 430.28 **23. a.** 1200; Yes **b.** $14; No; $14.00 **c.** 48; No; 49.3 **d.** 360; Yes **e.** 130; No; 135.3 **f.** 45; Yes **g.** 4; No; 4.65 **h.** $450; No; $439.45

Workshop 17 **Self-Check: 1.** $640 \div 80 = 8$; 8.07 **2.** $9000 \div 300 = 30$; 27.3 **3.** $\frac{1}{3} \times 900 = 300$; 324.1 **Problems: 5.** 700; 697 **7.** 110; 107 **9.** 40; 42.72 **11.** $900; $895.50 **13.** 30; $31\frac{11}{14}$ **15.** $50; $48.75 **17.** $4000; $3958.33 **19.** $80; $81.82 **21.** $33.00; $32.06 **23.** $2000; $2050 **25.** $70; $17,500; $18,200 **27.** $160 to $180; $153.85 to $173.08 **29.** $18; $17.88

Workshop 18 **Self-Check: 1.** $4 \times 600 = 2400$; 2332 **2.** $4 \times \$30 = \$120$; $119.09 **3.** $(3 \times \$2) + \$1 = \$6 + \$1 = \$7$; $6.34 **Problems: 5.** 2000; 2063 **7.** $160; $161.24 **9.** 36; 35.87 **11.** $17; $18.28 **13.** 3600; 3500.67 **15.** $18; $18.60 **17.** $9; $9.40 **19.** 2400; 2346 **21.** $36; $35.56 **23.** $92; $88.45

Workshop 19 **Self-Check: 1.** $3180 **Problems: 3.** $69.88 **5.** $8784 **7.** $444.50 **9.** Approximately 1495 rotations **11.** 224 miles **13.** 15 loaves

Workshop 20 **Self-Check: 1.** Problem cannot be solved. Need number of payments. **Problems: 3.** Cannot be solved. Need relationship between pints

and pounds. **5.** $24 **7.** Cannot be solved. Need weight of watermelons. **9.** 5' 11" **11.** Cannot be solved. Need cost of tennis balls. **13.** 19 **15.** Impossible to average 60 mph.

Workshop 21 **Self-Check: 1.** $14.38 **Problems: 3.** $\times$, $\times$, $+$, $216.60 **5.** $\times$; $+$; $+$; $\times$, compare, $+$; $29.60 **7.** $\times$; $\times$; $+$; $187.65 **9.** $-$; $\times$; $+$; 28.5 degrees C **11.** $+$; $-$; 223 **13.** $\times$; $+$; $-$; $2250 **15.** $\times$; $+$; $-$; $4.55

Workshop 22 **Self-Check: 1.** No; he used 93 instead of $0.93. Answer is $9.30. **Problems: 3.** No; Used 89¢ as $89; estimate: $4.50 **5.** Yes; No error **7.** No; Decimal point; estimate: $50 **9.** No; Added instead of subtracted; estimate: $30 **11.** About $500 **13.** About $1250 **15.** About $200 **17.** About $3 **19.** About 25

Workshop 23 **Self-Check: 1.** 7 bicycles; 4 unicycles **Problems: 3.** 6 bicycles; 3 tricycles **5.** 20 nickels; 20 quarters **7.** 25 **9.** 12 **11.** 4 pennies; 4 dimes; 1 quarter

Workshop 24 **Self-Check: 1.** 81; 243; 729 **2.** 16; 22; 29 **Problems: 3.** 34; 40; 46 **5.** 18; 15; 12 **7.** 25; 36; 49 **9.** 1093; 3280; 9841 **11.** 8 dimes; 12 quarters **13.** 33 **15.** 2:30 p.m. **17.** 4×4; 6×3 **19.** 11

Workshop 25 **Self-Check: 1.** 7; 8; 9 **Problems: 3.** 7 letters; 8 postcards **5.** All sums are 12. **7.** 2 **9.** Ford $= 10$; Chevy $= 8$; Dodge $= 4$ **11.** Sum is 27; Top: 8; 13; 6; Mid: 7; 9; 11; Bot: 12; 5; 10

Workshop 26 **Self-Check: 1.** 29 days **Problems: 3.** 5 **5.** $15,000 **7.** 56 dozen **9.** 62 **11.** 6

Workshop 27 **Self-Check: 1.** 11 **Problems: 3.** Cork: 5¢; Bottle: $1.05; **5.** 54 yards **7.** $7962 **9.** Holstein **11.** 7; 9; 11 **13.** $B = 3A - 2$; 118 **15.** Harry $= 197$; Jerry $= 210$; Darrel $= 192$

Workshop 28 **Self-Check: 1.** 3 **Problems: 3.** 12 **5.** none **7. a.** 182 **b.** 198

Workshop 29 **Self-Check: 1.** $(18" + 7") - 10"$ **Problems: 3.** 3 miles west and 6 miles north **5.** 8556.5 sq ft **7.** 15 **9.** $(4 + 7) - 10$ **11.** 342.26 sq ft

Workshop 30 **Self-Check: 1.** $42,390 \div 5280 = 8$ **Problems: 3. a.** 400 **b.** 225 **5.** 625 **7. a.** 352 ft **b.** 4400 ft **9.** Borrow 1 horse from neighbor; then Rick $= 9$; Mike $= 6$; and Peter $= 2$. **11.** There is no extra dollar.

1-1; pages 78–79 **Self-Check 1.** $304 **2.** $296.25
Problems 3. $288 **5.** $186 **7.** $453.13 **9.** $165
11. $812.50 **13.** $197.20 **15.** $444.06 **17.** $187
19. $215.63 **Maintaining Your Skills 21.** 0.50
23. 0.10 **25.** $136.13 **27.** $471.25 **29.** 140
31. 100 **33.** 14,400 **35.** 1000

1-2; pages 80–81 **Self-Check 1.** $441 **2.** $580.75
Problems 3. $240; $72; $312 **5.** $192; $28.80;
$220.80 **7.** $680; $442; $1122 **9.** $355.20;
$115.20; $470.40 **11.** $329 **13.** $229.50
15. $570.24 **Maintaining Your Skills 17.** $247.35
19. 1746.162 **21.** $441.60 **23.** $150 **25.** $125

1-3; pages 82–84 **Self-Check 1.** $7\frac{1}{2}$ **2.** $8\frac{1}{4}$
Problems 3. $7\frac{3}{4}$ **5.** $7\frac{3}{4}$ **7.** $8\frac{1}{2}$ **9.** $8\frac{1}{4}$ **11.** $8\frac{3}{4}$
13. 9 **15.** $49\frac{1}{2}$ **17.** $295.44 **19.** 8; $8\frac{1}{4}$; $7\frac{3}{4}$; 8; 8;
40; $300 **21.** S + S = 13 hrs; M − F = $33\frac{1}{2}$ hrs.
Maintaining Your Skills 23. $\frac{7}{8}$ **25.** $\frac{23}{24}$ **27.** $7\frac{3}{4}$ **29.** $3\frac{3}{4}$
31. 40,000 **33.** 68,000 **35.** 300

1-4; pages 85–86 **Self-Check 1.** $448 **2.** $224.10
Problems 3. $369 **5.** $269.70 **7.** $826.50
9. $263.86 **11.** $630 **13.** $401.25 **15.** $60.30
17. $69 **19.** $49 **21.** $1709.25 **Maintaining Your
Skills 23.** $264.00 **25.** $54.36 **27.** $128.00
29. $34.357 = $34.36 **31.** $1587.6 **33.** 456,000

1-5; pages 87–88 **Self-Check 1.** $1650 **2.** $758.33
Problems 3. $650 **5.** $1375 **7.** $1280.73 **9.** $240.38
11. $710 **13.** $2700; $623.08 **15.** $1450
17. $320.38 − $290.19 = $30.19 **19.** $3280
Maintaining Your Skills 21. $285.80 **23.** 13.01
25. 559.38 **27.** 1329.73 **29.** 104 **31.** 104

1-6; pages 89–91 **Self-Check 1.** $752 **2.** $7887
3. $160 **4.** $2125 **Problems 5.** $7840 **7.** $1535.68
9. $139.35 **11.** $2394.60 **13.** $1495; $1600
15. $364; $850 **17.** $3521.10; $3521.10
19. $274.69 **21.** $181.50; $181.50 **23.** $8363.38
25. $188.21 **27.** $244.63 **Maintaining Your Skills**
29. 2.13 **31.** 6.75 **33.** 9.25 **35.** $365.50
37. $331.20 **39.** 123.1 **41.** 521 **43.** 0.6 **45.** 0.8

1-7; pages 92–93 **Self-Check 1.** $3050 **2.** $559.20
Problems 3. $100; $360; $460 **5.** $90; $160; $40;
$290 **7.** $482 **9.** $243.63 **11.** $142.50 **13.** $673.75
Maintaining Your Skills 15. $1034.31 **17.** 1.960
19. 0.1025 **21.** 0.97 **23.** $105.00 **25.** $542.50

Reviewing the Basics, page 94 **Skills 1.** $45.67
3. $0.05 **5.** $172.26 **7.** 20.534 **9.** $2.48
11. 0.00021 **13.** 4660 **15.** $19\frac{1}{2}$ **17.** $10\frac{3}{4}$
19. $341.00 **21.** 0.5 **23.** 0.75 **25.** 0.125
27. 0.0125 **29.** 3.85 **Applications 31.** $8\frac{1}{2}$ hrs.
33. 104 **35.** 16

2-1; pages 100–102 **Self-Check 1.** $26.00 **2.** $34.00
Problems 3. $20 **5.** $33 **7.** $39 **9.** $4 **11.** $30

13. $45 **15.** $28 **17.** $46 **19.** $12 **Maintaining
Your Skills 21.** ones; 4 **23.** thousands; 6000
25. hundreds; 600 **27.** $27 **29.** $1.49 **31.** $276.10
33. $874.89

2-2; pages 103–104 **Self-Check 1.** $668 **2.** $1445
Problems 3. $280 **5.** $280 **7.** $3000 **9.** $2200;
$377.50 **11.** $1662.30 **13.** $4220 **Maintaining Your
Skills 15.** 65.87 **17.** 31.903 **19.** 192 **21.** 12.5

2-3; page 105–106 **Self-Check 1.** $33.75 **2.** $32.02
Problems 3. $3200 **5.** $677 **7.** $56.29 **9.** $22.82
Maintaining Your Skills 11. 0.547 **13.** $235.92
15. 16.54 **17.** 0.03 **19.** 12 **21.** 208

2-4; pages 107–108 **Self-Check 1.** $183.60 **2.** $10.25
Problems 3. $5.13 **5.** $10.67 **7.** $368.73 **9.** $0;
over $55,500 **11.** $86.51 **13.** $61; $36.11
15. November **Maintaining Your Skills 17.** 12.5
19. 0.15 **21.** 320 **23.** 23,980 **25.** 3800
27. 147,400

2-5; pages 109–110 **Self-Check 1.** $40.00 **2.** $18.88
Problems 3. $95 **5.** $44.75 **7.** $6.21 **9.** $35.00
11. $14.62 **Maintaining Your Skills 13.** 13.05
15. 86.50 **17.** 7.68 **19.** 7.89

2-6; pages 111–114 **Self-Check 1.** $215.32 **Problems
3.** 17.00; 23.27; 6.08; 4.56; 224.69 **5.** 47.00; 32.23;
9.64; 7.37; 303.01 **7.** 54.00; 35.23; 11.51; 8.06;
19.89; 331.81 **9.** $295.68 **Maintaining Your Skills
11.** 67.316 **13.** 253.23 **15.** 482.337 **17.** 2171.16
19. 63.06 **21.** 65.38

Reviewing the Basics; page 115 **Skills 1.** $239
3. $154.81 **5.** $40.10 **7.** $103.31 **9.** $24.01
11. $135.43 **13.** $19.00 **15.** $4.75 **17.** $4.76
Applications 19. $63.68 **21.** 24 **23.** 72

3-1; pages 126–127 **Self-Check 1.** $74.90 **2.** $105.00
Problems 3. $41.80; $41.80 **5.** $178.20; $168.20
7. $1275.95; $1240.95 **9.** $435.44 **11.** $599.22
13. $151.45 **15.** $1238.03 **17.** $108.03
Maintaining Your Skills 19. $396.05 **21.** $373.93
23. $507.00 **25.** $659.40 **27.** $653.32

3-2; pages 128–129 **Self-Check 1.** $215.32 **2.** One
hundred forty-three and $\frac{32}{100}$ **Problems 3.** Forty and $\frac{40}{100}$
5. Sixty-three and $\frac{74}{100}$ **7.** Thirty-four and $\frac{6}{100}$ **9.** One
thousand nine hundred seventeen and $\frac{00}{100}$ **11.** Two
hundred one and $\frac{9}{100}$ **13.** Five thousand three hundred
twenty-seven and $\frac{17}{100}$ **15.** No; Ninety-eight and $\frac{72}{100}$
17. Yes **19.** Yes **Maintaining Your Skills 21.** $35.15
23. Nineteen and $\frac{25}{100}$ **25.** Four hundred thirty-five and $\frac{00}{100}$
27. Five thousand two hundred seventy-four and $\frac{19}{100}$
29. Six thousand one hundred ten and $\frac{50}{100}$

3-3; pages 130–132 **Self-Check 1.** $1201.01 **2.** $1131.55
Problems 3. $401.43 **5.** $366.72 **7.** $367.33
9. $211.86 **11.** $162.66; $109.63; $2.33

13. $268.30; $218.30; $505.00; $378.65; $669.65
15. $313.88; $358.98; $224.32; $22.32; $161.72; $50.57; $20.57; $−28.35 overdrawn **Maintaining Your Skills** 17. $1245.96 19. $466.21 21. $816.55
23. $247.97 25. $77.09 27. $356.91

3-4; pages 133–135 **Self-Check** 1. $258.70 2. $63.90
Problems 3. $47.66 5. $1026.12 7. $15,533.36
9. $709.66 11. $401.24 13. 581.63; 2329.90; 1030.45; 7.17; 1888.25 **Maintaining Your Skills**
15. $1066.70 17. $47,082.74 19. $500.82
21. $1362.97 23. 846.35 25. 865.464

3-5; pages 136–140 **Self-Check** 1. $275.49 2. $916.33
Problems 3. $142.90; $142.90 5. $1446.89; $1446.89 7. $316.54; Yes 9. a. $44.89
b. $191.37 c. $794.83 d. $794.83 e. Yes
Maintaining Your Skills 11. $411.05 13. $93.78
15. $243.77 17. $171.92 19. $46.36

Reviewing the Basics; page 141 **Skills** 1. Twenty-five and $\frac{79}{100}$ 3. One thousand three hundred seventy-two and $\frac{35}{100}$ 5. $262.18 7. $621.78
9. $359.35 11. $211.68 **Applications** 13. $825.29
15. c 17. b 19. d

Unit 4   Savings Accounts

4-1, pages 146–147 **Self-Check** 1. $163 2. $101
Problems 3. $79.15; $79.15 5. $92.71; $62.71
7. $1032.58; $1012.58 9. $197.40 11. $935.28
13. $500.03 15. $256.48 17. $950.51 **Maintaining Your Skills** 19. $58.78 21. $697.97 23. $78.93
25. $43.33 27. $50.60

4-2, pages 148–149 **Self-Check** 1. Forty-five and $\frac{76}{100}$ dollars 2. $291.42 **Problems** 3. Seventeen and $\frac{35}{100}$ dollars 5. Forty-four and $\frac{93}{100}$ dollars 7. Four hundred six dollars 9. Seven thousand eight hundred fifty-two and $\frac{3}{100}$ dollars 11. a. 17594179 b. $831.95
c. Eight hundred thirty-one and $\frac{95}{100}$ dollars
13. a. 81-0-174927 b. $318.29 c. Three hundred eighteen and $\frac{29}{100}$ dollars 15. a. 06029175 b. $76.60
c. Seventy-six and $\frac{60}{100}$ dollars **Maintaining Your Skills**
17. Ninety-four and $\frac{78}{100}$ dollars 19. One hundred sixty-two and $\frac{5}{100}$ dollars 21. Four thousand two hundred eleven and $\frac{15}{100}$ dollars 23. $251.27
25. $25,696.29

4-3, pages 150–151 **Self-Check** 1. $851.00 2. $3191.30
Problems 3. $568.12 5. $113.08 7. $616.05
9. $1357.10 **Maintaining Your Skills** 11. $438.00
13. $1173.18 15. $175 17. $397.54 19. $6075.25

4-4, pages 152–154 **Self-Check** 1. $869.50 **Problems**
3. $429.50 5. $7529.03 7. $1947.25 9. $3832.44
11. $579.86 13. $18,219.75; $18,339.93; $17,939.93; $18,559.39; $18,159.39; $18,059.39; $18,179.24; $18,179.24 **Maintaining Your Skills** 15. $1267.52

17. $1047.62 19. $4488.90 21. $39.10
23. $2000.90

4-5, pages 155–156 **Self-Check** 1. $6.50 2. $56.25
Problems 3. $43.20; $10.80 5. $39.24; $19.62
7. $256.45; $64.11 9. $43.62 11. $32.84 13. $14.48
15. $21.67 **Maintaining Your Skills** 17. 0.5 19. 0.25
21. 6.25 23. 0.055 25. 0.095 27. 0.10625

4-6, pages 157–158 **Self-Check** 1. $2163.20 **Problems**
3. $2.01; $404.01 5. $342.38; $18,602.38; $348.79; $18,951.17 7. $824.18 9. $907.95 11. $9712.13
13. $1965.65 **Maintaining Your Skills** 15. 0.0525
17. 0.0575 19. 0.0725 21. $49.40 23. $1176.10
25. $362 27. 0.33 29. 0.60

4-7, pages 159–161 **Self-Check** 1. $2230.88
2. $4776.12 **Problems** 3. $1003.90; $103.90
5. $1555.41; $215.41 7. $4233.44; $361.77
9. $156.39 11. $55.22 13. $29.52 15. $896.04; $96.04 **Maintaining Your Skills** 17. 2% 19. 0.5%
21. 4.25% 23. 1.3125% 25. 1.875%
27. 1.625%

4-8, pages 162–163 **Self-Check** 1. $6022.62 2. $22.62
Problems 3. $80301.60; $301.60 5. $6549.08; $49.08 7. $15600.30; $279.30 9. $31.41
11. $384.84 13. $1440.63 **Maintaining Your Skills**
15. $4085.24 17. $554.99 19. $953.72 21. $7395.57
23. $94.18 25. $8057.49

Reviewing the Basics, page 164 **Skills** 1. Four hundred seventy-nine dollars 3. Three thousand ninety-one and $\frac{47}{100}$ dollars 5. $212.05 7. $734.96
9. $6723.06 11. 2.125 13. 1.685 15. 8.25
17. 21.75 19. 0.0675 21. 0.1995 **Applications**
23. 1.14339 25. 88 days 27. $\frac{3}{12} = \frac{1}{4}$ 29. $\frac{9}{12} = \frac{3}{4}$
31. 16 33.–37. Answers will vary.

Unit 5   Cash Purchase

5-1, pages 168–169 **Self-Check** 1. $21 2. $513
Problems 3. $2.44 5. $1.09 7. $59.70 9. $0.30
11. $4.61 13. $6.12 15. $1.79 17. $3.71
Maintaining Your Skills 19. $622.80 21. $1.20
23. 4000 25. 2,437,000 27. 35,230 29. 1130
31. $279.54 33. $2.86 35. 358.0 37. 521.7

5-2, pages 170–172 **Self-Check** 1. $260.10
Problems 3. $1.22 5. $12.47 7. $64.15 9. $63.62
11. $305.83 13. $132.23 15. $31.15 17. $139.92
19. $29.98; $119.95; $23.96; $26.94; $7.72; $208.55; $12.51; $221.06 **Maintaining Your Skills** 21. 142.6152
23. 0.0321 25. 27.187 27. 776.156 29. 1382.65
31. 126.25 33. 1251.846

5-3, page 173–174 **Self-Check** 1. 3.09¢ = 3.1¢
2. 12.375¢ = 12.38¢ **Problems** 3. 8.3¢ 5. 65.9¢
7. $0.19 9. $1.00 11. $44.61 13. $0.15 15. $3.75
17. a. $0.03 b. $0.75 c. $0.55 d. $0.20
Maintaining Your Skills 19. 4.62 21. 0.02 23. 20
25. 1350 27. 741.9 29. 1132.6 31. 87.16
33. 1000.00

5-4, pages 175–176 **Self-Check 1.** 20 oz at 3.1¢ **2.** 100 at 2.39¢ each **Problems 3.** 3.7¢; 4.0¢; small **5.** 66.1¢; 66.5¢; small **7.** b. **9.** 9 oz can **11.** 15-can box **13.** 40-oz jar **15.** 2.7¢; 2.6¢; 2.8¢; 3.1¢; Giant size 14-lb box **Maintaining Your Skills 17.** 21.37 **19.** 145.05 **21.** $1586 **23.** 5.1

5-5, pages 177–178 **Self-Check 1.** $1.09 **2.** $4.27 **Problems 3.** $2.14 **5.** $2.39 **7.** $1.58 **9.** $1.44 **11.** $25.95 **13.** $57.67 **15.** $82.04 **Maintaining Your Skills 17.** $1.30 **19.** $1.54 **21.** $0.86

5-6, pages 179–181 **Self-Check 1.** $580 **2.** $35.80 **Problems 3.** $70 **5.** $10.49 **7.** $2.99 **9.** $16.63 **11.** $2900 **13.** $0.57 **15.** $13.55 **17.** $87.48 **19.** $64 **21.** $7.00; $13.75; $14.75; $6.25; $19.50 **23.** White sale **Maintaining Your Skills 25.** 69.776 **27.** 10.467 **29.** $112 **31.** 1265

5-7, pages 182–183 **Self-Check 1.** $31.50 **2.** $175.20 **Problems 3.** $48; $72 **5.** $2.75; $8.24 **7.** $292.50; $157.50 **9.** $61.08 **11.** $12.56 **13.** $27.88 − $19.79 = $8.09 = 29% of $27.88 **Maintaining Your Skills 15.** $18.70 **17.** 25.76 **19.** 82.806 **21.** 0.0141 **23.** $4.49

Reviewing the Basics, page 184 **Skills 1.** 14.4¢ **3.** 17.25¢ **5.** $1.72 **7.** 422¢ **9.** $279.20 **11.** $71.20 **13.** $37.52 **15.** $77.24 **17.** 7.08¢ **19.** $3.41 **21.** $31.69 **Applications 23.** $279.98 **Terms 25.** e **27.** a **29.** b

Cumulative Review: Units 1–5, pages 189–190 **Skills 1.** Forty-seven and $\frac{59}{100}$ **3.** Two thousand thirty-five and $\frac{26}{100}$ **5.** $298 **7.** $33.18 **9.** $138.00 **11.** 15.4¢ **13.** 33.0¢ **15.** $22 **17.** $1809 **19.** $10,598 **21.** $1206 **23.** $664.83 **25.** $559.95 **27.** $112.91 **29.** $4135.35 **31.** $342 **33.** $0.20493 **35.** $0.222 **37.** $0.38 **39.** $134.44 **41.** $199.04 **43.** $325.15 **45.** $959.38 **47.** $422.29 **49.** 6 **51.** 0.5 **53.** 25.25 **55.** 0.0725 **57.** 0.125 **Applications 9.** $9.99 **61.** $25 **63.** 1.12005 **65.** 1.21946 **67.** 2 hrs. 40 min. **69.** 9 hrs. 28 min. **71.** 19 **73.** 99 **75.** $\frac{1}{4}$ **77.** $\frac{5}{6}$ **79.** 156 **81.** 18 **83.–100.** Answers will vary.

Unit 6   Charge Accounts and Credit Cards

6-1; pages 194–196 **Self-Check 1.** $6.96; $122.96 **2.** $5.00; $104.95 **Problems 3.** $10.08; $154.08 **5.** $8.42; $137.97 **7.** $84.09 **9.** $28.26 **11.** $6.27 **13.** $62.74 **15.** $113.29 **17.** $44.67 **19.** $402.79 **21.** $491.77 **23.** $82.05 **Maintaining Your Skills 25.** 0.08 **27.** 0.045 **29.** 0.092 **31.** $296.27 **33.** $87.27 **35.** 2.94 **37.** 7.052 **39.** $35.70

6-2; pages 197–199 **Self-Check 1.** $597.50 **2.** $270.78 **Problems 3.** $649.00 **5.** 337.66 **7.** 416.34 **9.** $375.66 **11.** $323.72 **13.** $796.35 **15.** $109.90; $188.73; $369.04 **Maintaining Your Skills 17.** $405.04 **19.** $306.69 **21.** $309.30 **23.** $287.84 **25.** $3297.34

6-3; pages 200–202 **Self-Check 1.** $2.70 **2.** $1.318 = $1.32 **Problems 3.** $0.90 **5.** $2.37 **7.** $5.03

9. $2.55; $191.55 **11.** $7.46; $460.41 **13.** $19.23; $1076.52 **15.** $1.80; $123.42 **17.** $99.79; $1.77; $253.35 **Maintaining Your Skills 19.** 43.16 **21.** 9.54 **23.** 3.00 **25.** $3.75 **27.** $0.44 **29.** 35.04 **31.** 0.528

6-4; pages 203–205 **Self-Check 1.** $300; $4.50; $374.50 **2.** $70; $1.05; $166.05 **Problems 3.** $400; $6; $486 **5.** $275; $4.13; $369.13 **7.** $416; $6.24; $644.74 **9.** $372.87; $5.59; $526.40 **11.** $1.32; $129.32 **13.** $1.95; $195.74 **15.** $374.29; $7.49; $461.09 **17.** $71.45; $307.68; $6.15; $130.48; $444.31 **Maintaining Your Skills 19.** $497.58 **21.** $410.93 **23.** $369.10 **25.** $77.47 **27.** 8.8 **29.** 19.5 **31.** 3.5 **33.** 14.89

6-5; pages 206–210 **Self-Check 1.** $5000 **2.** $400 **3.** $400 **4.** 19 **5.** $7600 **6.** 30 **7.** $13,000 **8.** $13,000 **9.** 30 **10.** $433.33 **11.** $2.48 **12.** $192.48 **Problems 13.** $350.00; 30; $1500.00; $50 **15.** $243.55; $3.65 **17.** $4.85; $253.36 **19.** $82.73; $1.65; $152.95 **21.** $122.21; $2.44; $146.34 **Maintaining Your Skills 23.** 2400 **25.** 825 **27.** 1578.5 **29.** 2163 **31.** 125 **33.** 36.45 **35.** 72 **37.** 283.4

6-6; pages 211–213 **Self-Check 1.** $3500 **2.** $600 **3.** $600 **4.** $3000 **5.** $450 **6.** $450 **7.** $450 **8.** $7200 **9.** 30 **10.** $14,750; average daily balance = $491.67 **Problems 11.** 12; $7200; 1; $740; 6; $4440; 1; $620; 10; $6200; 30; $19,200 $640.00 **13.** $147.76; $192.78; $3.86 **Maintaining Your Skills 15.** 1800 **17.** 3081.68 **19.** 1097.58 **21.** 1783.04 **23.** 42.90 **25.** 256.89 **27.** 421.17 **29.** 574.84

Reviewing the Basics, page 214 **Skills 1.** $173.14 **3.** $38.00 **5.** $537.38 **7.** $119.59 **9.** $32.11 **11.** $8320 **13.** $199.35 **15.** $1272.05 **17.** $68.71 **19.** $18.80 **21.** $735.89 **23.** 5.4321 **Applications 25.** 26 days **27.** 19 days **29.** 61.83 **31.** $70.60 **Terms 33.** c **35.** f **37.** a

Unit 7   Loans

7-1; pages 218–219 **Self-Check 1.** $15 interest; $615 maturity value **2.** $19.73 interest; $819.73 maturity value **Problems 3.** $913.50 **5.** $18.70; $868.70 **7.** $1349.21; $10,784.21 **9.** $403.20; $8803.20 **11.** $21,393.75 **13.** $48,670.63 **Maintaining Your Skills 15.** 0.40 **17.** 0.07 **19.** 0.1564 **21.** $\frac{1}{2}$ **23.** $\frac{1}{3}$ **25.** 0.667 **27.** 0.35 **29.** 0.60

7-2; pages 220–221 **Self-Check 1.** $272; $1088 **2.** $217.50; $507.50 **Problems 3.** $512 **5.** $443.50; $1330.50 **7.** $1422; $8058 **9.** $1012 **11.** $7115.68 **13.** $3216 **15.** $790.36 **17.** 25% **Maintaining Your Skills 19.** 0.32 **21.** 0.25 **23.** 0.20 **25.** 65 **27.** 142.50 **29.** $178.38 **31.** $17.63

7-3; pages 222–224 **Self-Check 1.** $82.335 = 82.34; $1976.16; $326.16 **Problems 3.** $111.07; $1999.26; $219.26 **5.** $155.25; $931.50; $31.50

7. $183.40; $200.80 **9.** $468.48 **11.** $134.76
13. 18%; $28.98 **15.** $687.68 **17.** 20%; $142.56
19. $614 **Maintaining Your Skills 21.** 20 **23.** 13.50
25. $53.13 **27.** 4158 **29.** 1209.19

7-4; pages 225–227 **Self-Check 1.** $18.87 **2.** $18.87;
$294.33 **3.** $294.33; $1214.97 **Problems 5.** $86.40;
$2313.60 **7.** $28.00; $57.51; $1622.49 **9.** $21.13;
$44.88; $930.12 **11.** $30 **13.** $30; $49.35; $1450.65
15. $100; $205.40; $5794.60 **17.** $58.67; $154.77;
$3045.23 **19.** $238.05; $112; $126.05; $8273.95
21. $200.75; $1035.78 **23.** $8.33; $204.79; $628.23
25. $4.21; $208.91; $212.48 **Maintaining Your Skills**
27. $600 **29.** $672 **31.** $610.71 **33.** $20.45
35. $864.29

7-5; pages 228–229 **Self-Check 1.** $800 + $8 =
$808 **2.** $1280 + $16 = $1296 **Problems**
3. $4848 **5.** $12.17; $1472.97 **7.** $59.87; $3325.74
9. $103.23; $8361.23 **11.** $28.20; $1908.10
13. $2034.82 **15.** $724.47 **Maintaining Your Skills**
17. 600.38 **19.** 118.0248 **21.** 550.6677 **23.** 320.571

7-6; pages 230–232 **Self-Check 1.** $3.08; 10.50%
2. $11.32; 10.50% **Problems 3.** 11.25%
5. $14.05; 17.00% **7.** $14.23; 10.50% **9.** 13.00%
11. 18.00% **13.** 10.00% **15.** 17.25% **17.** 14.50%
19. Less than 10% **21.** 14.75%; 18.25% **23.** $128;
12.25%; $2115 **Maintaining Your Skills 25.** $19.44
27. $40.31 **29.** 30 **31.** 56.21 **33.** 70.16 **35.** 1.25
37. 421 **39.** 1.92

7-7; pages 233–234 **Self-Check 1.** 23.33%
2. $58.97824 = $58.98 **Problems 3.** $18 **5.** 22.22%;
$77.77 **7.** 14.29%; $40.24 **9.** $84.19 **11.** $84
13. $23.06 **15.** $781.66; $95.99 **Maintaining Your Skills**
17. 421.07 **19.** 340.00 **21.** 40.8 **23.** 6.895

Reviewing the Basics, page 235 **Skills 1.** $732.51
3. $84.60 **5.** $117.76 **7.** $3239.99 **9.** $162.24
11. 0.1263 **13.** 0.3862 **15.** $198 **17.** 0.75 **19.** 0.583
21. 0.2 **23.** 0.12 **25.** 0.153 **27.** 15% **29.** 13.7%
31. 22.35% **33.** 45.4% **35.** 2051.2% **Applications**
37. $63.21 **39.** $\frac{5}{6}$ **41.** $\frac{7}{12}$ **43.** $\frac{185}{365}$ **45.** $\frac{90}{360}$
47.–51. Answers will vary.

Unit 8 Automobile

8-1; pages 240–241 **Self-Check 1.** $10,551 **2.** $12,497
Problems 3. $11,000 **5.** $13,145 **7.** $24,128
9. $18,335 **11.** $891.20 **Maintaining Your Skills**
13. 9064 **15.** 13,306 **17.** 9609.70

8-2; pages 242–243 **Self-Check 1.** $7200 +
$2250 + $340 = $9790 **Problems 3.** $17,512
5. $10,018 **7.** $12,460.75 **9.** Sticker price:
$10,760; est. dealer's cost: $9088 **11.** $13,643.80
Maintaining Your Skills 13. 56 **15.** 166.32 **17.** 4410
19. 6846.4 **21.** 6728.25 **23.** 5377.22

8-3; pages 244–245 **Self-Check 1.** $4445
Problems 3. $5350 **5.** $5000 **7.** $5100 **9.** $11,825
Maintaining Your Skills 11. 6045 **13.** 5125
15. 7975 **17.** 1980

8-4; pages 246–249 **Self-Check 1.** $233.20 +
$62.00 + $204.00 = $499.20 base premium;
$748.80 annual premium **Problems 3.** $576.00;
$921.60 **5.** $708.40; $2762.76 **7.** $712.00; $961.20
9. $657.60; $2038.56 **11.** $517.20; $956.82
13. $524.40; $1494.54 **15.** $239.76 **17.** $2054.92;
$160.72 **19.** $287.60 **Maintaining Your Skills**
21. 820.16 **23.** 1305.57 **25.** 185.96 **27.** 864
29. 2519.544 **31.** 7423.5 **33.** 1187.808

8-5; pages 250–252 **Self-Check 1.** $2600; $0.26/mile
2. $3628.50; $0.30/mile **Problems 3.** $2250; $0.25
5. $5900; $0.40 **7.** $6634.40; $0.32 **9.** $6400;
$0.26 **11.** $0.17 **13.** $0.31 **15.** $0.25 **17.** $0.46
19. $0.32; $1987.38 **21.** $3059.09; $0.36 **Maintaining**
Your Skills 23. 21.75 **25.** 4.40 **27.** 0.04 **29.** 1.07
31. 3.21 **33.** 0.09 **35.** 0.28 **37.** 0.48

8-6; pages 253–254 **Self-Check 1.** $8101
2. $16,215 **Problems 3.** $5256; $5719 **5.** $5712;
$7047 **7.** $22,440; $24,061 **9.** $7452 **11.** $15,560
13. $9852.00 + $851.50 = $10,703.50
Maintaining Your Skills 15. 169.5 **17.** 382.55

8-7; pages 255–257 **Self-Check 1.** $120.00 + 94.60
+ 18.90 = $233.50 **2.** $233.50 ÷ 430 = $.543
Problems 3. $0.39 **5.** $224.88; $0.36 **7.** $76.89;
$0.32 **9.** $183.02; $0.38 **11.** $274.34; $0.42
13. $621.95; $0.57 **15.** $147.87; $0.55 **17.** $135.49;
$0.97 **19.** $0.67 **21.** $491; $0.33 **23.** $320.35;
$0.76 **Maintaining Your Skills 25.** 508.35
27. 113.838 **29.** 23.82 **31.** 1.39 **33.** 1.47

Reviewing the Basics; page 258 **Skills 1.** $312.75
3. $223.11 **5.** $103.45 **7.** $69.54 **9.** $196.30
11. $131.04 **13.** $969.86 **15.** $117.90 **17.** $0.13
19. $0.11 **21.** $4195.80 **Applications 23.** $4200
Terms 25. e **27.** a **29.** b

Unit 9 Housing Costs

9-1; pages 266–267 **Self-Check 1.** $20,000; $60,000
2. $60,000; $140,000 **Problems 3.** $69,600
5. $24,700; $74,100 **7.** $9700; $38,800 **9.** $21,920
11. $195,415 **13.** $84,375 **15.** $70,000 **Maintaining**
Your Skills 17. 21,000 **19.** 70,000 **21.** 21,275
23. 76,000 **25.** 86,250 **27.** 100,400

9-2; pages 268–269 **Self-Check 1.** $869.40;
$208,656; $118,656 **Problems 3.** $737.80; $221,340;
$151,340 **5.** $2637; $632,880; $407,880 **7.** $784.80;
$235,440; $155,440 **9.** $209,640 **11.** $720.60;
$129,708 **13.** $34,506 **Maintaining Your Skills**
15. 15,523.2 **17.** 5133 **19.** 80,041

9-3; pages 270–271 **Self-Check 1.** $45 + $1200 +
$120 + $250 + $180 + $660 = $2455 **Problems**
3. $3645 **5.** $14,695 **7.** $2811.25 **9.** $26,407
Maintaining Your Skills 11. 3440 **13.** 1872 **15.** 360
17. 127.65 **19.** 815 **21.** 1728.13

9-4; pages 272–274 **Self-Check 1.** $799.77 **2.** $23.43
3. $79,953.37 **Problems 5.** $700; $37.80; $69,962.20

7. $1400; $92.80; $119,907.20 **9.** $460 **11.** $402.08;
$36.80; $38,563.20 **13.** $600; $32.40; $59,967.60
15. $345; $39.12; $35,960.88 **17.** $833.94; $785.83;
$48.11; $81,951.89 **19.** $5126.38; $195.45
21. **a.** $303.10; $310.00; $31,317.99. $300.13;
$312.97; $31,005.02 **b.** $8.72; $604.38; $305.33.
$308.26; $2.93; $305.33; 0 **Maintaining Your Skills**
23. 91,760.45 **25.** 78,879.02 **27.** 14,807.3

9-5; pages 275–276 **Self-Check** **1.** 0.0655 **2.** $28,000
3. $1834 **Problems** **5.** $23,920 **7.** $1302.30
9. $5967.85 **11.** $14,742.51 **13.** $124,500
Maintaining Your Skills **15.** 124.1556 **17.** 3.74928
19. 2289.544 **21.** 63,000 **23.** 302,496 **25.** 34,650

9-6; pages 277–278 **Self-Check** **1.** $54,000
2. $21,600 **3.** $10,800 **Problems** **5.** $95,000
7. $115,200; $23,040 **9.** $259,200; $129,600;
$51,840; $25,920 **11.** $376,000; $37,600; $188,000;
$75,200; $188,000; $3760 **Maintaining Your Skills**
13. $24,000 **15.** 63,000 **17.** 489,600

9-7; pages 279–280 **Self-Check** **1.** $481 **Problems**
3. $303 **5.** $293 **7.** $882 **9.** $1836.08
Maintaining Your Skills **11.** 112,000 **13.** 35,000
15. 10,710 **17.** 69,100

9-8; pages 281–284 **Self-Check** **1.** No; $3100 ×
35% = $1085 **2.** Yes; $3600 × 35% = $1260
Problems **3.** $385 **5.** $308 **7.** $1904 **9.** $688.92;
Yes; $693 **11.** $1584.80; No; $1505 **13.** $742.36;
No; $696.50 **15.** No; $1540; Total = $1605.73
17. Mort. = $1234.80; Ins. = $40.08; Taxes =
$357.85; Total = $1840.23; FHA; Yes; $1890
Maintaining Your Skills **19.** 74,846.50 **21.** 826.70
23. 378.47 **25.** 515.13 **27.** 947.14 **29.** 1840
31. 1781.40 **33.** 8946 **35.** 98.1

Reviewing the Basics, page 285 **Skills** **1.** $1615
3. $36,192 **5.** $13,600 **7.** $4993.81 **9.** $46.93
11. $373.72 **13.** 40,809.6 **15.** $1545.95 **17.** $2843.75
Applications **19.** $909.60 **Terms** **21.** b **23.** f **25.** e

Unit 10 Insurance and Investments

10-1; pages 290–291 **Self-Check** **1.** $676
2. $56.33 **Problems** **3.** $1064 **5.** $4200; $1050
7. $4500; $2700 **9.** $1570; $549.50; $22.90
11. $112.50 each **Maintaining Your Skills** **13.** 5%
15. 20% **17.** 201.85 **19.** 573.3

10-2; pages 292–293 **Self-Check** **1.** $1250 **2.** $450
Problems **3.** $580 **5.** $12,000; $2400; $2650
7. $3200; $890 **9.** $426.40 **Maintaining Your Skills**
11. $498 **13.** 0.9 **15.** 10,679 **17.** 128 **19.** 703,560

10-3; pages 294–295 **Self-Check** **1.** $55.50
2. $531.60 **Problems** **3.** $19.10 **5.** 25; $2.99;
$74.75 **7.** $240 **9.** $30.75 **11.** $325.55; $1360;
$1627.75 **Maintaining Your Skills** **13.** 544.714
15. 21.624 **17.** 18.4 **19.** 8.975 **21.** 3436.2
23. 0.12 **25.** 52 **27.** 73,200 **29.** 0.57

10-4; pages 296–297 **Self-Check** **1.** $816.90
2. $2237.40 **Problems** **3.** $728.50 **5.** $29.00
7. $1014.60; $119.40 **9.** $64.16; $5.04 **Maintaining
Your Skills** **11.** 3.05 **13.** 24.44 **15.** 275.85
17. 1461.54 **19.** 63.26 **21.** 371 **23.** 1749
25. 28,552.5

10-5; pages 298–299 **Self-Check** **1.** $680.70
2. $18,169 **Problems** **3.** $4823.37; $323.37
5. 1.375666; $12,380.99; $3380.99 **7.** $3271.24;
$271.24 **9.** $548.45; $48.45 **11.** Granite Trust;
$85.83 **Maintaining Your Skills** **13.** 140 **15.** 340
17. 800 **19.** 0.067 **21.** 4.641 **23.** 2191.06
25. 7610

10-6; pages 300–301 **Self-Check** **1.** 9.11% **2.** 9.31%
Problems **3.** $6446.97; $446.97; 7.45% **5.** 1.096524;
$5482.62; $482.62; 9.65% **7.** 9.58% **9.** $859.88;
8.60% **11.** 9.00%; $372.33; 9.31% **13.** 8.75%;
$827.45; 9.04% **Maintaining Your Skills** **15.** 8.91%
17. 2550.00% **19.** 30,000 **21.** 33,960

10-7; pages 302–303 **Self-Check** **1.** $8775
2. $3286.50 **Problems** **3.** $536.50 **5.** $36,625;
$36,943 **7.** $12,150; $12,272 **9.** $12,225;
$12,408.38 **11.** Ballon; $63.75 **13.** $24,118.17
Maintaining Your Skills **15.** 3.125 **17.** 14.375
19. $18,200 **21.** $22,950

10-8; pages 304–305 **Self-Check** **1.** $45; 5.32%
Problems **3.** 5.23% **5.** 7.63% **7.** 4.55%
9. $807.50; 5.39% **11.** $740; 6.14% **13.** $0.32;
$96; 0.85% **Maintaining Your Skills** **15.** 0.12
17. 65.22 **19.** 25.83% **21.** 3.126% **23.** 13.145%

10-9; pages 306–307 **Self-Check** **1.** $535.50 profit
2. $2098 loss **Problems** **3.** $4067.25 **5.** $5165.40;
$759.60 loss **7.** $201.60 loss **9.** $11,500; $2,312.50
profit **11.** $923.44; $1857.56 loss **13.** $1774.10
profit **Maintaining Your Skills** **15.** 381.6 **17.** $219.84
19. 14.92 **21.** 16.492 **23.** 108.4176

10-10, pages 308–310 **Self-Check** **1.** $60; $805;
7.45% **2.** $725; $9200; 7.88% **Problems** **3.** 8.42%
5. $9237.50; $850; 9.20% **7.** $40; $875; 4.57%
9. $375; $9437.50; 3.97% **11.** $5046.04; $652.50;
2009 **Maintaining Your Skills** **13.** $13,987.50 **15.** 1080
17. 1034 **19.** 626.56 **21.** 31.34 **23.** 3.42

Reviewing the Basics, page 311 **Skills** **1.** $74.14
3. $318.33 **5.** $42,337.50 **7.** $171,802.30
9. $2972.50 **11.** $227.04 **13.** $81.97 **15.** $10.80
17. 0.3125 **19.** 1.625 **21.** 1.25 **23.** 9%
25. 18.8% **27.** 179% **Applications** **29.** $576.40
31–37. Answers will vary.

Unit 11 Recordkeeping

11-1, pages 316–317 **Self-Check** **1.** $774 **2.** $1533.55
Problems **3.** $3440.00; $688 **5.** $8802.34; $1760.47
7. $545.45; $109.09 **9.** $131.23 **11.** $675.00
13. No; do not know their monthly net income.
Maintaining Your Skills **15.** $1537.42 **17.** $1213.50

11-2, pages 318–320 **Self-Check 1.** $1820 **2.** $1446.57
Problems 3. $751.50 **5.** $1742 **7.** $1796.67
9. Rent, life insurance, car insurance, car registration
11. $1008.75 **13.** $226.79 **15.** No **Maintaining Your**
Skills 17. $241.00 **19.** $25.74 **21.** $2014.61
23. $60.25 **25.** $6.44

11-3, pages 321–323 **Self-Check 1.** $8.90 less **2.** $24.42
more **Problems 3.** $15.05 less **5.** $4.79 more
7. Telephone bill; Water bill **9.** $100.00; More;
$16.70 **11.** Less; $12.86 **13.** No **15.** $1740.09;
less; $56.57 **17.** Electric, heating fuel, water bill
19. $162; less; $1.55 **21.** More; $31 **23.** No
Maintaining Your Skills 25. $174.85 **27.** $2231.61
29. $6.84 **31.** $15.07 **33.** $352.19

Reviewing the Basics, page 324 **Skills 1.** $56.91
3. $111.11 **5.** $580.38 **7.** $75.05 **9.** $336.97
11. $104.96 **13.** 24.7% **15.** 9.7% **Applications**
17. $225.44 **19.** $399.34

Cumulative Review: Units 6–11, pages 329–330 **Skills**
1. 4.7 **3.** 96.201 **5.** $219.68 **7.** $75.00 **9.** $110
11. $1292 **13.** $35 **15.** $69 **17.** $1851 **19.** $104.21
21. $202.31 **23.** $1008.29 **25.** $56.12 **27.** $43.33
29. $557.30 **31.** $33,658.20 **33.** $17,080.80
35. $59.50 **37.** $4.34 **39.** $8.04 **41.** $13,502.67
43. 0.875 **45.** 0.65625 **47.** 74.8% **49.** 71.4%
51. 0.45 **53.** 2.12 **55.** $381.12 **57.** $18.88
59. $5.04 **61.** $0.79 **Applications 63.** $60.05
65. 24 months **67.** 26 **69.** 48 **71.** 117 **73.** $\frac{1}{3}$
75. $\frac{1}{4}$ **77.** $\frac{182}{365}$ **79.** $8.54 **81.** $5.82
83–105. Answers will vary.

Unit 12   Personnel

12-1, pages 342–343 **Self-Check 1.** $6120 **2.** $23,795
Problems 3. $979.10 **5.** $2417.90 **7.** $12,880.70
9. $6128.65 **11.** $30,892.55 **Maintaining Your Skills**
13. $2325 **15.** $17,698 **17.** $5380.40 **19.** $13,791.58

12-2, pages 344–345 **Self-Check 1.** $21,035.70
Problems 3. $17,020 **5.** $30,299 **7.** $22,984
9. $40,457.60 **11.** $36,240.60 **13.** $14,872.00;
$15,600.73; $7.50 **Maintaining Your Skills 15.** $22,975
17. $798.30 **19.** $468.10 **21.** $2220.95

12-3, pages 346–347 **Self-Check 1.** 30%
Problems 3. 35% **5.** 32% **7.** $3807.11; 32.1%
9. a. $1180, $1104.48, $1180, $2347.02, $5811.50;
$452, $423.07, $452, $899.03, $2226.10; $572,
$535.39, $572, $1137.71, $2817.10; $288, $269.57,
$288, $572.83, $1418.40 **b.** All 19% **Maintaining**
Your Skills 11. $1854.97 **13.** $951.05 **15.** 40%

12-4, pages 348–351 **Self-Check 1.** $21,510; $1792.50;
2. $971 **Problems 3.** $20,000.00, $1666.67;
5. $20,563.20; $1713.60 **7.** $21,932.40; $1827.70
9. $1076 **11.** $1556 **13.** $1483 **15.** $47,985.30;
$3998.78 **17.** $1483 **19. a.** $26,380.73; **b.** $2198.39
21. $1556 **23.** $761 **Maintaining Your Skills 25.** 35
27. 39 **29.** 26 **31.** 31 **33.** $959.30 **35.** $525.17
37. $18,376.47

12-5, pages 352–353 **Self-Check 1.** $931.80 **Problems**
3. $16.80; $63.70 **5.** $37.80; $318.20 **7.** $130.83;
$939.62 **9.** $524.70 **11.** $664.52 **Maintaining Your**
Skills 13. $207 **15.** $30.45 **17.** $99 **19.** $149.80
21. $428.55 **23.** $347.60 **25.** $380 **27.** $600.10

12-6, pages 354–355 **Self-Check 1.** $1637 **Problems**
3. $915 **5.** $2838 **7.** $1455 **9.** $4640 **11.** $1339.60
Maintaining Your Skills 13. $580 **15.** $592 **17.** $434
19. $670 **21.** $1551 **23.** $1606.11

Reviewing the Basics, page 356 **Skills 1.** $15,304.64
3. $1287.50 **5.** $820.19 **7.** $2428.49 **9.** 32.3%
Applications 11. $30,300 **Terms 13.** c **15.** g **17.** a

Unit 13   Production

13-1, pages 364–365 **Self-Check 1.** $0.02 **2.** $0.22
Problems 3. $0.092 **5.** $0.014; $0.036 **7.** $0.003
9. $0.048 **11.** $0.02 **Maintaining Your Skills**
13. 21.5¢ **15.** 0.7¢ **17.** 7.2¢ **19.** $0.05 **21.** $0.077
23. $0.147 **25.** $0.038 **27.** $0.041 **29.** $0.023

13-2, pages 366–367 **Self-Check 1.** 175,000 **2.** 166,000
Problems 3. 95,000 **5.** 450,000 **7.** 19,281 **9.** 28,986
11. 323,688 **13.** 480,000 **Maintaining Your Skills**
15. $5.36 **17.** $6.75 **19.** $37.59 **21.** 190,000
23. 274,039

13-3, pages 368–370 **Self-Check 1.** 2.5% **2.** 6%
Problems 3. 4%; In **5.** 8%; Out **7.** 6%; Out
9. In **11.** 8%; Out **13.** 2%; 0%; 6%; 4%; 4%;
Out of control at 1 p.m. **15.** 0%, 4%, 4%; 4%, 0%,
0%; 4%, 8%, 8%; 0%, 0%, 4%; 4%, 4%, 12%
17. 5%; 4%; 3.75%; 4%; 5%; 4% **Maintaining Your**
Skills 19. 6% **21.** 3.3% **23.** 2.7% **25.** 1%
27. 6% **29.** 6.7% **31.** 5% **33.** Same

13-4, pages 371–372 **Self-Check 1.** 8 **2.** 30
Problems 3. 66 **5.** 50 **7.** 4 **9. a.** Avg. 1.7; 14.9;
7.1; 1.5; 1.2 **b.** 26.4 sec. **c.** 136 **Maintaining Your**
Skills 11. 18.4 **13.** 35.7 **15.** 800 **17.** 217

13-5, pages 373–374 **Self-Check 1.** 31.25% **2.** 18.75%
Problems 3. 37.50% **5.** 12.5% **7.** 31.25%
9. 37.33% **11.** 30% **13.** 14.44%; 34.44%;
24.44%; 5.56%; 3.33%; 6.67%; 11.11% **Maintaining**
Your Skills 15. 10 **17.** 9 **19.** 23.33% **21.** 21.25%

13-6, pages 375–377 **Self-Check 1.** $20\frac{1}{2}$ in **2.** 73 cm
Problems 3. 0.8 cm **5.** 12.8 cm **7.** 0.4 cm
9. 25.2 cm **11.** 0.6 cm **13.** 49.4 cm **15.** $19\frac{5}{8}$ in H $\times$
$20\frac{1}{8}$ in W $\times$ $24\frac{5}{8}$ in L **17. a.** $7\frac{1}{4}$ in H $\times$ $9\frac{1}{2}$ in W $\times$
$7\frac{1}{4}$ in L **b.** $7\frac{1}{4}$ in H $\times$ 5 in W $\times$ 14 in L
19. a. 14.6 cm H $\times$ 28.6 cm W $\times$ 9.6 cm L
b. Answers will vary. **Maintaining Your Skills**
21. $\frac{5}{16}$ **23.** $\frac{9}{16}$ **25.** $1\frac{5}{16}$ **27.** $\frac{3}{8}$ **29.** $1\frac{5}{16}$ **31.** $\frac{1}{2}$ **33.** $1\frac{1}{2}$
35. $3\frac{3}{4}$ **37.** 2 **39.** 2 **41.** $2\frac{1}{4}$ **43.** $8\frac{3}{4}$

Reviewing the Basics, page 378 **Skills 1.** $0.18
3. $147.72 **5.** 192.3 **7.** $17.55 **9.** $0.09
11. 816,524 **13.** $8\frac{7}{16}$ **15.** $2\frac{1}{4}$ **17.** $34\frac{1}{2}$ **19.** 32.1%
Applications 21. $4000 **23.** 16.54 **Terms 25.** d **27.** e

Unit 14 Purchasing

14-1, pages 384–385 **Self-Check 1.** $114 **2.** $266
Problems 3. $105 **5.** $140; $260 **7.** $55.68; $118.32
9. $8.53; $86.22 **11.** $127.02 **13.** Dis.: **a.** $3.99
b. $5.63 **c.** $9.79; Net: **a.** $15.96 **b.** $18.86
c. $25.16 **15.** $97.92 **Maintaining Your Skills 17.** $4.78
19. $83.32 **21.** $139.75 **23.** $43.02 **25.** $139.74

14-2, pages 386–387 **Self-Check 1.** 70% **2.** $266
Problems 3. $90 **5.** 65%; $17.55 **7.** 76%; $241.28
9. 92%; $384.46 **11.** $1590.60 **13.** $44.17; $16.40;
$83.78; $96.82 **15. a.** $129.45 **b.** $50.34
Maintaining Your Skills 17. 50% **19.** 78% **21.** 92%
23. 58% **25.** $26.28 **27.** $74.35 **29.** $2628.26

14-3, pages 388–389 **Self-Check 1.** $56 **2.** 40%
Problems 3. 20.0% **5.** $18; 30.0% **7.** $37.81;
33.6% **9.** $888.89; 17.3% **11.** $20; 20%; 23.6%
13. 35%; 25%; 0%; 20% **Maintaining Your Skills
15.** $23.40 **17.** $40.67 **19.** $287.47 **21.** 30%
23. 25.0% **25.** 10.0%

14-4, pages 390–391 **Self-Check 1.** $280 **2.** $210
Problems 3. $70; $280 **5.** $248; $992; $148.80;
$843.20 **7.** $868; $1302; $325.50; $976.50
9. $60.36 **11.** $46.93; $41.44; $35.02 **13.** $349.91;
$103.36; $40.24; $5.36 **Maintaining Your Skills
15.** $114 **17.** $8178.14 **19.** 51 **21.** $56.61
23. $425.91 **25.** $13,530.44 **27.** $167.61
29. $1025.27 **31.** $5614.77

14-5, pages 392–393 **Self-Check 1.** 63%; $352.80
2. 37%; $207.20 **Problems 3.** 44%; $272.80
5. $209.28; 52%; $226.72 **7.** $74.49; 49.6%; $73.31
9. $1644.75; $505.25 **11.** $101.11; $148.16; $20.24;
32.5% **13.** $31.77; $46.68; $181.85; $159.97; $35.35;
$23.02; $101.14; $46.72 **Maintaining Your Skills
15.** $114.56 **17.** $2481.58 **19.** 70% **21.** 75%
23. 90% **25.** 93% **27.** $35.29

14-6, pages 394–395 **Self-Check 1.** $19.20 **2.** $620.80
Problems 3. $725.20 **5.** 0; $348.64 **7.** $6447.17
9. $898.44 **11.** $227.31 **Maintaining Your Skills
13.** $889.20 **15.** $107.28 **17.** $95.72 **19.** $4.44
21. $7.15

14-7, pages 396–398 **Self-Check 1.** December 10
2. $6958 **Problems 3.** $602.70 **5.** $287.39;
$6897.34 **7.** $907.24 **9.** $723.78 **11.** $7889.74
Maintaining Your Skills 13. $1403.62 **15.** $16.42
17. $1011.45 **19.** $8356.07

Reviewing the Basics, page 399 **Skills 1.** $9187.76
3. $71.87 **5.** $12.78 **7.** $278.24 **9.** $42,212.39
11. $5.38 **13.** 5.96 **15.** $5622.98 **17.** 20.0%
19. 33.3% **Applications 21.** 70% **23.** 31.6%
25. 49.6% **Terms 27.** c **29.** a **31.** e

Unit 15 Sales

15-1; pages 404–405 **Self-Check 1.** $11.86 **2.** $272.73
Problems 3. $3299.40 **5.** $64.89 **7.** $181.95
9. $70.86 **11.** $48.11 **13.** $4.71 **15.** $3.85

17. $131.61 **19.** $6.60; $0.55 **Maintaining Your Skills
21.** $10.89 **23.** $86.78 **25.** $1.22 **27.** $3.25
29. $4406.32

15-2; pages 406–407 **Self-Check 1.** $14.94; 30%
2. $87.22; 35% **Problems 3.** 50.1% **5.** 36.7%
7. $80.75; 48.2% **9.** 32.8% **11.** 20.0% **13.** 30.3%
Maintaining Your Skills 15. $6.13 **17.** $35.79
19. $1309.22 **21.** 24.1% **23.** 100.0% **25.** 50%

15-3; pages 408–409 **Self-Check 1.** $84; $70; $14
2. $385.60; $337.40; $48.20 **Problems 3.** $8.00
5. $27.00 **7.** $33.45 **9.** $6.69 **11.** $29.96
13. $246,461.75 **Maintaining Your Skills 15.** $1.11
17. $0.46 **19.** $7.67 **21.** $26.39 **23.** $25.88
25. $16.73 **27.** $82.42 **29.** $5887.33

15-4; pages 410–411 **Self-Check 1.** $74.72; $47.80;
$26.92; 22.5% **2.** $121.90; $78.58; $43.32; 19.3%
Problems 3. $25; 50% **5.** $29.07; 19% **7.** $3976.62;
47% **9.** 22.0% **11.** 42.2% **13.** 16.2% **15.** 30.2%
17. a. 23.8% **b.** 20.2% **Maintaining Your Skills
19.** $1.96 **21.** $112.47 **23.** $3241.93 **25.** 41.4%

15-5; pages 412–413 **Self-Check 1.** $460 **2.** $2.00
Problems 3. $10 **5.** 50%; $173.48 **7.** 62.5%;
$76.54 **9.** $18.70 **11.** $3.95 **13.** $50; $20; $25
15. 0 **Maintaining Your Skills 17.** 80% **19.** 60%
21. 52.4% **23.** $60.00 **25.** $786.38 **27.** $3337.25
29. $189.05

15-6; pages 414–415 **Self-Check 1.** 66.7% **2.** 100%
Problems 3. $0.50; 40% **5.** $38; 80% **7.** $0.91;
700% **9.** 150% **11.** 205.1% **13.** 205.1% **15.** 150%;
66.7%; 100%; 185.7% **Maintaining Your Skills
17.** $89.69 **19.** $109.80 **21.** $5157.90 **23.** 66.7%

15-7; pages 416–417 **Self-Check 1.** $35; $85 **2.** $70;
$210 **Problems 3.** $81.00 **5.** $129.60; $216.00
7. $657.83; $1409.63 **9.** $18.81 **11.** $33.48
13. $295.70 **15.** $0.82 **17.** $2.29; 62.4%
Maintaining Your Skills 19. $129 **21.** $1919.49
23. $180.00 **25.** $22.32 **27.** $8.08 **29.** $126.53

15-8; pages 418–419 **Self-Check 1.** $20; 25%
2. $69.92; 40% **Problems 3.** $5.00; 20.0%
5. $97.88; 39.5% **7.** 71.3% **9.** 33.3% **11.** 25.3%
13. a. 36.0%; 29.2%; 31.9%; 35.7%; 34.5% **b.** 24
× 24; $2.30 **c.** Outside: 36.0% **Maintaining Your
Skills 15.** $60.00 **17.** $33.50 **19.** 20.0% **21.** 39.9%
23. 45.0% **25.** 55.1% **27.** 5.0% **29.** 33.3%

Reviewing the Basics, page 420 **Skills 1.** 10.2%
3. 147.9% **5.** $165.82 **7.** $972.85 **9.** $53.74
11. $1195.61 **13.** $763.11 **15.** 0.5 **17.** 0.125
19. 0.0525 **21.** $801.00 **23.** 20.0% **Applications
25.** 60% **27.** 98% **Terms 29.** f **31.** c
33. e

Unit 16 Marketing

16-1, pages 426–427 **Self-Check 1.** 75% **2.** 92%
Problems 3. 80%8 **5.** 60% **7.** 89% **9.** 9.5%
11. a. 405; 602; 451; 282 **b.** 23.3%; 34.6%

25.9%; 16.2% **c.** 13.1%; 37.3%; 31.1%; 18.5%
d. 25–34 **Maintaining Your Skills 13.** 56 **15.** 300

16-2, pages 428–429 **Self-Check 1.** 900,000
Problems 3. 2%; 36,000 **5.** 12%; 12,000,000
7. 900,000 **9.** 4,698,000 **11.** 1,800,000
Maintaining Your Skills 13. 0.45 **15.** 11.52

16-3, pages 430–431 **Self-Check 1.** 10% **2.** 2.2%
Problems 3. 40% **5.** 28% **7.** 6% **9.** 10%
11. 9.1% **13.** 36.5% **15.** 7.8% **17.** 4.8%
Maintaining Your Skills 19. 25% **21.** 111.1%
23. 23 **25.** 0.7 **27.** 13

16-4, pages 432–433 **Self-Check 1.** $5.5 million;
$6.5 million; $7.5 million **Problems 3.** $38,000
5. $44,000 **7.** $50,000 **9.** $14.4; $14.8; $15.2
Maintaining Your Skills 11. $44,286.00 **13.** 17.6
15. 103.675 **17.** Fifty-two **19.** Three thousand one
hundred four

16-5, pages 434–435 **Self-Check 1.** $441,000
2. $1,435,000 **Problems 3.** $400.000 **5.** $4,050,000
7. $324,000 **9.** $390,000 **11.** $130,224 **13.** 44,650
Maintaining Your Skills 15. 23.78 **17.** $17,220
19. $5 **21.** $2400 **23.** $18

16-6, pages 436–437 **Self-Check 1.** $503.88 **2.** $341.50
Problems 3. $523.50 **5.** $41.46; $1658.40 **7.** $45.54;
$227.70 **9.** $277.20 **11.** $663.36 **Maintaining Your
Skills 13.** 8.82 **15.** 0.032 **17.** 0.147 **19.** 16.5

16-7, pages 438–439 **Self-Check 1.** $9600
2. $227,500 **Problems 3.** $11,000 **5.** $345,000
7. $800 **9.** $244,900 **11.** $33,750 **Maintaining
Your Skills 13.** $\frac{1}{8}$ **15.** $3\frac{19}{56}$

16-8, pages 440–441 **Self-Check 1.** $390,000;
$468,000 **2.** $4.50 **Problems 3.** $120,000
5. $17.14; $38.14; $258,020 **7.** $0.53; $1.58;
$2,061,500; $0.71; $1.76; $1,918,000; $1.67; $2.72;
$1,734,000; $2.00; $3.05; $1,300,000; $3.75 selling
price, 950,000 units **Maintaining Your Skills 9.** 576.02
11. 45.2

Reviewing the Basics, page 442 **Skills 1.** $4.75
3. $87.70 **5.** $139,840 **7.** 230.35 **9.** 0.55
11. $9241.50 **13.** 6944 **15.** 30.6% **Applications**
17. 3771; 3076; 2870; 1051; 2025; 3938; 4805; 10,768

Cumulative Review: Units 12–16, pages 445–446 **Skills**
1. $7.85 **3.** 80¢ **5.** 7.5% **7.** 125.2% **9.** $199.99
11. $61.91 **13.** $4194.51 **15.** $33.16 **17.** $27.43
19. $1946.76 **21.** $7847.95 **23.** $297.96 **25.** $1.15
27. $61.11 **29.** $4.29 **31.** $25.97 **33.** $61.00 **35.** $\frac{5}{8}$
37. $3\frac{7}{8}$ **39.** 5 **41.**   **43.** $14 **45.** $28 **47.** 0.3
49. 0.1 **51.** 0.055 **53.** 0.003 **55.** $125.34
57. $247.14 **59.** $15.00 **61.** 25.0% **63.** 39.0%
Applications 65. $14.24 **67.** 89.7% **69.** 75%
71. 95% **73.** 1986 **75.** Approx. 7000 **77.** 13
79. 3.08 **81.** $30,960 **83–101.** Answers will vary.

Unit 17 Warehousing and Distribution

17-1, pages 450–451 **Self-Check 1.** 14,700 ft^3

2. 56,000 cm^3 **Problems 3.** 510 in^3; 255,000 in^3
5. 0.288 m^3; 144 m^3 **7.** 5.292 yd^3; 635.04 yd^3
9. 1350 ft^3 **11.** 1036.8 ft^3 **13.** 52,500 ft^3
15. 2275 ft^3 **17.** 22.5 m^3 **Maintaining Your Skills**
19. 30 ft^3 **21.** 192 in^3 **23.** 494.7 cm^3 **25.** 18,000 ft^3
27. 28,000 in^3

17-2, pages 452–454 **Self-Check 1.** 350 **2.** 240
Problems 3. 340 **5.** 320; 340 **7.** 95 **9.** 119; 102;
26 **11.** 188; 164; 129; 273; 230 **13.** 40; 184; 126
15. 90 **17.** Date 8/1, Bal. 47; Date 8/6, Out 24, Bal.
23; Date 8/16, In 40, Bal. 63; Date 8/27, Out 36, Bal.
27; Date 8/31, In 50, Bal. 77 **Maintaining Your Skills**
19. 195 **21.** 133 **23.** 24 **25.** 1878 **27.** 4398

17-3, pages 455–457 **Self-Check 1.** $3.93 **2.** $196.50
Problems 3. a. $2.15 **b.** $49.45 **5. a.** $280.00;
$263.20; 2060; $543.20 **b.** $0.26 **c.** $249.60
7. $9.61 **9. a.** $465.00; $306.75; $710.50; 310;
$1484.25 **b.** $465; $285; $750 **c.** $23.75; $710.50;
$734.25 **11. a.** FIFO; Low cost items go first.
b. LIFO; High cost items go first. **13.** No, only if
costs consistantly rise or decline. **Maintaining Your
Skills 15.** $1785 **17.** $20,409.12 **19.** $5696.53
21. $38,572.38 **23.** $15.67 **25.** $141.73
27. $48.19 **29.** $79.16

17-4, pages 458–459 **Self-Check 1.** $60,000 **2.**
$17,000 **Problems 3.** $240 **5.** $1170 **7.** $384.14
9. $4200 **11.** $3880.80 **13. a.** $2622.90; $4121.70;
$1498.80; $749.40; $374.70; $187.35; $187.35;
$9742.20 **b.** $4871.10 **Maintaining Your Skills**
15. $24,000 **17.** $43,950 **19.** $25,070.50
21. $193,792.50 **23.** $1468.25 **25.** $226,596.67

17-5, pages 460–461 **Self-Check 1.** $14.79
2. $516.69 **Problems 3.** $662.70 **5.** $365.24
7. $340.41 **9.** $472.00 **11.** $44.33 **Maintaining Your
Skills 13.** $73.01 **15.** $22.61 **17.** $892.04

17-6, pages 462–463 **Self-Check 1.** $40.10 **2.** $168.42
Problems 3. $934.69 **5.** $14.37; $212.10 **7.** $36.67;
$139.71 **9.** $443.37 **11.** $129.68 **13.** $144.48
15. $901.09 **Maintaining Your Skills 17.** $236.64
19. $460.34 **21.** $216.87 **23.** $365.50 **25.** $300.96
27. $27.67 **29.** $1224.35 **31.** $432.53 **33.** $147.33

Reviewing the Basics, page 464 **Skills 1.** 500
3. 1100 **5.** 111 **7.** 1301 **9.** 168 **11.** $161.57
13. $38,797.06 **15.** $86.98 **17.** $10.45 **19.** $0.84
Applications 21. $2177.51 **23.** $0.73 **25.** 1260 m^3
Terms 27. a **29.** c

Unit 18 Services

18-1, pages 470–471 **Self-Check 1.** 5000 ft^2; $1666.67
2. 5625 ft^2; $1640.63 **Problems 3.** 450 ft^2; $300
5. 1500 ft^2; $1250 **7.** $403 **9.** $1560 **11.** $1619
13. $1021; $1082 **Maintaining Your Skills 15.** $2150
17. $756 **19.** $18,071

18-2, pages 472–473 **Self-Check 1.** $680 **2.** $277.65
Problems 3. $168.00; $593.00 **5.** $220.50;

$1168.06 **7.** $257.76 **9.** $2260.33 **11.** $710.10
13. $156.98 **Maintaining Your Skills** **15.** $146.25
17. $95.06 **19.** $46.53

18-3, pages 474–475 **Self-Check** **1.** $203.52 **2.** $620.34
Problems **3.** $94.78; $394.28 **5.** $102.76; $873.46
7. $1231.65 **9.** $469.90 **11.** $2063.56 **Maintaining Your Skills** **13.** $25.80 **15.** $8.05

18-4, pages 476–477 **Self-Check** **1.** $27.76 **2.** $32.08
Problems **3.** $28.17 **5.** $2.10; $1.33; $45.78
7. $30.40 **9.** $35.94 **11.** $64.01 **Maintaining Your Skills** **13.** $36.81 **15.** $44.44

18-5, pages 478–480 **Self-Check** **1.** $1521 **Problems**
3. $2401 **5.** $2223.90 **7.** $1598.13 **9.** $2527.53
Maintaining Your Skills **11.** $87 **13.** $145.50

18-6, pages 481–482 **Self-Check** **1.** $31,000
2. $2512.50 **Problems** **3.** $1500 **5.** $5000
7. $150,000 **9.** $9130 **11.** $650,000
13. $526,912.50 **Maintaining Your Skills** **15.** $885.55
17. $102,250 **19.** $57,082.50

Reviewing the Basics, page 483 **Skills** **1.** $43
3. $4172 **5.** $1601 **7.** $27,163 **9.** 3237 **11.** 2631.18
13. 23.21 **15.** 4064.88 **17.** 7200 **19.** 7200
21. 2762.55 **23.** $348.17 **25.** $887.92 **27.** $8173.75
Applications **29.** 1200 ft^2 **31.** 18,000 ft^2
33. 509.25 ft^2 **Terms** **35.** d **37.** a **39.** b

Unit 19   Accounting

19-1, pages 484–491 **Self-Check** **1.** $285.96
Problems **3.** $114.60 **5.** $231.59; $743.41
7. $102.70; $416.30 **9.** $318.92 **11.** $62, $45.90,
$177.65, $422.35; $65, $44.75, $170.74, $414.26; $52,
$31.75, $119.92, $295.08; $68, $43.21, $171.62, $393.18;
$69, $41.12, $151.81, $385.61; $57, $47.37, $174.99,
$444.21; $82, $46.85, $173.91, $438.53; $3933.94,
$455, $300.95, $107.84, $68.85, $208.00, $1140.64,
$2793.20 **13.** $505.50, $53, $38.67, $5.06, $116.95,
$388.55; $536.40, $81, $41.03, $5.36, $148.85, $387.55;
$617.88, $63, $47.27, $6.18, $149.04, $468.84; $208.25,
$27, $15.93, $2.08, $53.34, $154.91; $261.80, $23,
$20.03, $2.62, $56.12, $205.68; $2129.83, $247, $93.07,
$162.93, $21.30, $524.30, $1605.53 **Maintaining Your
Skills** **15.** $247.63 **17.** $386.36 **19.** $1446.82

19-2, pages 492–493 **Self-Check** **1.** 7.5% **2.** 16.7%
Problems **3.** 10.0% **5.** 7.5% **7.** 66.7%
9. $131,704; 64.2% **Maintaining Your Skills**
11. $202,783 **13.** $449.18 **15.** 25.0%
17. 16.6% **19.** 2.0%

19-3, page 494–495 **Self-Check** **1.** $384 **2.** $1250
Problems **3.** $1250 **5.** $19,531.25 **7.** $17.25
9. $5760.00; $12,480.00 **11.** **a.** M: $1440.00; U:
$7200.00; S: $1152.00 **b.** M: $1038.00; U: $5190.00;
S: $830.40 **c.** M: $762.00; U: $3810.00; S: $609.60
Maintaining Your Skills **13.** 0.021 **15.** 0.005

19-4, page 496–497 **Self-Check** **1.** $500 **2.** $150
Problems **3.** $400.00 **5.** $1275.00 **7.** $8500

9. $10,947 **11.** $28,750 **13.** $360 **Maintaining
Your Skills** **15.** $286 **17.** $669 **19.** $735.90
21. $153.69 **23.** $80.85 **25.** $1.47

19-5, page 498–499 **Self-Check** **1.** $36,000
2. $16,000 **Problems** **3.** $2400; $2400; $1100
5. $100; $100; $180 **7.** $4665 **9.** $9472.34
11. $23,690, $23,690, $91,070; $23,690, $47,380,
$67,380; $23,690, $71,070, $43,690; $23,690,
$94,760, $20,000 **Maintaining Your Skills** **13.** $1650
15. $70.40 **17.** $102,600 **19.** $197,200
21. $7186.50

19-6, page 500–501 **Self-Check** **1.** $48,000
2. $36,000 **Problems** **3.** $2600 **5.**$\frac{10}{55}$; $20,000; $\frac{9}{55}$;
$18,000 **7.** $6000 **9.** $2720; $2040; $1360; $680
11. $28,800, $28,800, $51,200; $21,600, $50,400,
$29,600; $14,400, $64,800, $15,200; $7200, $72,000,
$8000 **Maintaining Your Skills** **13.** $8420 **15.** $5340
17. $1137

19-7, page 502–503 **Self-Check** **1.** $40,000; $40,000
2. $20,000; $20,000 **Problems** **3.** $3290 **5.** 33$\frac{1}{3}$%;
$14,000 **7.** $2400; $4800 **9.** $20,000, $30,000;
$32,000, $18,000; $7200, 39,200, $10,800; $4320,
$43,520, $6480; $2592, $46,112, $3888 **Maintaining
Your Skills** **11.** $540 **13.** $2970 **15.** $103,900
17. $426

19-8, page 504–505 **Self-Check** **1.** $3400.00;
$5440.00; $3264.00; $1958.40; $1958.40; $979.20
2. $6958 **Problems** **3.** $1848.00; $2956.80;
$1774.08; $1064.45; $1064.45; $532.22
5. $6968.00; $11,148.80; $6689.28; $4013.57;
$4013.57; $2006.78 **7.** Yr. 1, Dep. $6797.28;
Yr. 2, Dep. $11,657.24; Yr. 3, Dep. $8325.24;
Yr. 4, Dep. $5945.24; Yr. 5, Dep. $4250.68;
Yr. 6, Dep. $4250.68; Yr. 7, Dep. $4250.68;
Yr. 8, Dep. $2122.96; $47,600.00 **9.** $48,260.00,
$9,652.00, $9,652.00, $38,608.00; $48,260.00,
$15,443.20, $25,095.20, $23,184.80; $48,260.00,
$9,265.92, $34,361.12, $13,898.88; $48,260.00,
$5559.55, $39,920.67; $8339.33; $48,260.00;
$5559.55; $45,480.22, $2779.78; $48,260.00,
$2779.78, $48,260.00, ($0.00) **Maintaining Your Skills**
11. $28,310 **13.** $118

Reviewing the Basics, page 507 **Skills** **1.** $7151
3. $144 **5.** $197,478 **7.** $69.01 **9.** $4.28
11. $17,269.90 **13.** $74,000 **15.** $6800 **17.** $1198.80
19. $55,725 **21.** 0.065 **23.** 0.03 **25.** $1080
27. $15,531.25 **29.** $5.39 **31.** 15.9% **33.** 13.8%
Applications **35.** 2400 ft^2

Unit 20   Accounting Records

20-1, pages 512–513 **Self-Check** **1.** $132,000
2. $42,000 **Problems** **3.** $77,000 **5.** $36,250
7. $17,450 **9.** $104,015; $23,397.80; $80,617.20
11. $221,797; $142,257; $79,540 **Maintaining
Your Skills** **13.** $42,758.35 **15.** $20,422.38
17. $39,052.00 **19.** $21,624.05

20-2, pages 514–516 **Self-Check** **1.** $67,000
2. $67,000 **Problems** **3.** $33,000; $12,300; $20,700
5. $52,500; $22,500; $27,500 **7.** $99,800; $65,000;
$34,800 **9.** 80,467.50; 51,013.20; 29,454.30;
80,467.50 **11.** Total Assets $199,504,200; Total
Liabilities $142,599,410; Owner's Equity, Capital
$56,904,790; Total Liabilities and Owner's Equity
$199,504,200 **Maintaining Your Skills** **13.** $27,970
15. $5,072, 400 **17.** $9.9

20-3, pages 517–518 **Self-Check** **1.** $16,920
2. $12,404 **Problems** **3.** $96,600 **5.** $17,950
7. $58,127 **9.** $7871.68 **11.** $9921.32 **13.** $99.92
Maintaining Your Skills **15.** $16,596.26 **17.** $65,319.64
19. $4349.88 **21.** $95.76 **23.** $17.40

20-4, pages 519–521 **Self-Check** **1.** $61,500; $26,800
2. $289,000; $183,600 **Problems** **3.** $4685
5. $3675; $1935; $1107 **7.** $673,642; $255,726;
$159,307 **9.** $388,125; $157,125; $85,209 loss
11. NS: $3080; GP: $2209; NI: $871 **13.** GP: $58.3;
OE: $38.7; NI: $19.6 **Maintaining Your Skills**
15. $766.22 **17.** $14,709.46 **19.** $90.14 **21.** $5715

20-5, pages 522–524 **Self-Check** **1.** 1.4% **2.** 31.8%
3. 1.7 to 1 **4.** 1.3 to 1 **Problems** **5.** 50.0%
7. 44.5% **9.** 2.0:1; 1.2:1 **11.** $3553; 46.2%
13. 1.6:1; 1.3:1 **Maintaining Your Skills** **15.** 9.1%
17. 78.4% **19.** 1.0:1

20-6, pages 525–526 **Self-Check** **1.** 25%
2. −28.6% **Problems** **3.** 15.3% **5.** $23,090;
20.1% **7.** −$775; −16.7% **9.** $8943; 21.1%
11. **a.** −$7.5; −6.4% **b.** −$2.2; −3.9%
c. −$5.3; −8.8% **d.** $0.5; 1.4% **e.** −$5.8;
−23.4% **Maintaining Your Skills** **13.** 20.0%
15. 67.2% **17.** 140.4%

Reviewing the Basics, page 527 **Skills** **1.** $1,345,902
3. $9262 **5.** $35,885 **7.** $8459.51 **9.** $7723.96
11. $72,769.79 **13.** 19.5% **15.** 29.7% **Applications**
17. $19,325 **19.** $903,500 **Terms** **21.** c **23.** b

Unit 21 Financial Managment

21-1, pages 532–533 **Self-Check** **1.** $16,650 **2.** $11,250
Problems **3.** $5511.90 **5.** $64,823; $11,205.75
7. $194,884; $58,754.76 **9.** $7737.25 **11.** $13,838.40
Maintaining Your Skills **13.** $6.18 **15.** 507.5
17. 277.44 **19.** 44.88 **21.** $27,816

21-2, pages 534–535 **Self-Check** **1.** $7,568,000
2. $36,965,150 **Problems** **3.** $455,500; $6,344,500
5. $1,050,000; $1,104,650; $19,895,350
7. $6,978,133.75 **9.** $1,156,250 **11.** $2,525,737.50
Maintaining Your Skills **13.** $63,700 **15.** $44,580

21-3, pages 536–537 **Self-Check** **1.** $102,812.50
2. $155,541.67 **Problems** **3.** $1750; $71,750
5. $1312.50; $38,812.50 **7.** $1835.44; $39,485.44
9. $52,612.85 **11.** **a.** $117,725 **b.** $1,567,725
Maintaining Your Skills **13.** 3.4 **15.** 0.007

21-4, pages 538–539 **Self-Check** **1.** $95,600
Problems **3.** $2900; $97,135 **5.** $2194.38;
$82,835.62 **7.** $2365.72; $94,674.28 **9.** $92,816.12
11. $204,039.38 **Maintaining Your Skills** **13.** 60.548
15. 4.586 **17.** 0.007 **19.** 21.875

21-5, pages 540–541 **Self-Check** **1.** $866.67 **Problems**
3. $4120; $4095 **5.** $1836.25 **7.** $3003.75
9. $8342.55 **Maintaining Your Skills** **11.** 0.23 **13.** 0.34

21-6, pages 542–544 **Self-Check** **1.** $39,850
2. $142,250 **Problems** **3.** $12,595 **5.** $12,041
7. $8550 **9.** $2,777,290.40 **Maintaining Your Skills**
11. 600.712 **13.** 5.361

Reviewing the Basics, page 545 **Skills** **1.** $61,628.25
3. $980,400 **5.** $145,687.50 **7.** $2723.33
9. $4583.33 **11.** 0.095 **13.** 1.75 **15.** $8666
17. $1789.84 **19.** $0.22 **Applications**
21. $19,526.40 **Terms** **23–27.** Answers will vary.

Unit 22 Corporate Planning

22-1, pages 554–556 **Self-Check** **1.** 6.7%
2. $13,416.80 **3.** $25.50 **Problems** **5.** 1.5%
7. $18,086.49 **9.** $216,859.43 **11.** $3.93 **13.** 26.4%
15. $23,990.05 **17.** $0.59 **19.** $16,932.46
21. $0.77 **23.** **a.** −75% **b.** It is negative
c. Deflation **Maintaining Your Skills** **25.** $5.63
27. 5.0% **29.** 11.2%

22-2, pages 557–558 **Self-Check** **1.** $40,996,800
2. $1179.91 **Problems** **3.** $118.3 mil; $309
5. $448.9 bil; $6450.13 **7.** $2,652.3 bil; $1170.83
9. $732,672,000 **11.** $203.28 billion; $13,216.39
13. 461.7 billion; $18,320.96 **15.** $148.257 billion;
$15,457.67 **17.** **a.** France **b.** U.K. **c.** France **d.** U.K.

22-3, pages 559–561 **Self-Check** **1.** 250.0 **2.** $2680.83
3. $17.50 **Problems** **5.** 212.8 **7.** $666.76 **9.** $5.71
11. $163.80 **13.** $334.7 **15.** $118,462.75 **17.** $50.78
19. $188.36 **21.** 169.2 **23.** **a.** Less than 100 **b.** 0
c. No **Maintaining Your Skills** **25.** 1.3 **27.** 0.4
29. $59,409.95

22-4, pages 562–564 **Self-Check** **1.** Budget;
$332,000; $24,000; $20,000; $16,000; $8,000
2. Difference; +$28,000; −$4000; +$6000;
−$7000; −$5000 **Problems** **3.** $20,000; −$5000
5. $60,200; +$3800 **7.** $3,763,200; −201,200
9. $105,000; $30,000; $15,000 **11.** −$40,000;
−$30,000; +$10,000; −$60,000 **13.** $14,610,240;
$304,380; $304,380 **15.** Yes; +$244,000
17. **Budget** $27,000; $9000; $4500; $3600; $900;
total $45,000; **Difference** −$2500; −$700; −$400;
−$1200; +$300; total −$4500 **Maintaining Your**
Skills **19.** $3,175,743.50 **21.** $122,996.66
23. $9578.94

Reviewing the Basics, page 565 **Skills** **1.** $48.99
3. $2841.97 **5.** 25.0% **7.** 17.4% **9.** $6.44
11. $44.20 **13.** $534.74 **15.** $23.72 **17.** $37.09
19. $27.35 **21.** 2.9% **23.** $52.11 **Applications**
25. $1207.64 **27.** $2867.99 **Terms** **29–35.** Answers
will vary.

Cumulative Review: Units 17–22, pages 569–570 **Skills**

1. $179 3. $4218 5. 135 7. $1069 9. $229
11. $49 13. $184 15. $7756 17. $19.70 19. $85.12
21. $4039.95 23. $454.15 25. $512.46 27. $344
29. $7500 31. $32,175 33. $8.40 35. $22.57
37. $3168 39. $15 41. $106 43. $150.71 45. $2.55
47. $14.22 49. $566.43 51. $36 53. $36,735.30
55. $1516 57. 0.08 59. 0.7435 61. $496.48
63. $26.57 65. $14,551.53 67. 47.6% 69. 20.7%
Applications 71. over $100,000 73. $0–$50,000
75. $3466.20 77. $2.31 79. $0.57 81. 180 ft^2
83. 43,620 in^2 85. 1650 cm^2 87. 246.24 m^3
89. $38\frac{1}{2}$ ft^3

Skills File

1–pages 580–581 1. ones; 5 3. ten; 70 5. ones; 8
7. tens; 0 9. hundreds; 700 11. ones; 9 13. tenths;
0.3 15. thousandths; 0.009 17. hundredths; 0.02
19. ones; 4 21. hundredths; 0.02 23. thousandths;
0.007 25. thousandths; 0.003 27. thousandths;
0.001 29. Thirty-four 31. Seventy-eight 33. Eight
hundred forty-two 35. Five thousand twelve 37. One
hundred nineteen 39. One thousand two hundred
forty 41. One hundred ninety-four 43. Twenty-five
dollars 45. Seventeen and $\frac{9}{100}$ dollars 47. One
hundred twenty and $\frac{17}{100}$ dollars 49. Seven thousand
seven hundred fourty-nine dollars 51. Fifty cents or $\frac{50}{100}$
dollars 53. $71.64 million 55. 18.4 thousand 57. 171.6
thousand 59. 2.135 billion 61. $3,400,000
63. $11,200,000,000 65. 7,300,000,000 67. 400,000
69. $720,000,000 71. 54 73. 835 75. 211 77. 5763
79. 4444 81. 1236 83. 650 85. $101 87. $93
89. $742 91. $791,000 93. 2.0 95. 25.1 97. 58
99. 0.1 101. 0.5 103. 0.020 105. $10 107. $19
109. $600 111. $9521.39 113. $2000.00

2–page 582 1. 30 3. 70 5. 220 7. 8100 9. 600
11. 6000 13. 16,000 15. 1000 17. 0.1 19. 0.5
21. 0.47 23. 0.36 25. 0.601 27. 0.213
29. 137.492 31. 38,000 33. 48,000 35. 900
37. 30 39. 580 41. 36 43. 189.7 45. 0.1
47. 152.43 49. 0.179 51. 108.437

3–page 583 1. 106 3. 137 5. 1189 7. 799 9. 897
11. 488 13. 10,878 15. 8987 17. 71 19. 210
21. 897 23. 1542 25. 4638 27. 12,224 29. 12,201
31. 12,787 33. 3798 35. 12,318 37. 8152 39. 9484
41. 6777 43. 14,953 45. 33,587 47. 87,481
49. 14,999 51. 14,450

4–page 584 1. 331 3. 151 5. 3442 7. 3311
9. 1111 11. 8020 13. 6 15. 6231 17. 64
19. 99 21. 2258 23. 1166 25. 1909 27. 3667
29. 1275 31. 859 33. 2983 35. 1568 37. 9111
39. 1862 41. 2238 43. 2901 45. 16,526
47. 37,503 49. 126,403 51. 100,621

5–page 585 1. 478.99 3. 801.12 5. 791.19
7. 543.113 9. 816.659 11. 64.54 13. 177.349
15. 21.901 17. 24.460 19. 82.332 21. 10.122
23. 460.909 25. 502.175 27. 1064.683 29. 1.241
31. 21.901 33. 220.611 35. 772.652

6–page 586 1. 51.15 3. 134.89 5. 5.09 7. 25.51
9. 79.68 11. 203.84 13. 70.85 15. 84.32 17. 19.6
19. 8.07 21. 57.44 23. 377.63 25. 5.09 27. 19.983
29. 325.86 31. 204.354 33. 557.87 35. 5215.946
37. 1.1178 39. 778.029 41. 3.59806 43. 2.74502

7–page 587 1. 1533 3. 87,344 5. 100,152
7. 77,256 9. 959,373 11. 2808 13. 251,565
15. 4,268,236 17. 1665 19. 310,992 21. 1,662,024
23. 69,000 25. 28,182 27. 357,000 29. 36,060,000
31. 4,200,672 33. 43,955,000 35. 1,070,443
37. 169,130 39. 623,770 41. 1,309,622
43. 2,804,972 45. 15,805,077

8–page 588 1. 4.5 3. 87.822 5. 188.55 7. 230.8218
9. 43.9556 11. 287.352 13. 0.1008 15. 0.096
17. 0.584 19. 0.6156 21. 0.0567 23. 0.0861
25. 4116 27. 36 29. 876.403 31. 1212.86 33. 112.5
35. 81.9698 37. 161,713.902 39. 0.156462 41. 84
43. 0.07608533

9–page 589 1. $2\frac{9}{27}$ 3. $2\frac{2}{41}$ 5. $14\frac{12}{53}$ 7. $84\frac{17}{21}$ 9. $67\frac{19}{43}$
11. $59\frac{12}{23}$ 13. $71\frac{42}{81}$ 15. $16\frac{13}{114}$ 17. $15\frac{17}{321}$ 19. $8\frac{792}{892}$
21. $23\frac{22}{409}$ 23. $13\frac{617}{625}$ 25. $20\frac{831}{843}$ 27. 18 29. 16
31. 72 33. $14\frac{22}{621}$ 35. $12\frac{349}{549}$ 37. $5\frac{765}{842}$ 39. $181\frac{29}{473}$
41. $87\frac{527}{622}$ 43. $1600\frac{3}{51}$ 45. $252\frac{46}{88}$ 47. $208\frac{166}{417}$
49. $1345\frac{177}{514}$

10–page 590 1. 11.125 3. 21.2 5. 5.1 7. 19.75
9. 34.5 11. 22.8 13. 16.1 15. 206.8 17. 2.02
19. 10.865 21. 21.524 23. 8.71 25. 10.45 27. 31.13
29. 5.63 31. 17.5 33. 14.44 35. 230.05 37. 41.42
39. 25.54 41. 39.5

11–page 591 1. 4.6 3. 5.71 5. 17 7. 6.5 9. 42.91
11. 26 13. 31 15. 26 17. 8.96 19. 13.21 21. 1.41
23. 245 25. 1.21 27. 256 29. 67.48 31. 22.57
33. 1.75 35. 38.5 37. 98.86 39. 0.01 41. 17.13
43. 0.01

12–page 592 **1.** 10 **3.** 50 **5.** 24 **7.** 22 **9.** 51
11. 24 **13.** 35 **15.** 72 **17.** 20 **19.** $\frac{2}{3}$ **21.** $\frac{1}{2}$ **23.** $\frac{2}{3}$
25. $\frac{1}{3}$ **27.** $\frac{5}{6}$ **29.** $\frac{10}{27}$ **31.** $\frac{4}{7}$ **33.** $\frac{1}{2}$ **35.** $\frac{3}{4}$ **37.** $\frac{94}{183}$
39. $\frac{1}{2}$ **41.** $\frac{19}{36}$ **43.** $\frac{2}{13}$ **45.** $\frac{4}{25}$

13–page 593 **1.** $\frac{25}{4}$ **3.** $\frac{59}{8}$ **5.** $\frac{20}{3}$ **7.** $\frac{39}{5}$ **9.** $\frac{43}{10}$ **11.** $\frac{145}{32}$
13. $\frac{16}{3}$ **15.** $\frac{29}{10}$ **17.** $\frac{53}{16}$ **19.** $\frac{165}{32}$ **21.** $\frac{79}{11}$ **23.** $\frac{57}{5}$
25. $6\frac{1}{2}$ **27.** $5\frac{3}{4}$ **29.** $5\frac{1}{2}$ **31.** $2\frac{1}{3}$ **33.** $2\frac{1}{2}$ **35.** $2\frac{20}{21}$
37. $5\frac{5}{6}$ **39.** $3\frac{1}{6}$ **41.** $6\frac{1}{8}$ **43.** $6\frac{3}{7}$

14–page 594 **1.** 0.8 **3.** 1.13 **5.** 2.43 **7.** 1.37 **9.** 0.3
11. 4.38 **13.** 3.08 **15.** 0.03 **17.** 5.15 **19.** 2.12
21. 7.14 **23.** 0.44 **25.** $\frac{1}{10}$ **27.** $\frac{3}{10}$ **29.** $\frac{3}{100}$ **31.** $\frac{53}{100}$
33. $\frac{3}{1000}$ **35.** $\frac{8}{125}$ **37.** $\frac{111}{250}$ **39.** $\frac{469}{1000}$ **41.** $7\frac{1}{2}$ **43.** $1\frac{2}{25}$
45. $2\frac{83}{100}$ **47.** $\frac{21}{200}$ **49.** $\frac{1}{2000}$ **51.** $1\frac{402}{625}$ **53.** $\frac{1}{5000}$

15–page 595 **1.** $1\frac{2}{5}$ **3.** $\frac{3}{4}$ **5.** $1\frac{4}{7}$ **7.** $1\frac{4}{25}$ **9.** $1\frac{5}{16}$
11. $1\frac{8}{45}$ **13.** $11\frac{2}{3}$ **15.** $14\frac{2}{7}$ **17.** $12\frac{1}{3}$ **19.** $17\frac{1}{4}$
21. $9\frac{7}{8}$ **23.** $11\frac{2}{5}$

16–page 596 **1.** $1\frac{1}{10}$ **3.** $\frac{20}{21}$ **5.** $1\frac{1}{6}$ **7.** $1\frac{13}{110}$ **9.** $\frac{59}{60}$
11. $\frac{71}{150}$ **13.** $9\frac{1}{6}$ **15.** $16\frac{1}{3}$ **17.** $20\frac{19}{26}$ **19.** $28\frac{22}{63}$
21. $28\frac{43}{56}$ **23.** $29\frac{5}{48}$

17–page 597 **1.** $\frac{2}{9}$ **3.** $\frac{5}{8}$ **5.** $\frac{4}{7}$ **7.** $\frac{1}{8}$ **9.** $\frac{1}{5}$ **11.** $\frac{2}{5}$
13. $\frac{1}{4}$ **15.** $\frac{7}{20}$ **17.** $1\frac{1}{7}$ **19.** $3\frac{1}{4}$ **21.** $2\frac{1}{6}$ **23.** $5\frac{3}{16}$
25. $2\frac{7}{13}$ **27.** $8\frac{1}{2}$ **29.** $25\frac{2}{7}$

18–page 598 **1.** $\frac{3}{8}$ **3.** $\frac{3}{20}$ **5.** $\frac{1}{18}$ **7.** $\frac{1}{15}$ **9.** $\frac{3}{20}$ **11.** $\frac{1}{10}$
13. $4\frac{1}{4}$ **15.** $6\frac{5}{18}$ **17.** $5\frac{7}{60}$ **19.** $3\frac{1}{8}$ **21.** $2\frac{3}{20}$ **23.** $2\frac{1}{8}$

19–page 599 **1.** $1\frac{1}{2}$ **3.** $2\frac{1}{10}$ **5.** $2\frac{6}{11}$ **7.** $3\frac{1}{18}$ **9.** $12\frac{11}{12}$
11. $1\frac{37}{70}$ **13.** $2\frac{7}{8}$ **15.** $1\frac{3}{8}$ **17.** $\frac{1}{8}$ **19.** $3\frac{7}{9}$ **21.** $\frac{9}{20}$
23. $11\frac{5}{16}$ **25.** $14\frac{7}{9}$ **27.** $13\frac{5}{7}$ **29.** $12\frac{5}{32}$

20–page 600 **1.** $\frac{1}{3}$ **3.** $\frac{5}{8}$ **5.** $\frac{5}{12}$ **7.** 2 **9.** 9 **11.** $16\frac{2}{15}$
13. $58\frac{1}{3}$ **15.** $46\frac{1}{8}$ **17.** $\frac{1}{6}$ **19.** $\frac{11}{12}$ **21.** 6 **23.** $3\frac{3}{4}$
25. $19\frac{1}{2}$ **27.** $8\frac{8}{15}$ **29.** 64 **31.** $25\frac{29}{72}$ **33.** $97\frac{11}{12}$ **35.** $71\frac{1}{24}$

21–page 601 **1.** 2 **3.** $1\frac{7}{8}$ **5.** $7\frac{1}{2}$ **7.** 6 **9.** 16
11. $3\frac{7}{36}$ **13.** $2\frac{11}{12}$ **15.** $\frac{1}{4}$ **17.** $5\frac{5}{33}$ **19.** $4\frac{20}{133}$ **21.** $1\frac{1}{2}$
23. $1\frac{1}{2}$ **25.** $\frac{2}{3}$ **27.** 15 **29.** 6 **31.** 30 **33.** $\frac{43}{176}$
35. $\frac{143}{375}$ **37.** $6\frac{1}{2}$ **39.** 1

22–page 602 **1.** 1500:600 **3.** 600:1200 **5.** 600:1500
7. $\frac{1}{5}$ **9.** $\frac{1}{15}$ **11.** $\frac{1}{25}$ **13.** $\frac{2}{5}$

23–page 603 **1.** T **3.** F **5.** F **7.** T **9.** T **11.** F
13. 18 **15.** $7\frac{1}{3}$ **17.** $\frac{6}{11}$ **19.** $5\frac{5}{9}$ **21.** $12\frac{4}{13}$ **23.** 20
25. 5 **27.** 24 **29.** 64 **31.** 15 **33.** 50

24–page 604 **1.** $\frac{99}{12} = \frac{x}{7}$ **3.** $\frac{40}{5} = \frac{x}{17}$ **5.** $3.48 **7.** 117
pp. **9.** 1485 wheels **11.** $28 **13.** $33\frac{1}{3}$ minutes

25–page 605 **1.** 288 **3.** 80 **5.** 64; 640 **7.** A **9.** B
11. B

26–page 606 **1.** 10% **3.** 25% **5.** 82% **7.** 213%
9. 575.3% **11.** 39.1% **13.** 1210.4% **15.** 1082%
17. 10.6% **19.** 4% **21.** 50.3% **23.** 70% **25.** 710%
27. 1050% **29.** 1260% **31.** 2250% **33.** 3720%
35. 750% **37.** 1630% **39.** 620% **41.** 380%
43. 67% **45.** 880% **47.** 32% **49.** 300.3%
51. 21.87% **53.** 2500% **55.** 14% **57.** 51%
59. 186.8% **61.** 0.3% **63.** 26% **65.** 0.32%
67. 29% **69.** 342% **71.** 210% **73.** 167%
75. 254.9% **77.** 207.7% **79.** 19%

27–page 607 **1.** 25% **3.** 75% **5.** 12.5% **7.** 35% **9.** 14%
11. 62.5% **13.** 275% **15.** 666.7% **17.** 855.6%
19. 817.5% **21.** 722% **23.** 537.5% **25.** 58.3%
27. 1530% **29.** 2080% **31.** 1862.5% **33.** 93.8%
35. 1135.7% **37.** 77.8% **39.** 573.3% **41.** 1735%

28–page 608 **1.** 0.105 **3.** 0.40 **5.** 1.20 **7.** 0.067
9. 0.089 **11.** 1.19 **13.** 0.172 **15.** 1.00 **17.** 0.053
19. 0.007 **21.** 0.00625 **23.** 0.056 **25.** 0.068 **27.** 0.304
29. 0.5784 **31.** 0.7425 **33.** 0.1880 **35.** 0.1445
37. 0.00125 **39.** 0.0014 **41.** 0.0879 **43.** 0.0095
45. 0.00025 **47.** 0.22125 **49.** 0.252

29–page 609 **1.** $\frac{9}{20}$ **3.** $1\frac{3}{4}$ **5.** $\frac{117}{1000}$ **7.** $\frac{3}{2000}$ **9.** $\frac{53}{500}$
11. $\frac{3763}{10,000}$ **13.** $\frac{1}{2}$ **15.** $\frac{101}{1000}$ **17.** $\frac{809}{1000}$ **19.** $\frac{1}{16}$ **21.** $\frac{7}{16}$
23. $\frac{37}{400}$ **25.** $\frac{31}{200}$ **27.** $\frac{51}{200}$ **29.** $\frac{1}{6}$ **31.** $\frac{3}{10}$ **33.** $\frac{5}{16}$
35. $\frac{1}{8}$ **37.** $\frac{4}{15}$ **39.** $\frac{1}{9}$ **41.** $\frac{1}{2}$ **43.** $\frac{1}{80}$ **45.** $\frac{51}{400}$
47. $\frac{43}{800}$ **49.** $\frac{1}{15}$

30–page 610 **1.** 9 **3.** 3.84 **5.** 7.8 **7.** 14.24 **9.** 3.486
11. 45.359 **13.** 10 **15.** 40 **17.** 6 **19.** 2.25 **21.** 128
23. 8.19 **25.** 15 **27.** 2 **29.** 80 **31.** 9.144 **33.** 25
35. 6.375

31–page 611 **1.** 50% **3.** 20% **5.** 35.7% **7.** 89.2%
9. 69.9% **11.** 87.2% **13.** 150% **15.** 120%
17. 133.3% **19.** 129.2% **21.** 400% **23.** 513.3%
25. 80% **27.** 46.2% **29.** 55.1% **31.** 80% **33.** 120%
35. 25% **37.** 138.9% **39.** 200% **41.** 66.7%
43. 5% **45.** 146.7%

32–page 612 **1.** 72 **3.** 8048.4 **5.** 1600 **7.** 336.4
9. 60 **11.** 134 **13.** 160 **15.** 60 **17.** $90\frac{2}{3}$ **19.** $54\frac{86}{91}$
21. 75 **23.** 2.5 **25.** 52.5 **27.** 78 **29.** 180 **31.** 225
33. 55 **35.** $197\frac{1}{67}$

Applications File

Page 613 **Application A** **1.** 600 mi **3.** 448 m
5. 210 m **7.** 1216 ft^2 **9.** 360 km^2 **11.** 31.4

13. 23.4 **15.** $437.95 **17.** $1429.00 **19.** $5.25
21. $46.19 **23.** $50.16

Page 614 **Application B** **1.** 308 in **3.** 8775 yd
5. 8 h **7.** 30 s **9.** 30 **11.** 248 **13.** 20 **15.** 50

Page 615 **Application C** **1.** $1,200,000 **3.** $7,900,000
5. Bronx, Columbus, Dallas, Lincoln, San Diego
7. Express plus **9.** **a.** Express **b.** Mon.

Page 616 **Application D** **1.** 1 q., 1 do., 1 five do. **3.** 1 p.,
3 do., 1 five do. **5.** 2 p., 1 di., 1 q., 4 do. **7.** 2 di., 3 q., 4 do.,
4 do. **9.** 1 p., 2 di., 2 q. **11.** 1 q., 2 do. **13.** 1 di., 1 do.,
1 five do. **15.** 1 q., 2 do. **17.** 2 q., 1 do., 1 ten do.

Page 617 **Application E** **1.** 8:45 **3.** 7:00 **5.** 11:15
7. 12:15 **9.** 3:30 **11.** 8:45 **13.** 7:30 **15.** 9:45
17. 12:45 **19.** 6:00 **21.** 11:30 **23.** 10:15
Application F **1.** 2 h:46 min **3.** 3 h:15 min **5.** 51 min
7. 5 h:15 min **9.** 8 h:20 min **11.** 10 h:45 min

Page 618 **Application G** **1.** 137 days **3.** 212 days
5. 258 days **Application H** **1.** No **3.** Yes **5.** No
7. Yes **9.** Yes **11.** Yes **Application I** **1.** 79 days
3. 231 days **5.** 222 days

Page 619 **Application J** **1.** $\frac{3}{4}$ yr. **3.** 1 yr. **5.** 3 yr.
7. $1\frac{3}{4}$ yr. **9.** $1\frac{1}{6}$ yr. **11.** $\frac{5}{6}$ yr. **13.** $\frac{35}{73}$ yr. **15.** $\frac{50}{73}$ yr.
Application K **1.** 52 **3.** 16 **5.** 12 **7.** 6 **9.** 60

Page 620 **Application L** **1.** 0.85 **3.** 0.65 **5.** 0.20
7. 9% **9.** 53% **11.** 93% **13.** 96.5% **15.** 24.1%
Application M **1.** $1300 **3.** $2100 **5.** $900 **7.** $400

Page 621 **Application N** **1.** $8000 **3.** $0
Application O **1.** 650 **3.** 350

Page 622 **Application P** **1.** $0.27 **3.** $110,160
5. 108°; 54°; 36°; 18°

Page 623 **Application Q** **1.** $54\frac{1}{3}$ **3.** $6726\frac{3}{4}$
5. $1,235,563 **Application R** **1.** $148\frac{1}{2}$ **3.** 88 **5.** 0.09
7. 2.08 **Application S** **1.** $24 **3.** None **5.** 0.01

Page 624 **Application T** **1.** m **3.** mm **5.** m **7.** mm
Application U **1.** kg **3.** g **5.** t **7.** kL **9.** g

Page 625 **Application V** **1.** 10.5 cm **3.** 2.6 ft **5.** 162 mi
Application W **1.** 12.56 mi **3.** 2.512 m **5.** 4.71 ft
7. 56.52 ft **9.** 39.25 mi

Page 626 **Application X** **1.** 150 in^2 **3.** 0.675 m^2
5. $10\frac{1}{9}$ ft^2 **Application Y** **1.** 13.5 in^2 **3.** 200 m^2
5. 19.625 yd^2 **7.** 23,223.44 mm^2 **Application Z**
1. 519,552 m^3 **3.** 396 cm^3 **5.** $157\frac{1}{4}$ yd^3

CREDITS

Mechanical art by The Mazer Corporation.

Photographs: